ROBOTS
AND
MANUFACTURING
AUTOMATION

SECOND EDITION

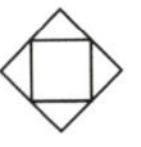

ROBOTS AND MANUFACTURING AUTOMATION

C. RAY ASFAHL
University of Arkansas, Fayetteville

JOHN WILEY & SONS, INC.
NEW YORK CHICHESTER BRISBANE TORONTO SINGAPORE

Acquisitions Editor *Charity Robey*
Production Manager *Linda Muriello*
Cover/Text designer *Karin Gerdes Kincheloe*
Production Supervisor *Micheline Frederick*
Manufacturing Manager *Lorraine Fumoso*
Copy Editor *Deborah Herbert*
Photo Researcher *Jennifer Atkins*
Cover Photo *Alex Pietersen*

Library of Congress Cataloging-in-Publication Data

Asfahl, C. Ray, 1938–
Robots and manufacturing automation / C. Ray Asfahl. -- 2nd ed.
p. cm.
Includes bibliographical references (p.) and index.
ISBN 0-471-55391-3 (cl.)
1. Robots, Industrial. I. Title.
TS191.8.A83 1992
670.42'7--dc20 91-32355
CIP

Printed in the United States of America

Printed and bound by the Hamilton Printing Company.

10 9

For your advice and encouragement, this book is gratefully dedicated to Morey Kays and Del Kimbler.

PREFACE

This book is about robots, but not *just* about robots. The industrial robot has become the standard bearer for the manufacturing automation movement. But, as sensational as industrial robots are, they are not the most significant development in manufacturing automation today. Behind the scenes, programmable controllers, microprocessors, process control computers, and industrial logic control systems are enjoying even more acceptance and application in manufacturing automation than are industrial robots. All of these devices should be considered members of the family of *flexible* automation equipment that is changing the way in which products are manufactured. This book describes the relationships among all of these devices in a manufacturing automated system.

This book is intended for the student's first course in manufacturing automation. In an industrial engineering curriculum, the ideal level for introduction of robots and manufacturing automation topics is at the junior level. Then the student who wants to specialize can take follow-on courses at the senior level on topics such as CAD/CAM, CIMS, FMS, and digital electronics, or specialized robotics courses, depending upon laboratory facilities. This book is also appropriate for upper division technology and industrial management curricula and has been used successfully in graduate level courses in industrial engineering and operations management.

The focus of this book is on how to *apply,* not how to *design,* robots and manufacturing automated systems. Hence, the book is appropriate for engineers, manufacturing technologists, and industrial managers. Calculus and computer programming backgrounds are not required for comprehension of this book, but some of the end-of-chapter exercises are designed to challenge students who possess such backgrounds. These exercises can be comfortably omitted with no loss of continuity in courses that are aimed at students who do not have technical backgrounds. The accompanying instructor's manual explains the end-of-chapter exercises and provides a basis for selecting classroom assignments suitable for the intended audience.

The revised edition of this text has been updated to recognize rapidly developing technology and its application to manufacturing automation. Two chapters—Chapter 8, "Machine Vision," and Chapter 16, "Computer Integrated Manufacturing"—are new in this edition. Included in Chapter 16 is up-to-date material on interfacing and networking, especially local area networks (LANs).

In the 1990s, industry is finding alternatives to the purchase of turnkey, commercially available robots. Today's students are learning to build manufacturing automation applications from basic components, such as sensors, switches, vision systems, and microprocessors. This book emphasizes these alternatives, especially programmable logic controllers (PLCs) and the application of industrial logic control systems. Additional design tools for analyzing logic control systems have been added to this revised edition. Safety has also been given increased emphasis, as have process control and quality. Case studies and exercises have been added for classroom use.

Accreditation bodies, in particular ABET, are placing increased importance upon design content in engineering and technology curricula. Design problems must challenge students to use their creativity to seek solutions to problems that are open-ended and have multiple solutions. This book and the accompanying *Instructor's Manual* identify several "design case studies" for strengthening the design content of courses in which this book is used. All examples, case studies, and design case studies are numerated as "examples" for consistency, but the "case studies," examples that are complex or are based upon real industrial applications, and the "design case studies," examples that have multiple solutions and design content, are further designated as such.

In addition to supporting ABET design criteria, the expanded *Instructor's Manual* that accompanies the revised text contains lecture outlines, audiovisual suggestions, projects for the laboratory, and sample examination questions with solutions. The questions and solutions are keyed to the appropriate page numbers in the book, so that examinations can be focused and solutions can be conveniently justified. In the case of design problems, sometimes multiple solutions are shown in the *Instructor's Manual*. Computer disks containing a limited version and tutorial for TISOFT™ and a tutorial for STAGE™ programming concepts are available to instructors who adopt this course for their classes.

This book can be used in a course either with or without a laboratory. Even without any equipment, the reader should be able to work all the exercises and understand each topic. If the course does have a laboratory, specific equipment manuals will be beneficial supplements to the textbook. Robots are expensive, but industrial-grade programmable controllers, microprocessors, and process microcomputers are not, making the provision of an interesting laboratory a feasible option. Low-cost, tabletop experimental robots are also available.

Many industrial readers want to know how to get started with robots and manufacturing automation, and there are some prerequisites. Chapter 1 describes product design and process control features that enhance automation potential. Sometimes the preparation for automation is more beneficial to the efficiency of the company than is the automation itself, and the text documents cases of this phenomenon. Chapter 2 uses simple language to describe the electrical and mechanical building blocks comprising the robots and automated systems presented in subsequent chapters. Mechanized parts feeding, handling, and orientation (Chapter 3) are essential for many robot applications, and the principles studied in this chapter can be applied to manual operations, too. Chapter 4 covers automated production and assembly lines and includes the analysis of the reliability of integrated automated systems. This analysis capability is needed for subsequent chapters on robotics. Chapter 5 is about NC machines, the forerunners to industrial robots and the originators of flexible automation, and this chapter includes material on CAD/CAM. Chapters 6 and 7 concentrate upon robotic manipulators, their characteristics, and their programming, and include a new section on AML. Chapter 8 is the new chapter on machine vision. Chapter 9 covers implementation, including safety. Chapter 10 is for robotic applications and includes analysis of economics. Chapters 11 and 12 describe and analyze industrial logic control systems, the most versatile and widely used tools in manufacturing automation today. These systems are even used as controllers for industrial robots themselves, though the public usually sees only the robot. Chapter 13 introduces programmable logic controllers, the most popular way of implementing industrial logic control systems. Upon completion of Chapter 13, readers should be able to configure custom robots in-house from standard components as an alternative to purchasing turnkey, commercially available robots. Process control computers (Chapter 14) are used to control banks of robot workcells, whole production lines, and even whole factories. At the opposite extreme are the decentralized systems controlled by tiny, low-cost microprocessors (Chapter 15), the technology of which has been the driving force behind the automation and robotics movement. Chapter 16 addresses computer-integrated manufacturing, and the new material in this edition on interfacing and networking is included in this chapter. The final chapter considers the ethical issues associated with the robotics and automation movement.

Many people contributed to the second edition. I acknowledge the conscientious reviews of Professors Alec Chang, Warren Liao, Michael Diesenroth, Y. S. Chadda, John Nazemetz, and James Rice. Michael Diesenroth provided many helpful suggestions for appropriate laboratory support in addition to his valuable suggestions for the text. John Nazemetz provided editorial assistance and pointed the way to significant material to be added.

Earnest Fant acted as my consultant on machine vision in electronics manufacturing, and Rajiv Mehrotra, Eric Webb, Sylvia Tran, and Gary Shepard were helpful in providing real examples for the new chapter on machine vision. David Boyster and Matthew Walker advised and assisted me in integrated circuits and microprocessors. Frank Broadstreet assisted with solutions to exercises. Other colleagues who were helpful are Robert Sims, William Boyd, and Eric Malstrom.

Thanks go to ARTRAN, Texas Instruments, Intel, General Dynamics, IBM, Singer, and a host of robot manufacturers not only for their hardware and software but also for the industrial case studies that enhance the second edition. Special thanks go to Margi Berbari, Marc Langston, Tom Jacoway, Kevin Price, Caile Spear, Rebecca Fant, Kay Fowler, Lois Giles, Genevie Payne and Dale Batson.

Finally, I acknowledge the assistance of my very capable secretary, Nancy Sloan, whose talents, thoughtfulness, and enthusiastic effort far surpassed her expected role.

C. Ray Asfahl

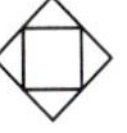

CONTENTS

CHAPTER 1

GETTING READY TO AUTOMATE

THE BOA CONSTRICTOR

James A. Baker, then executive vice president and technical systems sector executive of the General Electric Company, once likened the fierce competition in today's discrete-item manufacturing industries to the work of a boa constrictor squeezing its prey. In his words:

> It is common notion that this large snake simply drops out of a tree and crushes its prey to death, but the actual mode of operation is more subtle. The boa encloses a victim's chest tightly in its coils, and each time the prey relaxes and exhales, it takes up the slack. After three or four breaths, there *is* no more slack. The prey quickly suffocates and is swallowed by the boa [ref 12].

Baker sees the peaks and valleys of economic cycles as the breathing cycle of an industry under competitive siege. Every time sales are up, inventories are down, and profit margins return, it is easy to become complacent about production efficiency and exhale a sigh of relief. But this is when the boa constrictor of competition tightens its coils to take up the slack.

There was a time when the competitive struggles for industrial survival took place within a country's borders. Worldwide barriers to transportation, communication, and trade provided a measure of insulation between a country's industries and their foreign competitors. Even more important was the financial and technological advantage possessed by a privileged few industrialized nations that seemed impregnable to the leaders of industries of less developed nations. But the luxuries wrought for the rich nations by these competitive barriers have become their weaknesses—the chinks in the armor that formerly protected them from industrial competition. High wage rates, inefficient management, and obsolete factories are among these luxuries, which have allowed hungrier competition to break down the barriers and seize markets (see Figure 1.1) using low wages, determined management, and new factories that employ some of the latest technology developed by the very countries that are under economic siege.

Any nostalgic retreat to thoughts of domination by the historical industrial giants gives the boa constrictor cause to smile. Looking back to the glory days will not save an industry that is losing its competitive advantage.

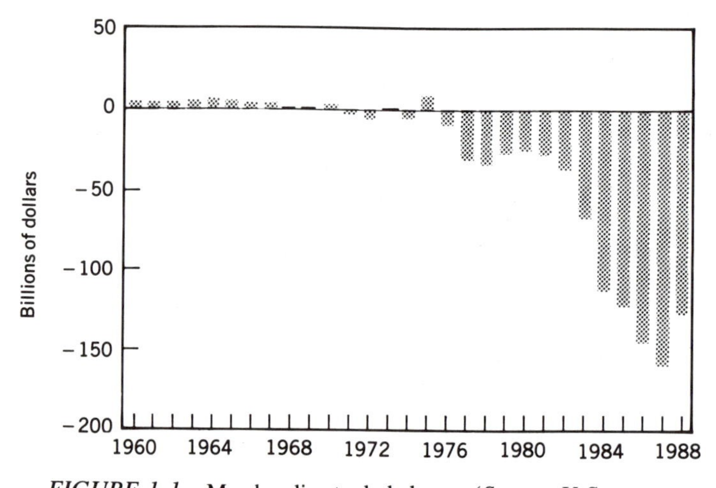

FIGURE 1.1 Merchandise trade balance. (*Source:* U.S. Department of Commerce, Bureau of Economic Analysis, *Business Conditions Digest, September 1989* [Washington, D.C.: U.S. Government Printing Office, September 1989], table 622.)

The United States

The United States of America has seen its position as world manufacturing leader under serious question in the decades of the 1970s and 1980s. Some have seen robotics as a possible savior to reverse the trend. Others have despaired that even robotics will not impede what they consider to be the inevitable demise of the industrial colossus that has been the United States. However, toward the end of the decade of the 1980s, some rays of hope had begun to shine through.

Although it is true that the rate of growth in U.S. productivity fell 60 percent from the levels of the 1950s and 1960s, that growth rate has slowed in other countries as well, including Japan, which is often thought to have replaced the United States as world manufacturing leader. The truth is that the United States is still the free-market industrial leader in the world, and Japan is not even a close second. In a 1986 study of the 24 member countries of the Organization for Economic Cooperation and Development (OECD), the United States ranked first in industrial output at 39 percent of the organization total. Furthermore, the United States' share has *grown* from 36 percent in 1973 [ref. 15]. These data suggest that, far from losing out, the United States is increasing its domination in world manufacturing competition. From 1962 to 1985 total United States employment in manufacturing grew from 24 percent of the OECD industrial employment total to 28 percent. During the same period Japan's growth in industrial employment was only slightly more impressive (from 15 percent to 20 percent). Corresponding to the increases in the United States and Japan, manufacturing employment share of total OECD industrial employment fell in Great Britain, Germany, and France.

Another despairing rumor declares that the American economy is degenerating into a purely service economy. While it is true that a growing percentage of the American workforce is employed in service occupations, the implication that manufacturing jobs are decreasing is false. What is happening is that the U.S. population is growing rapidly, while the populations of Britain, France, and Germany have remained relatively constant. Thus, with a growing population, the United States is able to seize a larger share of worldwide manufacturing while still increasing the percentage of service workers within its total economy. Even in the countries in which the populations have been stable, the numbers of service workers as a percentage of the total number of workers in those economies has increased *in every case*. In fact, the rate of service sector percentage growth in the nineteen countries for which data are available has been faster than the service sector percentage growth in the United States, faster in every other country except New Zealand. While service sector employment percentage in the U.S. increased 10 percent from 1965 to 1980, Japan's increased 31 percent.

Another source from the *Wall Street Journal* [ref. 45] characterizes the United States as somewhat paranoid about its economic future. Outsiders do not share the Americans' paranoia. In the words of Hubert Vedrine, national security adviser to French President François Mitterrand: "I do not believe at all in American decline. Americans adore frightening themselves." Perhaps the economic demise of the United States is not as imminent as the soothsayers of the early 1980s, including this author, were suggesting. Certainly there is hope, but only if the United States has the determination to apply itself to the task of meeting world manufacturing competition.

Despite intense wage rate competition and the commitment to quality of such industrialized countries as Japan, the United States has a tremendous advantage over other countries in manufacturing. This advantage is the presence of a very large and ready domestic market for its products. Despite the feats of flexible automation, a production lot of 100,000 identical products, or nearly identical products, is less costly per piece to manufacture than a lot of 500. Volume still is a factor, and the United States can more readily tap its large domestic market and achieve manufacturing economies before foreign competition has the opportunity to penetrate. Free trade diminishes this advantage, but does not eliminate it because the opportunity to restrict trade with import quotas or tariffs remains a possibility. Canada possesses nearly the same advantage as the United States because of its own market and its proximity and excellent relationship with the large market of its neighbor to the south. Europe is seeking a similar advantage of its own by creating the European Common Market. The combined economies of the countries of Europe make up a large and powerful market base on which to build volume production with the associated economies of scale that U.S. industries have enjoyed.

For those industries that qualify, automation offers opportunities for quantum advances in production efficiency—the kind of advances necessary to reverse trends, recapture markets, and break free from the boa's coils. Automation is certainly not new, and in a broad sense it can be traced back to the Industrial Revolution when machines first began to multiply vastly the productive capability of workers. The history of automation, however, has not been characterized by a steady progression; instead, it has been a series of breakthroughs. One breakthrough was interchangeable manufacture; another was Henry Ford's assembly lines. This book is about another such breakthrough, one that is still in the process of unfolding and that has the potential of becoming the greatest breakthrough to date. Although the word *robot* appears in the title of this book, industrial robots are not themselves the breakthrough; they are a *product* or result of the breakthrough. But robots have become the standard bearers of the current industrial automation movement and deserve the attention of any automation engineer whose is involved in discrete-item manufacturing.

Labor's Role

Automation certainly is not the only way to break the boa's clenching coils. When the chips are down, the response of labor has been remarkable. Demands for higher wages formerly seemed almost insatiable. In the United States, the labor leaders of the 1960s would have been incredulous had they been afforded a glimpse of the wage and benefits concessions made by labor unions in the 1980s.

In recent years it has sometimes appeared that labor has more determination to meet world competition than does management. In the 1980s and early 1990s large U.S. companies have become bankrupt or suffered severe cutbacks in operations or corporate mission. These cutbacks have resulted in numerous plant closings, sometimes accompanied by severe hardship, especially in small towns in which the plant was the major employer. So determined are labor and local management in these crises, that, in some cases, employee groups have mounted drives to buy the facility from the parent company and continue operations. This scheme has been successful in several instances, and labor has been able to accomplish what management has failed to do. One reason for the success of this strategy is that labor has been willing to make sacrifices for the survival of the firm (*their* firm), sacrifices that they were unwilling to make in their old role of negotiating with management for all they could get. In effect, worldwide competition has forced labor and management to lay down their arms and unify to assure survival against competition. The development has been beneficial and has brought U.S. labor and management closer to the spirit of cooperation that has proved so successful in the Japanese culture.

Another concept that is being used as a competitive weapon is participative management. The first arena for the display of this weapon was product quality, and the most popular term describing participative management is still *quality circles*. The Japanese were the first to embrace quality circles as a means to overcome the image of poor quality borne by their products in the 1950s. Quality circles solicited ideas from factory workers at all levels for ways to improve the quality of their manufactured products. The challenge was met, and the Japanese now enjoy a reputation for superior products at competitive prices. Today, the quality circles concept is being pursued in other countries, notably in the United States. The objective has been broadened to include not only quality but cost reduction and safety as well.

This book is a survey of the types of equipment and methods that can be used to achieve *factory automation,* which has been called "industry's survival kit" by Baker of General Electric. While pointing to ways to automate, the text also notes the disadvantages and pitfalls along the path to automation. One of the easiest mistakes to make is to leap into automation before assuring that the product and process are ready. The purpose of this chapter is to point to ways to get a factory ready for automation. Most of the preparations will benefit any plant, whether or not automation is achieved.

DESIGNING FOR AUTOMATION

The product itself must be designed for producibility and, in particular, "assemblability." The advent of the industrial robot has resulted in increased attention to product design to facilitate orientation, positioning, and mating of assembly components. The unexpected good news is that such attention to producibility in the product design phase, to make robot handling and assembly feasible, has resulted in more efficient manual production as well. In fact, in a substantial percentage of cases the result of a robotics project is a highly efficient, semi-automated process that may not even include the robot in the final analysis.

We will see in Chapter 3 that most automatic machines for handling and orienting piece parts operate by feel—that is, they use the *mechanical* characteristics of the part to achieve the desired orientation and position. Although some robots are being equipped with sophisticated vision systems for orienting and handling parts, the majority of applications will not support the additional cost of sophisticated vision systems. Therefore, the robot or automated system must work "blind," and it is good practice for automation engineers to close their eyes, too, when attempting to imagine how a machine will handle a part.

Symmetry

Curiously, in some cases part symmetry can be beneficial to automation, and in other cases symmetry can make automation nearly impossible. Consider the examples shown in Figure 1.2. In each of these examples, both ends were designed to be identical to make orientation unnecessary. In two of the examples, the achievement of symmetry requires additional details to be fabricated into the part, but the additional fabrication cost would be minimal compared to the cost of achieving parts orientation for the corresponding asymmetrical designs. Figure 1.3 presents a different problem, however. In these examples, important features on each part are difficult to detect mechanically, and the solution to the problem is to *remove* symmetry. On one part, a projection is added on one side of a basically disk-shaped part. On another, a flat is milled on a cylindrical surface to indicate the angular position of a hidden hole through the shaft. On the third, a pin is used for a similar purpose to locate around a disk the angular position of a hole whose axis is parallel to the center line of the disk. In each case, the part lost some of its symmetry in the redesign to make automation more feasible. Disk and circular objects are particularly good candidates for asymmetric design features, because without locating

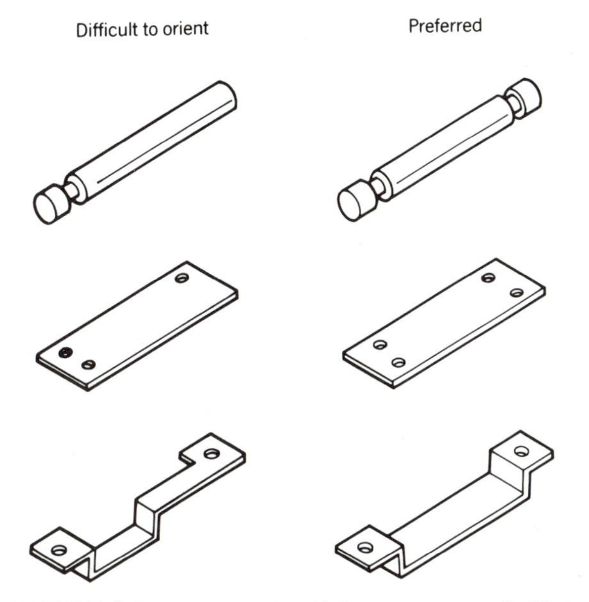

FIGURE 1.2 Examples in which part symmetry facilitates part orientation. (Reprinted with permission of the Society of Manufacturing Engineers, Proceedings of the 13th International Symposium of Industrial Robots and Robots 7, April 1983, p. 11–49.)

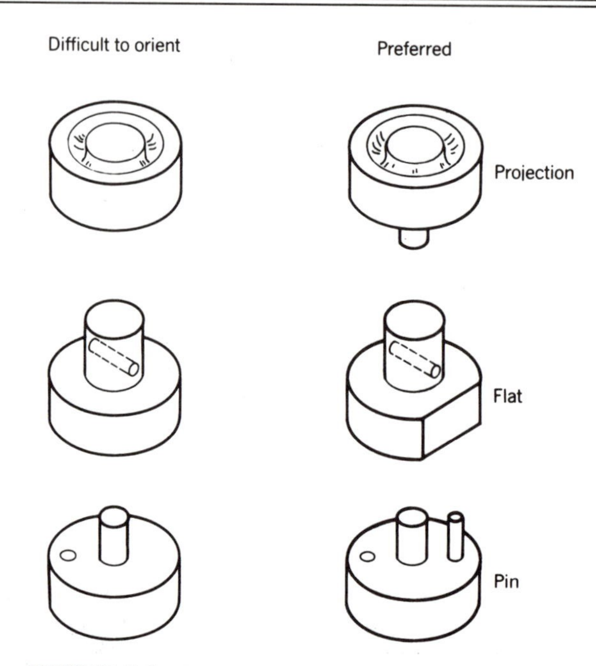

FIGURE 1.3 Examples in which parts asymmetry facilitates part orientation. (Reprinted with permission of the Society of Manufacturing Engineers, Proceedings of the 13th International Symposium on Industrial Robots and Robots 7, April 1983, p. 11–49.)

features they can assume an infinite number of rotational orientations. Rectangular shapes, however, usually benefit from symmetry because there are only a few feasible orientations, and frequently the designer can make all of these orientations workable.

Parts Tangling

Anyone who has sifted through a box of assorted springs can appreciate the need for eliminating features that cause parts to tangle or snarl. Figure 1.4 illustrates the solution to the tangled springs problem along with some other examples of part designs that reduce the tendency to tangle. Parts often have both holes and projections in which the functions of these features are irrelevant to each other and the projections are not intended to enter the hole. The relationship between hole size and part projection dimensions for such parts is important to prevent the projection from sticking into the hole and causing a tangle.

Design for Feeding

In Chapter 3, we shall see machines for feeding parts in single file along a track. Most of these machines utilize vibration or gravity, and force is transmitted from piece to piece as parts are pushed forward from the rear. The method is particularly good for parts that are fairly flat and stable in the desired orientation. But if parts are too thin or if edges are beveled, the parts will tend to "shingle," as shown in Figure 1.5. A similar problem is "wedging," caused when part ends are not orthogonal to the direction of travel, as seen in Figure 1.6.

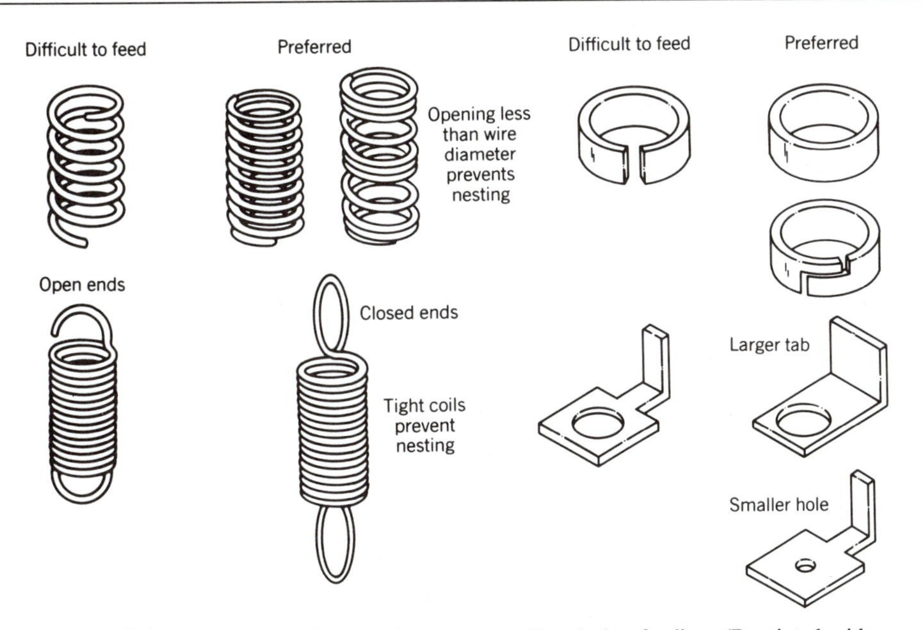

FIGURE 1.4 Avoid parts' features that cause tangling during feeding. (Reprinted with permission of the Society of Manufacturing Engineers, Proceedings of the 13th International Symposium on Industrial Robots and Robots 7, April 1983, p. 11–50).

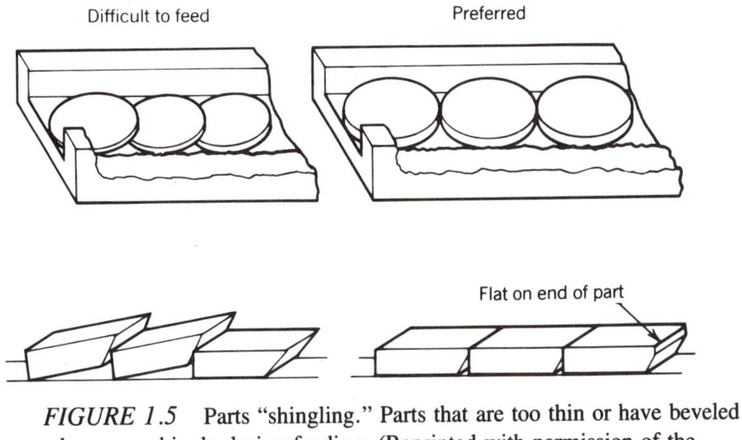

FIGURE 1.5 Parts "shingling." Parts that are too thin or have beveled edges may shingle during feeding. (Reprinted with permission of the Society of Manufacturing Engineers, Proceedings of the 13th International Symposium on Industrial Robots and Robots 7, April 1983, p. 11–49.)

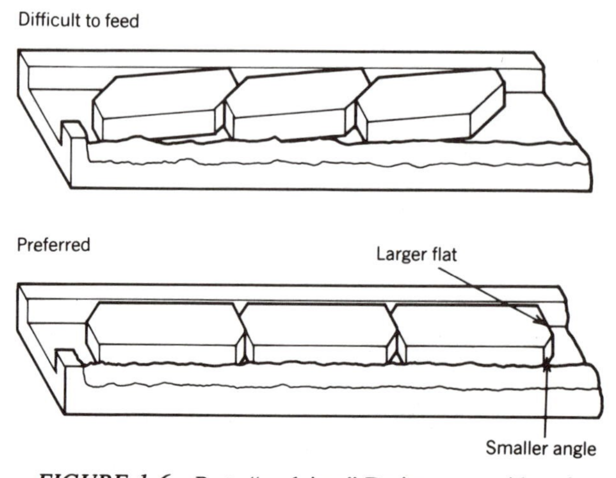

FIGURE 1.6 Parts "wedging." Design parts with ends that are "flat"—i.e., orthogonal to the direction of travel. (Reprinted with permission of the Society of Manufacturing Engineers, Proceedings of the 13th International Symposium on Industrial Robots and Robots 7, April 1983, p. 11–50.)

Designing for Insertion

Even if parts are oriented correctly, when the tolerances are close it is difficult to achieve the perfect alignment necessary to accomplish an insertion task with an industrial robot or other automatic machine. In Chapter 6, we shall discover some tricks for overcoming this difficulty, but it is frequently possible to design the mating parts to make the insertion

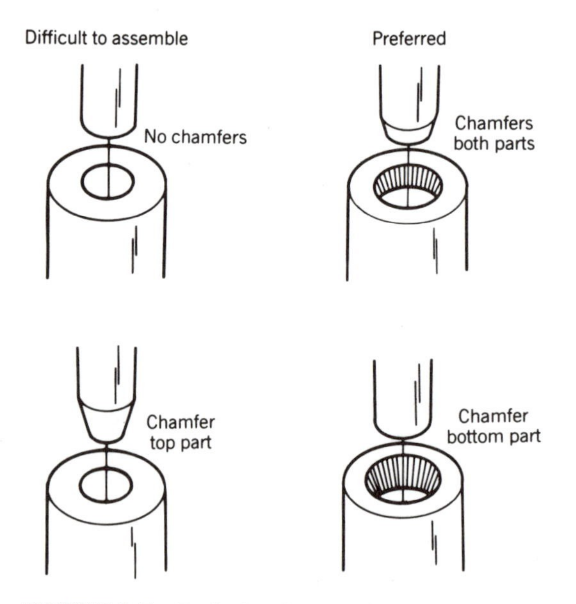

FIGURE 1.7 Designing for parts insertion using chamfering. (Reprinted with permission of the Society of Manufacturing Engineers, Proceedings of the 13th International Symposium on Industrial Robots and Robots 7, April 1983, p. 11–51.)

job easier. Figure 1.7 illustrates the use of chamfering to force the realignment of mating parts during insertion. Such minor automatic realignment is critical to the success of industrial robotics applications.

Fasteners

In Chapter 4, we shall discuss machines for driving screws, but it is wise to avoid screws and fasteners because of the complexity they add to the assembly process. It is often possible to do this with no compromise in quality or integrity of the assembly. Figure 1.8 illustrates a clip and slot insertion design that simply snaps into place. Figure 1.9

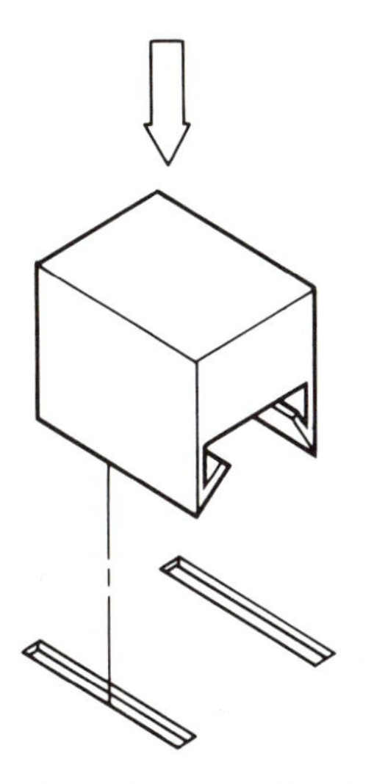

FIGURE 1.8 Clip and slot insertion design. (Reprinted with permission of the Society of Manufacturing Engineers, Proceedings of the 13th International Symposium on Industrial Robots and Robots 7, April 1983, p. 11–52.)

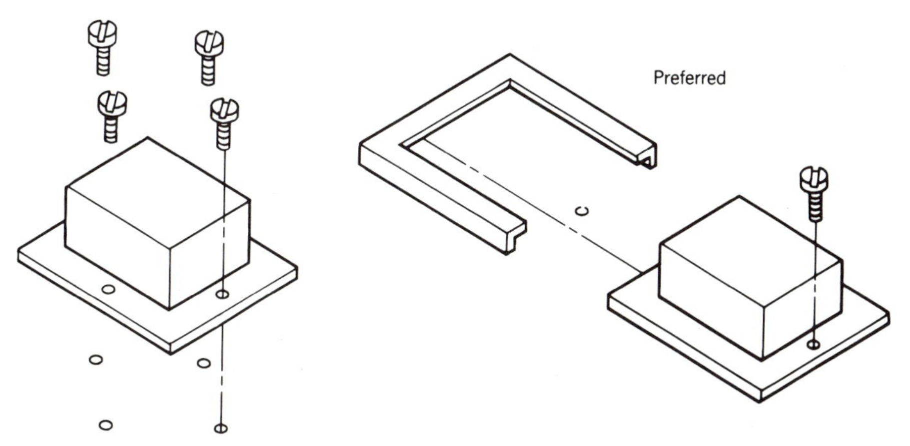

FIGURE 1.9 Minimize the number of screw-type fasteners required. (Reprinted with permission of the Society of Manufacturing Engineers, Proceedings of the 13th International Symposium on Industrial Robots and Robots 7, April 1983, p. 11–52.)

shows how attention to design principles for automation results in a design in which only one screw does a job that might have required four screws.

If fasteners must be used, automated processes will benefit from standardization of fasteners by type and size. It seems to violate engineering principles and economic good sense to use a larger, stronger, and more expensive screw than is necessary in a product application. But if such a screw is needed elsewhere in the assembly, it may be worthwhile to specify that same screw throughout to achieve uniformity just to benefit automation. The standardization of screw size and type may result in great savings in the purchase of fewer models of this type of automation equipment.

We conclude this section on product design for automation by examining two industry case studies that act as examples to show how adherence to basic design principles as described in this chapter can lead to dramatic savings in manufacturing cost.

EXAMPLE 1.1 (DESIGN CASE STUDY)
Design for Automated Assembly (IBM Printer Component)

The assembly is a mechanism for feeding single sheets of paper into a printer in production in an IBM plant in Austin, Texas. Figure 1.10(*a*) is the original design and has a total of 27 parts. In the redesign in Figure 1.10(*b*), the number of parts has been reduced to 14 and the motor can be installed with a push-and-twist motion. Note the reduction in the number of screws required [ref. 7].

EXAMPLE 1.2 (DESIGN CASE STUDY)
Design for Automated Assembly

In our second example of the principles of product design to facilitate automatic assembly we consider the assembly in Figure 1.11(*a*), the function of which is not disclosed [ref. 85]. The original design shows 21 distinct parts. Figure 1.11(*b*) displays the redesign, which reveals 16 distinct parts, a reduction of more than 20 percent. In addition, the assembly was greatly simplified. Note again the reduction in the number of screws required.

From this real example and from the illustrations of the principles of designing for automation, it should be apparent that many product design improvements to facilitate automation are of great benefit to the manual method of production. In fact, in the IBM printer mechanism redesign the manual assembly time became so short that the application no longer justified an industrial robot. Perhaps the greatest benefits to be achieved in the field of robotics and automation are the indirect benefits derived from plantwide preparations for automation.

Product Cycle Time

An additional benefit of designing for automation is the gains that can be made in reducing the product design cycle time, or the time required to launch a new product from concept, market analysis, product design, and process development. This cycle time is distinguished from the manufacturing cycle time, which refers to the production time required to manufacture an individual item of product, a concept that is also significantly affected by automation and will be discussed throughout this book, especially in Chapter 4.

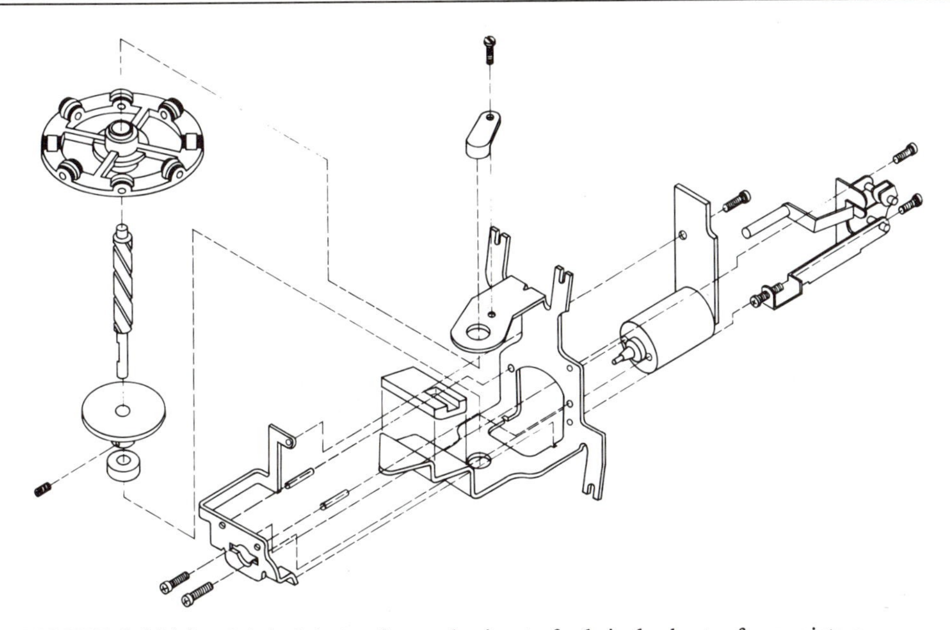

FIGURE 1.10(a) Original design for mechanism to feed single sheets of paper into a printer, manufactured in an IBM plant in Austin, Texas. (Reprinted with permission of the Society of Manufacturing Engineers, Proceedings of the 13th International Symposium on Industrial Robots and Robots 7, April 1983, p. 11–55.)

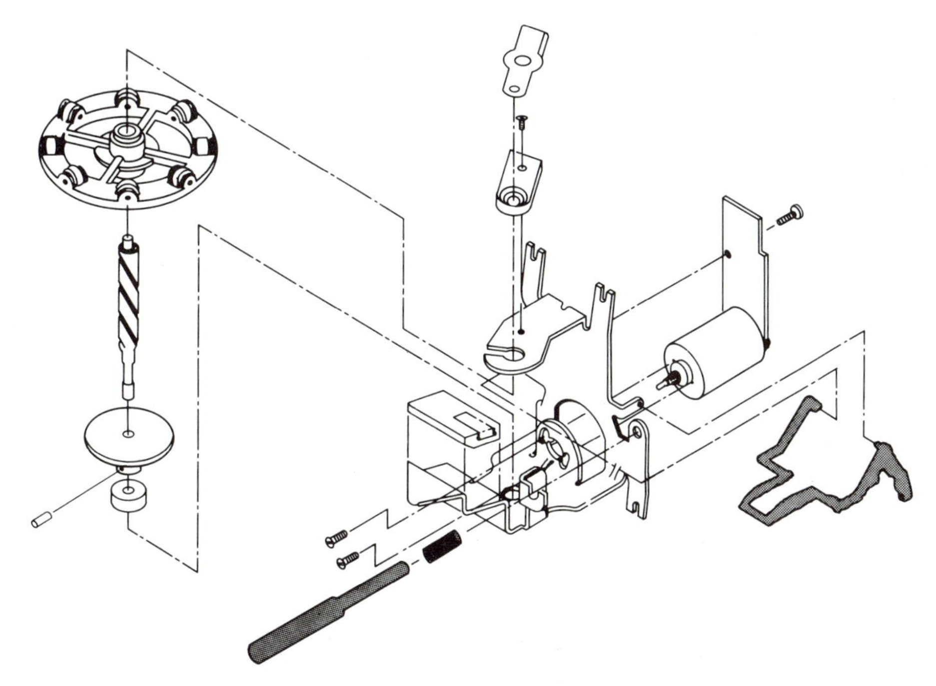

FIGURE 1.10(b) Revised design with changes to facilitate automated assembly. Details obscured at the request of IBM Corporation. (Reprinted with permission of the Society of Manufacturing Engineers, Proceedings of the 13th International Symposium on Industrial Robots and Robots 7, April 1983, p. 11–56.)

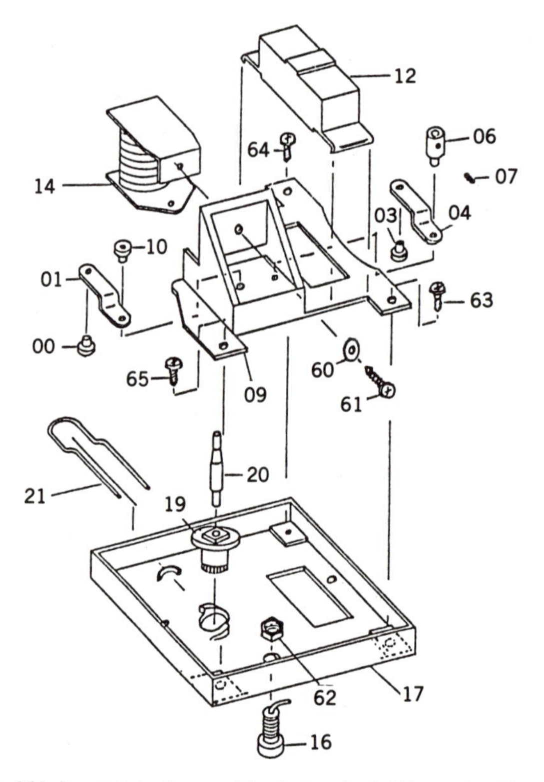

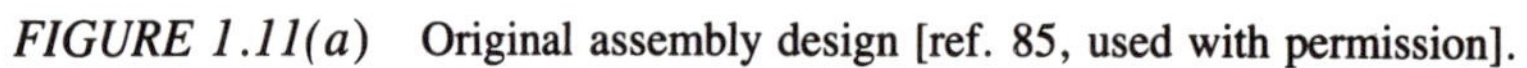

FIGURE 1.11(a) Original assembly design [ref. 85, used with permission].

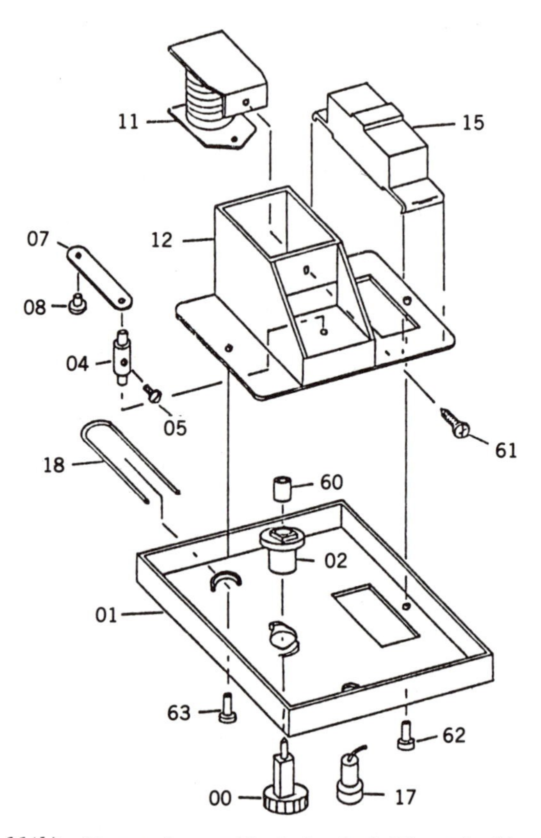

FIGURE 1.11(b) Proposed assembly design [ref. 85, used with permission].

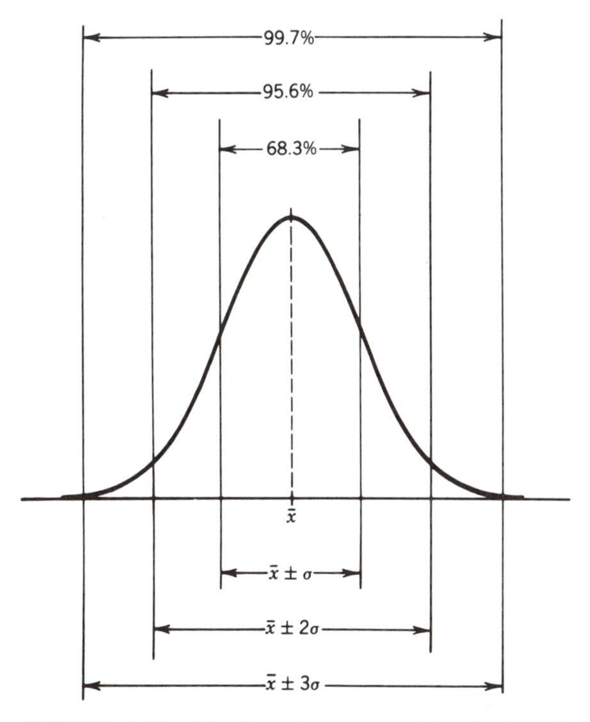

FIGURE 1.12 The Gaussian or normal distribution of a process variable.

A simple product design is easier to implement through process planning and into production. A fast product development cycle is crucial to the success of a company working in an industry that has very short product lives. This is most true of computer-related industries, in which the manufacturers of product clones steal the market from the original manufacturers in a very short time after product introduction. It is also true of many other industries, as highly efficient foreign competition is able to mimic the industries that have originated the product idea and have already carried out the market research. The best defense against the imitators is for the originators to take immediate advantage of their early research and to move quickly to establish themselves as market leaders before the imitators are able to get organized.

IBM's Dr. Low [ref. 56] stated it this way:

> Cycle time reduction is absolutely essential in business where a product is considered "mature" after two years and downright decrepit after five. As you all know, the only way to cut cycle time is to eliminate defects and simplify processes so we're doing both, and we're starting with simpler product designs.
>
> To give you a case in point, the power supply in our current DASD product is contained in two circuit cards—that's down from 12 cards in the previous model. Just think of how much faster we can assemble the new product, and let's not forget—the fewer the parts, the fewer the defects.

STABILIZING THE PROCESS

Although the flexibility of modern industrial robots and manufacturing automation equipment has decreased the importance of process stability, a relatively stable product and process are still prerequisite to automation. Even market demand should be stable or growing to justify significant capital investment in automation equipment. At the very least, frequent model changes mean additional "software" (programming) for robots and automated systems, and at the worst mean expensive retooling and obsolescence of expensive fixed automation equipment.

Statistical control techniques have been effectively used to monitor process variables, especially product quality, and to assess stability. The basic assumption underlying statistical control is that factory variables are not the result of a simple cause but are in fact the summation of a wide variety of independent contributing factors, each of which adds its own minor variations. Under such conditions, the central limit theorem [ref. 55] states that these summation variables have the familiar bell-shaped distribution known as the Gaussian distribution or simply the normal distribution. Figure 1.12 plots the bell-shaped distribution curve of the normal random variable. The curve approximates the frequency distribution plot of large samples of data points for normally distributed variables. The curve then is the theoretical ideal, and the vertical axis represents the probability density for any value of the variable along the horizontal axis.

Although some normal variables have wide variations and some concentrate their variations over a narrow range, any normal variable can be fitted to the curve of Figure 1.12 by appropriate scaling of the horizontal axis. The scaling is done by computing the standard deviation from a sample of data using the following formula:

$$s = \sqrt{\frac{\sum_{i=1}^{n} (x_i - \bar{x})^2}{n - 1}} \tag{1.1}$$

where x_i = sample observations

n = sample size

$$\bar{x} = \text{sample mean} = \frac{\sum_{i=1}^{n} x_i}{n}.$$

The sample standard deviation s is an estimate of the theoretical standard deviation σ of the normal random variable. The proportion of the normal probability density that lies between $\pm n\sigma$ deviations from the mean of a normal random variable is known. This is the key to the use of statistical control techniques: If an observation occurs outside the range $\bar{x} \pm 3\sigma$, it is typically presumed to represent a basic change in the process variable, not a random variation. The chances of a random observation of the same variable falling outside the $\bar{x} \pm 3\sigma$ range is extremely slim if $\bar{x}$ represents the true mean of the normally distributed variable. This point should be evident from Figure 1.12.

The relationship between part tolerance and process standard deviation is important to both product quality and the automation potential of a process. Figure 1.13 illustrates several possible relationships between part tolerances and normal process variation. Only the relationship shown in Figure 1.13 (d) demonstrates sufficient control of the process variable for factory automation. Even a process that fits Figure 1.13(d) quite nicely might find itself fitting Figure 1.13(c) when tighter tolerances are applied to satisfy the require-

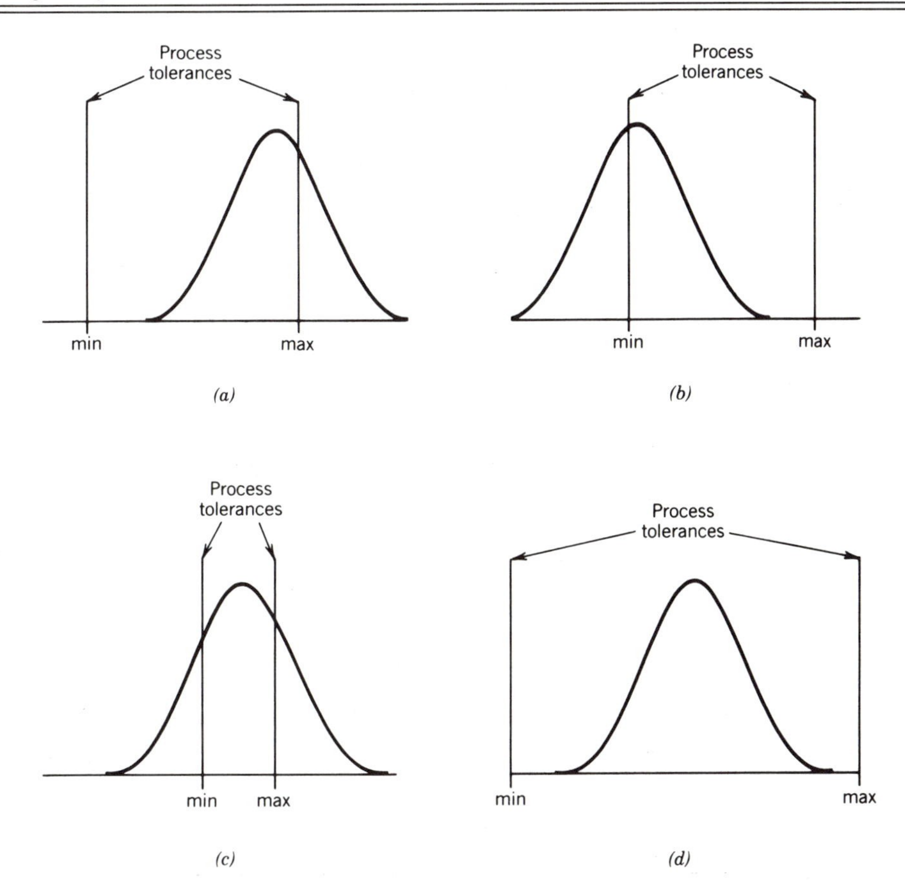

FIGURE 1.13 Relationships between normal process variation and process tolerance. (*a*) Process aimed too high to fall within tolerance. (*b*) Process aimed too low to fall within tolerance. (*c*) Process variation too large to meet required tolerance. (*d*) Process variation under control and within tolerance.

ments of an industrial robot or other automated manufacturing equipment. This is a possibility that the automation engineer should evaluate before plunging into a robotics or automation project.

Certainly there are ways to correct each of the three improper relationships between process variation and part tolerance that are shown in Figure 1.13. It is not at all unusual for an industry to purchase specially manufactured, high-quality screws and fasteners at additional cost to meet the quality requirements of automated assembly stations. The sole purpose of such purchases at added expense is to facilitate automated manufacturing, not to increase the quality of the product. Automation engineers should evaluate the control of their processes to determine whether the processes really qualify. Process means and standard deviations should be known or estimated and then compared with specified tolerances. The tolerances themselves should also be scrutinized to determine whether they will stand up to the rigorous demands of automation. The capabilities of automation equipment considered for purchase should be scrutinized closely to see whether tolerance for part variation matches the capability of the part fabrication process. An example will illustrate.

EXAMPLE 1.3

A particular industrial robot gripper can pick up parts within the following dimensional specification: 2 ± 0.010 in. The process standard deviation for this dimension is five thousandths. Is the process ready for this industrial robotics application? If the piece part dimension mean is exactly 2 inches, and the variation in the dimension is normally distributed, what percent of the piece parts will the robot be unable to pick up?

Solution

This process appears not to be ready for automation owing to the excessive variation in the process. A substantial percentage of the product will lie outside the feasible range of the robot gripper. The range of the gripper is

$$\begin{array}{r} 2.010 \\ -1.990 \\ \hline 0.020 \end{array}$$

which is

$$\frac{0.020}{0.005} = 4$$

that is, four times the process standard deviation. Even if the process mean were exactly 2.000 in., assuming the normal distribution, the upper and lower limits of the robot gripper are related to the process as follows:

$$\text{upper limit} = 2.000 + 0.010 = 2.000 + \frac{0.010}{0.005}\sigma = 2.000 + 2\sigma$$

$$\text{lower limit} = 2.000 - 0.010 = 2.000 - \frac{0.010}{0.005}\sigma = 2.000 - 2\sigma$$

Therefore the gripper range is $\bar{x} \pm 2\sigma$. According to Figure 1.12, the percentage of parts that lie outside the range $\bar{x} \pm 2\sigma$ is $100.0 - 95.6\% = \underline{4.4\%}$.

QUALITY AND AUTOMATION

An easily recognized principle at this point is that quality and automation go hand in hand. It is important to both quality and automation that processes be tightly controlled and that the consequences of piece-part variation be understood. Statistical process control methods were once thought to belong solely to the mission of product quality. They are now understood to have a significant impact upon the capability of the process to be automated.

The classic "three sigma limits" have been the standards of quality in decades past. Ninety-nine and seven tenths percent of the total has been considered an ideal goal for acceptable production. Only three defectives in a lot of 1000 has been considered reasonable in conventional manufacturing in most cases. But now the old three sigma limits have been challenged as not good enough for today's competitive marketplace. And now that automation is more important, extreme precision is being demanded for piece-parts. Consider the impact upon an automatic machine that jams and shuts down an entire integrated assembly line whenever an out-of-tolerance piece-part enters the feed track.

Even "four sigma limits" may not be good enough, especially when dealing with large volumes of product. Consider the U.S. Postal Service, for example. Nearly everyone blames the U.S. Postal Service for an error once in a while. But the U.S. Postal Service has a much lower tolerance for error than most manufacturing firms. It has been speculated [ref. 56] that if the U.S. Postal Service ever became satisfied with "four sigma limits" of performance, 20,000 pieces of mail would be lost every hour! By the same standard, there would be 5000 incorrect surgical procedures every week and 200,000 faulty prescriptions every year. The marketplace is demanding and getting higher quality than that. That is why some companies, such as IBM and Motorola, are reaching for even higher standards of quality, precision, and error-free performance. In 1990 Dr. Paul Low [ref. 56] stated that IBM is committed to reaching "six-sigma quality" by 1994.

Just what kind of target is "six-sigma quality"? Is it an attainable goal? Referring back to Figure 1.12, the tails of the normally distributed random variable reach to infinity. Table 1.1 tabulates the percentage of the area of the probability density function of the normally distributed random variable that lies within $\pm\, n\sigma$ of the mean. The area of the curve lying within 6 sigma limits of the mean is 99.9999998 percent [ref. 64]. Stated in terms of product quality, six-sigma quality would be represented by two defects in one billion. Does any industry ever achieve that kind of freedom from defects? The answer is yes, if the market demands it. For example, the market has demanded an extremely safe record for commercial air travel. Safe air travel requires thousands of individual actions by equipment, flight crew, and ground crew, many of which are critical to the safe completion of a given flight. In recent years the safety record of commercial airlines [ref. 1] has been generally fewer than one fatality in one billion passenger miles, and in one year (1986) there were only seven deaths in approximately 350 billion passenger miles. The record is even better for "large" commercial airlines, for which the National Safety Council [ref. 1] reports fewer than one fatal accident in one billion *aircraft* hours for most years in the decade since 1985. Since an aircraft hour on a large commercial jet airliner represents at least 50,000 passenger miles on the average, it is estimated that the per passenger mile rate of the large commercial airlines is fewer than one fatality in 50 trillion passenger miles. Such a record represents "six-sigma" performance. The airline record for baggage handling is not six-sigma, but it, too, is better than most people think, because of the enormous daily volumes involved.

Automation may not demand the same level of process stability that is required by the commercial airline traveler, but exacting specifications and removal of piece-part variation is a consideration not to be overlooked by the automation engineer. Consider the following case study.

TABLE 1.1 Areas of Probability Density of the Normally Distributed Random Variable

Region	*Area (%)*
$\mu \pm \sigma$	68.26
$\mu \pm 2\sigma$	95.44
$\mu \pm 3\sigma$	99.74
$\mu \pm 4\sigma$	99.994
$\mu \pm 5\sigma$	99.99994
$\mu \pm 6\sigma$	99.9999998

Ref. 64.

EXAMPLE 1.4 (CASE STUDY)

ARTRAN Automation, a world leader in the design of automated systems for food processing, invented a successful transfer system for poultry processing in 1989. In the initial test, which was conducted in one region of the country, the original design worked very reliably for chickens produced in that region. However, when the new system was installed in another region of the country and for a different customer, problems developed because the chickens processed in that plant were of varying size and the skin was drier and different in viscosity from the chickens in the test plant. The original design had to be altered in the field to make it work for the chickens processed in that region. Had it not been for the resourcefulness of the ARTRAN field engineer assigned to the project, this very successful automation project may well have ended in failure.

Example 1.4 points to the importance of considering the effects of piece-part variation and process instability upon the success of an automation project. There are few endeavors that are so dependent upon a multitude of seemingly tiny details—details that can make or break an automation project or even an entire company. Another of these details is considered in the section that follows.

ACHIEVING MACHINE RELIABILITY

Everyone wants reliable factory machines and processes, but automation adds a new urgency to this need. With automation comes processes that operate without operator attention. Automation also introduces a great deal of machine interdependence in sequential processing stations.

Chapter 4 will quantify some of the drastic consequences of machine malfunction and parts jamming, which can wreck an automation project in which the machines do not have sufficient prerequisite reliability. This is a complicated subject, but anyone who experienced one of the great northeastern power failures in 1965 or 1977 has a feel for the drastic consequences that are possible for automated systems that link equipment into an interdependent process. When one element fails, it can bring down the entire system if the interdependence between elements is critical. Manual processes are not immune to this characteristic either, but the problem intensifies when automation is introduced. Automation engineers need to evaluate factory machine and process reliabilities and to be sure that they meet minimums necessary for integration into an automated system. What those minimums are depends on the number of elements in series and the degree of dependence between elements. This will become clearer in Chapter 4 and in subsequent chapters of this book.

Closely related to reliability and of almost as much importance to automation is maintainability of equipment. What is important is *availability:* the percentage of total time the machine will be available for production. Availability of a machine is a function of how often the machine fails (reliability) and how long it takes to fix it when it does fail (maintainability). In a manual process operating in batch mode, it may be tolerable for a machine to be down 5 percent of the time (availability is 95 percent). The machine often can easily make up for lost time when it is back on line. But in a highly automated, interdependent system of machines, an availability of 95 percent can be intolerable, causing the automated process to be idle not 5 percent of the time but a majority of the time!

WHAT ABOUT VOLUME?

While it is still true that production volume is required to justify automation, it is not as true today as it was a decade or two ago. Automation always involves a capital investment that must be amortized through cost savings on each unit produced. But no longer do these units have to be identical for enormous production volumes because two basic types of automation are available to today's automation engineer.

Hard Automation

The classical type of automation, typical of the 1940s and 1950s, is fixed (hard) equipment, usually custom-made and designed to facilitate the manufacture of a specific product. Hard automation can achieve very high-speed production, but it is usually quite expensive. This expense can become very painful when a product model or design change is introduced. Fixed or hard automation, as the name implies, is not very adjustable or adaptable to product or process change. Therefore, at the very least, a certain amount of product and market stability is prerequisite to a decision to install hard automation equipment, and at the most, hard automation requires huge volumes to justify such a decision.

Although hard automation was the only type of automation available in the 1940s and 1950s, it is by no means obsolete today. Consider, for example, the production of electric light bulbs. General Electric alone produces approximately two billion light bulbs per year [ref. 12]. With this kind of volume, it is easy to justify specialized, high-speed, fixed automation equipment. In addition to their huge volumes, electric light bulbs have achieved a great deal of product stability. It appears unlikely that the incandescent light bulb will become obsolete for several decades. When it does so, millions of dollars of fixed, hard automation equipment will become obsolete with it.

Flexible Automation

Contrasted with fixed or hard automation is the newer type of automation available to today's automation engineer: flexible automation. We hesitate to call this type of automation "soft" because it consists of both hardware and software (computer programs and programmed operating systems). However, flexible automation is certainly soft compared to hard or fixed automation.

The salient and identifying feature of flexible automation equipment is that it is programmable, and therefore reprogrammable. Today, this usually means that the equipment has a digital computer as one of its components, but this was not always so. Early numerical control machines could be considered flexible automation, but they were not computer-based. The reprogrammability of flexible automation equipment gives it a key advantage over hard automation. Huge volumes are not necessary to justify flexible automation, because after the production run is complete the flexible automation equipment can be used again to produce something else. The equipment can be either reprogrammed or preprogrammed for a variety of tasks, calling upon its various learned routines by retrieving programmed software from a suitable storage medium such as magnetic tape or disk.

Flexible automation is the principal subject of this book. Flexible automation first appeared in the early 1950s with the introduction of numerically controlled (NC) machine tools. Later, NC machines evolved into computer numerical control (CNC) machines in which each machine had a small digital computer, usually a microcomputer, as one of its components. The topic of Chapter 5 is NC and CNC machines.

The industrial robot is a key example of flexible automation, and its development was a natural extension of the concepts of NC and CNC. Chapters 6, 7, 8, 9, and 10 discuss the subject of industrial robots. Chapter 6 is an introduction to industrial robots. Chapter 7 explains how an industrial robot can be programmed—the key to its flexibility. The responsibility for implementing a project in robotics is not without its pitfalls, and Chapter 9 is devoted to the topic of robot implementation. Chapter 10 uses case studies to illustrate and evaluate various types of industrial applications for robots.

Industrial robots and all flexible automation equipment must have systems for dealing with conditions as they arise, making logical decisions, and then taking appropriate actions. These systems are called industrial logic control systems and are the subject of Chapter 11. Chapter 12 explains how to diagram and analyze industrial logic control systems. The actual hardware for implementing industrial logic control systems is the programmable logic controller (PLC), the topic of Chapter 13. Programmable logic controllers are not well understood by the general public, and certainly they are overshadowed by the sensational industrial robots. But many industrial robots are themselves controlled by standard programmable logic controllers. It is even possible and practical in many cases for an automation engineer to design and build an industrial robot in-house, using a standard programmable logic controller for the robot's "brain."

Even before the development of programmable logic controllers, digital computers were employed to control industrial processes. Most such computers controlled continuous processes such as chemical plants, refineries, cement plants, and paper mills, but discrete-item manufacturing can also be controlled by process control computers. Most engineers are familiar with batch-mode data processing and computer programming, but process control computer programming and application have some key differences from conventional computer data processing. These differences are introduced in Chapter 14 so that the automation engineer can visualize the integration of industrial robots and flexible automation equipment into an overall computer-controlled factory system.

Finally, in Chapter 15 the driving force of the industrial robot and flexible automation revolution—the microprocessor—is isolated. The development of the microprocessor is one of the key technological breakthroughs of the century. Other potential candidates are the invention of the airplane, the discovery of nuclear power, and the invention of television, transistors, the computer, and the laser. The microprocessor is the most recent entry into this elite group, and, as will be seen in Chapter 15, there is reason to believe that its impact upon humankind will eclipse the others. At present, it is fueling the robot revolution, and it promises an exciting new era for manufacturing automation and the engineers who will create this era.

Chapter 16 introduces computer-integrated manufacturing (CIM), interfacing, flexible manufacturing systems (FMS), and Group Technology (GT). Chapter 17 closes with the issue of ethics and where we ought to be going regarding the relationship between robots, automation, and humankind.

SUMMARY

In this chapter, we have attempted to satisfy two objectives: (1) to enable the automation engineer to assess and appreciate the need for factory preparation for automation, and (2) to provide an outline to prepare the reader for the rest of this book.

Robots and manufacturing automation hardly need a sales pitch. Manufacturing industries, especially in industrially developed countries such as the United States, already feel a sense of urgency to automate due to growing worldwide competition. This urgency has been compared to that felt by the prey of a boa constrictor. The boa subtly tightens its coils every time the prey relaxes and exhales in an economic upturn with its usual wage increases and tolerance for production costs.

The terrifying urgency compelled by the boa constrictor of worldwide competition has caused some naive and impulsive engineers and manufacturing managers to plunge into automation before their products and processes are ready. Perhaps the greatest benefit to be gained from an examination of the feasibility for installing an industrial robot is the attention it places upon the design of the product, the stability of the process, and the reliability of existing process machines and equipment. This prerequisite design, analysis, and improvement of product and process may be so beneficial as to remove the need for an industrial robot or other automated system.

One prerequisite for robots and manufacturing automation needs special scrutiny: production volume. Huge volumes generally are considered essential to the success of manufacturing automation, but this old truth is beginning to lose ground. Recent advances have made possible the automation of medium-range production runs. These advances have introduced machines that are programmable and of sufficient flexibility to be used in a variety of applications. Flexible automation is the principal topic of this book, and the archtype of flexible automation is the industrial robot and the industrial logic systems that control it.

EXERCISES AND STUDY QUESTIONS

1.1. Why is a sense of urgency associated with the implementation of industrial robots and manufacturing automation?

1.2. Would you say that the industrial robot is a technological breakthrough? Why or why not?

1.3. What are some alternatives to automation in attempting to compete with industries worldwide?

1.4. Compose some product design rules to facilitate robots and manufacturing automation.

1.5. Does part symmetry help or hinder automation? Explain.

1.6. Consider the two product component designs in Figure 1.14 that are considered equal in terms of product quality, cost, and effectiveness. Which design would appear to be more appropriate for automation? Why?

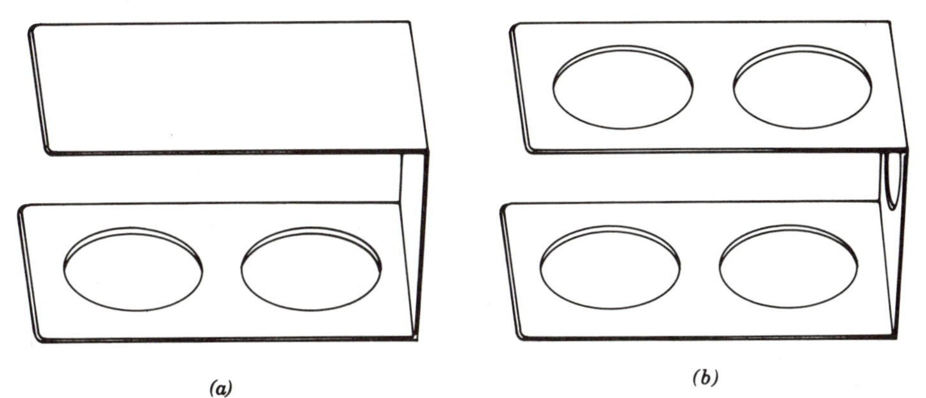

(a) (b)

FIGURE 1.14 Alternate product designs (Exercise 1.6).

1.7. An industrial process has a piece-part dimension as follows:

$$3.500 \pm .006 \text{ in.}$$

Example 1.3 & p. 14-16

The dimension as toleranced above can be held, provided the process is capable of being controlled to "three sigma limits." Without changing the process, what tolerance would have to be specified to accept "six-sigma" variation in this process dimension? Suppose that an industrial robot that is expected to pick up the piece-part has a tolerance for this dimension of $\pm .050$ in. Would you say that this process has sufficient stability for the employment of the robot described?

1.8. In Example 1.3, suppose that the piece part dimension mean shifted $^{1}/_{100}$ in. higher due to tool wear, while the balance of the process remained the same. With the new mean dimension of 2.01 in., what percentage of the piece parts will the robot be able to pick up?

Similar to above

CHAPTER 2

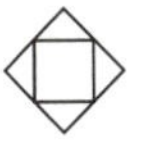

BUILDING BLOCKS OF AUTOMATION

The objective of this chapter is to identify basic devices for implementing an automation project, whether or not an industrial robot will be used in the final implementation. Successful productivity improvement projects usually isolate tiny bottlenecks or inefficiencies at the work station level, where people—not machines—are still playing the lead role. One by one, better methods are being found to sense, move, position, orient, fabricate, and assemble products using a wide variety of ingenious basic components—the building blocks of automated systems. Until a manufacturing work station has been analyzed thoroughly and fitted out with the basic components of automation, it usually is not ready for more exotic hardware such as industrial robots. Indeed, the industrial robots themselves are constructed of some of these same basic components of automation. The purpose of this chapter then is to examine and classify the basic components of automation at the work-station level, components that will appear in examples and case studies throughout the remaining chapters of this book. Some of the items examined in this chapter (for example, gears, switches, and motors) will be common and familiar to almost all readers but are listed here anyway to quicken the reader's awareness of the automation potential of these devices. Also, it is not the *design* of these components that we are studying here; that is another kind of engineering. Our objective is to recognize the *usefulness* of these items and to see how they can be integrated into systems for enhancing manufacturing productivity.

To make sense out of the diversity of automation components, some sort of crude classification is needed. The classification is rough because some of the most useful components find their way into several of the categories, depending upon how they are used. We can, however, talk about components as primarily belonging to one of the following four classes:

1. Sensors
2. Analyzers
3. Actuators
4. Drives

The approximate relationship of these four broad categories is shown in Figure 2.1. It should be pointed out that the operator in Figure 2.1 is a human, not a robot. The industrial

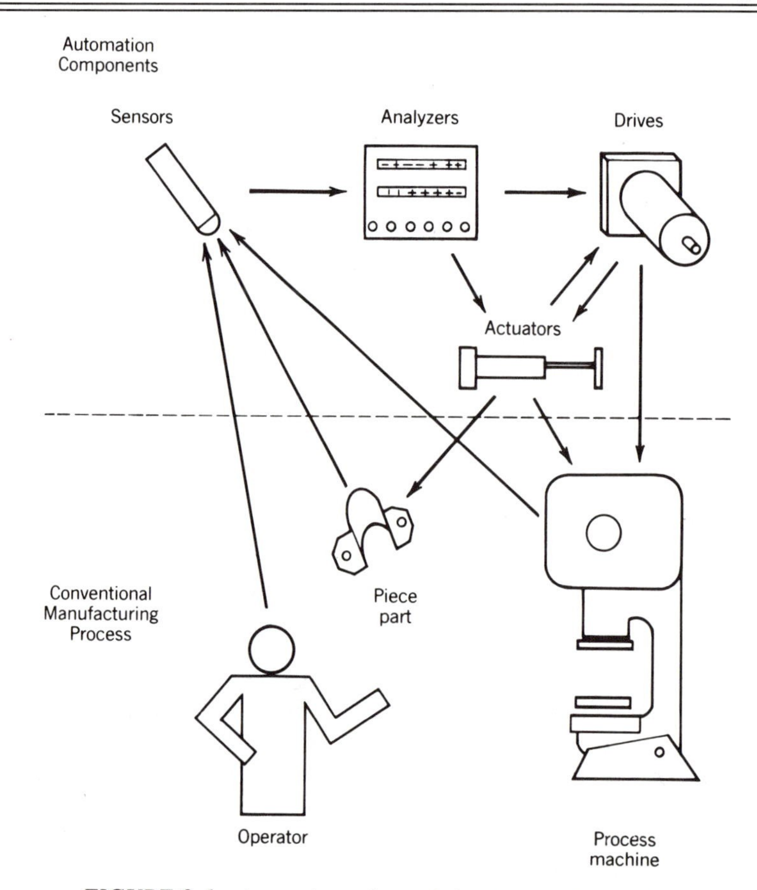

FIGURE 2.1 Approximate interrelationships of the four basic categories of automation components and their interface with the conventional process.

robot is part of the automated system (upper half of the figure). The industrial robot is actually an integrated system made up of all four of the basic automation component categories—sensors, analyzers, actuators, and drives.

SENSORS

Sensors are the first link between the typical automated system and the conventional process, as was indicated in Figure 2.1. Sensors convey information from the manufacturing process equipment, the piece part being manufactured, and from the human operator, if any. It may seem strange that the automated system senses the human operator, but this is without doubt the most important link between the automated system and the real world. The first sensor we shall examine is in that vital link.

Manual Switches

The most familiar sensor of all is the manual switch. Most people do not think of an electric lamp switch as a sensor, but the switch is the link between the lamp and the person who desires the lamp to be turned on or off. In the same fashion, an automation system is linked to the operator, who may desire to turn the system on or off or make adjustments to the automated cycle. Virtually all manual switches are electric, and we shall ignore any exceptions to this statement.

For a switch to be on means that the circuit it controls is closed or "made." When a switch is off, its circuit is open or "broken." Most switches have two stable states: on and off. However, many switches have only a single stable state. Such switches have a spring action that returns them to the normal state whenever they are released from an outside force. That normal state can be in either the open position or the closed position, which leads to the terms *normally open* (NO) and *normally closed* (NC) used to describe switches.

The ordinary wall switch is an example of a *toggle switch,* and in its simplest form it merely opens or closes a single circuit. Such switches are designated *single-pole, single-throw* (SPST) and are diagrammed as in Figure 2.2*a*. The term *single-pole, single-throw* implies that multiple poles and multiple throws are possible, and in fact these types must be examined in order to understand the simpler SPST designation. The toggle switch provides examples of each configuration. Toggle switches on equipment are usually two-position switches, as in the ordinary light switch, but may be three-position switches, with the center position designated as off. This makes it possible for the switch to complete two different circuits. Toggled to the left makes circuit *A;* toggled to the right makes circuit *B;* and the center position holds both circuits *A* and *B* open. Even when a toggle switch has only two positions, it may be wired to throw two circuits. Figure 2.2*b* illustrates such a switch in which one lead to the switch is common but the position of the switch determines whether circuit *A* or circuit *B* is made. Such a switch is designated *single-pole, double-throw (SPDT)*. The switch is single-pole because of the common lead on one side of the switch, but it is double-throw because it can complete either of two circuits on this same common pole. It is possible also to have two leads on both sides of the switch, which enables the switch to make two different circuits that do not have a common pole. In effect, such a switch is making two different circuits with a single mechanical throw. Such a switch is called a *double-pole, single-throw* (DPST) switch and is illustrated in Figure 2.2*c*. In still another configuration, each contact on one side of the switch can be connected to either of two contacts *each* on the other side of the switch. Such a switch is a *double-pole, double-throw* (DPDT) switch and is illustrated in Figure 2.2*d*.

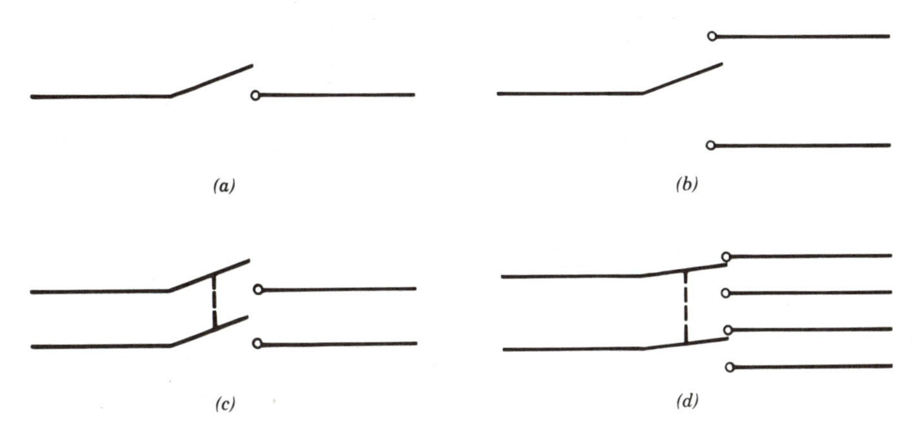

FIGURE 2.2 Example electric switch configuration. (*a*) Simple switch: single-pole, single-throw (SPST). (*b*) Single-pole, double-throw switch (SPDT). (*c*) Double-pole, single-throw switch (DPST). (*d*) Double-pole, double-throw switch (DPDT).

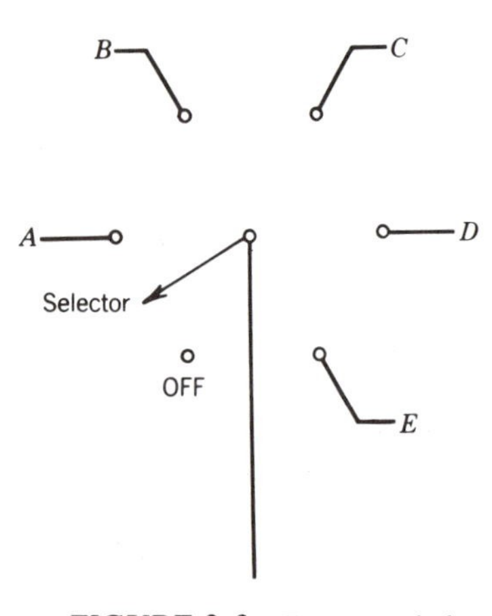

FIGURE 2.3 Rotary switch.

It is possible to add as many poles as desired, but to go beyond two throws the use of a toggle switch becomes impractical. For multiple throws, a *rotary switch* becomes appropriate. Figure 2.3 illustrates a rotary switch that could be classified as a single-pole, five-throw switch, although it is usually designated simply as a six-position rotary switch. A mechanical detent is placed at each of the six positions. Note that there are six positions but only five throws, because one detent is used for the off position to represent "all circuits broken." A manual control knob without detents is usually a continuous control used for some variable in the automated process that can take on any value over a continuous range. Such control knobs may vary voltage, resistance, or capacitance in an electrical circuit and thus vary such quantities as speed, position, or intensity in the automated process. However, for the purpose of applying robots and automation to manufacturing systems, the understanding of switches and other discrete selection devices is vastly more important than the understanding of continuous control devices.

The safety of robots and other automated systems demands that most operator control switches have only one stable state: off. Thus, a positive operator action must be maintained to keep the switch on. The advantage of single-stable-state switches is that control can be exercised whenever and wherever it is necessary to change the operational state of the system. To understand this point, consider the folly of designing a large, complicated automated system that is controlled by a single toggle switch in which both the on and off positions of the switch are stable. To kill the system in an emergency, the operator would have to find that lone switch and turn it off. But with a *momentary switch,* the system could be switched on at one location and then switched off by other momentary switches at any of several convenient locations around the machine. In the automation industry, these momentary off switches located at various points about the equipment are called *emergency power offs* or simply *EPOs*.

A point that was given little attention in the foregoing discussion was that positive operator action must be maintained to keep the switch on. Such a mode of operation may seem impractical, but note that we are referring to the switch—not to the system itself. Through logic circuitry it is possible to hold the system on and to use additional logic to turn if off immediately by any of several EPOs. How to accomplish this and other feats of industrial logic control is the subject of Chapter 11.

The most popular physical configuration of a momentary switch is the familiar *push button*. Push buttons can be either "making" or "breaking"—that is, normally open or

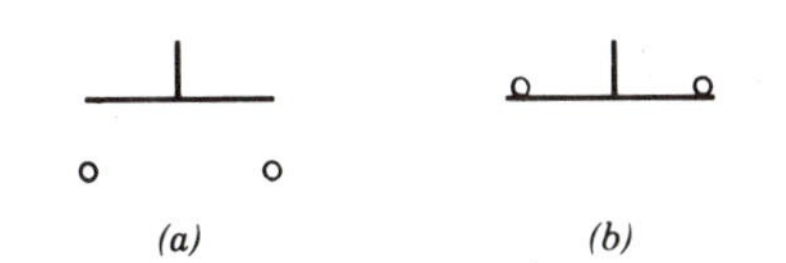

FIGURE 2.4 Momentary (spring-return) push button. (*a*) Normally open. (*b*) Normally closed.

normally closed, respectively. Figure 2.4 illustrates the standard momentary push-button diagram in both configurations. There is another type of push-button switch that has two stable states; the action of the push button is to switch the circuit to the opposite state, whatever that state might have been. Such a switch resembles a toggle switch in function but may have a disadvantage in failing to display which of the two feasible circuit states (open or closed) currently exists.

Limit Switches

Like manual switches, *limit switches* are actuated mechanically, but limit switches are automatic inputs from the manufacturing process, the material, or the automated system itself, without intervention by the operator. There are literally thousands of styles and models of limit switches, even more varieties than of their cousins, the manual switches. The reason for this is that limit switches must be designed to be exactly correct in size, lever travel, force of actuation, and ruggedness for the specific automation application desired out of the myriad of automation applications that might be feasible. By comparison, manual switches are designed for human operators who have relatively similar physical characteristics. In addition, many limit switches are used in situations for which they must be "industrially hardened" or built to withstand severe industrial environments to which a human operator might not be exposed. Limit switches are actuated by levers, toggles, push buttons, plungers, rollers, "cat whiskers," and just about anything else the inventor can devise to make an automation application feasible. Figure 2.5 illustrates a wide variety of limit switches used in automated systems.

Robot systems employ limit switches both in the construction of the robot itself and in the peripheral equipment. Limit switches can be used to limit the travel of a robot arm on any of its axes of motion. When the limit is reached, a circuit is opened (or closed) that removes power from that axis of motion either directly or via the robot controller. As an example, in the peripheral equipment a limit switch can be used on a gate in the perimeter barrier (see Figure 2.6) to act as an interlock (an automatic EPO) whenever someone enters the gate.

Proximity Switches

Some switches do not require physical contact or light radiation to "feel" or sense an object. Such switches are called *proximity switches* because they can sense the presence of a nearby object without touching it. Proximity switches are thus capable of performing a feat that is in the realm of the supernatural if attempted by human operators. As such, they can be used on robots to give the robot certain advantages over human operators.

Figure 2.7 illustrates two types of proximity switches, the first of which acts only upon presentation of a *ferrous* metal object within its sensing range (Figure 2.7*a*). The other type of switch is capable of sensing both ferrous and nonferrous metal objects

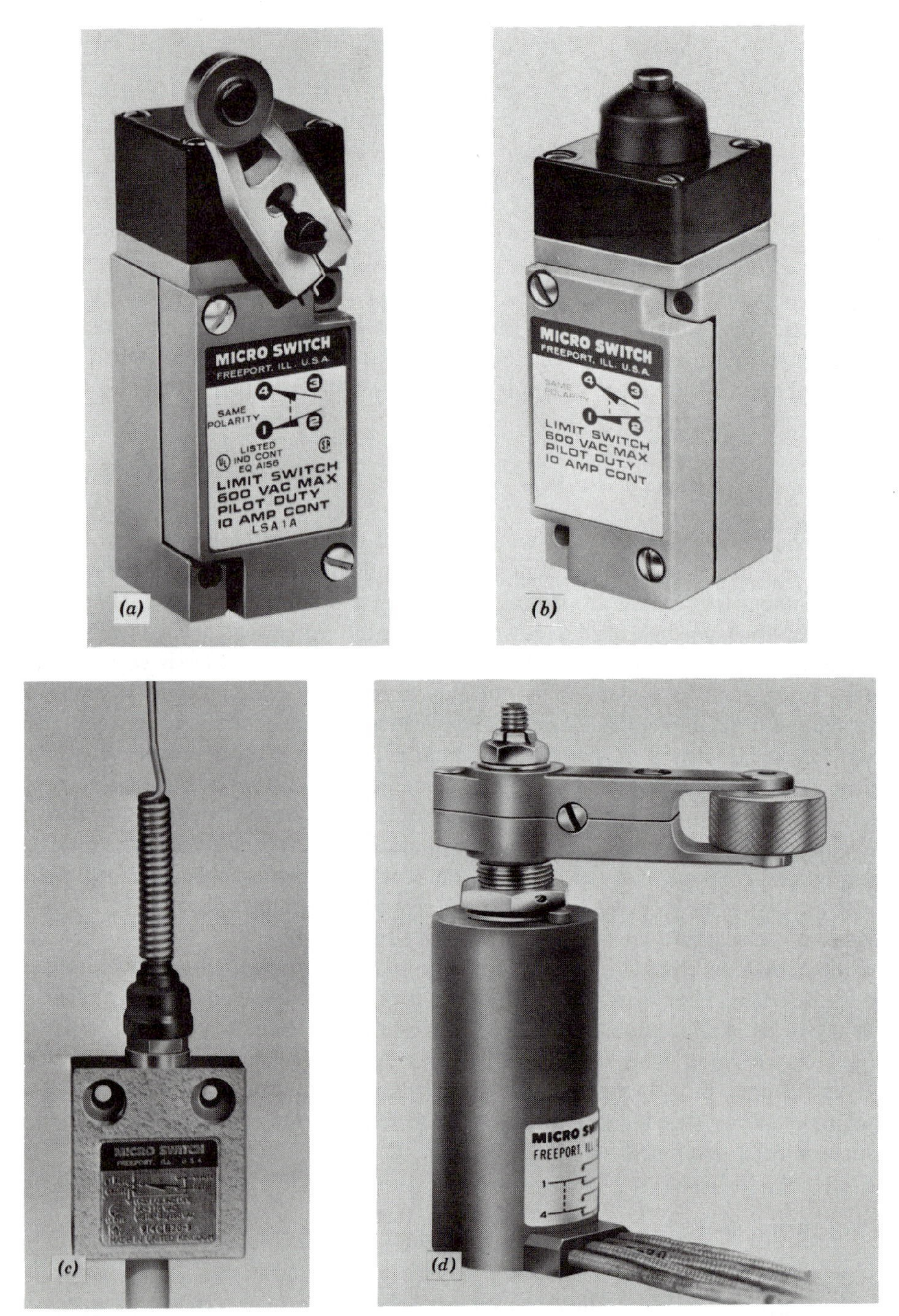

FIGURE 2.5 Examples of popular models and some of the more unusual styles of limit switches. (*a*) Standard side rotary limit switch. (*b*) Standard top plunger limit switch. (*c*) Miniature limit switch. (*d*) Aircraft-type limit switch. (Reprinted by permission of Micro Switch, a Honeywell Division, Freeport, Ill.)

FIGURE 2.6 Robot system equipped with interlocked barrier. (Reprinted by permission of Prab Robots, Inc., Kalamazoo, Mich.)

(Figure 2.7*b*). At first it might seem that the robotics and automation engineer would always prefer the switch that functions with all metals. However, consider a sorting application where items are similar, but one has more ferrous metal content. Another advantage to the ferrous-metal-only proximity switch is that nonferrous metal barriers become transparent to the switch so that the switch sensor can "see through" the barrier to act upon the presence of some ferrous metal object on the other side of the barrier. Figure 2.8 illustrates the use of a ferrous-only proximity switch sensing ferrous metal objects behind an aluminum surface.

There are physical bases for proximity switches that can respond to any object—metal or nonmetal. One type uses an electromagnetic (radio frequency) antenna specially designed and placed to fit the application. The antenna receives a signal transmitted by another strategically placed antenna, but the reception of the signal is disturbed by the intrusion of any object into the field. This disturbance is detected by the antenna that trips a switch when the disturbance reaches a specified level. Unfortunately for some applications (and fortunately for others), the sensitivity of the antenna is related to the electrical properties of the material of the object being detected. The size of the object to be detected also plays a role. The system can be tuned to be somewhat selective to specific items.

Another type of proximity switch that works for nonmetallic objects is the sonar type. Sonar systems transmit and receive reflections of pressure waves to detect object presence. These pressure waves are commonly called sound waves when their frequencies are within the audible range. Most sonar systems, however, use ultrasonic radiation, which has frequencies higher than audible pressure waves.

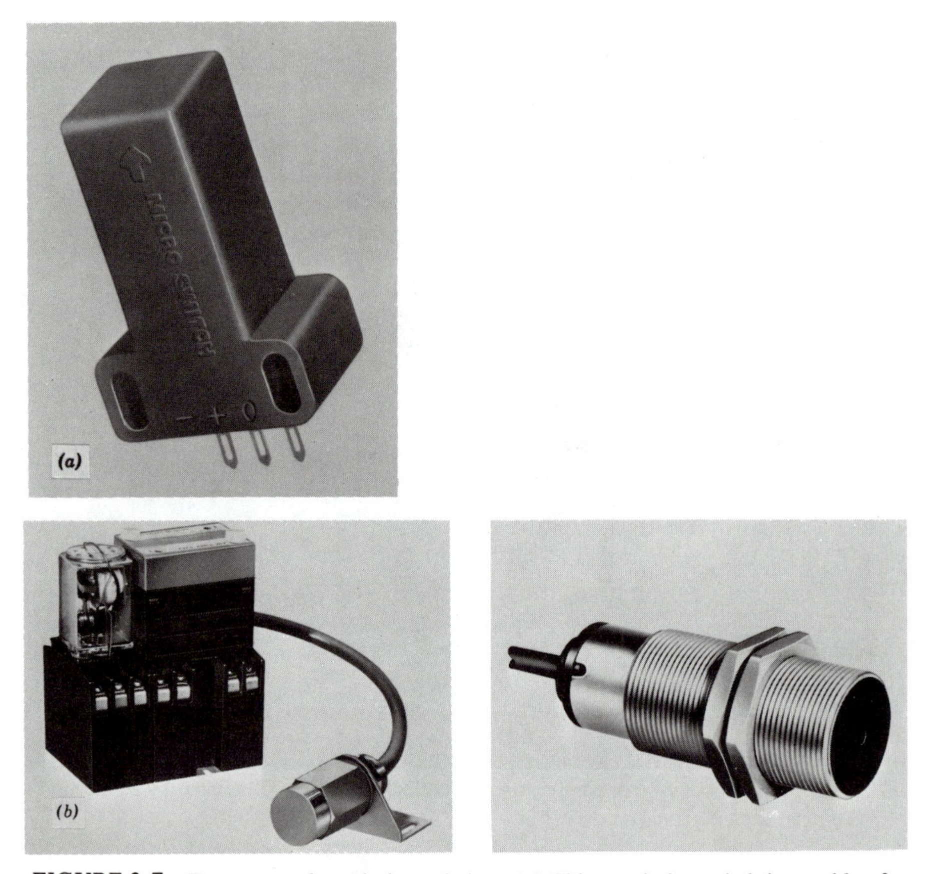

FIGURE 2.7 Two types of proximity switches. (*a*) This proximity switch is capable of sensing only objects containing ferrous metals. (*b*) These proximity switches are capable of sensing any metal object regardless of ferrous content. (Reprinted by permission of Micro Switch, Inc., a Honeywell Division, Freeport, Ill.)

A sophisticated proximity switch system employs the *Hall effect,* in which a small voltage is generated across a conductor carrying current in an external magnetic field. The amount of Hall voltage is proportional to the flux density of the magnetic field, which is perpendicular to the flow of current. This proportionality enables Hall effect proximity switches to detect not only presence but also relative distance to a sensed object.

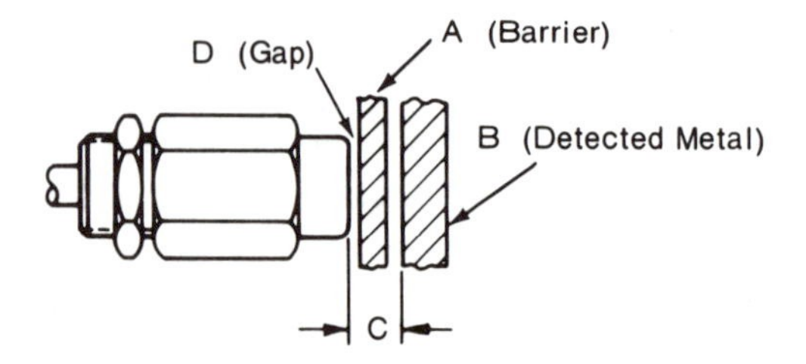

FIGURE 2.8 Nonferrous metal surface is "transparent" to the ferrous-only proximity switch that is detecting ferrous objects behind the nonferrous barrier. (Reprinted by permission of Micro Switch, Inc., a Honeywell Division, Freeport, Ill.)

Photoelectric Sensors

In wider use than proximity switches are sensors that are sensitive to light radiation: *photoelectric sensors*. Two basic approaches for employing photoelectrics are in use. The first approach merely uses a photocell to detect the presence of light radiating naturally from some object in the process. A good example is the use of photocells to turn on lighting systems automatically at dusk and to turn them back off again at dawn. Increasing emphasis upon energy costs has generated additional interest in this kind of automated system.

The second approach to photoelectrics employs a beam of light emitted by an artificial light source. The principal purpose of this approach is to detect the presence or absence of objects in the path of the beam. The beam emitter can be a separate unit or can be incorporated into the sensor. The combination variety requires some type of natural or artificial reflector to direct the light beam back to the emitter/sensor.

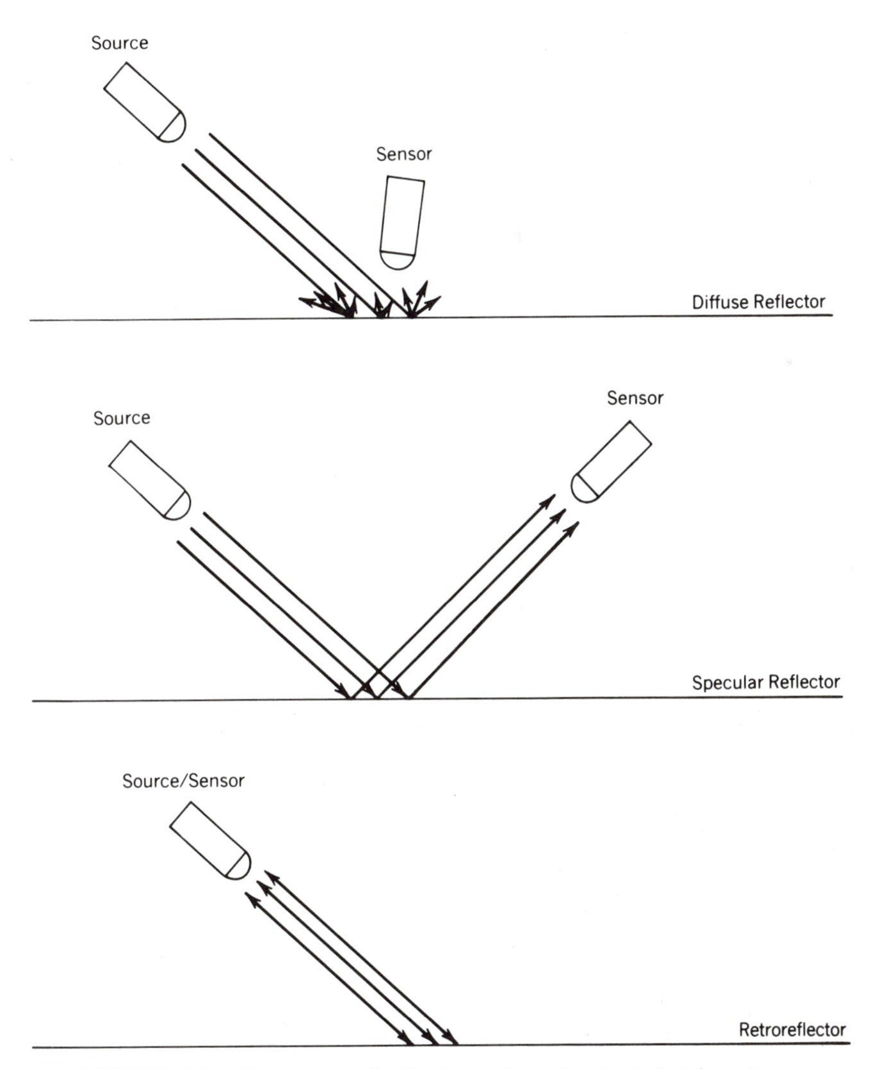

FIGURE 2.9 Three types of reflective surfaces for photoelectric systems.

Reflective surfaces for photoelectric systems are of three types: diffuse, specular reflective and retroreflective, as shown in Figure 2.9. The diffuse reflective surface is the lowest in cost and describes most reflective surfaces. Even an ordinary white object acts as a diffuse reflective surface in that it reflects light but not images. Diffuse reflectors scatter so much light that only a small fraction makes its way back to the photoelectric sensor. Therefore, the savings in the diffuse surface may be lost in the necessary provision for a more sensitive and perhaps more sophisticated sensor. Photoelectric systems that use diffuse reflective surfaces are also more susceptible to stray signals.

Specular reflective surfaces are most often associated with the word *reflective* and include mirrors and very shiny surfaces. Specular reflective surfaces obey the physical law that the angle of incidence equals the angle of reflection. It is obvious that the source and sensor must be more closely aimed for specular reflective surfaces than for diffuse surfaces. For systems in which the emitter and sensor are mounted in the same unit, the plane of the specular reflective surface must be perpendicular to the direction of the incident beam or the reflected beam will be lost. Once again, this can be either a disadvantage or an advantage, depending upon the objective of the inventor or automation engineer.

Retroreflective surfaces are the most complex and expensive of the three types. Retroreflectors are capable of reflecting back to the source a large percentage of the light beam regardless of the angle of incidence. Basically, the retroreflective surface violates the physical principle that angle of incidence equals angle of reflection, except when the plane of the surface is perpendicular to the incident beam.

A common example of a retroreflective surface is the red reflector on the rear of a bicycle. Although the reflector is not illuminated internally, it still glows brilliantly from the retroreflected light of the headlamp of a vehicle behind it. The light is concentrated in a beam back to the source regardless of its direction within certain operational limits. Certain highway signs, highway paint stripes, reflective tape strips, and movie screens are coated with a surface of tiny beads that cause a portion of incident light to be bounced back to the source. At least partially, these surfaces can permit photoelectrics to take advantage of their retroreflective properties. Obviously, alignment is not as important as for specular reflective surfaces. Sensitivity of the photocell can be lower than for diffuse reflective surfaces, so retroreflective surfaces somewhat combine the advantages of specular and diffuse reflective surfaces, but at a price—the retroreflective surface is the most expensive of the three types.

As useful as photoelectrics are, the automation engineer should beware of conditions that can foul, confuse, or even ruin the photoelectric system. Stray ambient light, even from worker-carried flashlights, can be a problem in confusing the system. Stray light from welders' arcs can be even more destructive, perhaps permanently damaging the sensor. High temperatures from manufacturing processes may also damage equipment. If the manufacturing environment presents a dirt or vapor problem, dust or condensation may accumulate on lenses or mirrors. If manual cleaning of these surfaces is necessary, some of the benefits of automation may be defeated. Vibrations from adjacent machines must be anticipated and dealt with in automation planning. Vibrations may cause source-sensor misalignment problems and may lower lamp reliability, especially for incandescent lamps.

There are ways to overcome most problems with photoelectrics, especially if the automation engineer plans for these problems from the beginning. For example, ambient light can be shielded and mirrors can be used to redirect the beam. Dust and condensate problems can be alleviated by automatic air jets. Another trick is to use overdesign when adverse environmental conditions are expected. For example, if the beam system is rated at 8 to 10 feet, use it at distances of 2 to 3 feet. To solve problems of vibration-caused misalignment, a retroreflective system may be the answer. Other problems may be alleviated by use of infrared beams or fiber optics, as will be discussed in the sections that follow.

Infrared Sensors

Sometimes it is useful to detect electromagnetic radiation outside the visible range. Infrared sensors respond to radiation in the range of wavelengths just beyond the visible spectrum at the red end. Hot objects emit infrared radiation, and thus infrared sensors are useful for locating heat sources in a process. Such applications in which "natural" infrared radiation is sensed are useful for monitoring systems to detect malfunctions.

Infrared sensors are also very useful when used with artificial beams to detect the presence or absence of objects, even more so than are photoelectric systems. Since infrared radiation is invisible, there are some advantages of using infrared beams and receivers instead of ordinary photoelectrics. Also, infrared sensors are virtually unaffected by stray ambient light—with obvious advantages.

A strategy that is gaining popularity is the use of a modulated infrared beam, in which the source is pulsed to provide much greater intensity and the sensor is modulated to receive at the same frequency. The modulated beam permits a much greater range for otherwise weak beam sources such as LEDs (light-emitting diodes). LEDs are solid state and have great advantages over incandescents in terms of power requirements and reliability.

Fiber Optics

A convenient supplement to photoelectric or infrared sensor systems are fiber optic light tubes, which are flexible pipes of glass or plastic that can be used to bend light beams around corners. When bundles of fibers are used together, whole images can be transmitted. However, the typical automation application is to use one fiber to transmit a light beam that is sensed by the system as either present or absent.

An advantage of fiber optics is their surprising efficiency. Fiber optics are so efficient that it becomes worthwhile for the telephone industry to convert communications circuits from electrical signals to modulated light signals for transmission via fiber optics and subsequent reconversion at the receiving end.

Lasers

Before leaving the subject of photoelectric sensors, laser light should be considered. Lasers are concentrated, amplified beams of collimated light. They are capable of delivering over a distance a large amount of energy into a tiny spot and thus have obvious industrial applications. In automated systems, the laser is useful in providing very long, precise light beams. Figure 2.10 shows the precision available from laser systems for providing very narrow, concentrated beams of light. The precision of these beams makes them excellent for detecting tiny objects that are capable of breaking the beam at large and varying distances. The presence or absence of a continuous beam then can be used as a logic input to an automated control system. Such precision also makes the laser a good tool for dimensional measurement.

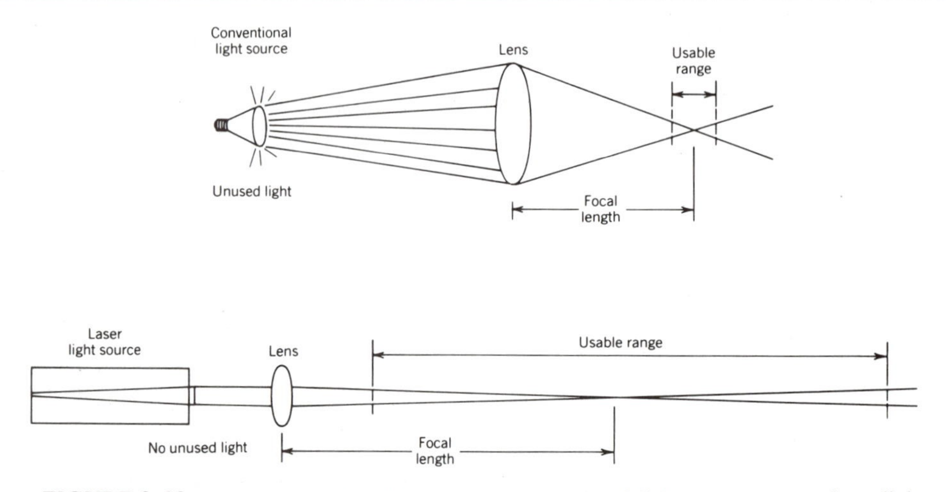

FIGURE 2.10 Comparison of usable range of conventional light sources versus laser light sources for the purpose of locating objects at a distance.

ANALYZERS

Once information is sensed by an automated system, it must be registered and analyzed for content, and then a decision must be made by the system as to what action should be taken. This function can be quite complex, and the system components that perform it are generally too complicated to discuss in detail in this chapter. But some of the components deserve mention here to enable the reader to understand the components of NC machines, robots, programmable controllers, and other manufacturing automation devices discussed in the remaining chapters.

Computers

Digital computers are the primary means of analyzing automation system inputs. Computers are extremely versatile in that the ways they can be programmed to manipulate data are limitless. The continuing miniaturization of computer circuits along with decreasing costs made possible by technological breakthroughs has created a continuing increase in the number of feasible applications of manufacturing automation. No other development has had more impact upon the growth of industrial robots. Computers are so important as analyzers that they are discussed in depth in Chapters 14 and 15. The importance of computer application to robots and other automation equipment is discussed throughout this book.

Counters

It is frequently useful for an automated system to determine how many of various items are present or pass through an automated system. This function can be handled either internally by a computer or programmable controller or externally by a separate device called a *counter*. The counter can be mechanical, but most automatic systems employ

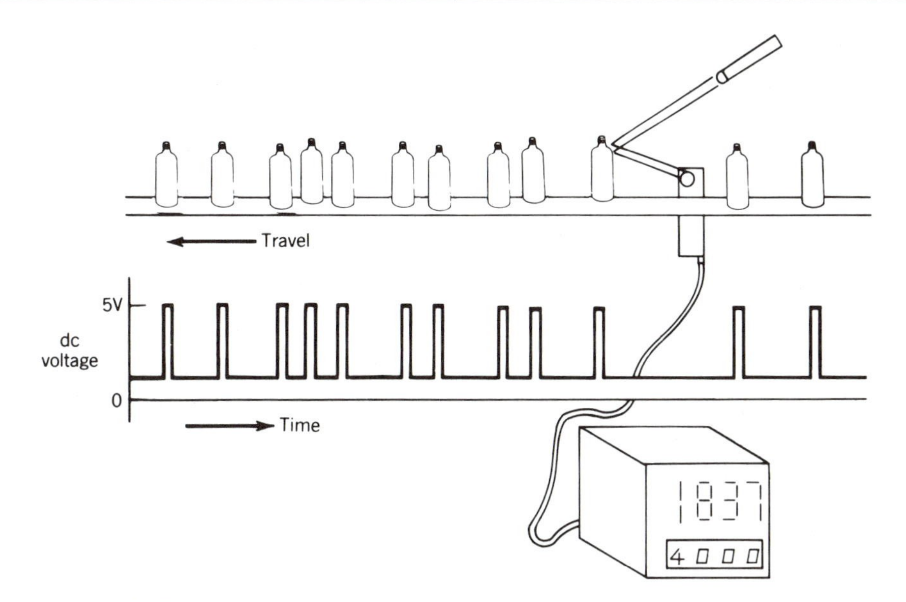

FIGURE 2.11 Automatic counting system using photoelectrics to count bottles coming down a conveyor line.

solid-state electronic counters. If the counter is a separate unit, it will usually have a display to report current status of the count in progress.

The quantity counted is usually a series of voltage pulses that have been generated by a sensor detecting some physical quantity of importance to the automated system—for example, glass bottles coming down a conveyor line. The curved glass would be somewhat specular, and at a precise position the angle would be exactly right to reflect a pulse of light from a positioned source upon a photoelectric sensor, as shown in Figure 2.11. The sensor would convert the light pulse to a voltage pulse, which would be transmitted to the counter.

Note in Figure 2.11 that the spacing of the voltage pulses is not uniform. It is also possible for the peaks themselves to be of varying width (time) and amplitude (voltage). This is entirely satisfactory—within limits, of course—because most industrial counters are capable of detecting peaks and peak intervals of less than 100 μs duration. There is also some tolerance in the voltage required to generate the count.

In Figure 2.11, the counter shown has two display registers. The top register is an electronic display (LED type) and represents the current count in progress. The bottom register is mechanical and represents a preset number that acts as a target to trigger an output signal when the current count register reaches the target value. This feature enables the counter to cause something to happen in the manufacturing process. A good example would be the automatic batching of manufactured products into preset lot sizes. Another useful feature of many counters is *bidirectionality,* which enables them to accept two kinds of inputs: count up and count down. This can be useful in automatic industrial quality control applications and in material handling.

The counter shown in Figure 2.11 is a separate unit, but in Chapter 13 we shall see that the counter can be an internal function of a programmable logic controller. Counters can also be devised through computer software (programming) to permit counting internal to an on-line process control computer.

Timers

If precise clock pulses are available, a counter that counts these pulses becomes a *timer,* basically a clock. An industrial timer is more like an alarm clock than an ordinary clock. When elapsed time becomes equal to a preset value, an output signal is generated—hence, the likeness to an alarm clock. Like counters, industrial timers can be bidirectional—that is, time up and time down.

Timers often have the additional feature of being interruptible—that is, they can be cumulative in summing the various periods of voltage up time interrupted by various periods of voltage down time. This concept will become clearer when it is demonstrated in Chapter 12, which discusses the diagramming of industrial logic control systems. Timers are very useful devices in industrial logic control systems.

The applications of industrial timers to robots and manufacturing automation is even greater than that of their cousins, the industrial counters. Besides being available as separate units, industrial timers can be internal to programmable logic controllers and on-line process control computers.

Bar Code Readers

Although it can be considered a sensor, a *bar code reader* is an analyzing system that incorporates a conventional photoelectric or laser scanner along with timers and counters. Successive bars of varying width, as seen in Figure 2.12, are scanned and counted. The scan is orthogonal to the bars, and thus voltage pulses from the photoelectric sensor can be compared to determine individual bar widths. The sequence and width of bars is then analyzed to decode the bars and translate them to an alphameric data string for processing by the automated system, as shown in Figure 2.13.

Bar code scanners usually employ lasers because the concentrated coherent laser light can retain its focus over a larger depth of field, as shown in Figure 2.10. What this is saying is that the bar code label does not have to be precisely positioned at a fixed distance from the scanner. The laser depth of field permits various-sized boxes or items to pass under the beam at varying distances and still be scanned by the scanner head. This point is further illustrated in Figure 2.14. Note also in Figure 2.14 that the items to be scanned have random orientations and present nonorthogonal surfaces, especially the footballs and baseballs. The zigzag scan pattern enables the laser scanner to seek out the bar code label, regardless of orientation.

Alternatives to bar code readers can be easier for humans to read but present more problems in manufacturing automation systems. Figure 2.12 shows two optical character recognition formats that are used in some automated systems. Other alternatives are Mark-Sense, widely used in scoring examinations and questionnaires, and magnetic ink printing, used almost exclusively in the banking industry on imprinted checks. Unfortunately, the automatic recognition of these characters is a much more delicate operation than is needed for bar codes. Alignment is important, and some of the characters have tiny differences. Consider the problem in designing a scan procedure to distinguish between the letters E and F, O and Q, or P and R. Multiple scans are essential, and defective type or smudges on the labels can easily result in a misread. Scanning programs contain checks for inconsistencies that can prompt a rescan if necessary to assure a valid read. After repeated attempts with no success, the scanner can be programmed to report "INVALID READ." Some label deficiencies, however, can trick the scanner into an *erroneous read,* a problem that is usually worse than an *invalid read.*

The probability of obtaining an erroneous read from a bar code scanner is made extremely low by the design of the code. A typical code employs a series of five bars

FIGURE 2.12 Examples of standard bar code and optical character recognition formats. (Courtesy, Metrologic Instruments.)

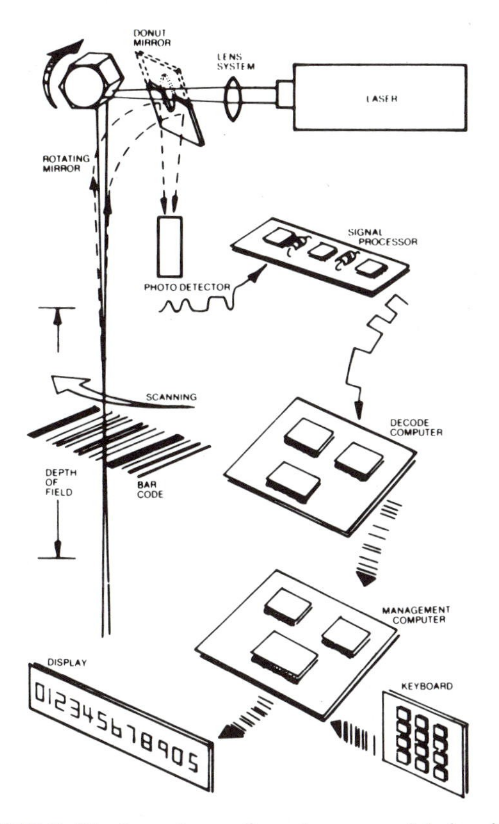

FIGURE 2.13 Laser bar code reader scans a label and processes the code to produce a numeric string for use by an automated data system. (Courtesy of Metrologic Instruments.)

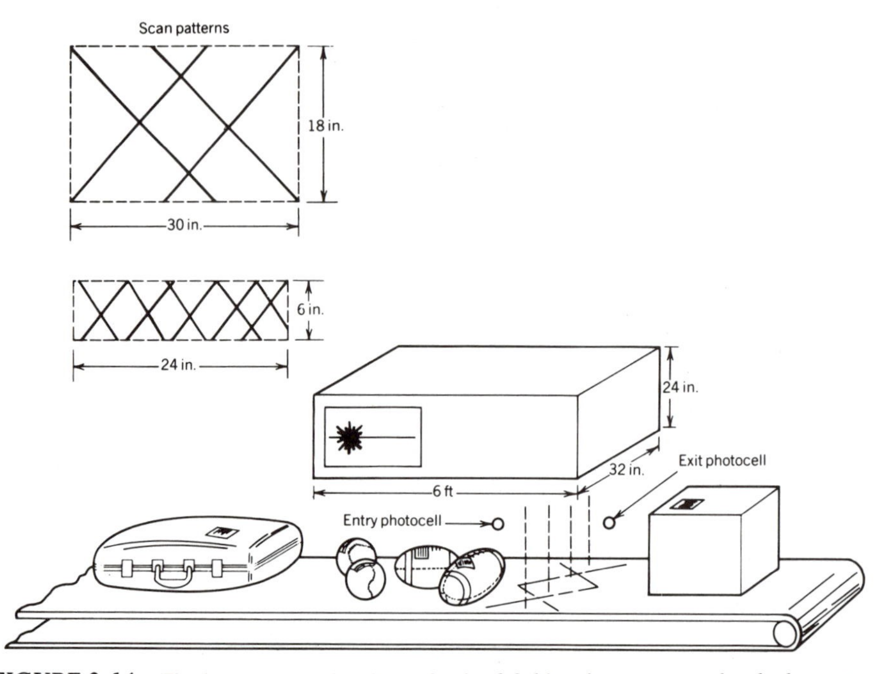

FIGURE 2.14 The laser scanner has large depth of field and uses a cross-hatched scan to identify bar code labels of random orientation and distance from the scanner. (Courtesy of Rexnord, Inc., Milwaukee, Wisc.)

and four intervening spaces for a single alphanumeric character. Each bar and each space can have either of two widths, which gives the code as many readable character combinations as a nine-digit binary number. Since $2^9 = 512$, and since fewer than fifty alphanumeric and special characters are utilized, the vast majority of the 512 bar combinations will be recognized by the scanner as invalid, prompting a rescan.

Further enhancing the reliability of bar code readers is the fact that some label errors will not cause a problem because the label will still be interpreted correctly. Consider the situation in which a narrow bar is adjacent to a wide space, as shown in Figure 2.15*a*. With the bar edge damaged, as shown in Figure 2.15*b*, no problems will result because the damage only makes the bar read narrower and the space read wider. If the edge damage is so great that a bar is completely broken (reads as a space) or two bars touch (space reads as a bar), the scan will invariably be detected as invalid because of compound errors.

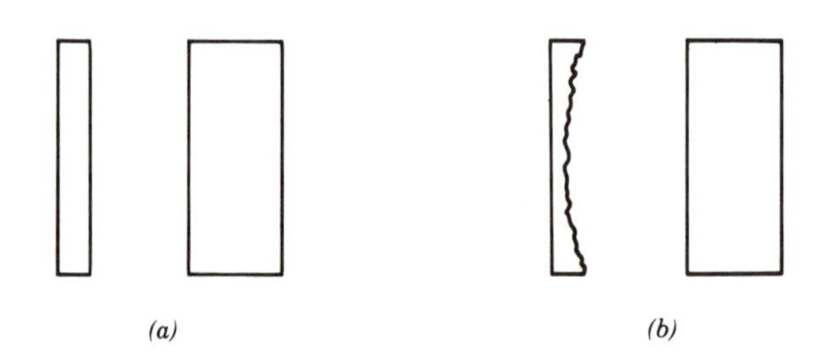

FIGURE 2.15 Example bar code series in which a narrow bar is adjacent to a wide space. The edge damage shown in (*b*) does not cause an erroneous read. (*a*) Narrow bar adjacent to a wide space. (*b*) Edge damage on a narrow bar.

If the bar code scanner shows a large proportion of refusal scans (invalid code detected), the chance of misreading a code combination increases because large numbers of refusal scans mean increased frequency of system faults in the equipment, the label, or the method. But even if the first scan reject rate is as much as 10 percent, the character substitution read error is tiny: less than one in several hundred thousand.

EXAMPLE 2.1
Code Design

Suppose that a machine is capable of imaging six different character configurations, as shown in Figure 2.16. As can be seen in Figure 2.16, the designer has decided not to use two of the feasible character configurations. These two unused characters are useful in detecting errors due to erroneous read or faulty code printing. Thus, if one of the two unused configurations is read by the machine, a false scan is indicated and a rescan is necessary to obtain a valid reading. In the given application a total of 30 distinguishable product codes is necessary. How many digits are required in this coding system? In the event of a reader error on the first pass, what is the probability that a wrong code will be interpreted as a valid identification of one of the 30 products?

For all six characters, assume equal likelihood of being misread and equal likelihood of being the result of any misread.

Solution

Each digit position provides four valid character configurations. The coding system is thus a base-4 numbering system. To distinguish between 30 different product codes, multiple digits are required. Two digits would permit $4^2 = 16$ distinguishable codes. Three digits would permit $4^3 = 64$ distinguishable codes. Since 16 is less than 30, but 64 is greater than 30, three digits are required.

Using three digits permits 64 unique combinations when only 30 are required. Thus over one half of the combinations are unused, but these are not wasted because they can be used for checking for read errors. The 34 unused combinations are interpreted as invalid, and thus are interpreted as erroneous and a rescan is signaled. In fact, since two of the feasible configurations (the all-black and the all-white squares) were also unused, a much larger set of invalid combinations lies within the scanning capability of the system. Considering all six configurations as scannable, using a three-digit system it would be feasible to read as many as $6^3 = 216$ combinations, of which only 30 are considered valid. In such a system the remaining ($216 - 30 = 186$) combinations would represent invalid code that would generate a rescan. Thus, in the event of a reader error, there are 215 possible interpretations ($216 - 1$). Of these 215 interpretations, 29 would be interpreted as valid code (although the code would be for the wrong product), and the rest ($215 - 29 = 186$) would be correctly interpreted as a reader error and signal a rescan. Therefore, the probability that a reader error would result in a wrong code being interpreted as a valid product identification would be

$$P = \frac{29}{215} = 0.139$$

It should be noted that the calculated probability is conditional upon a reader error occurring.

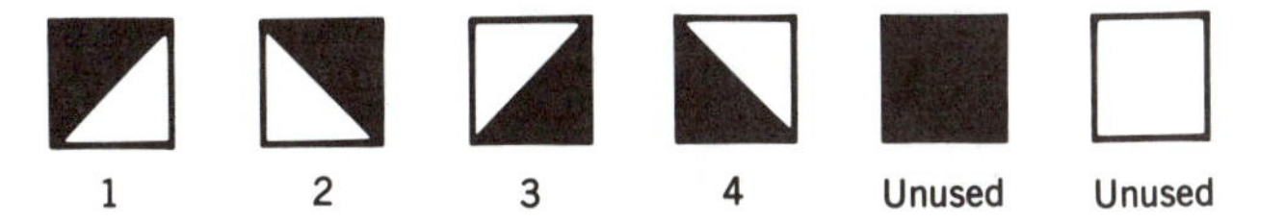
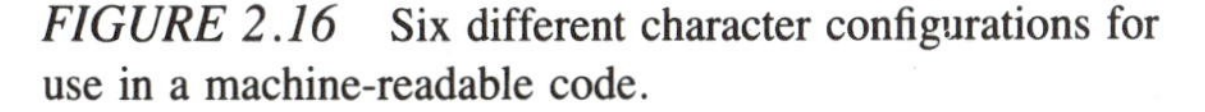

FIGURE 2.16 Six different character configurations for use in a machine-readable code.

Optical Encoders

The capability of rapidly scanning a series of bars makes possible additional automation opportunities when light and dark bars are placed in concentric rings on a disk. Figure 2.17 shows a portion of such a disk that can be rigidly attached to a shaft and housed in an assembly consisting of optical sensors for each ring. The assembly is called an *optical encoder* (see Figure 2.18) and is useful for automatically detecting shaft rotation. The

FIGURE 2.17 Sample portion of optical encoder disk. (Reprinted by permission of BEI Electronics, Inc., Little Rock, Ark.)

FIGURE 2.18 Absolute optical encoder for monitoring shaft rotation on robots or other automation devices. (Reprinted by permission of BEI Electronics, Inc., Little Rock, Ark.)

shaft rotation information can be fed back into a computer or control mechanism for controlling velocity or position of the shaft. Such a device has application for robots and numerical control machine tools.

Optical encoders can be either incremental or absolute. The incremental types transmit a series of voltage pulses proportional to the angle of rotation of the shaft. The control computer must know the previous position of the shaft in order to calculate the new position. Absolute encoders transmit a pattern of voltages that describes the position of the shaft at any given time. The innermost ring switches from dark to light every 180°, the next ring every 90°, the next 45°, and so on, dependent upon the number of rings on the disk. The resulting bit pattern output by the encoder reveals the exact angular position of the shaft, as can be seen in Example 2.2.

EXAMPLE 2.2
Absolute Optical Encoder

An absolute optical encoder disk has eight rings and eight LED sensors, and in turn provides 8-bit outputs. Suppose the output pattern is 10010110. What is the angular position of the shaft?

Solution

	Encoder Ring	*Angular Value (degrees)*	*Observed Pattern*	*Computed Value (degrees)*
(innermost)	1	180	1	180
	2	90	0	
	3	45	0	
	4	22.5	1	22.5
	5	11.25	0	
	6	5.625	1	5.625
	7	2.8125	1	2.8125
(outermost)	8	1.40625	0	
			Total	210.94

Stated as an equation the computation becomes

$$A = \sum_{i=1}^{n} m_i A_i \tag{2.1}$$

where i = ring number

m_i = 0 if ring i is white (clear)

= 1 if ring i is black (opaque)

A_i = angular value for ring i

n = total number of rings

Twelve-ring optical encoders are available that have resolutions of over 4000 counts per turn. Figure 2.19 on page 42 illustrates a sample low-resolution optical encoder in its actual use as a means of monitoring the position of the wheel for a mobile robot.

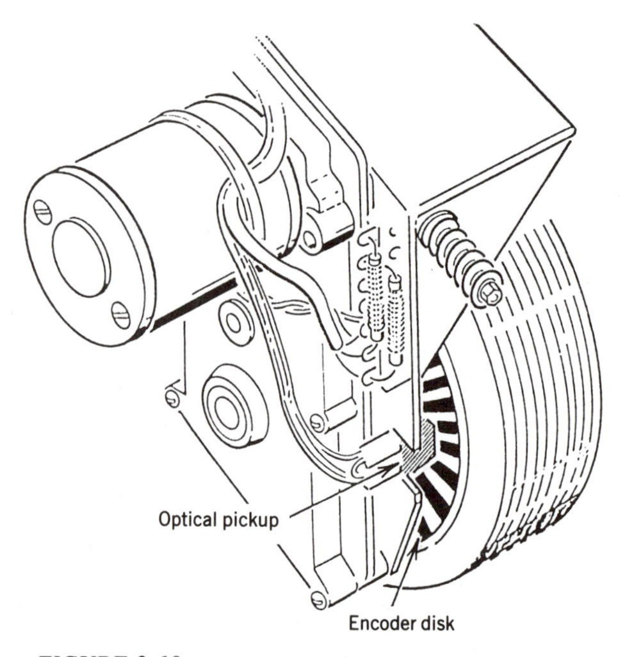

FIGURE 2.19 Sample optical encoder to monitor wheel rotation on a mobile robot. [ref. 44]. (Copyright Heath Company, reprinted with permission.)

ACTUATORS

Once a real-world condition is sensed and analyzed, something may need to be done about it. It is at this point that the automation of many systems ceases because it is believed that a human operator must intervene and apply judgment for taking some kind of physical action. Such systems may be called "process monitoring" if they merely sense and display or record data or "on-line assist" if they also analyze data and give advice or prompts to the operator suggesting specific actions to be taken. However, more and more automated systems are closing the loop by taking physical actions automatically without operator intervention.

Actuation may be a direct physical action upon the process, such as a sweep bar that sweeps items off a conveyer belt at the command of a computer or other analyzer. In other cases, an actuator is simply a physical making of an electrical circuit, which in turn has a direct effect upon the process. An example would be an actuator (relay) that turns on power to an electric furnace heating circuit.

Cylinders

When a linear movement is required in an automation application, a cylinder usually is chosen to accomplish it. The most popular are the pneumatic types because of the convenience of piping compressed air throughout a manufacturing plant. Shop air is generally regulated to the range 80–100 psi, which is adequate for most grippers, movers, positioners, and tool-stroking devices. Figure 2.20 shows the use of pneumatic cylinders in an automatic work station. The control of air cylinders is accomplished by valves that may be driven by electrical impulses or by air logic devices.

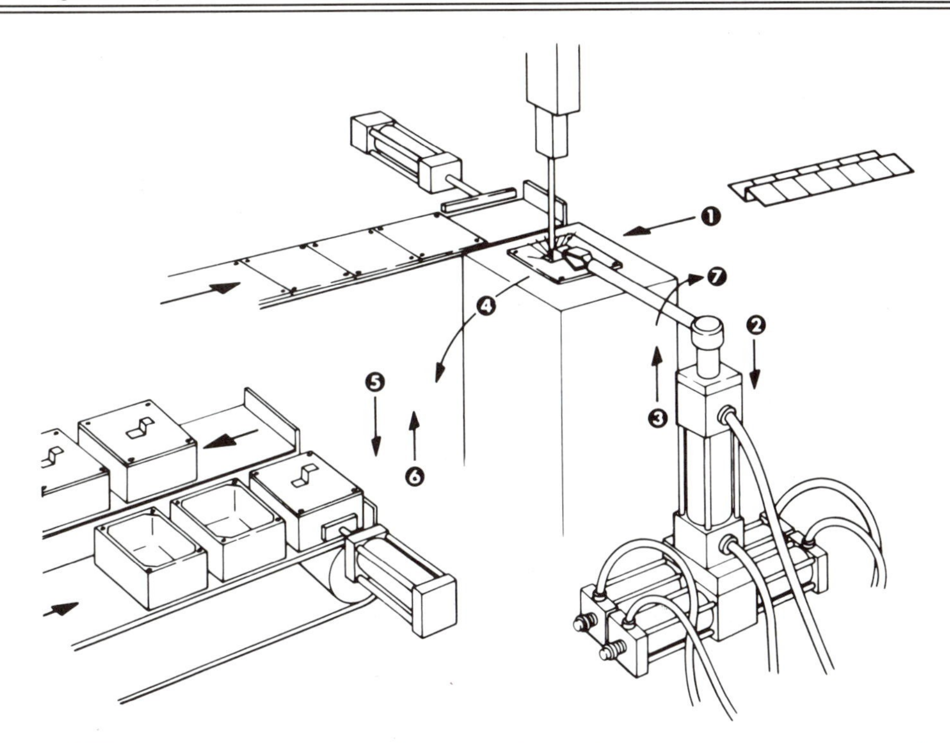

FIGURE 2.20 Use of pneumatic cylinders to achieve robotic motions in an automatic work station. (Reprinted by permission of PHD, Inc., Fort Wayne, Ind.).

When the manufacturing process requires forces to be applied automatically in excess of 200 pounds, the more powerful hydraulic cylinder is usually selected over the pneumatic cylinder. Hydraulic pressures in excess of 2000 psi are readily available; compare these pressures with the 80–100 psi commonly used in pneumatic systems. Given the mechanical advantage of a large enough cylinder, pneumatics can deliver as large a force as the hydraulics, but space and convenience tend to favor hydraulics for the large forces. The most powerful industrial robots are driven by hydraulic actuators.

Besides being powerful, hydraulic cylinders have the advantage of being well controlled throughout the stroke. In addition, they are quiet, although the pump and reservoir can be quite noisy. Disadvantages are high initial cost, maintenance, and problems from leaking cylinders.

A caution to observe in the design of either pneumatic or hydraulic actuators is that both pressure and volume requirements must be met. A system may have sufficient pressure to actuate cylinders or other actuators, but may not be able to maintain that pressure during high-speed operations. This mistake has been observed especially in partially automated factories in which pneumatic systems are used to power mechanized screwdrivers, staplers, and handling equipment. System design that anticipates the demands that will be placed upon the pneumatic or hydraulic equipment and actuators during peak periods will avoid this drawback.

Solenoids

When a small, light, quick linear motion is desired in an automated system, an electrical solenoid is a logical selection. In basic physics, we learned that the principle of the solenoid operation is the creation of a magnetic field set up by passing an electrical current through a coil. Thus, the core of the solenoid can be selectively drawn into the coil in response to an electrical current. In the absence of the coil current, the core can be automatically returned by spring action. The stroke motion of a solenoid is not very controlled in comparison, for example, with a hydraulic cylinder—but many automation applications require only a short, quick, discrete action, not a smooth, controlled stroke.

Relays

The most popular solenoid of all is one that is used to switch an electrical circuit—that is, the common relay. Switching-type circuits usually operate at lower voltages and especially at lower amperage than power circuits. The output of the switching logic network then can be used to trip one or more relays to close or open a power circuit. Figure 2.21 shows the use of relays to close electrical circuits under the automatic control of process sensors. Compare the logic of the two circuits shown. In Figure 2.21*a*, relays from *both* sensors must be energized to make the power circuit. In the arrangement in Figure 2.21*b*, the action of either relay *A* or *B* is sufficient to make the circuit. The myriad ways in which relays can be combined in switching networks form the basis for the classical approach to automating manufacturing systems. Chapters 11 and 12 will show how the requirements of automated systems can be satisfied by means of industrial logical circuits. Such circuits can utilize banks of relays, solid-state integrated logic circuits, or

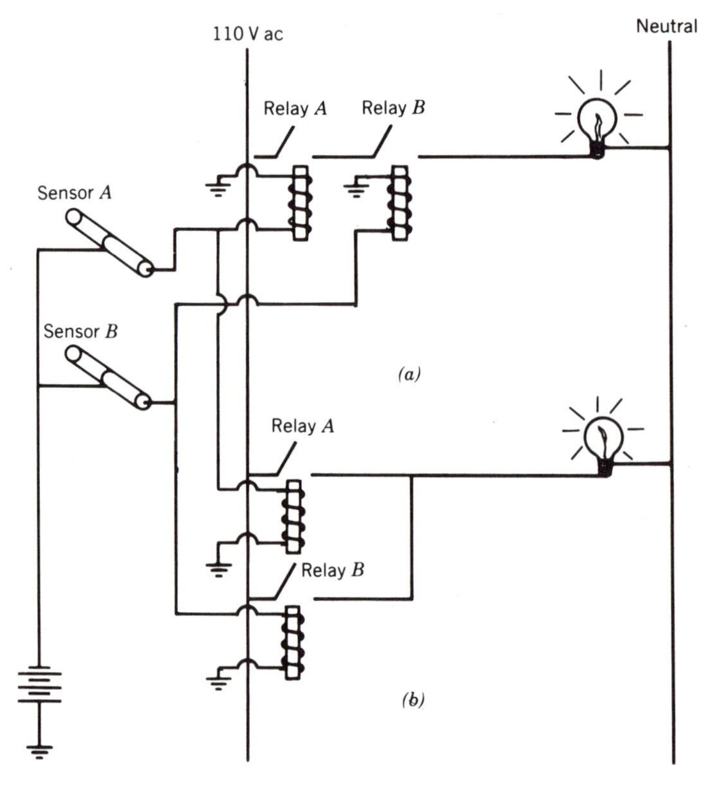

FIGURE 2.21 Using relays to automatically switch power circuits. In circuit (*a*), *both* sensors *A* and *B* must supply voltage to their respective relays to close the power circuit. In circuit (*b*), a voltage at *either* sensor is sufficient to close the power circuit.

programmable logic controllers to accomplish identical functions. Understanding of the concepts of these systems depends upon an understanding of basic relays.

A relay can be described as either "latching" or "nonlatching," a comparison of which is shown in Figure 2.22. A latching relay needs only an electrical *impulse* to pull and hold the power circuit closed. Another impulse is needed on a different switching circuit to release the latch. Nonlatching relays hold only while the switching relay is energized and thus require a *continuous* electrical signal. The discontinuation of that signal permits the relay to release the switch immediately.

Thus far, we have described relays that *make* circuits when energized, but relays can also *break* circuits when they receive an electrical signal. When the energization of a relay coil makes a circuit, the relay is designated "normally open." (The relays of Figure 2.21 are normally open.) Conversely, the relay that breaks a circuit when energized is designated "normally closed." It follows, then, that the normal state of an electric relay is the deenergized state. Figure 2.23 compares a normally open relay with a normally closed relay in a single power circuit. In Figure 2.23*a*, the sensor must provide an energized control input in order to make the power circuit. In Figure 2.23*b*, the control input must be *deenergized* to make the power circuit.

The typical relay and solenoids in general operate on low-voltage direct current. But the convenience and availability of 110-volt alternating current (ac) have given rise to the ac relay and ac solenoid. Another advantage to the higher voltage ac solenoids is their relative immunity to induced voltages from power conductors in the manufacturing environment. A sensitive low-voltage solenoid can be easily tripped inadvertently by a stray voltage induced by a powerful current in an electrical cable that passes close to a relay circuit, especially if the conductor paths are adjacent, parallel, and long.

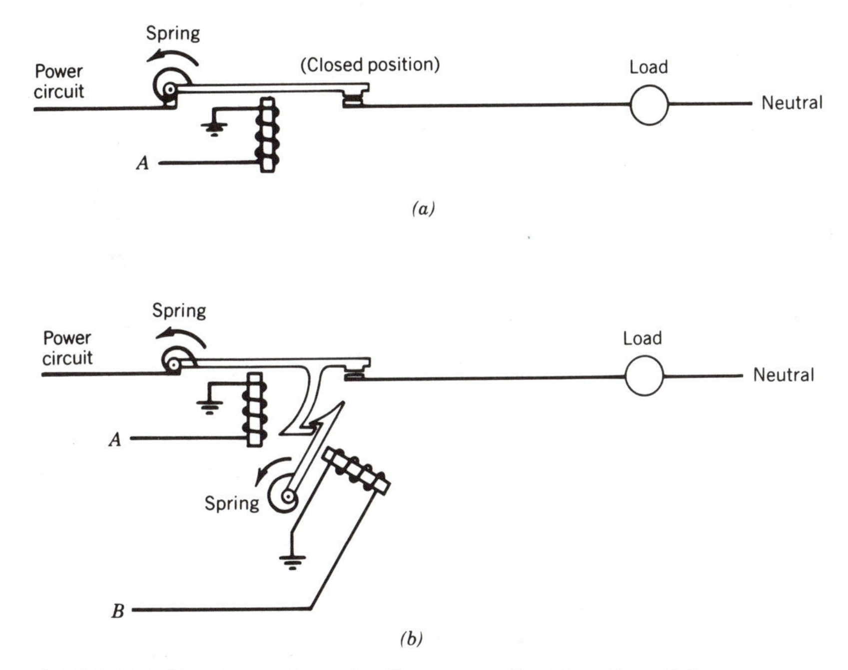

FIGURE 2.22 Comparison of ordinary versus latching relays. (*a*) Ordinary, nonlatching relay is closed only when current is applied to the coil via control input *A*. (*b*) Latching relay stays closed by a mechanical latch after an impulse at *A* pulls it in. An impulse at *B* releases the latch.

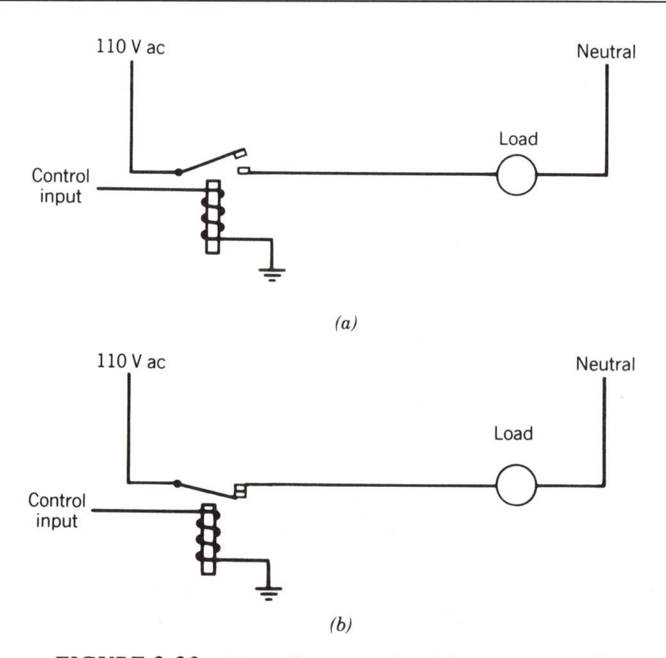

FIGURE 2.23 Normally open relay (*a*) versus normally closed relay (*b*).

As the amperage level of the power circuit increases, the nomenclature for relays changes. Small motors, handling devices, and automated tool actuators usually are served by power circuits of less than 10 A, and the relays that trip them are indeed called "relays." However, as the amperage ranges upward between 10 and 30 A, the nomenclature is *power relay*. At higher amperage, the relay may be called a *contactor*. Still, the basic principle of the simple relay is being employed, and the automation engineer should not be confused by these terms.

A special need for a relay is in the tripping of power circuits for electric motors. The automation engineer will hear reference to "motor starters"; these devices are either contactors or relays that in addition provide overload protection to open the motor circuit if a heavy mechanical load begins to cause the motor to carry too much current. A familiar period of overload is when the motor is first turned on, but a motor starter should permit a tolerable overload during the startup period. For this reason, the overload protectors are usually thermal, allowing them to be somewhat forgiving of a temporary overload.

DRIVES

Like actuators, drives take some action upon the process at the command of a computer or other analyzer. For purposes of classification, the distinction being made here between actuators and drives is that actuators are used to effect a short, complete, discrete motion—usually linear—and drives execute more continuous movements typified by, but not limited to, rotation. Actuators may turn drives on and off, and drives may provide the energy for the movement of actuators. Some automation devices, such as genevas and walking beams, seem to belong to both categories.

Motors

The quintessential drive is the motor. The automation engineer must have a broad perspective of the term *motor* to include not only electric motors but hydraulic and pneumatic motors as well. Internal combustion engines, also called motors, are relatively insignificant in automated manufacturing. Hydraulic and pneumatic motors are the converse of their corresponding pumps. When hydraulic or pneumatic pressure is conveniently available throughout a manufacturing system or automated work station, it may make more sense to tap this pressure to impart motion to a subsystem by means of a hydraulic or pneumatic motor. Hydraulic motors are capable of delivering a large amount of power in a confined space. Compared to hydraulic motors, pneumatic motors are more noisy and less powerful, but they may be more practical in many automation applications, especially considering the availability of compressed air in an automated manufacturing environment. Both pneumatic and hydraulic motors have some advantages over electric motors in systems in which electric motors may be hazardous either from an electrocution standpoint or from the ignition of flammable vapors or gases.

Even when referring to electric motors, automation engineers need to know to which type they are referring. The majority of service motors are ac motors, and a description of the myriad sorts of ac motors is beyond the scope of this book. There are two special types of dc electric motors used for drives for robots and manufacturing automation machines, however, and these deserve special emphasis.

Stepper Motors

For several reasons, the *stepper motor* is a very useful drive in automation applications. For one thing, it is driven by discrete dc voltage pulses, which are very convenient outputs from digital computers and other automation control systems. The stepper motor also is ideal for executing a precise angular advance as may be required in indexing or other automation applications.

Stepper motors are ideal for open-loop operations where the control system gives a specific output command and expects the system to react properly without monitoring results in a feedback loop. Some industrial robots use stepper motor drives, and stepper motors are useful in numerically controlled machine tools. Most of these applications are open-loop, but it is possible to employ feedback loops to monitor the position of the driven components. An analyzer in the loop compares actual position with desired position, and the difference is considered error. The driver can then issue voltage pulses to the stepper motor until the error is reduced to zero.

DC Servo Motors

DC servo motors are useful in numerically controlled machine tools and industrial robots for the control of motion. By using a feedback loop, the controller can deliver to the motor dc voltage that is proportional to the observed error. When the error is reduced to zero, the voltage goes to zero and the motor stops. More sophisticated automatic servos can supply dc servo-motor voltage proportional to the rate of change of error and/or to the summation of accumulated error over time.

One important characteristic of the dc servo motor that is also true of the stepper motor is that both hold their torque when they come to rest under power. Therefore, the power is useful not only for rotating the shaft but also for holding it motionless when no movement is desired. Numerically controlled machine tools and industrial robots are examples of automation equipment that must be able to hold axis positions between commands to change positions.

Kinematic Linkages

When applying automation at the work station of a manufacturing machine, it is easy to forget that an ample and ready source of mechanical power usually exists from the machine itself. For example, automation engineers will sometimes employ separate electric motors to feed a machine automatically in situations in which an off-machine linkage would be cheaper and more practical. Furthermore, there is a distinct advantage of off-machine linkages that is easy to overlook: synchronization. When a machine speed is increased, the kinematic linkages attached to it speed up right along with the machine. This can be a convenience when attempting to balance an assembly line. It is essential that handling systems at the work station coordinate with process machine actions, and off-machine linkages are usually the best way to achieve this coordination. Gears, cams, levers, and ratchets are the components of the kinematic off-machine linkages. The purpose of this book is not to explore the design of these general-purpose mechanical devices, but rather to recognize their usefulness in transferring power from manufacturing process machines to the automated systems that handle and position material in support of the process machines. Two mechanical linkages, however, are of particular importance to manufacturing automation in that they impart intermittent motion to automated flow lines; they are genevas and walking beams.

Genevas

A geneva mechanism is used to drive an indexing table intermittently. Figure 2.24 reveals the drive mechanism (usually on the underside) of an indexing table. Note that there are two wheels: a driver and the driven indexing table. The actual driving force is delivered by a pin on the driver, which slides in a series of slots in the driven indexing table. The driver rotates continuously, imparting motion to the table while passing through angle B of its motion (see Figure 2.25). During the remainder of the revolution of the driver, the driven indexing table is at rest. This at-rest (or dwell) period is the time during which work is accomplished at each work station of the indexing table.

In Figure 2.24, note that the path of the driver pin is precisely radial (orthogonal to tangential) to the driven wheel both at the times it enters and exits the slot. This is necessary to impart a smooth start and complete stop to the indexing table. In fact, any coasting of the table during the station execution (wheel-at-rest) phase will cause the driver pin to miss its next entry slot. This is a serious problem that will jam the drive for the machine and produce other mishaps that may occur on top of the table. Note also that the index and dwell time are dependent upon the driver speed and are constant from cycle to cycle.

It is possible to have various numbers of stations in a geneva indexing table setup, although the number cannot be less than three. More than eight stations should be considered rare. Figure 2.25 describes the relationships between number of stations, dwell time, index time, and driver speed. Note that at four stations angle A = angle B, and driver and driven wheel diameters are the same. An example will now illustrate calculations for a typical geneva.

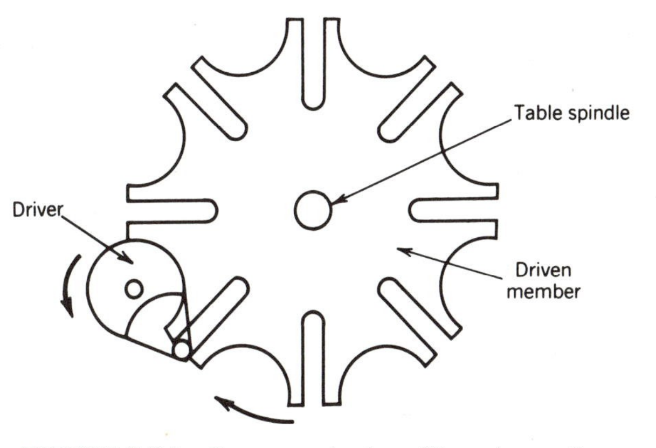

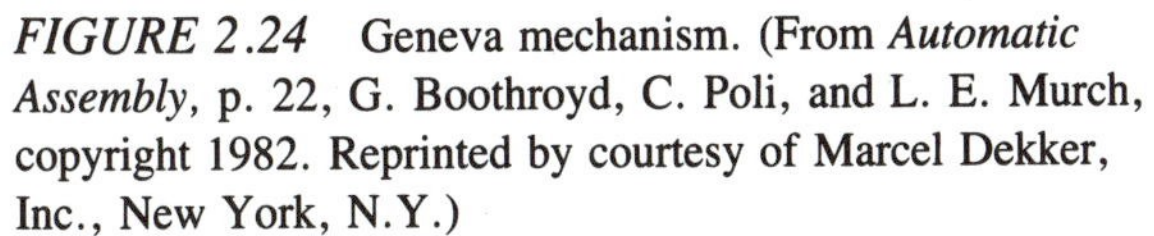

FIGURE 2.24 Geneva mechanism. (From *Automatic Assembly*, p. 22, G. Boothroyd, C. Poli, and L. E. Murch, copyright 1982. Reprinted by courtesy of Marcel Dekker, Inc., New York, N.Y.)

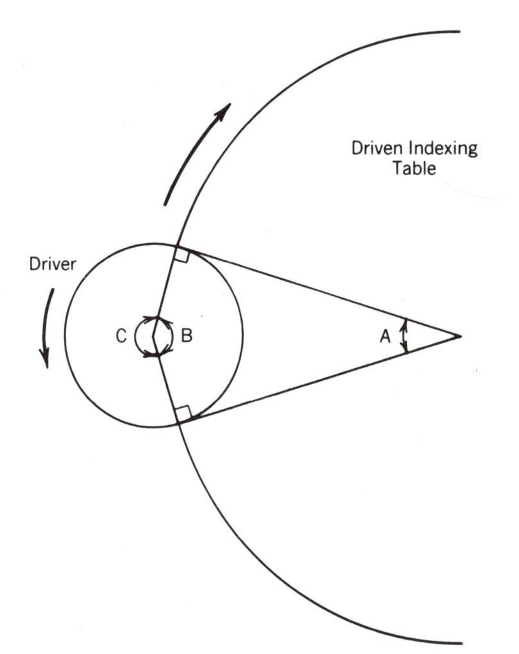

$$\text{angle } A + \text{angle } B = 180^\circ$$

$$\text{angle } B + \text{angle } C = 360^\circ$$

$$\text{Index time} = \frac{\text{angle } B}{360^\circ} \times \frac{1}{\text{driver rpm}} \text{ (minutes)}$$

$$\text{Dwell time} = \frac{\text{angle } C}{360^\circ} \times \frac{1}{\text{driver rpm}} \text{ (minutes)}$$

$$\text{Cycle time} = \text{index time} + \text{dwell time} = \frac{1}{\text{driver rpm}}$$

FIGURE 2.25 Geneva-driven indexing table calculations.

EXAMPLE 2.3
Indexing Table Calculations

An indexing table driven by a geneva mechanism has six stations and a driver speed of 12 rpm. Calculate

(a) index time
(b) dwell time
(c) ideal production rate per hour

Solution

$$\text{angle } A = \frac{360°}{6} = 60°$$

$$\text{angle } B = 180° - \text{angle } A = 120°$$

$$\text{angle } C = 360° - \text{angle } B = 240°$$

$$\text{index time} = \frac{\text{angle } B}{360°} \times \frac{1}{\text{rpm}}$$

$$= \frac{120°}{360°} \times \frac{1}{12 \text{ rev/min}} \times 60 \text{ sec/min}$$

$$= 1.67 \text{ sec}$$

$$\text{dwell time} = \frac{\text{angle } C}{360°} \times \frac{1}{\text{rpm}}$$

$$= \frac{240°}{360°} \times \frac{1}{12 \text{ rev/min}} \times 60 \text{ sec/min}$$

$$= 3.33 \text{ sec}$$

$$\text{cycle time} = \text{Index time} + \text{dwell time}$$

$$= 1.67 \text{ sec} + 3.33 \text{ sec}$$

$$= 5 \text{ sec}$$

$$\text{production rate (ideal)} = \frac{1}{5 \text{ sec}} \times 3600 \text{ sec/hr}$$

$$= 720 \text{ pieces/hr}$$

Walking Beams

Walking beams are a means of intermittently indexing a linear type of automated line and are thus analogous to the genevas used to drive rotary indexing tables. A walking beam is driven by an actuating cylinder, as shown in Figure 2.26 but has an advantage over the geneva in permitting arbitrary setting of index and dwell times by varying the cylinder stroke and return times. A constantly rotating crank also can be used to operate the beam, which would fix the relationship between index and dwell.

Walking beams have a reputation for being somewhat noisy and can damage the manufactured product by dropping the line onto the rests every time the line is indexed. However, the walking beam is more flexible than the geneva in the number of work stations it can conveniently accommodate. Linear-type automated lines also have plant layout advantages over rotary indexing tables, especially when the number of stations is large.

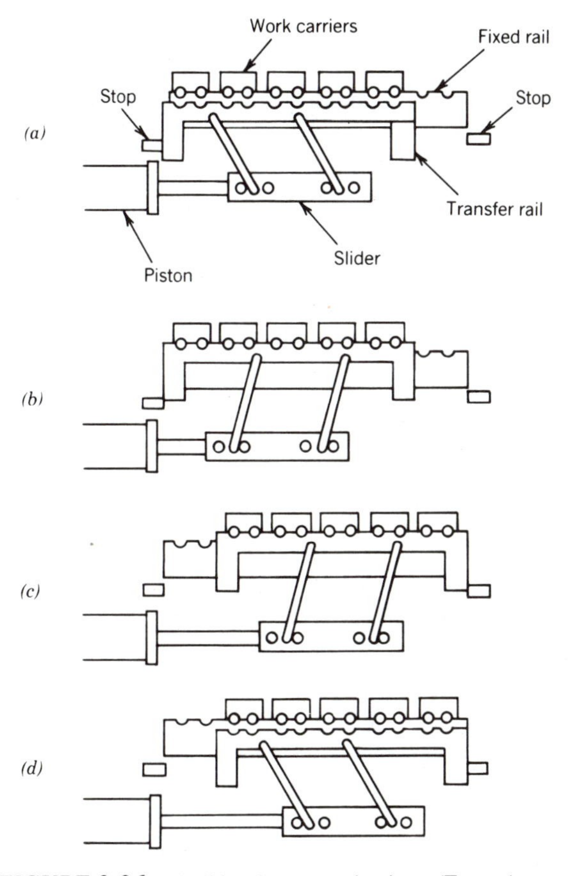

FIGURE 2.26 Walking beam mechanism. (From *Automatic Assembly,* p. 14, G. Boothroyd, C. Poli, and L. E. Murch, copyright 1982. Reprinted by courtesy of Marcel Dekker, Inc., New York, N.Y.)

SUMMARY

This chapter has provided the "nuts and bolts" of mechanization and automation. Automation begins with the simple and sometimes not so simple mechanization of portions of the operation of individual work stations. Although it is wise to keep the systems approach in mind for total automation of a factory, in reality most factories are automated a piece at a time.

It is difficult to classify the components of mechanization and automation, but four broad categories have been identified in this chapter: sensors, analyzers, actuators, and drives.

The application of switches and sensors for specific automation objectives usually provides advantages that cannot be seen until future automation objectives are added to the system. To a lesser degree, the same can be said for actuators and drives that execute the output phase of the automated system.

By describing types of integrated automated systems and their utility in manufacturing, the remainder of this book will build upon the various components of automation and mechanization discussed in this chapter.

EXERCISES AND STUDY QUESTIONS

2.1. What would be the most likely physical configuration of a single-pole, six-throw switch?

2.2. In Figure 2.27, the first line of bar code reads –609–. By using this information and comparing the two lines of code, you should be able to decipher the second line of code.

(a) What does the second line of code read? (Hint: Each character consists of five bars.)

(b) Write out the code for the digit 6.

FIGURE 2.27 Two lines of bar code for Exercise 2.2.

2.3. A photoelectric system is intended to detect the presence or absence of $^{3}/_{8}$-in. stove bolts that pass by in a correct orientation on a track. The bolts must completely break a beam that has a diameter of $^{1}/_{2}$ in. Is the photoelectric system technically feasible? What problems do you foresee? What are possible solutions to the technical problems presented, if any?

2.4. The switch in Figure 2.28 has how many poles? How many throws?

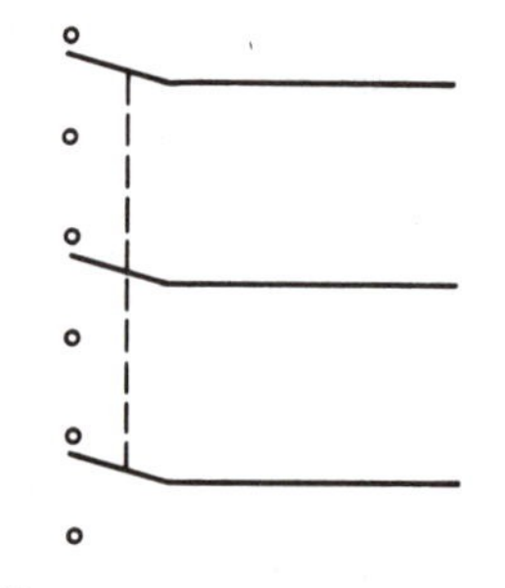

FIGURE 2.28 Switch configuration for Exercise 2.4.

2.5. An absolute optical encoder has 4 rings on its optical disk. The bit pattern output reads 1001. What is the absolute angular position of the shaft?

2.6. Suppose a label code has 500 feasible character configurations but there are only 50 valid characters recognized by the code. Would you say that the other 450 feasible characters were wasted? Explain your reasoning.

2.7. In the code of Exercise 2.6, if a random reader error occurs, calculate the approximate chance that a wrong character will be substituted.

2.8. Suppose you are setting up an automated system to employ the code described in Exercise 2.6 to identify warehouse pallets by means of identification labels.

The company has 600 such pallets, and you have the responsibility of selecting how many digits long all the labels will be. What minimum number of digits can feasibly handle the 600-pallet system?

2.9. In Exercise 2.8, suppose a random reader error occurs. What is the approximate chance that the system would result in a wrong pallet interpretation instead of a rescan given that the labels have two digits.?

2.10. Suppose company operating supervisors complain about the reliability of the automated system described in Exercise 2.9. How much improvement in reduction of wrong pallet interpretation could be achieved by using four digits instead of two?

2.11. Note in Figure 2.12 that the OCR-A code for the letters F and R is somewhat peculiar in that the top half of the letters is smaller and appears squeezed upward. Can you explain this seeming irregularity in the OCR-A format?

2.12. A certain bar code character consists of five bars and four spaces. Each five–four set has two wide bars and three narrow ones and has one wide space and three narrow ones. How many unique combinations exist? (Note that since the machine must see exactly two wide bars, three narrow bars, one wide space, and three narrow spaces, it would be virtually impossible to read a wrong character; instead, the machine would indicate an "invalid read.")

2.13 A familiar coding system has six valid characters as shown in Figure 2.29. The coding system of Figure 2.29 uses the presence or absence of a black dot in any of seven possible positions on the face of the die. Thus it can be seen as a seven-digit binary number system in which only six binary numbers are considered valid code. What is the probability that a reader error will cause a wrong code to be substituted instead of generating a rescan?

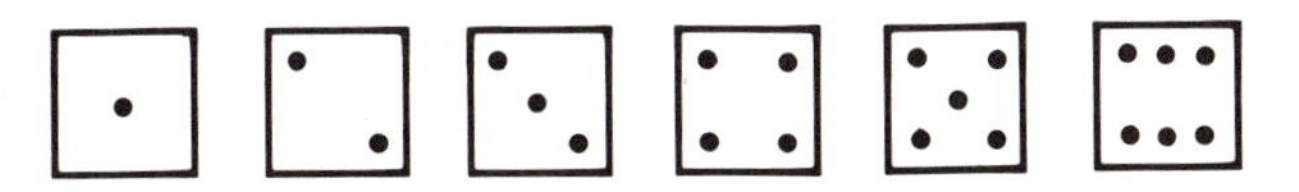

FIGURE 2.29 Six feasible characters represented by the faces of a die (as in a pair of dice).

2.14 In the dice-style coding system of Exercise 2.13 note that it would have been possible to place three rows of three dots each on a single die face as shown in Figure 2.30. Thus there would be nine positions on the face of the die, upon each of which a dot could be present or absent. Explain how this expanded system could be used to improve coding system reliability while retaining the original six-character code.

FIGURE 2.30 Expanded format for dice-style coding system.

CHAPTER 3

MECHANIZATION OF PARTS HANDLING

A prerequisite to most robotics applications is the mechanization of parts handling. Humans may be adept at selecting pieces in the random orientations in which they may be found in a storage bin, but most robots need a carefully positioned piece-part in a consistent orientation. Some sophisticated robots equipped with machine vision are able to accomplish what is known in the industry as "bin-picking," but these robots are typically not economically feasible for most robot applications. Failure to recognize the need for mechanized parts handling and orientation has resulted in total failure of many robotics applications, a reality now recognized by many automation engineers in industry who have rushed in to buy a robot before dealing with the subjects covered in this chapter.

The automated handling and positioning of piece-parts can be extended to include automated assembly. The automation engineer should consider assembly, too, when attempting to solve a handling problem, a point that will be stressed in this chapter. It will be found that many of the same devices used for parts handling can also be used to assemble the parts to a subassembly.

PARTS FEEDING

Fabricated piece parts, whether they be forgings, castings, machined parts, plastic injection moldings, wood milled products, electrical components, or rubber moldings, must be transported, selected, oriented properly, and positioned for assembly or subsequent operations. The automation engineer whose plant receives these components aligned, positioned, and ready for assembly is fortunate. However, whether shipped from a subcontract supplier or from another department within the plant, components usually are received jumbled into a large hopper. This fact leads to the first principle of automating an assembly station or any work station after the initial parts fabrication: parts source compatibility.

Parts Source Compatibility

When automating the assembly station, the first place the automation engineer should look is where the components were before they came to the assembly station. The engineer

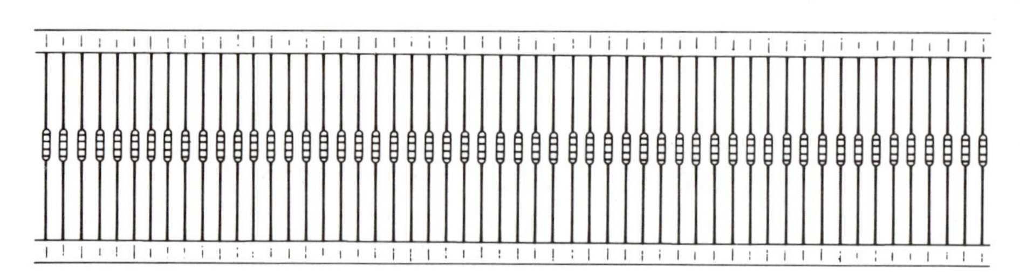

FIGURE 3.1 Continuous feed ribbon of electrical resistors.

should ask, "Was it really necessary for the components to arrive for assembly at this station all jumbled and piled into a hopper?" Even if the components were purchased outside the plant, it is sometimes permissible to specify that they be shipped in magazines or continuous feed strips, or at least compartmentalized containers. Witness the common industry practice to make electrical resistors and nails available in continuous strip form (see Figure 3.1), certainly a boon to the automated processes that use these components.

When writing specifications for purchased parts and when qualifying vendors, the packaging and orientation of parts should be a consideration. The material handling and in-house production cost may justify the likely higher cost of purchasing components in magazines, strips, or presorted form. Sometimes the additional cost of presorting is surprisingly low, and it is even possible for parts purchased this way to be cheaper. Consider the case of plastic injection moldings in Figure 3.2. The supplier may be using manual labor to separate the piece parts from the mold skeleton, increasing supplier cost. However, notice the symmetrical arrangement of the moldings on the skeleton in Figure 3.2. By purchasing the moldings with the skeleton intact, the moldings might be separated from the skeletons in-house, either manually or by a robot, thus preserving the correct orientation. It can be seen that automation engineers have a valid input to the parts purchasing process, and this principle will become even more evident in subsequent sections of this chapter.

Motion and Transfer

If the worst case is indeed the actual case and the parts are received in a jumbled mass in a hopper, the first requisite for automatic feeding of these parts is to get them to start

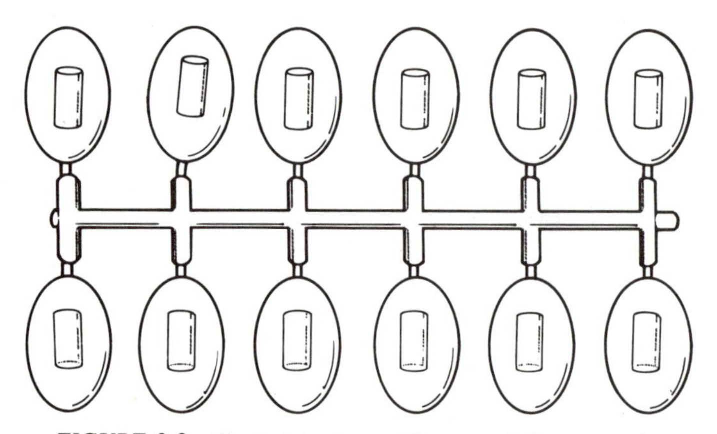

FIGURE 3.2 Plastic injection molding parts before separation from the mold skeleton.

moving. This is where mechanical engineers excel, and some outstanding research in this area has been reported by Boothroyd [refs. 20, 21, 89] and by den Hamer [ref. 30].

Motion can be imparted to parts by gravity, centrifugal force, tumbling, air pressure, or vibration. The power to drive these mechanisms can be either electrical or off-machine, the off-machine devices having synchronization advantages, as was pointed out in Chapter 2. The most versatile small parts feeding machine is the electrically driven *vibratory bowl* pictured in Figure 3.3. The most amazing feature of a vibratory bowl is that it literally causes piece parts to travel *uphill* as they vibrate up inclined ledges, or tracks, spiraling around the inside of the bowl. The physics creating this phenomenon is the variation in the acceleration of motion during the vibration cycle causing the parts to be tossed upward farther than they slip backward before the next vibration cycle tosses them farther upward. The direction of the vibration motion is canted to provide two components, one parallel to travel in the track and one perpendicular to the track (i.e., vertical). The frequency of vibration can be varied with increasing frequency resulting in faster feed rates up to a limit beyond which increasing frequency results in slower feed rates. The automation engineer will usually buy these commercially available bowls, not attempting to design and build them in-house. One of the principal features of vibratory bowls is recycling, the restart of components that either fall off the tracks or have been rejected by the parts orientation mechanism along or at the top of the track.

FIGURE 3.3 Vibratory bowl feeder. (Reprinted by permission of Moorfeed Corp., Indianapolis, Ind.)

FIGURE 3.4 Vibratory spiral elevator. (Reprinted by permission of Moorfeed Corp., Indianapolis, Ind.)

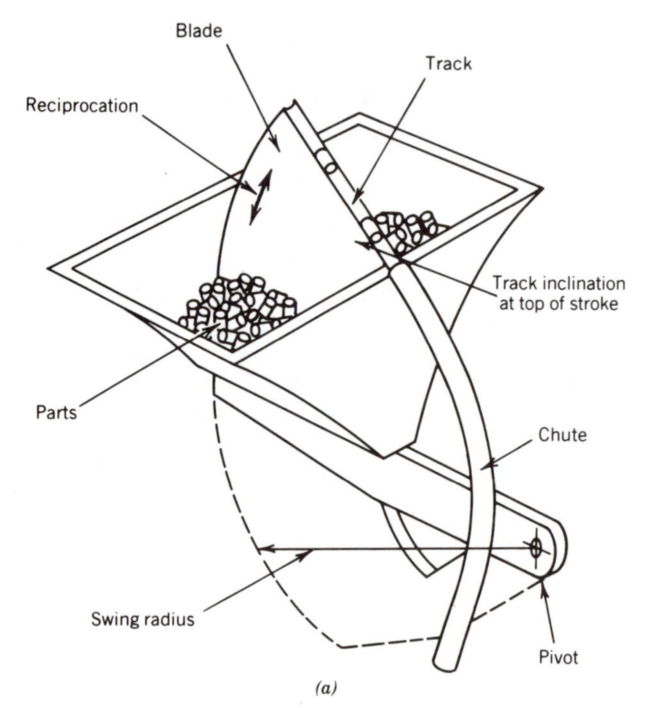

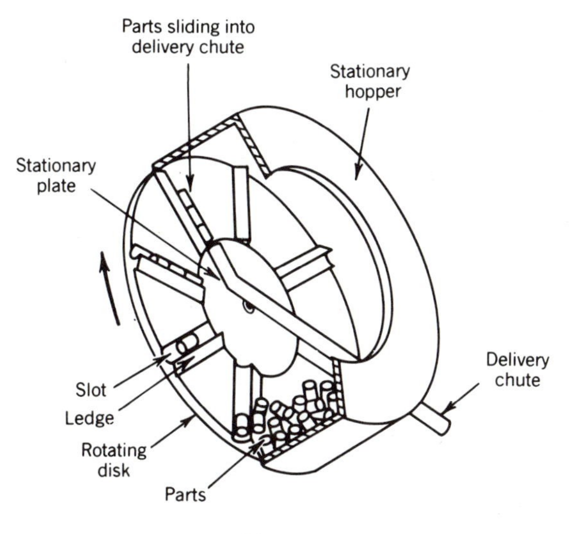

(b)

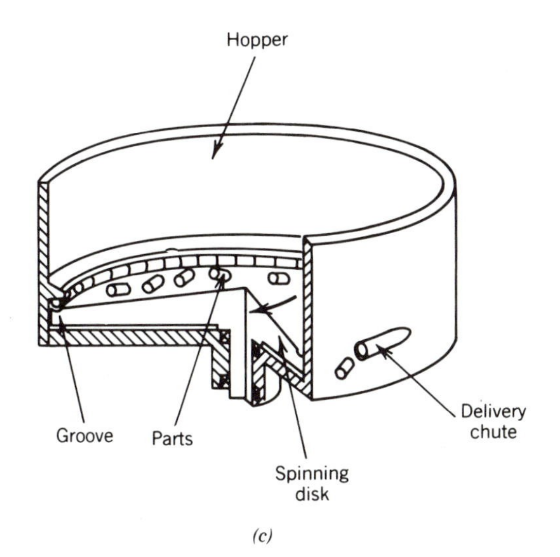

(c)

FIGURE 3.5 Three ways to feed cylindrical parts. (*a*) Centerboard hopper. (*b*) Rotary disk feeder. (*c*) Centrifugal hopper. (Adapted from *Automatic Assembly,* pp. 57, 73, 77, G. Boothroyd, C. Poli, and L. E. Murch, copyright 1982. Reprinted by courtesy of Marcel Dekker, Inc., New York, N.Y.)

A close relative of the vibratory bowl is the *vibratory spiral elevator,* shown in Figure 3.4. The spiral elevator has some advantages when the objective is strictly feeding, not orienting or rejecting misoriented parts. The unit shown in Figure 3.4 is used in a heat chamber to bring electronic parts to the required temperature during their 56-foot spiral travel up the track before being discharged into a test operation.

Three ways to feed cylindrical parts are shown in Figure 3.5. The centerboard hopper uses a reciprocating motion. The other two use rotation, but for different purposes. The centrifugal hopper uses the rotation to provide centrifugal force, while the rotary disk feeder merely uses it to tumble the parts around until they fall by gravity into one of the orientation slots. For more gentle feeding of delicate cylindrical parts, a rotating base hopper can be used, as shown in Figure 3.6.

Two more feeders for cylindrical parts, the bladed wheel hopper and the tumbling barrel hopper, are shown in Figure 3.7. The bladed wheel bears a resemblance to the centerboard hopper, but both feeders are really quite different. The bladed wheel does not capture parts; it merely stirs up the mass and kicks misaligned parts back up out of the way for unimpeded passage of those parts that have fallen in the slot. The radius of the bladed wheel is such that it does not touch parts that have already correctly fallen into the slot. The bladed wheel hopper is essentially a gravity feed device. The tumbling barrel, like a clothes dryer, has vanes around the drum to pick up parts as the drum rotates so that the parts will fall upon the track in the middle. Too much rotation speed will obviously defeat the purpose of this device.

The rotary centerboard hopper (Figure 3.8) appears similar to the bladed wheel hopper, but note in the figure that the blade is actually picking up the parts, not kicking them back. The part is U-shaped and travels down and around by gravity on the slowly turning blade to the point at which it is dropped upon the stationary delivery track.

For disk-shaped parts, a revolving hook hopper can be used (Figure 3.9). The appearance is similar to the rotating base hopper of Figure 3.6, but note that the base in Figure 3.9 is stationary. The purposes are quite different, however, as the revolving

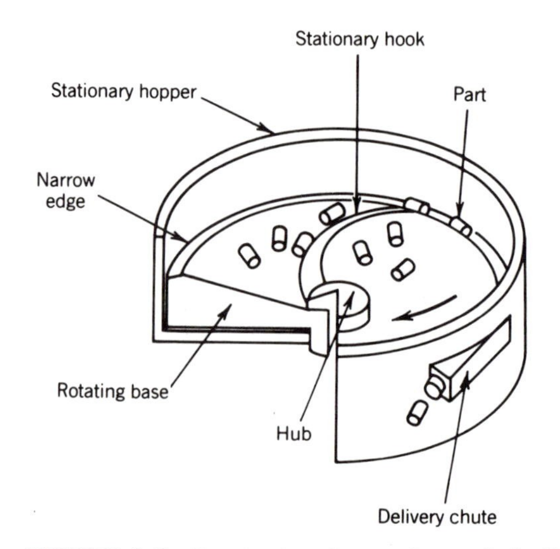

FIGURE 3.6 Rotating base hopper for gentle feeding of delicate cylindrical parts. (From *Automatic Assembly,* p. 82. G. Boothroyd, C. Poli, and L. E. Murch, copyright 1982. Reprinted by courtesy of Marcel Dekker, Inc., New York, N.Y.)

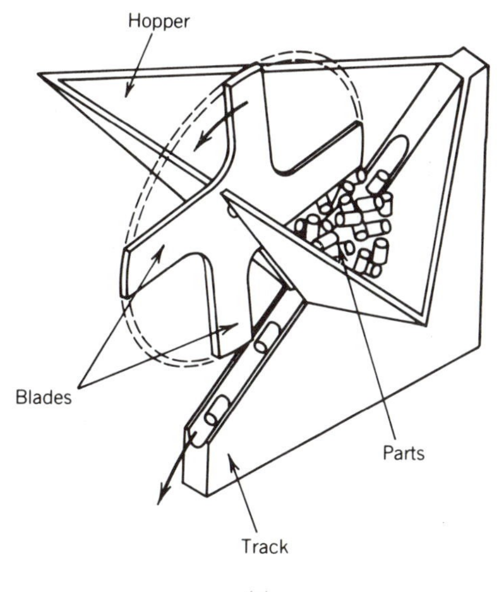

(a)

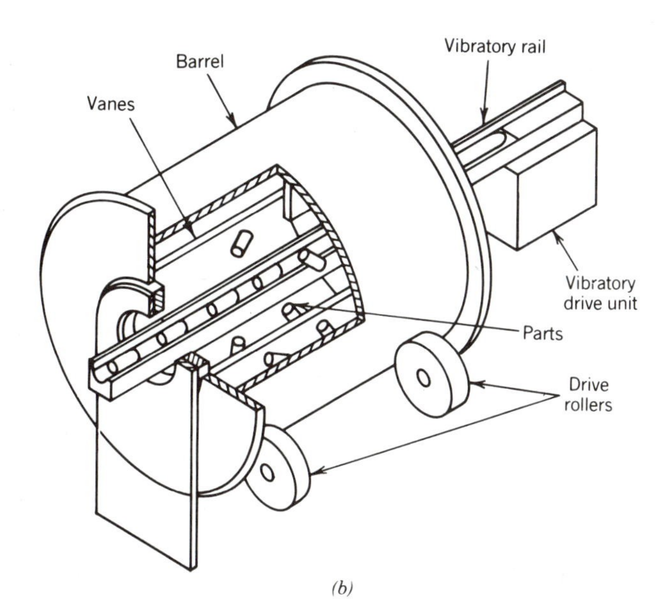

(b)

FIGURE 3.7 Bladed wheel hopper and tumbling barrel hopper for the feeding of durable cylindrical parts. (*a*) Bladed wheel hopper. (*b*) Tumbling barrel hopper. (Adapted from *Automatic Assembly,* pp. 87, 89, G. Boothroyd, C. Poli, and L. E. Murch, copyright 1982. Reprinted by courtesy of Marcel Dekker, Inc., New York, N.Y.)

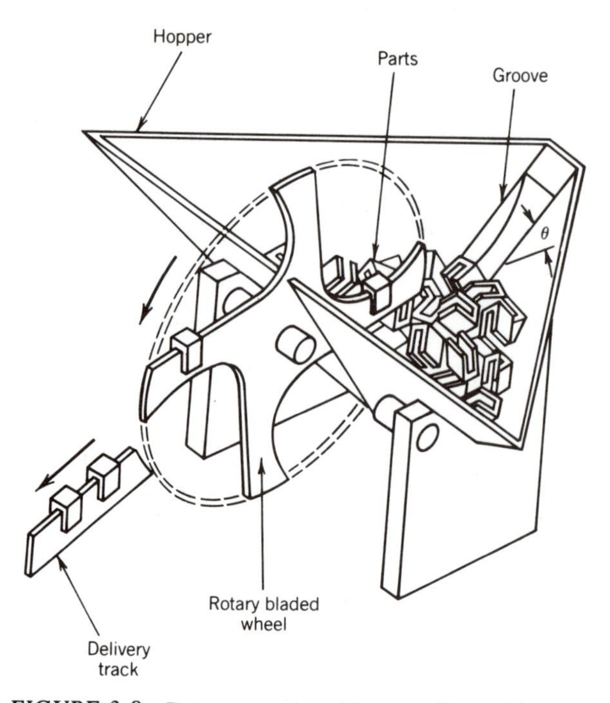

FIGURE 3.8 Rotary centerboard hopper. (From *Automatic Assembly,* pp. 93, G. Boothroyd, C. Poli, and L. E. Murch, copyright 1982. Reprinted by courtesy of Marcel Dekker, Inc., New York, N.Y.)

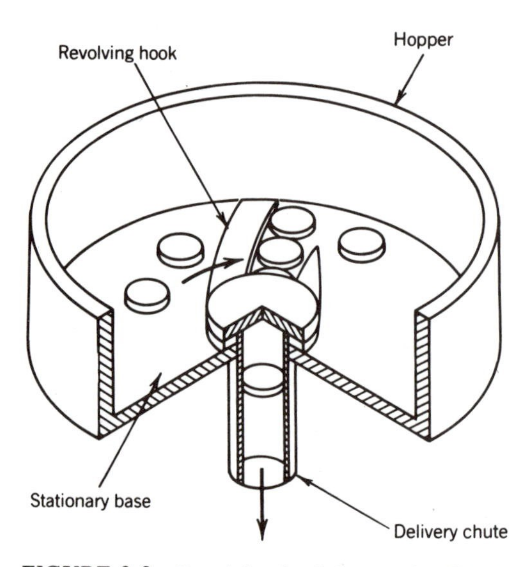

FIGURE 3.9 Revolving hook hopper for disk-shaped parts. (From *Automatic Assembly,* p. 80, G. Boothroyd, C. Poli, and L. E. Murch, copyright 1982. Reprinted by courtesy of Marcel Dekker, Inc., New York, N.Y.)

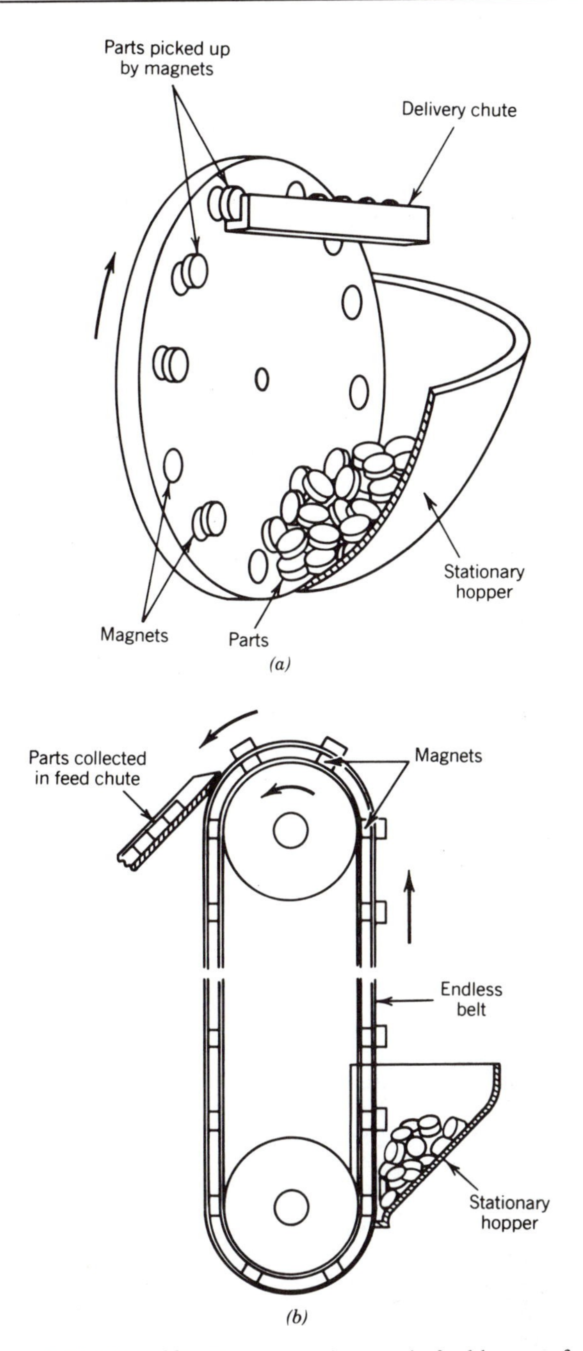

FIGURE 3.10 Two types of magnetic feed hoppers for ferrous parts that are relatively flat. (*a*) Magnetic disk feeder. (*b*) Magnetic elevating hopper feeder. (From *Automatic Assembly*, pp. 95, 99, G. Boothroyd, C. Poli, and L. E. Murch, copyright 1982. Reprinted by courtesy of Marcel Dekker, Inc., New York, N.Y.)

hook, which is of less thickness than the disks, slides along the floor of the hopper, gathering the disks into a channel that drops them into the center of the hopper. If the disks are ferrous, they may be fed using a magnetic feeder. Figure 3.10 shows two types of magnetic feeders.

Orientation Selection and Rejection

It can be seen that many of the transfer mechanisms that impart motion to the parts also orient them before discharging them onto a track or into a chute. However, the figures shown thus far have illustrated simple shapes such as cylinders and disks. The cylinders can be tapered or perhaps slightly compound, but for headed screws and other more complicated shapes a more sophisticated means of orientation is usually needed.

Returning to the vibratory bowl, the most versatile of feeding mechanisms, along the top of the bowl will be found some subtle irregularities in the inclined track that have been designed to recognize the geometry of the particular part being fed. These irregularities are designed to push, dump, or otherwise release incorrectly oriented parts to fall back down into the bowl for a retry at correct orientation. The headed screw or rivet is a good first example. Figure 3.11 shows several clever tricks to obtain the desired orientation of the rivets into the slot on the left, rivet head up. Taking advantage of the fact that the rivet is longer than the diameter of its head, a wiper blade diverts all rivets standing on their heads. Next, a pressure break causes the line of rivets to buckle if the delivery slot is too full to accept new rivets at the rate they are being pushed up from below. The pressure break also is a track-narrowing device that assures that the line of rivets will arrive single-file at the delivery slot. However, even at its narrowest point, the track accommodates the rivet turned either shank or head forward. Now note the slot, which is wide enough for the shank but not the head. The shank will fall into the slot whether the head or the shank is trailing. If the rivet is turned somewhat diagonally on the track so that the shank misses the slot, the downward slope on the inside shoulder of the slot causes the rivet to fall back into the bowl for a retry.

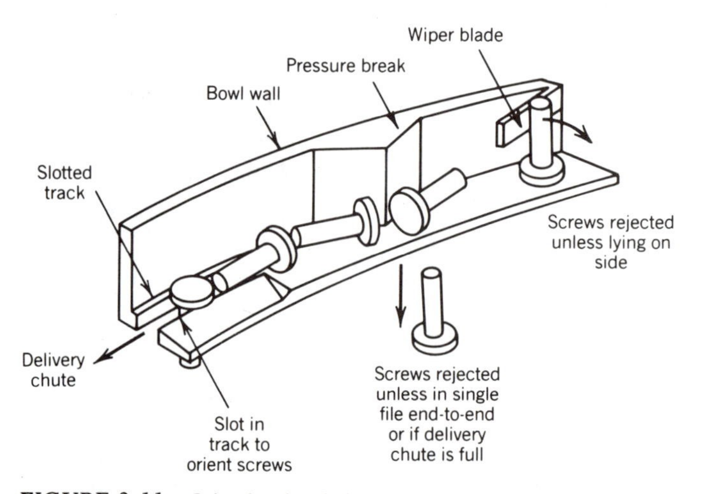

FIGURE 3.11 Orienting headed screws or rivets at the top discharge portion of a vibratory bowl. (From *Automatic Assembly*, p. 102, G. Boothroyd, C. Poli, and L. E. Murch, copyright 1982. Reprinted by courtesy of Marcel Dekker, Inc., New York, N.Y.)

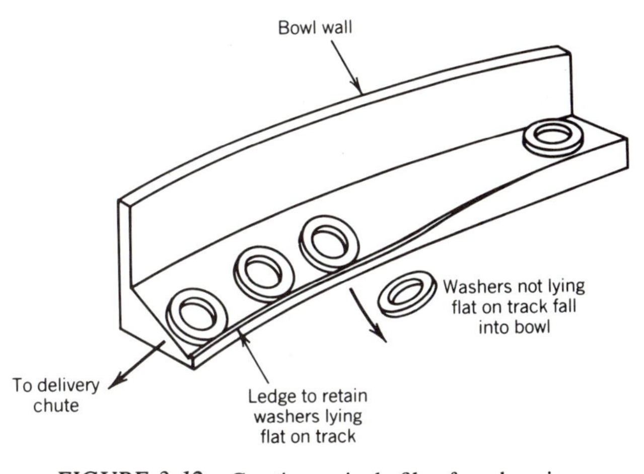

FIGURE 3.12 Creating a single file of washers in a vibratory feeder. (From *Automatic Assembly,* p. 103, G. Boothroyd, C. Poli, and L. E. Murch, copyright 1982. Reprinted by courtesy of Marcel Dekker, Inc., New York, N.Y.)

For washers or other disks, a device is needed to create a single file by dumping extras that may have climbed onto the backs of other washers. A sloped portion of the track with a tiny lip thinner than the washer does the trick, as shown in Figure 3.12.

For bottle caps or other cup-shaped objects, a wiper blade will work if the caps are *shorter* than their diameters (see Figure 3.13). Note that this rule is the opposite of the rule used earlier for screws and rivets. However, in the case of the caps the desired

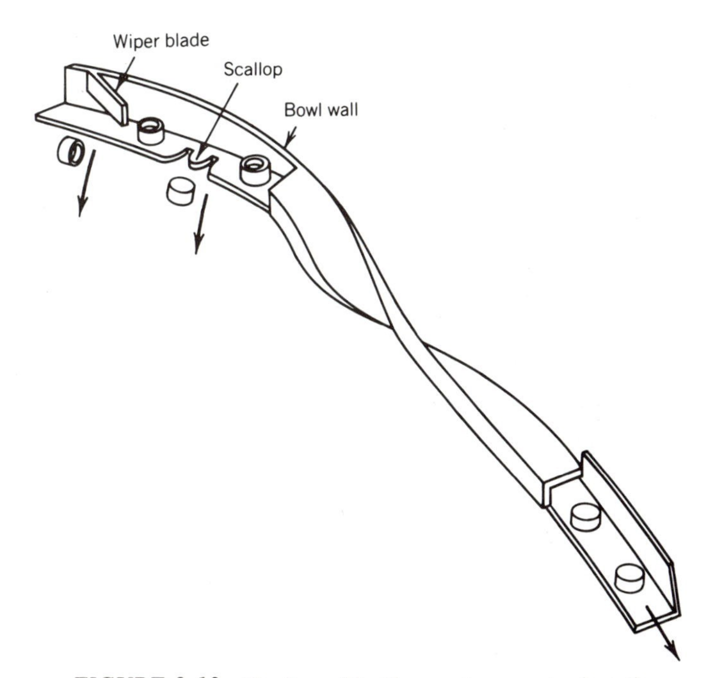

FIGURE 3.13 Feeding of bottle caps in correct orientation.

orientation happens to be open-end up. The scallop dumps any cap standing open-end down. Finally, a closed section of track can be used to invert the cap to a final useful orientation for capping oriented in the final, correct orientation, but some caps got that way too early in the process. The caps must be uniformly upside down when they reach the closed inversion track.

The systems described thus far seem to apply primarily to small geometric shapes of metal. Such piece parts are typical, but the methods described here are not limited to the metal products industries. For the sake of variety, consider Figure 3.14, which illustrates a rotating drum for orienting plastic bottles. Consider the billions of plastic bottles consumed each year in the United States alone, and the motivation for automation comes into focus. In Figure 3.14, note that there are two chutes, one for bottles coming out cap end first and the other for bottom end first. One orientation is desired, but there is no reason why one line cannot be inverted and then merged with the other to produce a continuous line of correctly oriented bottles.

Due to the bulkiness, lack of rigidity, and nonferrous characteristics of plastic bottles, the manufacture, sorting, and orientation of plastic bottles may present more of a challenge than for metal parts. However, the challenge these products present is not nearly as formidable as is the processing and sorting of food products. Fruits and vegetables must be sorted, oriented, and then chopped, peeled, or sliced before canning and freezing. Fruits and vegetables are variable in size and shape and are fragile and perishable. Attempts to use standard sorting and orienting machines can be completely frustrated. It is, however, possible to be somewhat scientific about the design of automation equipment for orienting all types of piece-parts and products, including food products, as will be seen in the section that follows.

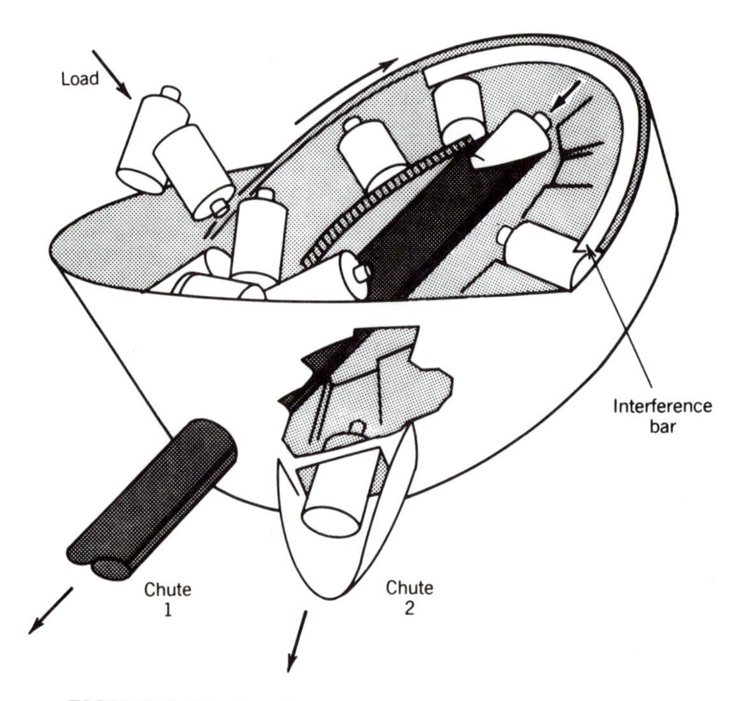

FIGURE 3.14 Rotating drum for orienting plastic bottles. (Excerpted from *Automation,* Copyright, Penton/IPC Inc., January 1970.)

PARTS FEEDING RESEARCH

We begin this section with a case study that considers the precise design parameters for a parts feeding and orientation system for a typical part geometry.

EXAMPLE 3.1 (CASE STUDY)

This case study is taken from the files of P. F. Rogers and Geoffrey Boothroyd [ref. 89], recognized authorities on the design of parts feeding and orientation equipment. Consider the asymmetrical part geometry illustrated in Figure 3.15. The feeder is vibratory so that parts slowly feed across the slot where they are intended to fall before being transferred to the next operation. The assignment is to design a selector slot of the proper width so that the part will fall heavy end first and thus be oriented correctly for the subsequent operation. The problem is that parts can feed across the slot in two feasible orientations, either heavy end first or light end first. The graphs plot slot widths versus experimental frequencies for the two ways that the parts fall. The shaded areas show that certain ranges of slot widths do not permit the part to fall at all. For some of these slot widths the part simply passes over the slot; for others the part is bounced back upstream, perhaps blocking the flow of additional parts. There are two graphs, one showing behavior of parts that happen to enter the system heavy end first, and the other showing parts entering light end first. The two graphs are aligned so that a vertical line drawn through both represents identical slot widths. The

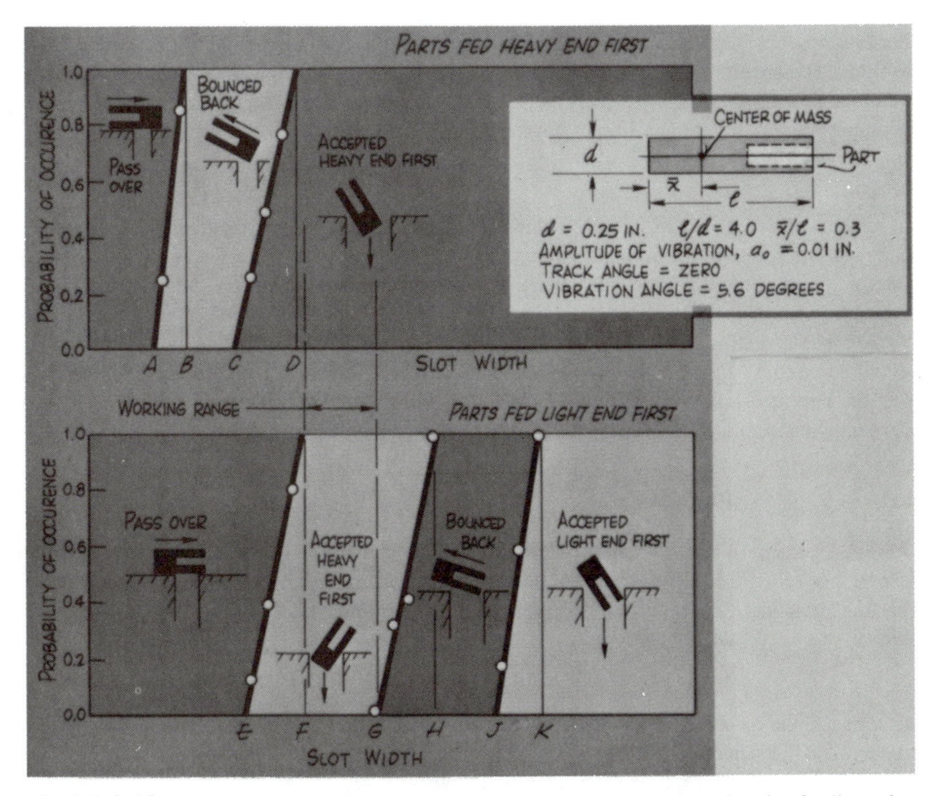

FIGURE 3.15 Experimental research to determine the "working range" for slot feeding of a given geometric shape. (Excerpted from *Automation*, Copyright, Penton/IPC Inc., January 1972.)

interesting characteristic revealed in this study is that there is a "working range" of slot widths for which all the parts will fall correctly, regardless of input orientation. An example will be used to further clarify the use of the experimental research diagram in Figure 3.15.

EXAMPLE 3.2
Slot Feeder Performance

Suppose that the actual numerical values for slot width points A through K in Figure 3.15 were found to be as follows:

$$A = .12 \text{ in.}$$
$$B = .19 \text{ in.}$$
$$C = .29 \text{ in.}$$
$$D = .43 \text{ in.}$$
$$E = .31 \text{ in.}$$
$$F = .45 \text{ in.}$$
$$G = .59 \text{ in.}$$
$$H = .73 \text{ in.}$$
$$I = .85 \text{ in.}$$
$$J = .96 \text{ in.}$$

Further suppose that the actual production slot feeder is set at slot width 0.40 in. If the input stream consists of 80 percent "heavy end first" and 20 percent "light end first," what will be the composition of the output stream? What percent of the product will pass over the discharge chute and not reach the output stream? If the input stream flows at 2000 pieces per hour, what will be the output flow?

Solution

The production feeder slot at 0.40 in. is between points C and D in the diagram. Therefore all the heavy-end-first product in the input stream will fall correctly into the discharge chute, although some of the pieces may be bounced back several times before eventually falling correctly. The light-end-first product will not all fall, since the diagram shows that a percentage will pass over the discharge chute for slot widths between points E and F. If the production slot width had been exactly equal to E, all the light-end-first product would have passed over the slot. For slot width F all the light-end-first product would have fallen correctly. Noting the straight line relationship between points E and F on the diagram, linear interpolation can be used as follows:

$$\frac{.40 - .31}{.45 - .31} = \frac{.09}{.14} = .643$$

Thus 64.3 percent of the light-end-first product would enter the discharge chute (correctly). It can thus be concluded that all product that actually falls into the discharge chute will fall correctly. The small amount of the input stream that passes over the slot without falling can be calculated as follows:

$$(1 - .643) \times 20\% = .357 \times 20\% = 7.1\%$$

In summary, the composition of the output stream is 100 percent heavy-end-first (correctly falling) product, and 7.1 percent of the product will pass over the discharge chute and not reach the output stream. The output flow rate is calculated as follows:

$$\begin{aligned} \text{output flow} &= (2000 \text{ pcs/hr}) \times (1.0 - .071) \\ &= 2000 \times .929 = 1858 \text{ pcs/hr} \end{aligned}$$

The slot feeder design study has shown that parts feeding design can be more scientific than most people think. Of course, changing the dimensions and center of mass of the part means that a new working range of slot widths must be identified. However, for standard shapes such as cylinders, disks, cubes, prisms, screws, and bolts of given materials, formulas can be derived that express slot width limits as a function of part geometries.

Of course, some part geometries will not even have a working range. In Figure 3.15, for example, if the correct orientation had been light end first, there would have been no working range at all. Experiments will show some parts to be impractical for feeding and orientation, and these parts would perhaps be a completely undesirable design for automation. The development of parts feeding research has reached the point that entire classes of part geometries can be identified that are either ideal candidates for automation, average candidates, or mediocre candidates.

Boothroyd and P. Dewhurst have developed an interactive computer program that acts as a decision algorithm for classifying and analyzing components and assessing their "assemblability." The computer software is equipped to assess costs and compute efficiences for both automatic and manual assemble [ref. 20].

Selector Efficiency

A general criterion for evaluating a mechanical parts selector is "efficiency," which is defined as follows:

$$\text{efficiency} = \frac{F_O}{F_I} \tag{3.1}$$

where F_O = output feed rate of correctly oriented parts

F_I = total input feed rate

It is possible to calculate the efficiency from experimental data after a selector system has been designed and fabricated. However, a growing body of scientific data has made possible the synthesis of efficiencies from available information on the individual selector mechanisms employed by the system. For each selector mechanism there is a "transition probability matrix" that reveals the probability that a piece will change to orientation y when it passes through a selector device when the part was in orientation x prior to encountering the device.

The most popular mode of selector device is the *rejector,* which simply discards an incorrectly oriented piece back into the center of the vibratory bowl or other hopper. The *pure rejector* is a device for which the transition probability from a given entry state to the rejected state is 100 percent. A device can be a pure rejector for some entry states and not others. If a device is a pure rejector for all entry states, it can be called a "total rejector," meaning that it rejects every part that encounters it, regardless of orientation. The concept of "total rejector" is introduced here merely to clarify the term "pure rejector." In a real automation application, pure rejectors would have utility; total rejectors would not.

The general case for a parts orientation device recognizes a probability of transition from any feasible orientation to any other feasible orientation, including the possibility of rejection. Therefore, an n by $n + 1$ matrix of probabilities is generated, in which n represents the number of feasible orientations. For example, a rectangular block has six feasible orientations, as shown in Figure 3.16. Thus, there are 42 probabilities in a 6×7 matrix used to describe completely the general orientation characteristics of a device that orients rectangular blocks. With most real devices, most of the probabilities in the

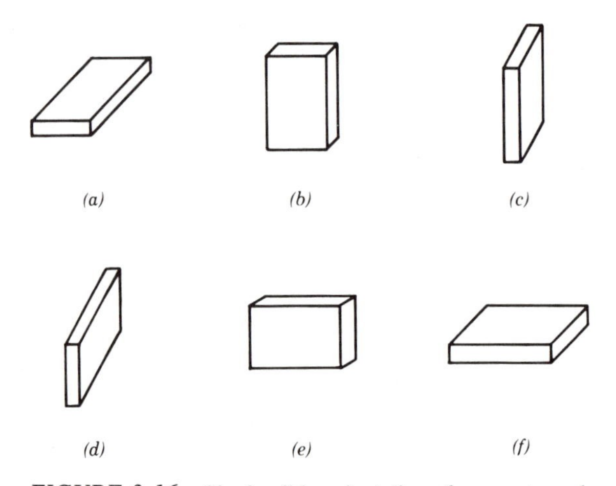

FIGURE 3.16 Six feasible orientations for a rectangular block. (*a*) Flat lengthwise. (*b*) Erect crosswise. (*c*) Erect lengthwise. (*d*) On-edge lengthwise. (*e*) On-edge crosswise. (*f*) Flat crosswise.

matrix are either zero or one (0 or 100 percent). Now let us consider an example combination of orientation devices and perform probability calculations to determine the distribution of output orientations.

EXAMPLE 3.3 (CASE STUDY) Orientation of Rectangular Blocks

Figure 3.17 illustrates a three-phase orientation device for feeding rectangular blocks. The desired orientation is on-edge lengthwise, as shown earlier in Figure 3.16*d*. The objective of this case study is to calculate efficiencies and output feed rates for a given input orientation distribution and a given input feed rate, given an assumed table of transition probabilities for each device in the selector system. Table 3.1 represents the transition probability matrix for the three phases (devices) of this selector system. For this example let us assume an input feed rate of 2000 pieces per hour and an input distribution of the following orientations:

Orientation	*Percent*
a	30
b	10
c	5
d	5
e	10
f	40

Solution

Before proceeding with a quantitative analysis of the probabilities associated with each selector device, the general characteristics of each selector can be observed in Table 3.1. The phase one wiper can be seen to be a pure rejector for the "erect lengthwise" orientation, an intuitively reasonable characteristic when the table is compared with the diagram in Figure 3.17. The wiper is only a 90

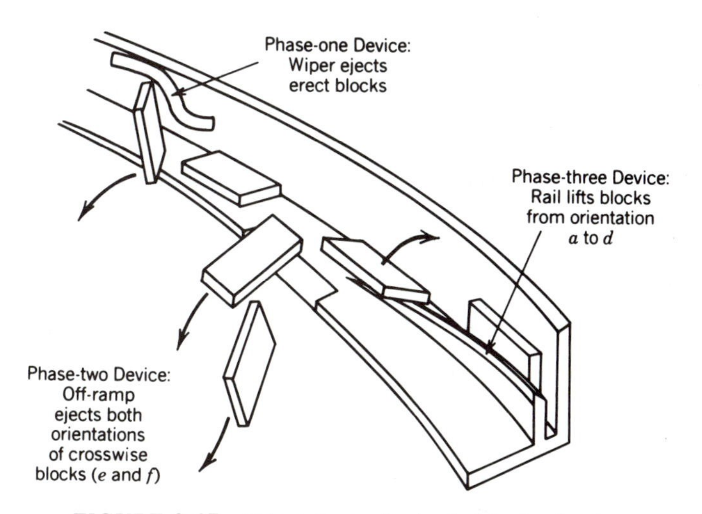

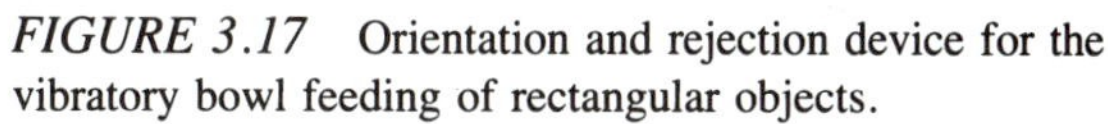

FIGURE 3.17 Orientation and rejection device for the vibratory bowl feeding of rectangular objects.

TABLE 3.1 Probability Matrix for Case Study 3.3

				Phase One Selector—Wiper			
From				*To*			
Orientation	*a*	*b*	*c*	*d*	*e*	*f*	x^a
a. Flat lengthwise	100	0	0	0	0	0	0
b. Erect crosswise	10	0	0	0	0	0	90
c. Erect lengthwise	0	0	0	0	0	0	100
d. On-edge lengthwise	0	0	0	100	0	0	0
e. On-edge crosswise	0	0	0	0	100	0	0
f. Flat crosswise	0	0	0	0	0	100	0
				Phase Two Selector—Off-ramp			
a. Flat lengthwise	100	0	0	0	0	0	0
b. Erect crosswise	0	100	0	0	0	0	0
c. Erect lengthwise	0	0	100	0	0	0	0
d. On-edge lengthwise	0	0	0	100	0	0	0
e. On-edge crosswise	0	0	0	0	0	0	100
f. Flat crosswise	0	0	0	0	0	0	100
				Phase Three Selector—Inclined Rail			
a. Flat lengthwise	0	0	0	80	0	0	20
b. Erect crosswise	0	0	0	0	0	0	100
c. Erect lengthwise	0	0	100	0	0	0	0
d. On-edge lengthwise	0	0	0	100	0	0	0
e. On-edge crosswise	0	0	0	0	0	0	100
f. Flat crosswise	0	0	0	0	0	0	100

[a]x is the "rejected state."

TABLE 3.2 Summary of Distribution Computations

	Distribution			
Orientation	*Input*	*Phase One*	*Phase Two*	*Phase Three (Output)*
a. Flat lengthwise	30	31	31	0
b. Erect crosswise	10	0	0	0
c. Erect lengthwise	5	0	0	0
d. On-edge lengthwise	5	5	5	29.8
e. On-edge crosswise	10	10	0	0
f. Flat crosswise	40	40	0	0
x (rejected)		14	50	6.2

percent rejector for the "erect crosswise" orientation; the other 10 percent are switched to "flat lengthwise," which is still an incorrect orientation. The four remaining input orientations are shown by the table to be unaffected by the wiper. The phase two off-ramp is shown in Table 3.1 to be a pure rejector for orientations e and f, just as was indicated in the diagram, and has no effect on the other four orientations. The final device is the phase three inclined rail, which has the function of lifting the "flat lengthwise" blocks to "on-edge lengthwise," the correct orientation. From Table 3.1 it can be seen that the rail is only 80 percent successful in this function, the other 20 percent being rejected from the track.

Having enumerated all probabilities for each of the three transition matrices, it is possible to calculate directly the output orientation distribution for this combination of three devices for any given input distribution and also to calculate the system efficiency. The method of the calculation is to multiply the probabilities of each input orientation by the appropriate transitional matrix probability to determine the output distribution of orientation. This is done successfully by phases until the overall system output is computed. Multiplying by the phase-one transition probabilities results in the following distributions at the end of phase one.

Orientation	*Calculation*	*New Distribution (%)*
a	30% × 1.0 + 10% × .1	31
b	10% × 0	0
c	5% × 0	0
d	5% × 1.0	5
e	10% × 1.0	10
f	40% × 1.0	40
x (rejected)	10% × 0.9 + 5% × 1.0	14

Continuing the calculation through the remaining phases results in the distribution calculation summarized in Table 3.2. Any such summary can be checked by adding up the first two columns, which should both total to 100. The total output (last column) plus the total number rejected (bottom row) also should total 100. From the final summary in Table 3.2, it can be determined that the system is 29.8 percent efficient.

The output feed rate calculation is based on the definition of efficiency:

$$\begin{aligned} \text{output feed rate} &= \text{input feed rate} \times \text{efficiency} \\ &= 2000 \text{ pcs/hr} \times 29.8\% \\ &= 596 \text{ pcs/hr} \end{aligned}$$

This output feed rate could then be matched to the next assembly or fabrication process in the automated or semiautomated process.

Efficiency Versus Effectiveness

A 29.8 percent efficiency might seem to be a little disappointing, but remember that the objective in Example 3.3 was to orient piece parts automatically into a single correct orientation when presented with six possible input orientations, and that the input distribution of those six orientations was very unfavorable, with only 5 percent of the input stream appearing in the correct orientation to begin with. The selector system of Example 3.3 had a challenging assignment, and most real industrial applications of automatic assembly present similar challenges in parts orientation. It is fascinating to watch an industrial vibratory bowl at work, and it is not uncommon at all to observe a large quantity of rejected orientations being tossed back into the bowl for a retry.

Another way to view the selector of Example 3.3 is in terms of "effectiveness" instead of efficiency. Selector effectiveness can be defined as follows:

$$\text{effectiveness} = \frac{F_O}{F_T} \tag{3.2}$$

where F_O = output feed rate of correctly oriented parts (the same as defined earlier for efficiency)

F_T = total *output* feed rate

Selector effectiveness is best visualized as the percentage of the output stream that is correctly oriented. The Example 3.3 selector had an effectiveness of 100 percent since all of the output stream was correctly oriented. To be feasible for use in real automated manufacturing applications, a parts selector and orientation system generally must be virtually 100 percent effective. Any deviation from the ideal of 100 percent effectiveness in a real system generally results in part jamming, a problem with unexpectedly severe consequences, as will be seen later, in Chapter 4.

Although an overall orientation system is usually nearly 100 percent effective, the selector stations that make up the system usually are much less than 100 percent effective. In fact, the phase-one selector in Example 3.3 was only 6 percent effective for the given input distribution, computed as follows:

$$\begin{aligned} F_O &= 30\% \times 0 + 10\% \times 0 + 5\% \times 0 + 5\% \times 1.00 \\ &\quad + 10\% \times 0 + 40\% \times 0 \\ &= 5 \\ F_T &= 30\% \times 1.00 + 10\% \times .10 + 5\% \times 0 + 5\% \times 1.00 \\ &\quad + 10\% \times 1.00 + 40\% \times 1.00 \\ &= 86 \\ \text{effectiveness} &= \frac{5}{86} = 6\% \end{aligned}$$

This figure could have also been calculated directly from column 2 of Table 3.3 by taking the ratio of outputs for orientation *d* to the sum of all orientation outputs from phase one, excluding, of course, the rejected parts, which would not appear in the output stream.

Phase-two effectiveness at 14 percent is hardly better than phase one. The beauty of the integrated system of parts selector/rejectors is the synergistic effect achieved by placing them in series. Taken alone, none of the three phases of Example 3.3 would have been very effective, but together they are virtually 100 percent effective.

Effectiveness is an interesting criterion for analysis, especially for individual selector stations. However, since most overall selector systems approach the 100 percent ideal for effectiveness, the more useful measure of the overall parts selector system is efficiency, not effectiveness.

Part Wear and Damage

A final consideration may be that the parts selector will overwork the parts, oscillating and vibrating them and kicking many of them back for retry after retry. If the effectiveness is virtually 100 percent, it is possible to use the system efficiency to calculate the chances that a part will be tossed back k times before reaching an acceptable orientation. The formula is:

$$p_k = \left[\frac{E}{100}\right]^1 \left[1 - \frac{E}{100}\right]^k \tag{3.3}$$

where p_k = probability that the part will be kicked back k times
E = efficiency
k = number of kickbacks

For Example 3.3:

$$p_0 = \left[\frac{E}{100}\right]^1 = \frac{29.8}{100} = 0.298$$

$$p_1 = \left[\frac{E}{100}\right]^1 \left[1 - \frac{E}{100}\right]^1 = (0.298)(0.702) = 0.209$$

$$p_2 = \left[\frac{E}{100}\right]^1 \left[1 - \frac{E}{100}\right]^2 = (0.298)(0.702)^2 = 0.147$$

$$p_3 = \left[\frac{E}{100}\right]^1 \left[1 - \frac{E}{100}\right]^3 = (0.298)(0.702)^3 = 0.103$$

$$p_{10} = \left[\frac{E}{100}\right]^1 \left[1 - \frac{E}{100}\right]^{10} = (0.298)(0.702)^{10} = 0.009$$

Thus, nearly one out of a hundred parts will be kicked back ten times before achieving an acceptable orientation. The automation engineer must decide whether or not this kind of treatment will damage the product.

The average number of kickbacks $\bar{k}$ for every part entering the 100 percent effective selector system is also a function of efficiency and is as follows:

$$\bar{k} = \sum_{i=0}^{\infty} i p_i$$

$$= \sum_{i=0}^{\infty} i \left[\frac{E}{100}\right]^1 \left[1 - \frac{E}{100}\right]^i \tag{3.4}$$

The infinite series of Eq. (3.4) may be recognized as being equivalent to the following closed form:

TABLE 3.3 Tabulation of Kickback Probabilities for Example 3.3

Number of Kickbacks k	*Probability of k Kickbacks* $\left[\frac{E}{100}\right]^1\left[1-\frac{E}{100}\right]^k$	$k\,p_k$	$\sum_{i=0}^{k} i\,p_i$
0	0.298	0	0
1	0.209	0.209	0.209
2	0.147	0.294	0.503
3	0.103	0.309	0.812
4	0.072	0.289	1.102
5	0.051	0.254	1.356
6	0.036	0.214	1.570
7	0.025	0.175	1.745
8	0.018	0.141	1.886
9	0.012	0.111	1.997
10	0.009	0.087	2.083
.			
.			
.			
∞	0	0	$2.356 = \bar{k}$

$$\bar{k} = \frac{1 - \dfrac{E}{100}}{\dfrac{E}{100}} \tag{3.4a}$$

For Example 3.3, $\bar{k}$ can be computed directly from Eq. (3.4a) to be 2.356, or Eq. 3.4 can be used to tabulate the calculations as in Table 3.3, displaying all of the kickback probabilities.

EXAMPLE 3.4

Stainless steel stampings are used to jacket and support pulleys used in the assembly of blocks of various designs used on sailboats. The stainless steel has a polished appearance and the quality of the surface finish is a factor in the acceptability of this costly product. Suppose these stamping are oriented for subsequent assembly using a bowl feeder having an effectiveness of 100 percent and an efficiency of 50 percent. What are the chances that a given stamping will be tossed back into the bowl three times before a correct orientation will present it for assembly? What is the average number of times a typical stamping will be tossed back into the bowl for this operation?

Solution

$$p_3 = \left[\frac{E}{100}\right]^1\left[1 - \frac{e}{100}\right]^3 = (.5)(.5)^3 = 0.625$$

$$\bar{k} = \frac{1 - \dfrac{50}{100}}{\dfrac{50}{100}} = \frac{.5}{.5} = 1$$

Having completed a discussion of the rather precise ways in which a parts selector can be analyzed, it should be acknowledged that the mechanization of parts handling and orientation is as much an art as it is a science. The fabrication of selector mechanisms and escape devices is usually done in a very experimental way in which the skilled artisan cuts and tries many different configurations until an effective design is perfected. Although many scientific principles govern the behavior of parts orientation and selection processes, trial-and-error experimentation can be used to develop practical, working systems without the science. The primary benefit of the analysis is to understand the dependence between successive steps in the orientation process and the importance of efficiency and effectiveness to the overall success of the automation project.

Food Product Orientation

In the canning industry, the sorting, orienting, and feeding of food products can present more complex problems of analysis due to the random variables, which cannot be avoided. Automation research by Kwei at the University of Arkansas [ref. 50] has led to the development of an optimum in-feeder design for the chopping of vegetables in a frozen food process (see Figure 3.18). At the point where the carrot was attached to the top of the plant is a tough portion that is considered undesirable. Mechanical choppers are capable of chopping off the tops if the proper end is fed to the chopper. Figure 3.18 shows an in-feed sorting and conveying mechanism, the objective of which is to advance the carrot top-end-forward to the mechanical choppers. The carrots are dumped from bulk conveyors onto a "shaker table," which is full of holes. The top end is heavier and will likely fall through the holes first regardless of carrot orientation on the shaker table. What happens after the carrots fall depends principally upon the relationship between carrot length and the size of the gap between the in-feed conveyor and the shaker table. Other factors are the speed of the in-feed conveyor and the downward slant of the entire assembly. If the carrots could be counted upon to be of constant length, the problem would be much easier. Analysis of the feed system design must consider the random variables of carrot size. Case Study 3.5 illustrates the computations for a simplified case.

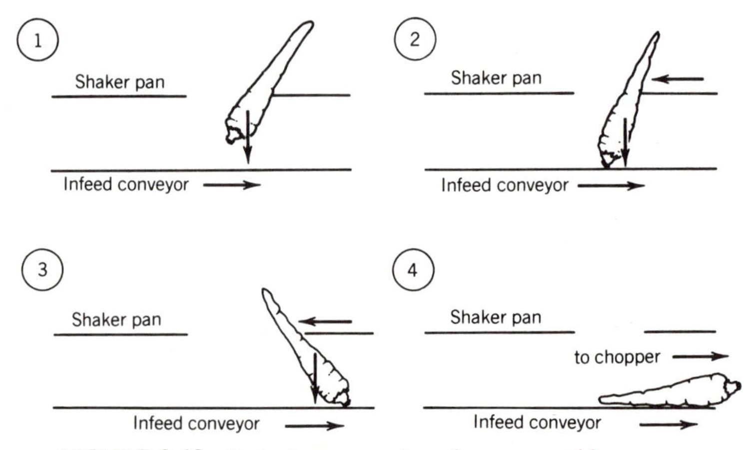

FIGURE 3.18 In-feed conveyor for a frozen vegetable process.

EXAMPLE 3.5 (CASE STUDY)
Design of a Frozen Vegetable In-Feeder

Figure 3.18 shows part of an automated system for preparing frozen carrots. In the diagram, a carrot is falling through a hole in the shaker pan onto the in-feed conveyor with a correct orientation. However, if the carrot is too short for the gap between shaker pan and in-feed conveyor, it is clear that the carrot might fall such that the point of the carrot will be forward and the cap to the wrong end. On the other hand, if the gap is too small, the carrot will not feed and will clog the system. Clogging causes disruption of the production line and is considered twice as serious a difficulty as "wrong end" orientation. As an approximation to the true situation, assume that clogging always will occur if the gap is one inch less than carrot length and that clogging will never occur if the gap is larger than one inch less than carrot length. Also, as an approximation, assume that a carrot will always fall incorrectly if the gap is greater than the carrot length. Assume that carrot length is normally distributed with mean 3 in. and standard deviation ½ in.

(a) What size gap should be specified?

(b) What percent of the incoming carrots would clog the in-feeder?

(c) What percent of the carrots would be chopped incorrectly?

Solution

Let L = length of carrot (normally distributed, $\mu = 3$ in., $\sigma = ½$ in.)

$F[L]$ = cumulative distribution function of L

G = gap in inches

G^* = optimum gap

$Q(G)$ = penalty function

= Prob [carrot will be chopped wrong end] + 2 Prob [carrot will clog]

$$Q(G^*) = \min_G Q(G)$$

$$\begin{aligned} Q(G) &= \text{Prob}[G \geqslant L] + 2(\text{Prob}[G \leqslant L - 1]) \\ &= \text{Prob}[L \leqslant G] + 2(\text{Prob}[L \geqslant G + 1]) \\ &= F[G] + 2(1 - F[G + 1]) \end{aligned}$$

Let $Z = \dfrac{G - \mu}{\sigma}$ = standard normal random variable

$$\phi(Z) = \int_{-\infty}^{Z} \frac{1}{\sqrt{2\pi}} e^{-z^2/2}\, dz = \text{cumulative distribution function of } Z$$

$$z_1 = \frac{G - 3}{1/2} = \text{the point at which the carrot will be chopped wrong end}$$

$$= 2G - 6$$

$$z_2 = \frac{G + 1 - 3}{1/2} = \text{the point at which the carrot will clog}$$

$$= 2G - 4$$

$$\begin{aligned} Q(G) &= \phi[z_1] + 2(1 - \phi[z_2]) \\ &= \phi[z_1] + 2\phi[-z_2] \end{aligned} \tag{3.5}$$

This function can be minimized either by standard search techniques or differential calculus. Intuitively, the gap should be greater than 2 in. because clogging is a more serious problem than wrong-end chopping. The minimum for the penalty function is found to lie in the vicinity of a 2.7 in. gap. Therefore,

$$G^* = 2.7$$

Using tables of the standard normal cumulative distribution function:

$$\% \text{ carrots chopped wrong end} = 27.43\%$$

$$\% \text{ carrots clogging} = 8.08\%$$

$$Q(G^*) = 0.2743 + 2(0.0808) = 0.4359$$

The carrot processing case study required some understanding of random variables and statistical theory for analysis, but even for those readers without a background in these techniques, the similarities between such a problem and machine parts feeding problems should be evident. Just as in the slot feeding problem, there is a *working range* to be established. For a nominal 3-in. carrot the in-feed conveyor can be set anywhere from 2 in. to 3 in. with no problem. The complication is that carrots vary in size, and a working range for a 3-in. carrot is not a working range for a 5-in. carrot. Fortunately, most real-world statistical distributions have a central tendency, and automation system designs can be aimed for the bulk in the mid-range of the distribution. This was done in Case Study 3.5, as shown in Figure 3.19, except that the upper tail of the distribution was favored somewhat over the lower tail because of the higher penalty associated with clogging the in-feeder.

Partial Automation

Figure 3.19 seems to be begging another question. So many of the carrots in the two tails of the distribution are in the trouble areas (clogging or wrong-end chopping) that it appears that the automated in-feeder is not doing a satisfactory job. In the real processing plant from which Case Study 3.5 was drawn, this was precisely the case (except that the vegetable was okra, not carrots). The variability of the product forced the process engineers to compromise and use a partially automated work station. Space for human operators on both sides of the shaker tables was provided so that vegetables that are seen to fall incorrectly could be turned over manually before entering the chopper area.

Partial automation is not limited to food processes, where product variability is the chief problem. Orientation and alignment of various kinds of products and machine parts may require intelligent vision or some sophisticated capability, which in a given manufacturing process may be infeasible either technically or economically. Partial automation, where the worker and machine each do their parts according to their capabilities, is an alternative that should not be overlooked by the automation engineer.

This chapter has considered the problem of moving the material or piece parts into position for processing or assembly. Although the operations have been seen to be complicated and some even ingenious, none has resulted in useful change to the product itself. Parts feeding and orientation was without doubt important, but we are now ready to consider the automation of machines that process or assemble these parts into useful products.

SUMMARY

The success of an assembly automation project depends not so much on the highly visible motions of industrial robots handling product parts as it does on the tiny subtleties of positioning and orienting of piece parts for assembly. Sometimes these detailed problems

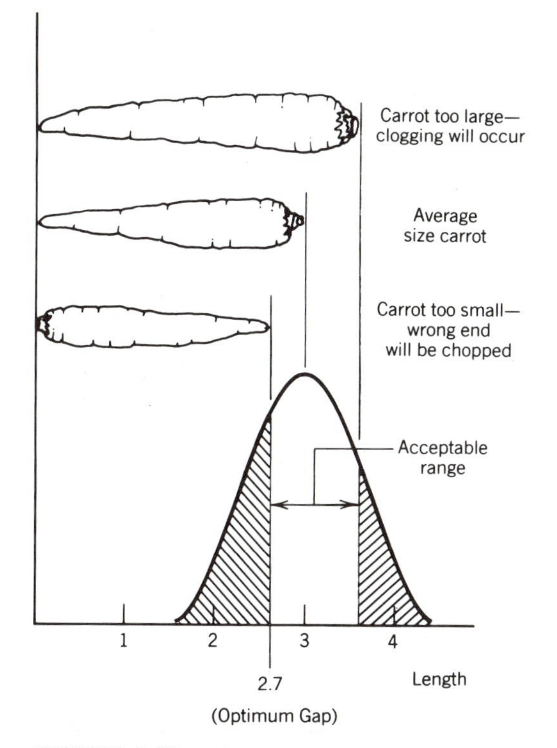

FIGURE 3.19 Aiming for the central bulk in automated systems dealing with random variables (Case Study 3.5).

can be avoided by preserving parts orientations to begin with, either during their manufacture or at the vendor's site for purchased parts.

Given that parts orientation must be accomplished, clever mechanical devices are available for this purpose with the principal representative being the vibratory bowl with parts selectors in the discharge track. Those parts that the bowl system is unable to select and orient correctly are rejected back into the bowl for a retry. The rejected parts are not lost; generally, the only things lost are time and production rate. The amount of decrease in production rate from rejection of misoriented parts is related directly to the efficiency of the vibratory bowl and part selector track. The overall efficiency of the system can be calculated from the matrix of individual rejector probabilities along the selector track. If higher production rates are necessary, vibration rate can be speeded up with correspondingly higher feed rates up to a point. The science of orientation and feeding applies not only to small metal parts, but also to plastic bottles, moldings, castings, and even food products. Food products present special problems for automation engineers due to the random variation in item sizes and shapes.

EXERCISES AND STUDY QUESTIONS

3.1. Describe circumstances in which purchased piece parts can be specified to be shipped in the correct orientation for a robotic assembly operation and still be cheaper than if purchased in the usual way (jumbled together in a box).

3.2. Name some methods of imparting motion to component parts in preparation for orientation and assembly. Which is the most popular?

3.3. Both the rotating base hopper and the centrifugal hopper feed cylindrical parts. What is the difference?

3.4. In the experimental slot feeding diagram of Figure 3.15, suppose the objective had been to slot feed the light end first. Determine the working range, if any.

3.5. In Example 3.2, regarding slot feeder performance, suppose the distribution of the input stream had been reversed, that is, 20 percent of the pieces were heavy-end-forward, and 80 percent were light-end-forward. Calculate effectiveness, efficiency, and output flow rate.

3.6. For Example 3.2, regarding slot feeder performance, use a slot width of 0.65 in. and an input distribution of 70 percent heavy-end-first and 30 percent light-end-first to calculate effectiveness, efficiency, and output flow rate.

3.7. In the preceding exercise, reverse the input flow distribution to 30 percent heavy-end-first and 70 percent light-end-first and rework the exercise.

3.8. Rework Exercise 3.6 using a slot width of 0.50 in. instead of 0.65 in.

3.9. Table 3.4 tabulates a transition probability matrix for an orienting device in a vibratory bowl selector track.

(a) Is the device a rejector?
(b) If so, is it a pure rejector?
(c) Is it a total rejector?
(d) What is the desired orientation?
(e) Would it help to add a second identical orienting device further up the selector track?

3.10. If the actual input orientations for the selector described by Table 3.4 are equally distributed among the four feasible possibilities shown, what is the efficiency of this selector station? What is the effectiveness of the selector station?

3.11. Rework Exercise 3.10 calculating system effectiveness for a system of two identical selector stations configured in series.

3.12. Is it possible to achieve 100 percent effectiveness with a series of selectors that performs as described by Table 3.4?

3.13. **(Design Case Study)** Suppose it is essential to achieve 100 percent effectiveness in the orienting system described by Table 3.4. This, of course, is impossible, but one thing we could do is add another selector that is a pure rejector for one or more states. This pure rejector could be placed either upstream or downstream from the present selector. For which input state(s) should the additional selector be a pure rejector to achieve an overall system effectiveness of 100 percent? Where should the rejector be placed—upstream or downstream from the selector? Compare the system efficiencies of these two alternatives.

TABLE 3.4 Part Selector Transition Probability Matrix for Exercises 3.9 through 3.13

	a	*b*	*c*	*d*	*x*
a	100	0	0	0	0
b	20	0	0	0	80
c	20	0	10	0	70
d	10	0	0	30	60

3.14. Fill in the missing entries in the part selector transition probability matrix of Table 3.5, and then calculate the selector effectiveness for the following distribution of input orientations if the desired orientation is orientation *a*.

Input Distribution

(a) 50%
(b) 10%
(c) 30%
(d) 5%
(e) 5%

3.15. Assuming equal distribution among the five input orientations, for which output orientation is the selector described by Table 3.5 (as completed in Exercise 3.14) most effective? What is the selector effectiveness for this orientation? What is the selector efficiency for this orientation?

TABLE 3.5 Part Selector Transition Probability Matrix for Exercises 3.14 and 3.15

	a	*b*	*c*	*d*	*e*	*x*
a	40	30	20	10		
b	10	10	10	10	10	
c						100
d	20	20	20	20		20
e	30	25		20	0	10

3.16. A 100 percent effective vibratory parts selector system has an efficiency of 80 percent. What are the chances that a part will be kicked back into the bowl five times before escaping the system? What is the average number of kickbacks per part for this system?

3.17. A design parameter of a certain parts feeding system is that the probability that a part will be kicked back ten times or more be kept to below one percent. The system is virtually perfect with respect to effectiveness, but is only 40 percent efficient. Will the system meet the design parameter?

CHAPTER 4

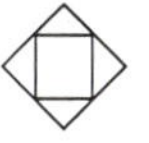

AUTOMATIC PRODUCTION AND ASSEMBLY

The purpose of parts orientation and feeding, studied in Chapter 3, is to prepare the parts for automatic processing or assembly. Production lines and assembly machines use a variety of pneumatic and electric drives and actuators of the general types described in Chapter 2. Although this chapter emphasizes the assembly cycle, many of the principles apply to other automatic processes such as transfer presses, gang drills, and automatic indexing lathes.

ASSEMBLY MACHINES

Assembly machines are so tailored to the individual assembly configurations that it is difficult to classify them in this text. Some operations are common to most assemblies, however, and one of these, the driving of screws, will be examined here.

Automatic Screwdrivers

Most of us think of an automatic screwdriver as a power-driven tool that the operator holds in his or her hands to facilitate manual driving of screws. But there are machines that do the entire operation. Screws are dumped in bulk into vibratory bowl hoppers, where they are oriented and fed down a discharge track to the point at which a power-driven screwdriver is guided automatically into position to drive the screw. Figure 4.1 shows a design reported by Cameron [ref. 23] in which the machine has many characteristics identical to its human competitors in the driving of screws. The mechanical jaws or "fingers" receive the screw from the feed track, hold it erect with the screw centerline aligned with that of the screwdriver, and then travel down with the screwdriver continuing to hold the screw until the automatic driver has engaged the slot and turned the screw several turns. Note the similarity between this operation and that of a human hand holding a screw to be driven manually.

The actual driver can be powered either electrically or pneumatically with advantages for each as discussed in Chapter 2. Because the driver must disengage within a given torque range, a clutch or other method must be used to achieve this disengagement.

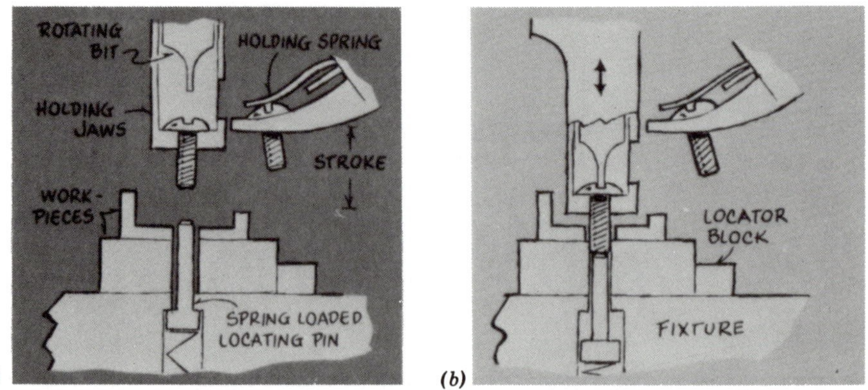

FIGURE 4.1 Automatic driving of screws. In (*a*) holding jaws receive the screw from the feeding and orientation system. In (*b*) the holding jaws move down with the tool until the screw enters the hole and contacts the locating pin. As the rotating driver engages the screw and torques it to specification, the holding jaws open. (Excerpted from *Automation,* Copyright, Penton/IPC, Inc., January 1972.)

Pneumatic motors may simply stall at approximately the correct torque if the air pressure is regulated correctly. If more air pressure is needed to achieve necessary spindle speeds to meet production requirements, or if electric drives are used, friction or ratchet clutches can be used.

Lack of uniformity either in the parts to be assembled or in the screws that fasten them can lead to problems such as improper torquing, undriven screws, stripped threads, and marred screw heads. This is a quality-control matter, and in general, quality control of components becomes more important when assembly is automated. Even with quality components, automatic screwdriving and other automatic assembly operations demand accurate fixturing to hold the parts properly during assembly.

Nut and bolt assembly is quite similar to screw assembly operations. Bolts can be positioned with robots or linear actuators. Nuts can be attached using mechanisms similar to the automatic screwdriver pictured in Figure 4.1. Larger components and housings can usually benefit from fixtures unique to the assembly. Industrial robots with special grippers are useful for these components also.

Indexing Machines

Automatic screwdrivers, nut runners, and other automatic assembly tools are useful aids to automation, but they are not known as assembly machines. The term *assembly machine* usually implies multiple station operation with or without storage between stations. The identity of the assembly machine is associated principally with the method of "indexing" or transferring the partially completed assembly from station to station.

The two principal types of indexing machines are in-line (Figure 4.2) and rotary indexing. The archetype of the in-line style is the automobile assembly line, although automobile assembly lines retain a large number of manual operations. In-line arrangements are more flexible in that the method can accommodate any number of stations. Perhaps even more important is the convenience of adding in-process storage capability along the line. In-process storage capability is the key to increasing the productivity of lines that have a large degree of variability in individual station cycle times.

The rotary-indexing assembly machine is more popular for small parts assemblies. These machines are sometimes called *dial indexing machines*. A circular table has several

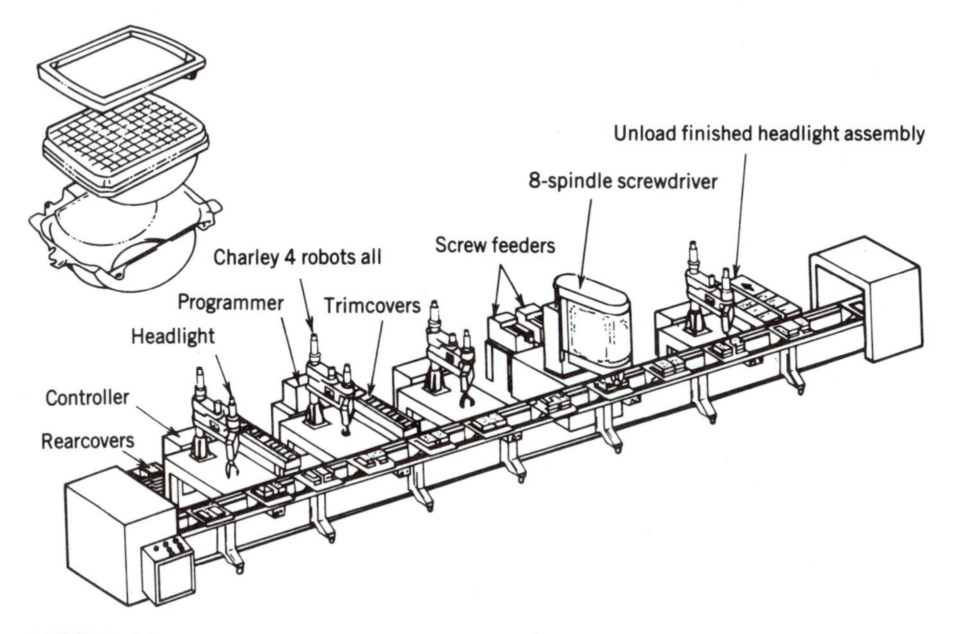

FIGURE 4.2 In-line, nonsynchronous, automatic transfer line for assembly of automobile headlamps. The process employs four Charley 4(TM) assembly robots. (Courtesy VSI Automation.)

stations that hold the product in various stages of its assembly as it advances around the circle. It is possible for the motion of the rotary table to be continuous, as in automatic bottling machines, but this requires the assembly apparatus to travel with the table. Most factory assembly machines advance in discrete intermittent steps. One station is used for the mounting of the main housing, base, or subassembly onto the rotary table. This step can be accomplished by a human operator or by an industrial robot. After the table makes nearly a complete circle, the completed assembly is removed at a station adjacent to the input station, or sometimes a single station serves both the loading and unloading functions.

PRODUCTION AND THROUGHPUT

The automatic assembly machine, whether it be rotary or in-line, produces a completed assembly every time the machine indexes, regardless of the number of stations in the assembly process. To compute the (ideal) production rate of an automatic assembly machine, one needs to know only the indexing cycle time; the number of stations is immaterial. It is true that throughput time, the time required to assemble a given assembly from start to finish, is dependent upon the number of stations and the index time, but production rate is not.

EXAMPLE 4.1
Production and Throughput Time

An automatic assembly machine is of the dial-indexing configuration, has eight stations, and is driven by a geneva mechanism in which the driver has a rotational speed of 30 rpm. What is the production and throughput time of this machine?

Solution

From Chapter 2, we understand the principle of the geneva mechanism in which every revolution of the driver constitutes one indexing of the assembly machine. Therefore the production rate is 30 units per minute.

$$\begin{aligned}\text{production time} &= \frac{1}{\text{production rate}} = \frac{1}{30 \text{ units/min}} \times \frac{60 \text{ sec}}{\text{min}} \\ &= 2 \text{ sec per unit} \\ \text{throughput time} &= \text{production time} \times \text{number of stations} \\ &= 2 \text{ sec} \times 8 \text{ stations} \\ &= 16 \text{ sec}\end{aligned}$$

We have been focusing upon ideal production rates in which the assembly machine functions without jamming, but this ideal is never completely achieved. Station malfunctions on an automatic assembly machine can have disastrous effects upon production, as will be seen in the section that follows.

Machine Jamming

It is easy to overlook the pitfalls of automation, and one of the most notorious examples can result from the setting up of automatic multistation assembly machines without due consideration to the potential effects of station malfunction or jamming. Consider an eight-station rotary indexing machine driven by a geneva mechanism in which the index time is three seconds and the dwell time is five seconds. The reader may want to check the validity of the ratio of index to dwell time, using the principles studied in Chapter 2. Under ideal operating conditions (no malfunctions), this eight-station rotary indexing machine will produce a completed assembly every eight seconds and will achieve a corresponding production rate of 450 units per hour.

In a more realistic case, suppose each station malfunctions on the average of once every 100 cycles, a seemingly tolerable rate of work stoppage. One or more station malfunctions will immediately jam the indexing machine, requiring an operator to make adjustments to restart the machine. For our example, let us say that this adjustment and restart process requires a mere ten minutes. But even with such minor applications of "Murphy's Law," the result is a drastic reduction in productivity of the automated assembly machine, as can be seen in the following series of calculations.

If the chance of a station malfunction is one in 100, the chance that a given station will *not* malfunction in a given cycle is 99 percent or 0.99. But all eight stations must operate without malfunction to produce a completed assembly successfully. So the probability of no malfunction in a given cycle is the product of the chances of no malfunctions at each station during that cycle. Multiplying the station success probabilities:

$$0.99^8 = 0.9227$$

Therefore, out of 10,000 machine cycles, 9227 assemblies will be produced without malfunction, with a cycle time of eight seconds per assembly. This will consume

$$9227 \times 8 \text{ sec} = 73{,}816 \text{ sec} = 20.50 \text{ hr}$$

In the other 773 cycles (10,000 − 9227), at least one station will malfunction and require a 10-minute repair. This will consume

$$773 \times 10 \text{ min/breakdown} = 7730 \text{ min} = 128.83 \text{ hr}$$

The total time to produce the 9227 assemblies is then

$$20.50 \text{ hr} + 128.83 \text{ hr} = 149.33 \text{ hr}$$

for the total of operating time plus malfunction downtime. But note which figure is the *larger!* The percent downtime is

$$128.83/149.33 = 0.863 = 86.3\%$$

of the total production time! The production rate has been reduced from the ideal of 450 units per hour calculated earlier to

$$9227 \text{ units}/149.33 \text{ hr} = 61.8 \text{ units/hr}$$

The efficiency of the assembly machine would be the ratio of the actual production rate to the ideal production rate, calculated as

$$\text{efficiency} = \frac{61.8 \text{ units/hr}}{450 \text{ units/hr}} = 0.137 = 13.7\%$$

an amazing 86.3 percent drop in efficiency from introduction of only a slight (one percent) station malfunction rate. Such is the world of automation, and the automation engineer should be prepared to deal with the realities of system malfunction and downtime when planning for a new automated assembly machine.

Component Quality Control

The previous section discussed the disastrous effects of station malfunction in an automatic assembly machine, but failed to explain why these malfunctions might occur. The predominant cause of assembly station malfunction is some random variation in the components being assembled—variation of a magnitude that cannot be handled by the assembly machine. If tighter specifications can be applied or closer quality control can be exacted upon the components produced to existing specifications, the automation engineer may be able to achieve astonishing improvements in assembly machine production rates. Suppose in the example in the previous section that 90 percent of the assembly malfunctions were due to faulty components. Elimination of the component quality problem would then reduce the station malfunction rate from one out of 100 cycles to one out of 10,000 cycles. Such station reliability corresponds roughly to "four-sigma" performance, as introduced in Chapter 1. Let us now refigure the production rate and percentage downtime with the component-quality problem eliminated:

Probability of no station malfunction in a given typical cycle:

$$\text{probability [no station malfunctions]} = 0.9999^8 = 0.9992$$

thus, in 10,000 machine cycles, 9992 assemblies will be produced with no malfunction with a cycle time of eight seconds per completed assembly. This will consume

$$9992 \times 8 \text{ sec} = 79{,}936 \text{ sec} = 22.20 \text{ hr}$$

In the other 8 cycles (10,000 − 9992), at least one station will malfunction and require a 10-minute repair time. This will consume

$$8 \times 10 \text{ min} = 80 \text{ min} = 1.33 \text{ hr}$$

Now the total time to produce 9992 assemblies is

$$\text{total time} = 22.16 \text{ hr} + 1.33 \text{ hr} = 23.49 \text{ hr}$$

for both operating and downtime. But the percentage downtime has been reduced from 86.3 percent to

$$1.33 \text{ hr}/23.49 \text{ hr} = 5.66\%$$

of the total production time. The production rate is back up to

$$9992 \text{ units}/23.49 \text{ hr} = 425 \text{ units/hr}$$

which is

$$425/450 = 0.944 = 94.4\%$$

of the ideal production rate.

A production efficiency of 94.4 percent is not ideal but it is considerably better than the 13.7 percent achieved without component quality control.

It can be seen that the control of component quality can easily become the difference between success and failure of an automated assembly system setup. A case study will now serve to review the essential concepts of the previous two sections.

EXAMPLE 4.2 (CASE STUDY)
Automatic Assembly with Station Malfunctions and Parts Jamming

An in-line automatic transfer and assembly machine has 30 consecutive assembly stations. The line is under control of a walking beam with an index time of four seconds and a dwell of 20 seconds. Each station along the line will operate without malfunction with a reliability of 0.999 when its hopper is supplied with quality components. Any defective component will cause a station to jam, which in turn will precipitate a line jam because there is no provision for in-process assembly storage along the line. A jam or malfunction requires ten minutes to correct.

(a) No Station Malfunction or Jamming

Assuming no station jamming or malfunction at all, what is the ideal production capability of this line? What is the throughput time?

Solution

Cycle Time

$$T_C = 4 \text{ sec index} + 20 \text{ sec dwell} = 24 \text{ sec}$$

Ideal Production Rate

$$R = \frac{3600 \text{ sec/hr}}{24 \text{ sec/unit}} = 150 \text{ units/hr}$$

Throughput Time

$$T_T = 30 \text{ stations} \times 24 \text{ sec/station} = 720 \text{ sec}$$
$$= 12 \text{ min}$$

(b) Assuming Station Malfunction

Assuming ideal component quality, what is the production rate considering station malfunction? What is the percent downtime? What is the throughput time?

Solution

$$\text{Prob [a cycle will not malfunction]} = 0.999^{30} = 0.9704$$
$$\text{Prob [a cycle will malfunction]} = 1 - 0.9704 = 0.0296$$

Time to Produce 9704 Assemblies

$$\begin{aligned}\text{total time} &= \text{total successful cycle time} + \text{malfunction correction time}\\ &= 9704 \text{ units} \times \frac{24 \text{ sec}}{\text{unit}} \times \frac{1 \text{ hr}}{3600 \text{ sec}}\\ &\quad + 296 \text{ malfunctions} \times \frac{10 \text{ min}}{\text{malfunction}} \times \frac{1 \text{ hr}}{60 \text{ min}}\\ &= 64.69 \text{ hr} + 49.33 \text{ hr}\\ &= 114.02 \text{ hr}\end{aligned}$$

Production Rate

$$\begin{aligned}R &= 9704 \text{ units}/114.02 \text{ hr}\\ &= 85 \text{ units/hr}\end{aligned}$$

Percent Downtime

$$D = 49.33 \text{ hr}/114.02 \text{ hr} = 0.43 = 43\%$$

Throughput Time

$$\begin{aligned}T_T &= 30 \text{ stations} \times \frac{1}{85 \text{ units/hr}} \times \frac{60 \text{ min}}{\text{hr}}\\ &= 21.18 \text{ min}\end{aligned}$$

(c) Assuming Defective Parts Jamming

Assuming component lot quality is at the level of $^1/_2$ of one percent defective, what is the effect of defective components upon production rate?

Solution

We now have two ways that a station can jam:

1. General station malfunction (reliability = 0.999)
2. Defective component ($\frac{1}{2}$ of one percent = 0.005)

For the situation in which neither of these possibilities occurs we must compute the product of probabilities:

$$\begin{aligned}\text{Prob}\begin{bmatrix}\text{a given station will not jam}\\ \text{at a given cycle}\end{bmatrix} &= (0.999) \times \text{Prob [no defective component]}\\ &= (0.999)(1 - 0.005) = (0.999)(0.995)\\ &= 0.994\end{aligned}$$

Since the system has 30 stations, any one of which is capable of bringing down the entire line, we must compute the product of probabilities that individual stations will not jam to obtain the overall system success probability for a given cycle:

$$\text{Prob [a cycle will not jam]} = 0.994^{30} = 0.8349$$

And the probability that the system *will* jam in a given cycle is the complement:

$$\text{Prob [a cycle will jam]} = 1 - 0.8349 = 0.1651$$

Time to Produce 8349 Assemblies

$$\begin{aligned}\text{total time} &= \text{total successful cycle time} + \text{unjam time}\\ &= 8349 \text{ units} \times \frac{24 \text{ sec}}{\text{unit}} \times \frac{1 \text{ hr}}{3600 \text{ sec}}\\ &\quad + 1651 \text{ jams} \times \frac{10 \text{ min}}{\text{jam}} \times \frac{1 \text{ hr}}{60 \text{ min}}\end{aligned}$$

$$= 55.66 \text{ hr} + 275.16 \text{ hr}$$
$$= 330.82 \text{ hr}$$

Production Rate

$$R = 8349 \text{ units}/330.82 \text{ hr}$$
$$= 25.24 \text{ units/hr}$$

Percent Downtime

$$D = 275.16 \text{ hr}/330.82 \text{ hr} = 0.83 = 83\%$$

Efficiency

$$\text{E} = \frac{\text{actual production rate}}{\text{ideal production rate}} = \frac{25.24 \text{ units/hr}}{150 \text{ units/hr}} = 0.17 = 17\%$$

Throughput Time

$$T_T = 30 \text{ stations} \times \frac{1}{25.24 \text{ units/hr}} \times \frac{60 \text{ min}}{\text{hr}}$$
$$= 71.32 \text{ min} = 1.19 \text{ hr}$$

Thus, a component quality level of only $^1/_2$ of one percent defectives in this case reduces the production rate from 95 units/hr to 25.24 units/hr, increases downtime from 43 percent to 83 percent, and throughput time increases from a little over 20 minutes to over an hour.

So important is the quality and uniformity of components to the success of assembly automation that some firms pay several times the standard price for such items as screws, nuts, and bolts to purchase the finest quality available for use in automated assembly. The prohibitive cost of machine jamming due to faulty components easily justifies the added cost for quality components in these cases.

Defective Component Assembly

One way to deal with the component-quality problem is to use automatic in-process inspection to prevent costly assembly operations from being wasted on defective components. This is a good strategy, but the results of Case Study 4.2 suggest that once the components reach the assembly machine a better strategy may be to assemble the defective components *intentionally* rather than stop the line. For some assemblies—electronic circuit boards, for instance—it may be more practical to screen out defective components after the assembly is complete, even though this entails the added cost of partial disassembly. The benefits of this strategy will be illustrated in Example 4.3.

EXAMPLE 4.3
Savings from Defective Component Assembly

A 20-station automatic assembly line operates on a two-shift basis, including maintenance, 350 days per year. Preventive maintenance is performed on the third shift and during normal operation. The line is stopped upon demand for machine repair or defective component replacement. The component quality level is one percent defective, and the current strategy is always to detect and remove defective components before assembly, even though such removal requires stopping the line an average of 15 minutes each time to remove defective components. The automatic assembly machine has an equivalent annual cost of $490,000 and achieves an ideal cycle time of ten seconds.

A strategy is proposed to keep the line running without stopping to remove defective components, even though this will entail rework at a cost of 0.6 labor hours at $20 per hour per defective assembly. Is the savings in production time worth the expensive rework?

Solution

The analysis of this strategy consists of comparing the typical cost of line downtime (alternative A):

$$\text{Cost } A = 15 \text{ min/stoppage} \times \frac{1 \text{ hr}}{60 \text{ min}} \times \frac{1 \text{ day}}{16 \text{ hr}} \times \frac{\$490{,}000/\text{yr}}{350 \text{ days}}$$

$$= \$21.875$$

With the typical rework cost (alternative B):

$$\text{Cost } B = 0.6 \text{ hr/rework} \times \$20 \text{ hr} = \$12$$

The better alternative of the two is to assemble defective components intentionally.

Note in Example 4.3 that production rate and cycle time did not enter the analysis. The lower cost alternative in this case study was to assemble defective components and rework later, but other alternatives should be considered. The assembly machine downtime for component removal is not really lost if it can be made up on the third shift. The equivalent annual cost of the machine itself would remain essentially the same since most of this cost would consist of capital recovery of the original investment in the machine. Two other alternatives would be: (1) automatic preinspection screening of the components to remove defectives before they reach the assembly machine, and (2) improvement of the source quality of the components during their manufacture.

Example 4.3 was a concoction of circumstances that would make the intentional automatic assembly of defective components a viable alternative. Usually, however, this alternative will not be the best one. If there is any idle capacity on the machine, even on the nightshift, it will usually be advantageous to use that idle capacity to make up for downtime rather than incurring rework costs due to intentional assembly of defective components.

In case the reader considers the circumstances of Example 4.3 to be too farfetched, it should be mentioned that even in conventional manufacture the intentional acceptance of known defectives is a calculated strategy. For instance, in the garment industry, fabric flaws are often detected by the cutting-room personnel during the spreading of "lays" before the cut is made in thicknesses of up to 250 ply. The flaws are left in the fabric to avoid disrupting the laying process and to prevent damage to other pieces that will be cut from the defective ply. It is usually hoped that later the defective pieces will be discovered and removed, either in the sewing room prior to stitching or even in final inspection where rework will be the unfortunate consequence. Either alternative may be preferable to removing the flaw in the cutting room.

In Chapter 10 we shall study a robotics application in which the robot detects a defect during automatic assembly and, without interrupting the assembly cycle, presses a "memory button" on the assembly pallet to alert a downstream station to correct the defect.

BUFFER STORAGE

If a large degree of random variation exists in the operation of the individual stations along an automatic assembly line, a degree of independence between the stations must be built in to achieve even a modicum of efficiency. This independence is achieved by placing buffer storage between stations. Such a system usually dictates a nonsynchronous material transfer system because a rigid, intermittent transfer system does not accommodate the variations in station operation time. Nonsynchronous systems are said to be "power-and-free."

To see the advantage of buffer storage between stations, consider a line in which the ideal cycle time is one minute. If each station on the line malfunctions once every 60 cycles and one hour is required to resolve each malfunction, each station will be available for production approximately 50 percent of the time. Hypothetically, if all of the station malfunctions could be synchronized, the line could theoretically achieve a production of one-half of 60, or 30, units per hour. But the nature of malfunctions is that they occur at random points in time. This characteristic greatly reduces the productivity of the line, because if any station along the line is down, the entire line is down—as was seen earlier in the analysis of indexing assembly machines. In the situation described above in which each station is available only 50 percent of the time, the availability of the entire indexing line is 0.5^n, where n represents the number of stations along the line. Such a line with only five stations would be down 97 percent of the time! The difference between this percentage and the hypothetical percentage of availability if the malfunctions could be synchronized:

$$97\% - 50\% = 47\%$$

represents the production loss due to interference between the stations. The interference is called "blocking" when station i cannot release its part to station $i + 1$ and is called "starving" when station i cannot obtain a part from station $i - 1$. In either case, station i is idle during this period even if it is not malfunctioning and would be able to produce if it had a part upon which to perform its operation.

The provision of buffer storage between stations permits each station to produce "to stock." The first station is started first and produces approximately one-half the capacity of its storage before the second station is started. The second station then begins, working from the stored supply produced by the first station while the first station continues to produce. Subsequent stations are added to production as each sequential storage reaches approximately one-half its capacity, provided stations are reasonably similar in operating characteristics and downtime. When all storages are thus loaded, the line can function with some resilience due to the buffer storages between stations. If the buffer storages are large enough to prevent blocking and starving, the production of the line described earlier can be back up to 30 units per hour.

EXAMPLE 4.4
Buffered Assembly Line Production

Example 4.2 considered the devastating effects of station malfunctions and parts jamming upon automatic assembly. Suppose the same 30-station operation were performed on a buffered assembly line with the same 20-second productive cycle. (Note that the four-second index time is ignored in this case because each buffered station will be producing to stock.) Calculate the improvement in line productivity.

Solution

Prob [a given station will not jam] = 0.994 [from solution to Example 4.2(c)]

Time to Produce 994 Assemblies

$$\text{production time} = 994 \text{ units} \times \frac{20 \text{ sec}}{\text{unit}} \times \frac{1 \text{ hr}}{3600 \text{ sec}} + 6 \text{ jams} \times \frac{10 \text{ min}}{\text{jam}} \times \frac{1 \text{ hr}}{60 \text{ min}}$$

$$= 5.52 \text{ hr} + 1 \text{ hr}$$

$$= 6.52 \text{ hr}$$

Production Rate (Buffered)

$$R_B = 994 \text{ units}/6.52 \text{ hr}$$

$$= 152.4 \text{ units/hr}$$

Production Rate (Unbuffered)

$$R = 25.24 \text{ units/hr [from solution to Case Study 4.2(c)]}$$

Productivity Improvement

$$\frac{R_B}{R} = \frac{152.4}{25.24} = 6.04$$

Thus, Example 4.4 shows how the addition of adequate buffer storage could increase the productivity of the line sixfold! This is a sobering thought because it is an indictment of the typical automatic-indexing assembly line, the epitome of automation. In fact, fully buffered assembly lines are really a degeneration to batch-type manufacturing, typical of a plant prior to automation.

Before the reader despairs, however, let us consider the drawbacks of buffer storage between stations. Example 4.4 omitted a calculation of throughput time, which goes up drastically when buffer storages are provided. The amount of the increase is dependent upon the amount of buffer storage provided, as will be illustrated in Example 4.5.

EXAMPLE 4.5
Effect of Buffer Storage upon Throughput Time

A 30-station assembly line with buffer storage between stations that averages 50 units each has a production rate of 152.4 units per hour (the same as in Example 4.4). Calculate throughput time.

Solution

Throughput time must consider the wait time each unit spends in the buffer storages between stations. The average wait time is the product of the average station cycle time (including downtime) *times* the average number in storage. Thus, for an average storage quantity of 50

$$\begin{array}{l}\text{average wait time} \\ \text{(between any two stations)}\end{array} = \frac{1}{152.4 \text{ units/hr}} \times 50$$

$$= 0.328 \text{ hr/buffer}$$

There would be $n - 1$ or 29 buffer storages in a 30-station buffered assembly line so the total wait time for units in the buffered assembly line would be

$$\begin{aligned}\text{total average wait time} &= (\text{average wait time in each buffer}) \\ &\quad \times \text{number of buffers} \\ &= 0.328 \text{ hr/buffer} \times 29 \text{ buffers} \\ &= 9.51 \text{ hr} \\ \text{total throughput time} &= \text{production time} + \text{wait time} \\ &= \left[\frac{1}{152.4 \text{ units/hr}} \times 30 \text{ stations}\right] + 9.51 \text{ hr} \\ &= 0.20 \text{ hr} + 9.51 \text{ hr} \\ &= 9.71 \text{ hr}\end{aligned}$$

Alternative Solution

Another way to calculate throughput time for the buffered assembly line is to recognize that there are

$$30 + (29 \times 50) = 1480$$

positions through which a unit must advance through the system at a rate of 1/152.4 hours per position:

$$\text{throughput time} = [30 + (29 \times 50)] \frac{1}{152.4} = 9.71 \text{ hr}$$

The throughput time of 9.71 hours calculated for Example 4.4 should be compared with the throughput time calculated for Case Study 4.2(c), which considered station malfunctions and jamming. That throughput time was calculated to be approximately one hour. Thus, although adding buffer storage increased line productivity sixfold (Example 4.3), it increased throughput time almost tenfold (Example 4.4). This conclusion is based upon the assumption that an average buffer storage of 50 units would be ample to prevent any blocking or starving of stations in the buffered assembly line.

Besides causing a gross deterioration in throughput time, adding buffer storage increases the complexity of parts handling and transfer between stations. In typical batch manufacturing, the degenerated state of a buffered assembly line, all of the parts positioning maneuvers studied in Chapter 2 are lost again at each station as each station dumps completed subassemblies into a bin for the next station.

It can be seen that the general problem of engineering a manufacturing sequence for discrete-item processing or assembly presents a dilemma: Maximizing productivity tends to defeat attempts to minimize handling cost and throughput time. But this is where the industrial robot enters the scene. The industrial robot is capable of preserving parts orientation and position and can intelligently permit production to stock or in assembly-line sequence as conditions arise. This is only one of the benefits of highly flexible industrial robots, which will be explored in depth in later chapters. But first we shall complete our study of automated processes by examining in Chapter 5 automatic numerically controlled machine tools, the forerunners of the industrial robot.

SUMMARY

Once machine parts have been correctly oriented and fed to the machine, they are ready for automatic processing or assembly. Operations such as screwdriving and nut running can be completely automated. Assembly machines index between stations arranged in-line or around a dial. Both types can present severe problems of inefficiency due to system downtime whenever one of the stations jams or malfunctions. The drop in efficiency and in production rate is more dramatic as the number of assembly stations in the machine increases.

The problem of station malfunction and defective component jamming can be alleviated somewhat by the provision of in-process buffer storage at stations along the line. While such buffer storages can effect a marked improvement in line productivity, a trade-off is the drastic deterioration in throughput time. Also, the insertion of buffer storage can present a challenge in preserving the costly orientation and positioning of parts for automatic assembly.

The importance of quality of components for assembly increases when automation is employed. The principal reason is that a defective component can jam its work station, bringing down the entire line. This problem can be so detrimental that in certain situations it becomes advantageous to assemble defective components and remove them at the end of the assembly process rather than to stop the line.

EXERCISES AND STUDY QUESTIONS

4.1. A 12-station automatic assembly line has a station cycle time of 18 seconds, including index time. Calculate ideal line production rate per hour and throughput time.

4.2. A six-station dial indexing machine is driven by a geneva with driver rotation at 18 rpm. What are the production and throughput times of this machine?

4.3. An eight-station geneva with driver rotation 24 rpm is compared to a six-station geneva with driver rotation 20 rpm. Which system has the faster production rate? Which has the faster throughput?

4.4. An assembly system has five sequential stations with no storage between stations. Ideal production rate for the system is 150 units per hour. With station malfunctions, the actual production rate deteriorates to 120 units per hour. What is the percent downtime? What is the ideal throughput time? What is the actual throughput time?

4.5. A dial-indexing geneva assembly machine with an ideal cycle time of four seconds has a throughput time of 20 seconds. How many stations are in the system? What is the rpm of the driver?

4.6. Assume a station reliability of .99 for each of the stations in the dial-indexing system in Exercise 4.5, that is, on the average a malfunction will occur approximately once in 100 cycles. Every malfunction stops the entire machine and requires ten minutes to clear. What is the actual production rate of the system, taking into consideration downtime to clear malfunctions?

4.7. A six-station assembly machine has zero storage between stations and a breakdown rate of 2 percent for each station. Average time to clear a breakdown and restart the line is ten minutes. With an ideal cycle time of ten seconds for this machine, including index time, what is its actual production rate? Calculate the production efficiency. Calculate the ideal and actual throughput times.

4.8. In Exercise 4.7, suppose the station breakdown rate could be cut in half by introducing a component-quality improvement project that includes a more stringent receiving inspection of purchased parts, vendor rating, and closer specified tolerances on both purchased parts and parts fabricated in-house. How much would this improve the efficiency of the assembly machine?

4.9. A ten-station automatic assembly line with no buffer storage has a normal cycle time of three seconds when no malfunctions occur. To allow for parts jams, which require two hours to repair, the system is designed to tolerate a total average cycle time of ten seconds. For the sake of discussion, let us assume that the only reason that a cycle will malfunction is when an out-of-tolerance piece part jams it. Will the ten-second cycle time limit be met if the piece-part quality is "three sigma"?

4.10. In Exercise 4.9, suppose the piece-part quality is "four sigma." What would be the total average cycle time and production rate?

4.11. In Exercise 4.9, what piece-part quality would be required to hold system downtime to less than half the total average cycle time?

4.12. In Exercise 4.9, what piece-part quality would be required to hold system downtime to 10 percent of total average cycle time?

4.13. An eight-station automatic assembly line has buffer storage between stations of capacity 50 units each. The eight stations are reasonably similar in operating characteristics and downtime. How many units would you recommend to be in each buffer storage when the system is operating normally?

4.14. What procedure would you recommend for initial startup and loading of the eight-station automatic assembly line described in Exercise 4.13?

4.15. Calculate average throughput time for the eight-station automatic assembly line described in Exercise 4.13.

4.16. (**Design Case Study**) Suppose a 50-station transfer line with no storage or slack between stations is being planned for a major automation thrust in the manufacturing plant where you are employed as the automation engineer. A cycle time objective has been set at 12 seconds in order to meet production quotas of 2000 units per eight-hour shift. The plant manager is very excited about this new major automation project and plans to invite top management from corporate headquarters to observe the fruits of automation as soon as you are able to get the operation on-stream. Would you share in the plant manager's optimism about the project? Do you think the eight-hour shift production quota of 2000 units is realistic? Is the production quota compatible with the cycle time objective? If all preventive maintenance is performed on another shift so that theoretically the line could operate during a full eight-hour shift, what would be the ideal production rate for this line? What is the maximum average station breakdown rate per cycle to be tolerated if the average time required to service a line breakdown and restart the line is six minutes?

CHAPTER 5

NUMERICAL CONTROL AND CAD/CAM

The forerunner to industrial robots was the numerically controlled (NC) machine tool, an automated version of the conventional machine tool. Conventional machine tools such as milling machines, lathes, drilling and boring machines, and grinders are manually guided into position by handwheels and cranks as is indicated in Figure 5.1. The tool cutter is held precisely by the machine tool to effect an accurate cut upon the workpiece, usually metal. Conventional machine tools are still appropriate for very-small-quantity production or for toolroom use where one-of-a-kind tools are fashioned by the skilled tool-and-die maker before being used subsequently in machine tools for high production.

The first tack taken by automation engineers was to use mechanical cams and linkages to automate mechanically a sequence of standard machine tool operations to produce a particular product. This strategy worked well as conventional machine tools were speeded up and labor was saved by performing the machine control operations mechanically. Another advantage of the automatic machine tool approach was that the operation became more stable and repeatable, and thus quality was enhanced. The mechanical cam and linkage type of automation for machine tools is exemplified by the automatic lathe (Figure 5.2) and reached its apex with the development of the automatic screw machine (Figure 5.3). The automatic screw machine, named after only one of the products it makes, is capable of automatically performing all the cuts on a workpiece such as a screw or other items shown in Figure 5.4 and achieving very high production rates by simultaneously performing six operations as the workpieces index on six spindles. This type of machine was the ultimate in automation in the 1940s.

Dramatic as the automatic screw machine and other mechanically sequenced machines are, something was still missing in the capability of automatic machines in the 1940s. Some production demands were for quantities too small to justify the investment in mechanically fixing an automatic sequence of operations. The only answer for these products was to employ skilled machinists to operate engine lathes, conventional milling machines, and other machine tools that required the operator to set all cuts and machine adjustments manually, despite the drawbacks in efficiency and quality. The aircraft and aerospace designs of the early 1950s demanded very highly skilled machinists to machine the large and surprisingly intricate aluminum structural members of these then-advanced vehicles.

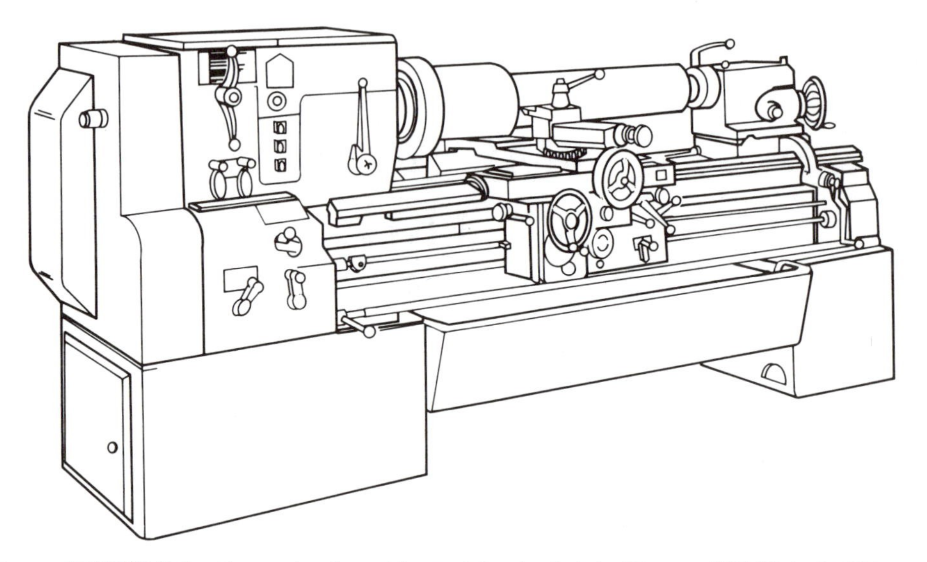

FIGURE 5.1 Conventional machine tool (engine lathe). (*Source:* NIOSH [ref. 60].)

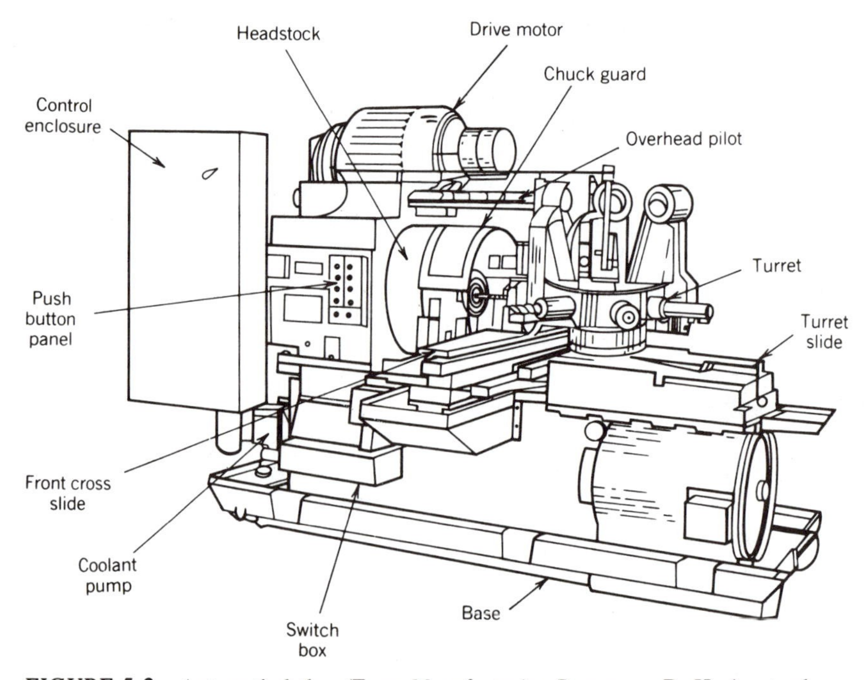

FIGURE 5.2 Automatic lathe. (From *Manufacturing Processes,* B. H. Amstead, Phillip F. Ostwald, and Myron L. Begeman, 7th ed., copyright 1979 by John Wiley & Sons, Inc., New York, N.Y. Reprinted by permission.)

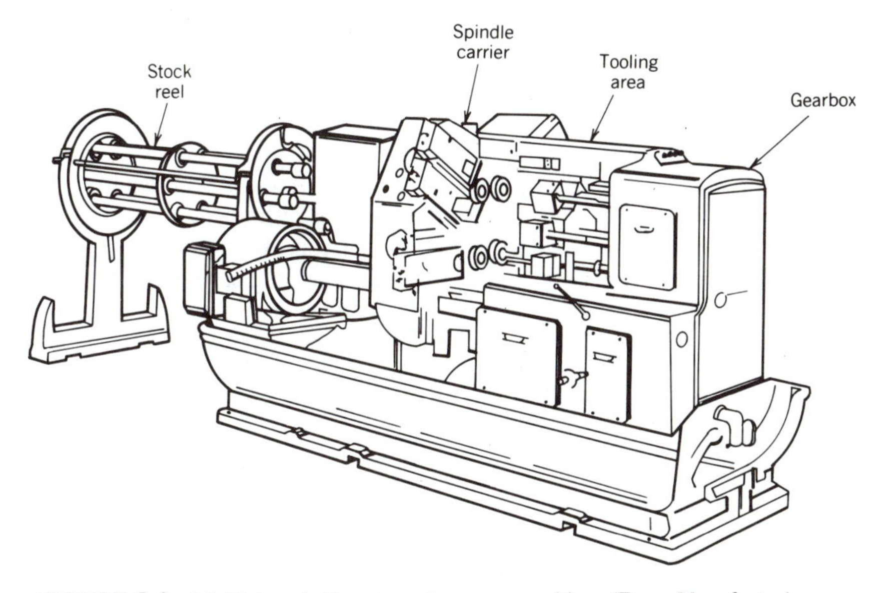

FIGURE 5.3 Multiple-spindle automatic screw machine. (From *Manufacturing Processes,* B. H. Amstead, Phillip F. Ostwald, and Myron L. Begeman, 7th ed., copyright © 1979 by John Wiley & Sons, Inc., New York, N.Y. Reprinted by permission.)

FIGURE 5.4 Items produced on an automatic screw machine. (Reprinted by permission of Davenport Machine Tool, Rochester, N.Y.)

Among other skills, blueprint reading was a necessary expertise of the machinist who would mill a large landing gear strut or wing spar from a solid block of aluminum. So diverse were the many cuts required to produce a single part such as the one illustrated in Figure 5.5 that the operation was difficult to break down into prescribed operation settings, and the alternative was to rely upon the skill of the machinist to devise an effective machining sequence directly from the blueprint. Another skill required of these master machinists was facility with precision measuring instruments. In a process dependent upon the judgment and skill of the operator, frequent checks were necessary to assure that the desired level of precision was being achieved. The need for such checks was aggravated by the low volume required for each complicated workpiece.

It was this setting that spawned a project in 1949 sponsored by the United States Air Force in which the Massachusetts Institute of Technology developed a prototype machine tool to produce wing spars and skins for military aircraft. The concept was that the machine's movements could be controlled by a numerical code that was machine-readable. The advantage would be the capability of storing a complicated sequence of machining operations on a suitable storage medium to be retrieved and rerun the next time that particular design or model workpiece was needed again.

Some writers point back centuries to textile looms and other early machines controlled by punched cards to mark the beginning of numerically controlled machines. It is true that earlier machines operated under such control concepts, but they were not known as NC machines, and they were of a different evolutionary era than the NC machine tools developed in the 1950s for the aerospace industry.

FIGURE 5.5 Machining a trunnion component for the NASA Space Laboratory on a milling and boring machine. (Reprinted by courtesy of Maho Machine Tool Corporation, Naugatuck, Conn.)

ADVANTAGES OF NC

The preceding background treatise cites the aerospace industry as the birthplace of modern numerical control machines, but why aerospace? The answer to this question is important because it provides insight into the principles and objectives of using NC machines. Aerospace is the ideal application for the use of NC machine tools for just about every reason or advantage one seeks by using NC machines. These advantages will now be examined using the aerospace industry as the ideal for comparison with other industries that may also be excellent candidates for NC machine tools.

Flexibility

Many products are made in quantities of 100,000 or even 1,000,000 for a single model run, but not airplanes or spacecraft. The much smaller production volumes characteristic of the aerospace industry call for a type of automation that can be easily modified to convert production from one model to another and even one workpiece to another on the same processing machine. Such a requirement is costly for the inflexible type of hard automation in use prior to 1950. Numerical control provided the flexibility to answer to this need, and it has since been applied to other industries once aerospace paved the way.

Another way to view the flexibility advantage afforded by NC machines is to recognize it as a reduction in setup time. Setup times ties up facilities, personnel, and machines just as does production operation time, but setup time produces no product. In the *first* production run of a product, NC machine setup times are comparable to those of hard automation setups because NC programming and checkout is required. But in the *re-*setup of the second and subsequent production lots, the NC machine setup time is much less than the conventional setup time.

Capability for Complex Workpieces

A long and complicated sequence of machining operations becomes difficult or impossible to perform automatically using hard automation. The hard automation approach uses special-purpose machines equipped with complex cams and mechanical linkages to perform a few, perhaps six, highly repetitive operations. By contrast, NC machines are very general purpose and can be programmed to do six operations in an automated sequence, or just as easily, continue on to do 66 or even 666 operations. Upon completion, the same NC program can be repeated on another workpiece, or a different NC program can be loaded, and, in a few minutes, the same machine can be manufacturing an entirely different workpiece requiring several hundred operations not at all similar to the operations of the previous program. Again, this is just what aerospace needed with its complicated wing spars and landing gear struts; other industries have followed with similar needs, not the least of which has been the machine tool industry itself.

Facility with Large Workpieces

Special-purpose automatic machines usually are employed to perform a few operations upon a workpiece, and then the workpiece is transported to another machine to continue the manufacturing process. This is also true somewhat for NC machines, but the difference is in the number of machines to which the workpiece must be transferred. Rather than pass the workpiece to another machine, the NC machine is more likely to change its own tool and continue processing. When the workpiece is large and awkward to handle, as in airplane parts, this advantage of NC machines can save manufacturing costs.

Reduced Jig and Fixturing Costs

The ability to leave a workpiece in the same machine for continued processing with the same or other tools holds another important advantage for NC machines: reduced tooling and fixturing costs to hold the workpiece. This is probably even more important than the advantage of dealing easily with large, awkward workpieces. The flexibility and versatility of the NC machine enables it to approach the workpiece from many angles and with various speeds and tools to accomplish a large number of manufacturing operations without moving the workpiece to additional fixtures.

Quality

It is logical that a machine that performs many operations in a single fixturing of a workpiece should be able to achieve greater uniformity and quality than a series of special-purpose machines that require several refixturings of the workpiece. Add to this the elimination of human variation as well as methods variation, both of which introduce some process irregularity. The elimination of these sources of variation made possible by the use of NC machines results in higher quality.

In addition to the above advantages for NC machines, it has been argued that NC equipment reduces skill levels required. It is true that the execution of a correctly programmed NC machine requires less machinist-type skill and experience than does the performance of the same operations without the benefit of numerical control. But other skills are required to program and maintain the NC machines. A familiar pattern is associated with the adoption of NC machine tool processes as occurs with other forms of flexible automation: Machine-operational skills and crafts-oriented skills are replaced by computer-oriented skills.

Even at the machine-operational level, it would be unwise to assume that the employment of NC machine tools will reduce the skill level required and thus the labor cost of machine tool operators. The writer of this text was astonished to discover from personal experience that in aircraft factories of the early 1960s NC machine tool operators were among the most experienced and most highly paid machinists in the plant. The reason for this anomaly was likely to be based more upon traditions for seniority than upon a genuine need for machinist experience. Also, the degree of responsibility may have been a factor in assigning experienced machinists to numerical control machines. Although there is less for the operator to do physically at a numerical control machine, both the machine and the workpiece it is processing are of an order of magnitude more expensive than is true for conventional machine tools.

MACHINE CODE

Numerical codes for machines to read are more efficiently expressed in binary-based systems than in decimal code. Figure 5.6 compares a binary-based numerical code, punched paper tape, with a decimal-based numerical code, represented by the now obsolete computer card. Both examples in Figure 5.6 represent 80 characters of code. Both formats are capable of being read either by machines or by humans. For humans to read the code, they must be proficient in binary-coded decimal (BCD) in the case of punched paper tape

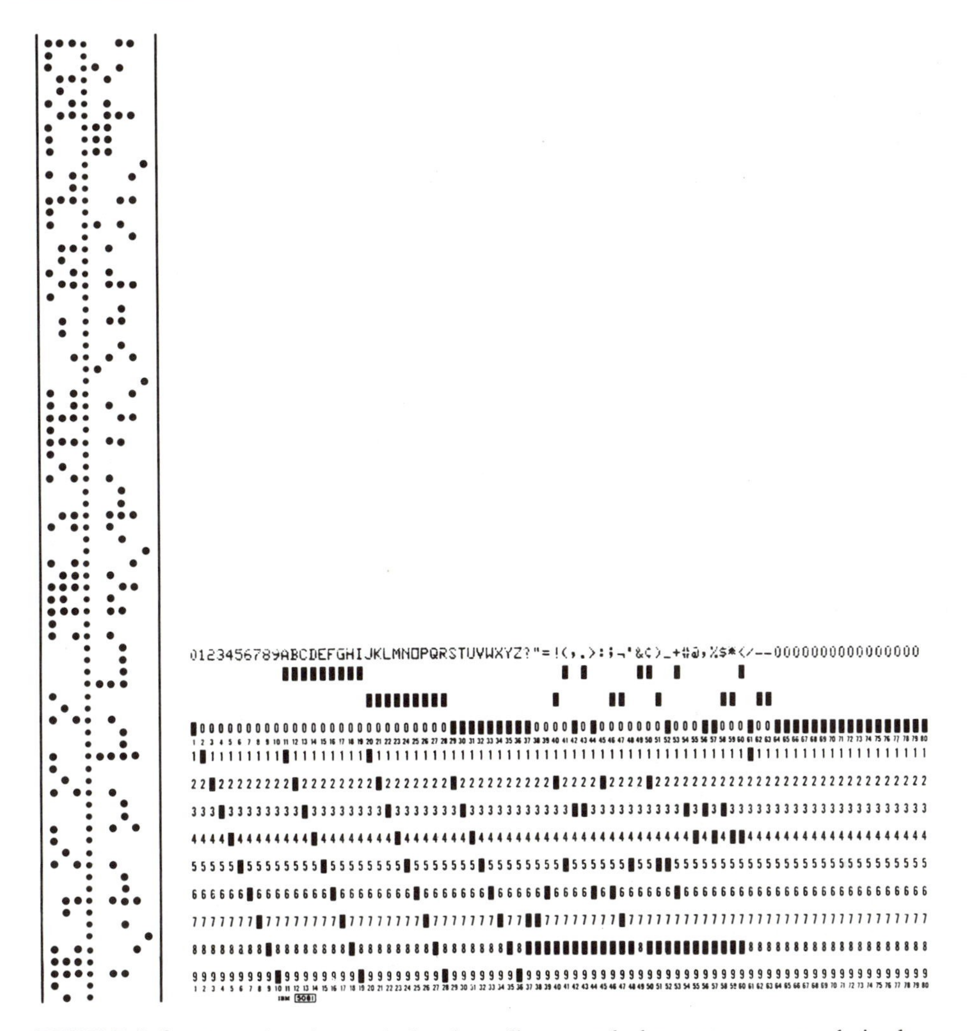

FIGURE 5.6 Binary-based numerical code medium, punched paper tape, versus decimal-based computer card. Both represent 80 characters of data.

or Hollerith code in the case of the computer card. The trend, however, is toward NC machine programs that are stored on more compact media in which the code is invisible to the human eye. The principal categories of these newer media are magnetic cassette tape and computer random access memory (RAM).

BCD

Binary-coded decimal code is important, not only because it is used in NC punched paper tape, but also because it is used in the invisible media of magnetic cassette tape and computer random access memory. Normally, an automation engineer will not find it necessary to read BCD physically without the help of a tape reader, but it is worthwhile to be able to read BCD and also to understand the structure of the numerical code used by NC machines.

A close examination of Figure 5.6 reveals that the one-inch wide punched paper tape has eight regular punch "channels" (rows of holes lengthwise of the tape) plus one channel of smaller holes near the center. The smaller holes are sprocket holes for feeding the tape on a machine, sometimes called a "flexo-writer," for punching and/or reading the tape. Note that the sprocket hole channel is positioned between channels three and four, not in the exact middle of the tape between channels four and five. This was a clever design decision to prevent inverting the tape when loading it into a machine or when reading it visually. Figure 5.7 deciphers two standard BCD character-coding schemes for NC machines. Each character is represented by eight possible punches in a horizontal row across the tape. The traditional paper tape code, EIA (Electronic Industries Association) 244A, shown on the right is the code most likely to be encountered if one is called upon to read a portion of a punched paper tape visually.

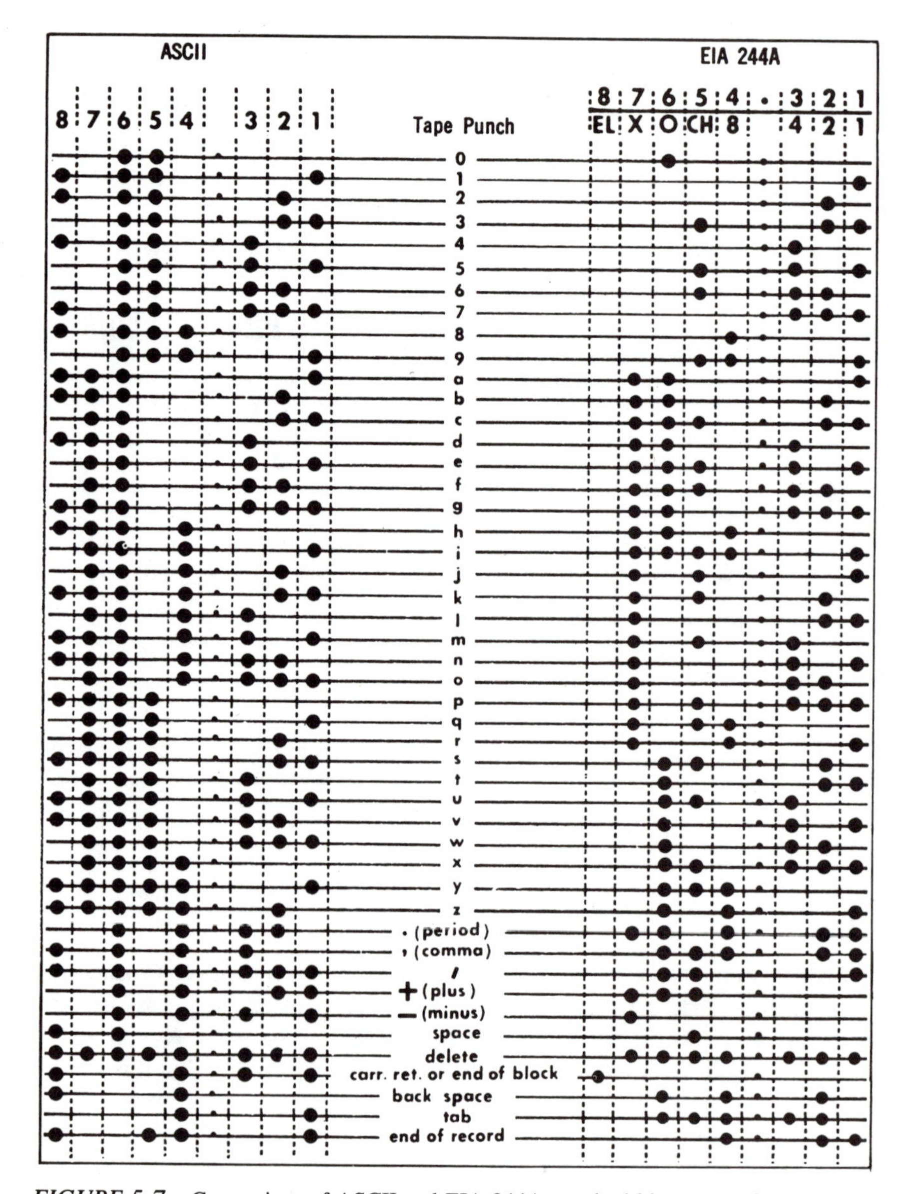

FIGURE 5.7 Comparison of ASCII and EIA 244A standard-bit patterns for NC code.

TABLE 5.1 Allocation of BCD Channels to the Representation of Decimal Digits

Tape channel number	1	2	3	4
Binary digit value	2^0	2^1	2^2	2^3
Decimal equivalent	1	2	4	8

For example, to represent decimal number 9 the following binary number would be written in channels 1 through 4: 1001.

Four binary digits are required to represent all ten decimal digits. These four binary digits occupy NC tape channels one through four, as summarized in Table 5.1. The BCD approach *dedicates* the four binary digits to a single decimal digit in order to make the code easier to interpret, even though this is somewhat wasteful of binary digits. Table 5.2 compares decimal with BCD equivalents and with the pure binary equivalents. Note that BCD and binary are the same up to the value decimal 10. At the value decimal 10, BCD begins a new binary digit (as does decimal) and starts over again with the lower-order digits. But pure binary continues with only four binary digits until all feasible combinations are exhausted. Thus, pure binary is more efficient but is difficult to read. If the reader understands the scheme at this point, he or she should be able to write down the BCD equivalent of any decimal number immediately, but not the pure binary equivalent. Stop reading for a moment and try to do this for the decimal number 1439. Construction of the BCD equivalent is illustrated in Figure 5.8. Note that four BCD digits are allocated to each decimal digit. The pure binary equivalent of the decimal number 1439 would be shorter in length than the BCD number and would be more difficult to derive. Decimal to *pure* binary conversions for large decimal numbers require compu-

TABLE 5.2 Decimal Versus BCD Versus Pure Binary Numbers

Decimal	*BCD*	*Binary*
0	0	0
1	1	1
2	10	10
3	11	11
4	100	100
5	101	101
6	110	110
7	111	111
8	1000	1000
9	1001	1001
10	10000	1010
11	10001	1011
12	10010	1100
13	10011	1101
14	10100	1110
15	10101	1111
16	10110	10000
17	10111	10001
18	11000	10010
19	11001	10011
20	100000	10100

Decimal number 1439 =
BCD number 1010000111001
(3 leading zeroes deleted)

Construction:

BCD	0001	0100	0011	1001
Decimal	**1**	**4**	**3**	**9**

FIGURE 5.8 Constructing the BCD equivalent of the decimal number 1439.

tations that are beyond the scope of this text and are not really needed by most automation and robotics application engineers. For comparison, however, the pure binary equivalent of decimal 1439 is 10110011111. This can be verified by finding the decimal equivalent to each of the binary ones and adding these decimal equivalents together to reach the composite decimal value of the entire binary number. The decimal equivalent to a binary digit is equal to 2^{n-1}, where n is the *position* of the digit in the binary number counting from the right. Counting from the right in our example, binary 10110011111 has binary digits 1 in positions 1, 2, 3, 4, 5, 8, 9, and 11. The decimal equivalent can then be computed as in Table 5.3.

Alpha Characters

Channels 6 and 7 are used in conjunction with the numeric channels to represent alphabetic and special characters. Figure 5.7 reveals that the NC codes employ a systematic procedure for using channels 6 and 7 for alphabetic letters, although the ASCII procedure is somewhat different from the EIA 244A procedure. The ASCII procedure signifies "alphabetic" by means of a bit in both channels 6 and 7 and then simply counts to decimal 26 in pure binary using channels 1 through 5, since there are 26 letters in the alphabet. The EIA

TABLE 5.3 Computation of Decimal Equivalent of Binary 10110011111

Digit Position	*Binary Digit*	*Decimal Equivalent*	*Computed Decimal Equivalent*
1	1	2^0	1
2	1	2^1	2
3	1	2^2	4
4	1	2^3	8
5	1	2^4	16
6	0	0	0
7	0	0	0
8	1	2^7	128
9	1	2^8	256
10	0	0	0
11	1	2^{10}	1024
Total			1439

244A scheme retains the BCD structure in the alphabetic letters by dividing the alphabet into three groups of nine letters each. The groups are denoted as follows:

Alphabetic letters A through I: bit in both channels 6 and 7
J through R: bit in channel 7 but not 6
S through Z: bit in channel 6 but not 7

Within groups, the code merely counts to decimal 9 in binary.

The careful reader has probably already recognized something wrong with the foregoing description of EIA 244A representation of the alphabet. The alphabet consists of twenty-six letters, not twenty-seven, so in "three groups of nine letters each" one code combination of bits is left over in the system. Can you determine from Figure 5.7 what the missing combination is and from where in the alphabet it was omitted?*

Parity Bits

Channel 5 in the EIA 244A code and channel 8 in the ASCII code are reserved for checking the reliability of the machine that produces the tape or the machine that reads the tape. A rigid discipline is followed by NC codes that requires that the number of bits in each horizontal row always be kept *odd* (in the case of the EIA 244A code) or *even* (in the case of the ASCII code). This discipline is called *parity* or *odd parity* or *even parity,* depending upon the code. Since some ordinary BCD characters will present an odd number of bits and some will present an even number, the parity channel is used to add an extra bit or omit it to maintain an odd (or even) number of bits for each character. The reason for this peculiar discipline will now be explained.

There are four basic mechanical errors that can occur in the production and subsequent interpretation of an NC tape:

1. The writer can write an extra bit that was unintended.
2. The writer can fail to write a bit in a position where it should have written one.
3. The tape reader can mistakenly interpret that a bit is present in a position in which a bit is really not present.
4. The tape reader can fail to detect a bit that is really present.

Each one of these errors should be extremely rare, but each can feasibly occur. If one instance of the above four errors occurs alone, the number of bits in the horizontal row in which the error occurs will violate parity. Thus, if a hardware device counts the number of bits in each row while the tape is written or read, a violation of parity can be immediately detected as a hardware error and the machine will stop. Of course, if two errors occur in the same horizontal row of the tape, parity will not be violated and the error may go undetected. "Two wrongs do not make a right," but the machine might mistakenly think so because parity would not be violated. The parity concept is to assume that the probability of two errors occurring in the same row is so remote as to be neglected as inconsequential.

Only one channel remains to be explained: channel 8 of the EIA 244A code. Channel 8 is abbreviated EL, which means "end of line" and is used to separate blocks of data. The channel 8 marker is recognized easily because end of line is the only meaning associated with this bit. This is particularly convenient for visual inspection of an NC punched paper tape.

*The missing combination is the set 6-5-1. It is omitted between the letters R and S in the alphabet.

AXES OF MOTION

The principal function of an NC instruction to a machine is to describe the motion to be taken by the tool or the relative motion between the tool and the work. This is a key concept and extends from machine tools in this chapter to industrial robots in Chapter 6.

Machine-tool motions are generally described in *x-y-z* Cartesian space, making the *x*-, *y*-, and *z*-axis coordinates the heart of the typical NC command. For illustrating these axes, we will call upon the archetype of all NC machine tools: the vertical milling machine. Figure 5.9 diagrams a vertical mill and labels the *x, y,* and *z* axes. A similar scheme is used in other machine tools as summarized in Figure 5.10. It can be seen in these diagrams that there is some conformity to a consistent rationale in naming the *x, y,* and *z* axes, but there are exceptions. Usually, the *x* and *y* axes describe a plane that is orthogonal to the penetration axis of the tool. The tool penetration axis is usually the *z* axis with negative

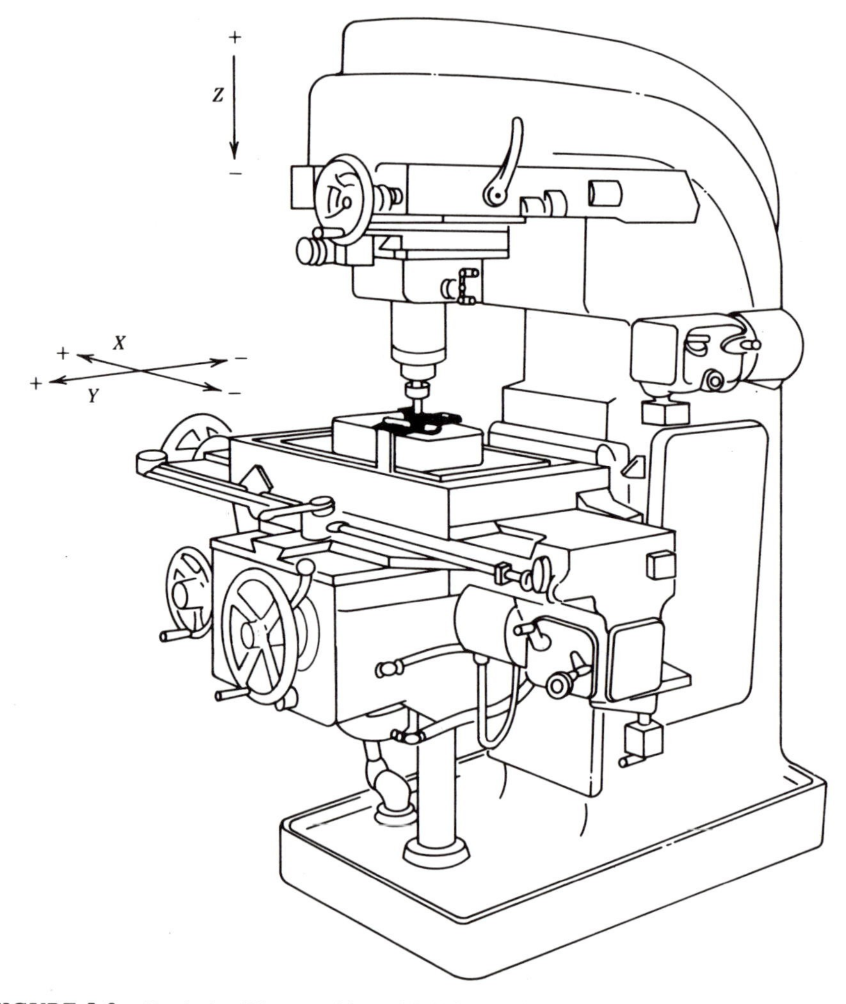

FIGURE 5.9 Vertical milling machine with NC axes labeled. The *z* axis is the vertical motion of the spindle. The *x* and *y* axes are motions of the table, which holds the workpiece. For example, a table move to the left causes the tool to make a cut to the right and achieve an *x*-axis cut in the + direction. (*Source:* NIOSH [ref. 60].)

z representing penetration and positive z representing withdrawal. In the x-y plane, the major axis (the axis of greatest range of travel) is usually taken as the x axis, and the minor axis is then the y axis. All of these rules are met by the vertical mill in Figure 5.9 but are violated by some of the example tools in Figure 5.10.

As the complexity of the machine tool increases, the simple x-y-z coordinate system becomes inadequate to describe the motion. Note in Figure 5.10 that additional axes are named besides x, y, and z. For instance, axis w is sometimes specified to refer to a reflected z axis in which the table or workpiece support moves in opposition to the tool penetration axis. The effect of the w axis in these cases is to achieve z axis motion, but

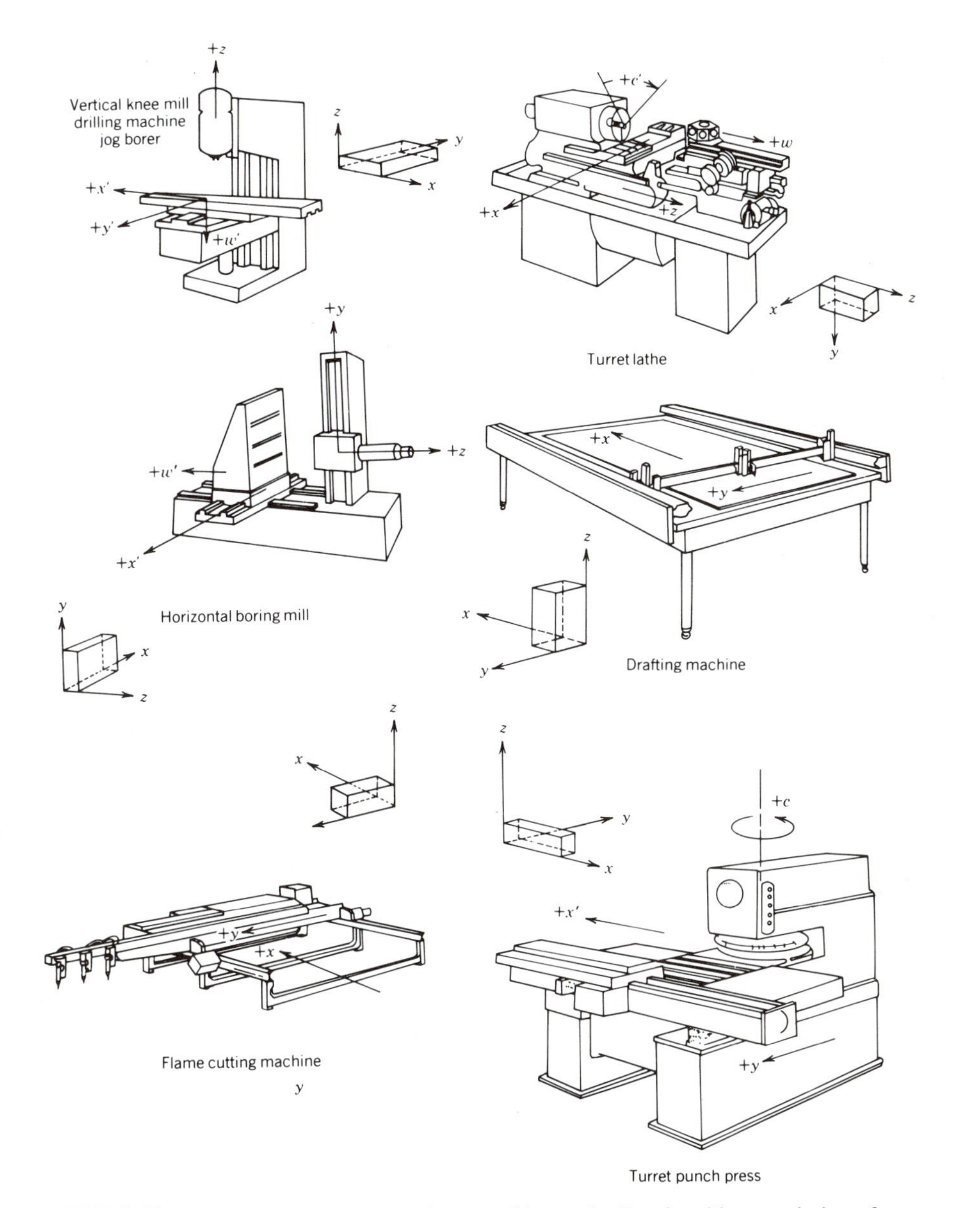

FIGURE 5.10 Example NC axes on various machine tools (Reprinted by permission of Electronic Industries Association, Washington, D.C. from EIA Standard RS-267-B, pp. 6, 8, 9, 11, 13, 14, 18, 26–29.)

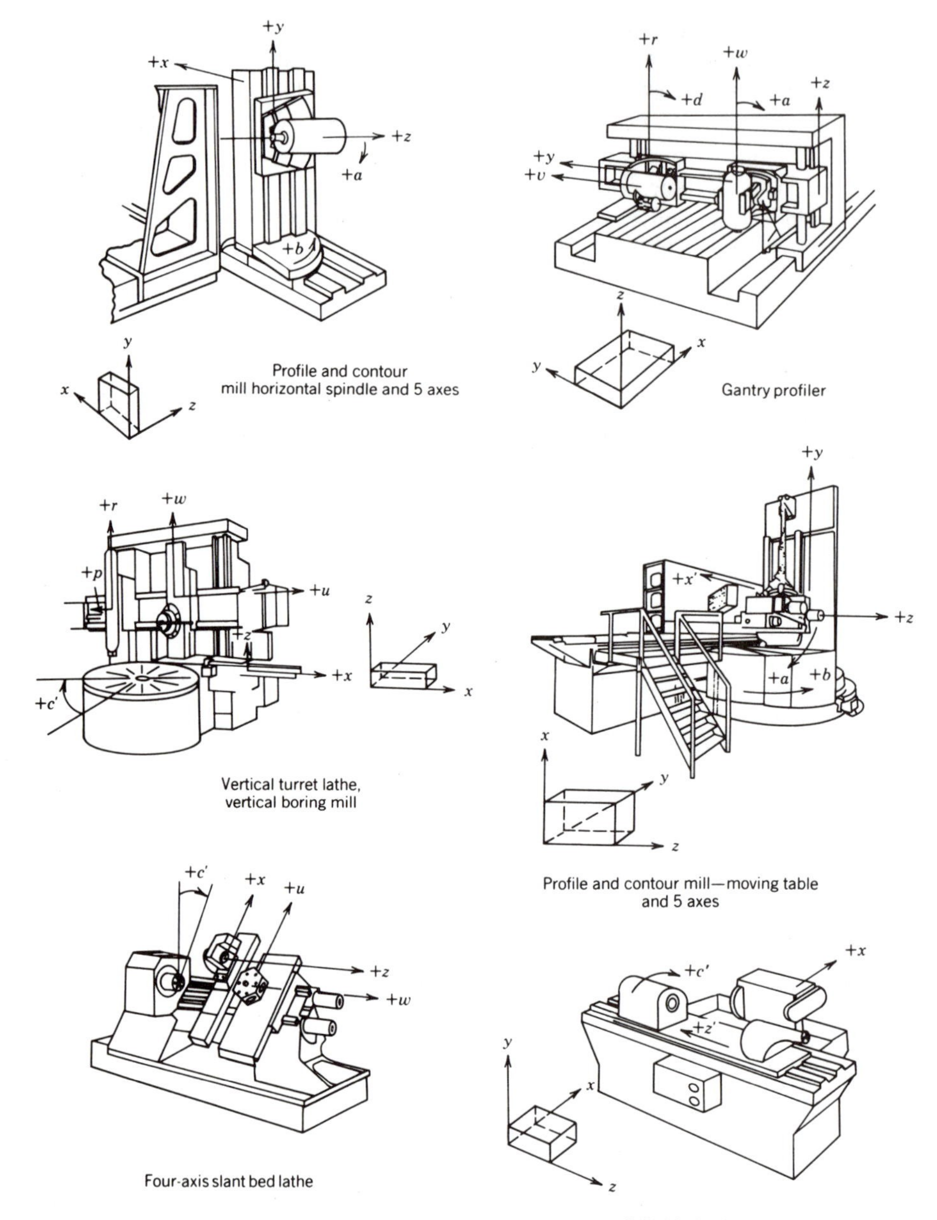

FIGURE 5.10 (continued)

it is achieved in a more practical way than by motion of the tool head. In some of the diagrams of Figure 5.10 it can be seen that the tool head has more sophisticated motion capability than simple penetration. The turret punch press, for example, has a planetary motion, designated axis c. The profile and contour mill, used for milling large structural components for spacecraft, has the capability of varying the horizontal pitch of the tool head (axis a) in addition to the penetration axis z. The gantry profiler is shown with two toolheads and seven axes of motion.

The machines shown in Figure 5.10 include some of the more unusual configurations, and the student of numerical control should not be overly concerned with memorizing every axis combination. What is important is the understanding of the wide variety of ways machine tools can move, and the relationship of these motions to the world coordinate system of x, y, and z. These principles carry over from CAD/CAM to robotics, as will be seen in the next chapter of this book.

INCREMENTAL VERSUS ABSOLUTE

There are two principal conventions for instructing the NC machine tool (or anything for that matter) to execute a motion. The first convention is to command the machine to move its tool to a new position in space that is a given x, y, and z distance away from its current position in space. This type of programming is generally designated as "incremental" in the NC industry. The other kind of NC programming is "absolute" and has a home position from which to reference all of its motion commands. An example will compare the incremental and absolute conventions in specifying a machine tool's position. Figure 5.11 illustrates a three-hole drilling sequence laid out in an x-y plane. For simplicity, the z axis is ignored in this example.

Most NC machine tools can operate in either incremental or absolute modes at the specification of the programmer. This was not always true but has been made possible by the continued development of flexibility in automation. Thus, the choice of incremental versus absolute can be made to suit the application, the product being manufactured, the dimensioning scheme adopted by the draftsman, or simply the convenience of the programmer.

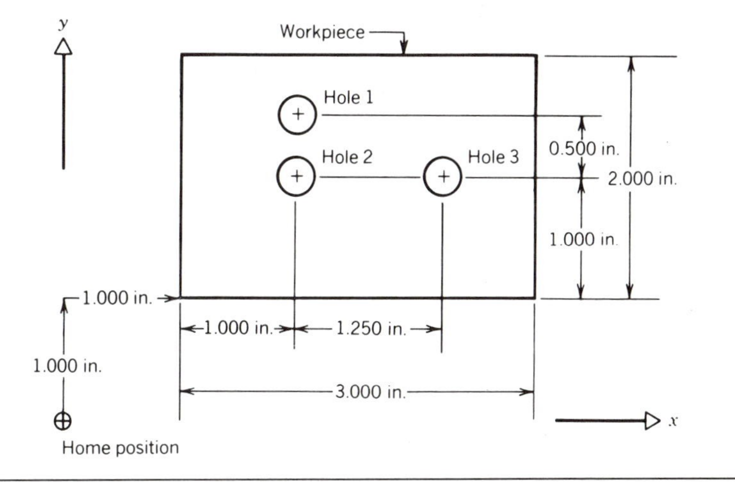

An NC machine is at home position and is commanded to drill three holes in the sequence shown in the diagram. Axis positions for each hole are described as follows:

Incremental	***Absolute***
Hole 1: $x = 2.000, y = 2.500$	Hole 1: $x = 2.000, y = 2.500$
Hole 2: $x = 0 \quad , y = -.500$	Hole 2: $x = 2.000, y = 2.000$
Hole 3: $x = 1.250, y = 0$	Hole 3: $x = 3.250, y = 2.000$

FIGURE 5.11 Example NC machine tool position sequence comparing the incremental versus absolute conventions for specifying the axis positions.

MOTION CONTROL

The degree of control over the path of the tool is a significant determinant in the cost of an NC tool, and many machining operations require much less control than others. The automation engineer needs to understand this difference to make rational decisions for investment in capital equipment. Even the economic feasibility of an automation application can hinge on the degree of motion control needed in the NC tool.

Point-to-Point

Hole drilling is an important application for machine tools, but no machining takes place while the tool positions itself from the previous operation to a point over the next intended hole to be cut. Then only one axis, usually the z axis, needs to be controlled during the drilling operation. This degree of motion control is called *point-to-point* (PTP). Milling a slot parallel to the x or y axis is another machining operation that can be performed in PTP. Figure 5.12 shows the path of a machine tool as it moves in PTP to locate over a hole to be drilled. Both x and y axes are advanced equally until the y axis motion is complete. Motion continues along the x axis until the final position is reached.

Contouring

The machining of a curved surface requires control of the relative velocities of tool motion in two or more axes. This is made possible by varying the voltages to the dc servo motors that drive the NC axes during the cutting operation. If one is willing to program thousands of tiny point-to-point movements, a PTP machine can be programmed to simulate a contouring operation by tiny straight line approximations. But a contouring machine would handle the complex calculations necessary to generate the contoured surface automatically.

NC PROGRAMMING

The objective of NC programming is to use one of the BCD codes to issue a series of commands to be executed by the machine. But within the standard BCD codes, various programming formats or schemes have been devised to communicate the commands to the machine. Some of these communication schemes are called "formats," because they specify the commands directly according to a rigid format. Higher level communication schemes are similar to computer languages and are called NC "languages." One thing to be noted before going any further is that regardless of the format or language selected to communicate the desired operations to the NC machine, the execution of these operations by the machine is independent of the format or language selected. In general, it is not possible to discern what language has been used to program the NC machine simply by watching the machine in operation.

Block Structure

Certain essential ingredients must be in every NC command; other ingredients are optional, and some are rare. When all of these ingredients are organized together, the resultant command is usually called a "block." The principal ingredient, the tool position, has already been covered in the section entitled "Axes of Motion." Another essential ingredient is an identifying number or "sequence number" to maintain order and to permit future

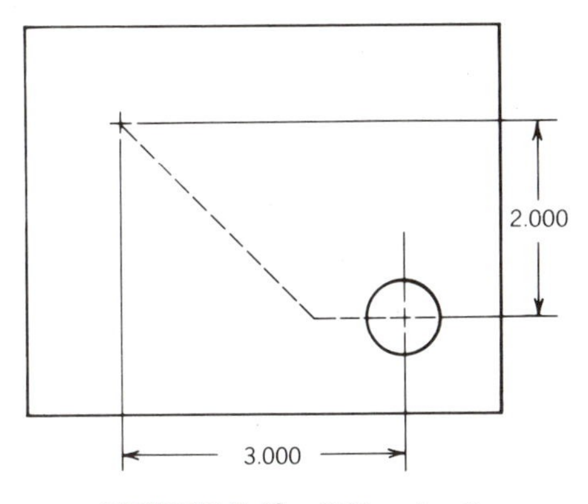

FIGURE 5.12 PTP tool path.

reference to a particular command in the program. Feed rates in inches per minute, millimeters per minute, or inches per turn of the spindle may need specification. If the tool spindle is to be turned on, it may be necessary to specify another ingredient, the spindle speed in rpm. Miscellaneous functions such as "turn on coolant" or "turn off coolant" may be specified. Each of the elements or ingredients of the NC block as described above is designated as an NC "word." So NC blocks are a collection of NC words separated by some kind of end-of-block character.

Block Formats

The most basic way of communicating with an NC machine is to specify the NC words in the block structure directly according to one of several prescribed, conventional formats. Certain purposes must be achieved by all NC formats, but the ways in which the formats achieve these purposes can be quite different. One format may be compact and efficient, another easier to interpret. Some of the differences are a matter of arbitrary style. Evolution also can be observed, just as in spoken languages, and some formats have become obsolete along with the machines that were designed to understand them. Following are some of the most important formats:

1. *Fixed sequential.* This was one of the earliest NC formats used. A specific sequence is required for every word in the block, and all words are required whether or not they specify meaningful change in the NC machine's state. For example, the word specifying the x-axis position must be included even if the x-axis position does not change.
2. *Block address.* In this format, a block code is used at the beginning of the block to inform the machine which words of the block are going to be specified in this particular command. Unspecified elements are left unchanged. Block address code is very compact. Table 5.4 provides some typical block address codes.
3. *Tab sequential.* This format is somewhat similar to the fixed format, except the words to be skipped are designated by a "tab character." The distinctive bit pattern of the tab character (6-5-4-3-2) makes it easy to recognize visually. Tab sequential is quite compact, although not as compact as block address.
4. *Word address.* In this format, an alphabetic character is used as a code at the beginning of each word to denote what word is being specified. As in tab

TABLE 5.4 Block Address Code

Change	*x*	*y*	*x and y*	*z*	*No Position Change*
Position only	00	10	20	30	
Feed	01	11	21	31	41
Speed	02	12	22	32	42
Coolant	03	13	23	33	43
Feed and speed	04	14	24	34	44
Feed and coolant	05	15	25	35	45
Speed and coolant	06	16	26	36	46
Feed, speed, and coolant	07	17	27	37	47

Spindle Speed Code

rpm	*Code*
0	0
77	1
154	2
308	3
424	4
583	5
847	6
1165	7
1659	8
2330	9

Feed Rate Code

Inches per Revolution (ipr)	*Code*
0	0
0.0025	1
0.0050	2
0.0075	3
0.0100	4
0.0150	5
0.0200	6

Coolant Code

Coolant	*Code*
OFF	0
ON	1

sequential and block address, unneeded words can be omitted, making word address a compact format. Word address is a very important block-type format because it has made the transition to modern CNC equipment to be discussed later in this chapter. Most word-address dialects have a required sequence for words specified, although words can be omitted. Some of the more common word-address codes are shown in Table 5.5. Included are some sample

TABLE 5.5 Word Address Codes

N	identification number
G	preparatory command
X	x axis
Y	y axis
Z	z axis
F	feed rate
T	tool select
S	spindle speed
M	miscellaneous
EOB	end of block
	Sample G-codes
G0	rapid transverse
G1	linear interpolation
G7	go home
G17	axis selection x-y
G20	axis selection z
G90	absolute programming
G91	incremental programming
	Sample M-codes
M0	no M code
M2	end of program
M3	start spindle rotation
M5	stop spindle

G-codes and M-codes that are widely used and will be needed for understanding Example 5.1, which follows.

The following case study illustrates and compares these four block-oriented formats.

EXAMPLE 5.1 (DESIGN CASE STUDY) NC Programming on a Milling Machine

A keyway and two holes are to be cut in Figure 5.13, using an end mill on a vertical milling machine. The end mill is a versatile milling tool that can be employed to cut metal using either its cylindrical sides, its end, or both simultaneously. Illustrate the NC program to accomplish this machining objective using the fixed sequential, block address, tab sequential, and word-address NC formats.

Solution

The solution to this case study is contained in Figures 5.14 through 5.17.

Although the various block formats are different from each other, they have a consistency in sequence and procedure in presenting data to the NC machine. They all require that any change in tool position among the three cartesian axes be specified as part of an NC block. This consistency is convenient for NC machine-tool builders, but the personnel who must program the NC machine may prefer a more user-oriented means of communication. Toward this end, NC languages have been developed and are described in the next section.

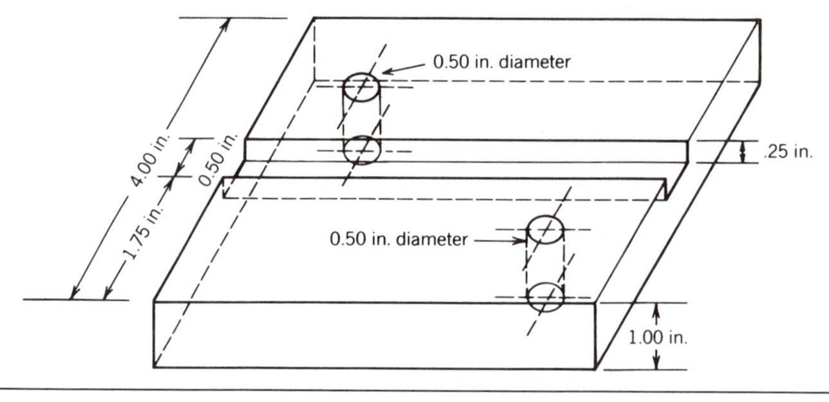

Zero plane for the z axis is 0.25 inches above the workpiece.

Word definitions

G0	rapid transverse	M0	no M code
G1	linear interpolation	M2	end of program
G7	go home	M3	start spindle rotation
G17	axis selection x-y	M5	stop spindle
G20	axis selection z		
G90	absolute programming		
G91	incremental programming		

FIGURE 5.13 Part to be machined using NC in Case Study 5.1.

Fixed Sequential

N	X	Y	Z	F*	S†	M
001	1.00	3.00	0.00	2.87	000	0
002	1.00	3.00	−0.75	2.87	573	3
003	1.00	3.00	0.00	2.87	573	3
004	−0.50	2.00	0.00	2.87	573	3
005	−0.50	2.00	−0.50	2.87	573	3
006	5.25	2.00	−0.50	2.67	382	3
007	5.25	2.00	0.00	2.67	382	3
008	3.00	1.00	0.00	2.67	382	3
009	3.00	1.00	−0.75	2.87	573	3
010	3.00	1.00	0.00	2.87	573	3
011	3.00	1.00	0.00	2.87	000	5
012	3.00	1.00	0.00	2.87	000	2

*Sample calculation for feed rate:

$$\begin{aligned} F &= (\text{recommended feed per rev}) \times (\text{spindle speed}) \\ &= 0.0050 \text{ in./rev} \times 573 \text{ rev/min} \\ &= 2.87 \text{ in./min} \end{aligned}$$

†Sample calculation for spindle speed:

$$\begin{aligned} S &= \frac{(\text{recommended cutting speed})}{\pi \times (\text{dia. of tool})} \\ &= \frac{75 \text{ ft/min} \times 12 \text{ in./ft}}{\pi \times .5 \text{ in.}} \\ &= 573 \text{ rev/min} \end{aligned}$$

FIGURE 5.14 Fixed sequential NC program for Case Study 5.1.

Block Address

N	Block Address Code	Data		
01	21	1.00	3.00	2*
02	32	−0.75	5†	
03	30	0.0		
04	20	−0.50	2.00	
05	30	−0.50		
06	04	5.25	2	4
07	30	0.00		
08	20	3.00	1.00	
09	34	−0.75	2	5
10	30	0.00		
11	42	0		
		end of program		

* From Table 5.4, Feed Rate Code

† From Table 5.4, Spindle Speed Code

FIGURE 5.15 Block address NC program for Case Study 5.1.

Tab Sequential

01	TAB	1.0	TAB	3.0	TAB	TAB	2.87	EOB		
02	TAB	TAB	TAB	−0.75	TAB	TAB	573	TAB	3	EOB
03	TAB	TAB	TAB	0.00	EOB					
04	TAB	−0.50	TAB	2.0	EOB					
05	TAB	TAB	TAB	−0.50	EOB					
06	TAB	5.25	TAB	TAB	TAB	2.67	TAB	382	EOB	
07	TAB	TAB	TAB	0.00	EOB					
08	TAB	3.00	TAB	1.00	EOB					
09	TAB	TAB	TAB	−0.75	TAB	2.87	TAB	573	EOB	
10	TAB	TAB	TAB	0.00	EOB					
11	TAB	TAB	TAB	TAB	TAB	000	TAB	5	EOB	
12	TAB	TAB	TAB	TAB	TAB	TAB	2	EOB		

FIGURE 5.16 Tab sequential NC program for Case Study 5.1.

Word Address

```
N01  G0 G17 X1.00 Y3.00 EOB
N02  G20 Z-0.75 F2.87 S573 M3 EOB
N03  G20 Z0.00 EOB
N04  G0 G17 X-0.50 Y2.00 EOB
N05  G20 Z-0.50 EOB
N06  G17 X5.25 F2.67 S382 EOB
N07  G20 Z0.00 EOB
N08  G0 G17 X3.00 Y1.00 EOB
N09  G20 Z-0.75 F2.87 S573 EOB
N10  G20 Z0.00 EOB
N11  M5 EOB
N12  M2 EOB
```

FIGURE 5.17 Word address NC program for Case Study 5.1.

NC Languages

For many NC programmers, it is more convenient to write programs in an abbreviated English-language format, using verbs such as GO TO, RUN, and EXECUTE. This is the style used in general-purpose computer languages such as BASIC, FORTRAN, and C. Persons who have programmed in one or more of these general-purpose computer languages will find familiarity in such NC languages as APT and COMPACT, which use the abbreviated English-language format.

There is another, more important advantage for using computer-type languages for NC machines, and that is the feature of *symbolic addressing*. In an NC language such as APT, the programmer can define a point in Cartesian space and then recall or return to that point later in the program. This saves much detailed specification of *x*-*y*-*z* coordinates in thousandths or ten-thousandths of an inch or in hundredths of a millimeter.

The programming of contoured paths for NC machines that have continuous path capability is impractical without the use of computer-style NC languages. Contoured

APT Program

Columns 1–6	Column 10	Comments (not part of program)
PARTNO	THE 2 HOLES AND KEYWAY PGM	Identify program
	MACHIN/MILL,2	Identify tool used
	CUTTER/.500	Define cutter diameter
PT0	POINT/0,0,0	Define point locations
PT1	POINT/1.0,3.0,0	
PT2	POINT/−.25,2.0,0	
PT3	POINT/5.25,2.0,0	
PT4	POINT/3.0,1.0,0	
	FEEDRAT/2.87	Set feed rate
	SPINDL/573	Set spindle speed
	FROM/PT0	Starting position
	GOTO/PT1	Above hole 1
	GODLTA/0,0,−0.75	Drill hole 1
	GODLTA/0,0,+0.75	
	GOTO/PT2	Position to cut keyway
	GODLTA/0,0,−0.5	
	FEEDRAT/2.67	Set speeds
	SPINDL/382	
	GODLTA/5.75,0,0	Cut keyway
	GODLTA/0,0,+0.5	At point 3
	GOTO/PT4	Above hole 2
	FEEDRAT/2.87	Set drilling speeds
	SPINDL/573	
	GODLTA/0,0,−0.75	Drill hole 2
	GODLTA/0,0,+0.75	
	GOTO/PT0	Return to start position
	SPINDL/0	Stop spindle
	FINI	Finish (of program)

FIGURE 5.18 APT program for Case Study 5.1.

surfaces are described mathematically as portions of standard geometric shapes or by combinations of these shapes. The NC language is used to specify the shapes to be contoured by the tool, and the computer software takes care of the interpolations necessary to command the machine to generate the shape. Imagine trying to achieve the same purpose with a block format in which the *x-y-z* position of each interpolated point must be calculated by the programmer and specified in detail. To be fair, we must say at this point that facility for arcs and simple geometric shapes has been provided also in the block formats for the convenience of the programmer, but the programming of contours is certainly not the forte of the NC block formats.

It was stated earlier that some NC formats are compact and efficient, while others are easier to interpret. Similar differences can be seen among NC languages. Some evolution of NC languages has been observed, the same as for NC formats. Families of NC languages have also evolved, giving rise to the term *dialect* to refer to an NC language that is closely related to another, more well-known NC language.

It is not the purpose of this text to provide an in-depth familiarity with each of the many NC languages or even with a single selected language used in each machine-tool industry. That objective should be pursued in a separate course, perhaps in-house for specific equipment in use or anticipated for purchase. The purpose here is to provide the automation engineer with a broad understanding of languages, their strengths and weaknesses, and their purpose in the overall scheme of robots and manufacturing automation.

A program in APT is now listed and annotated with remarks for comparison with the block formats used in Case Study 5.1. The reader is not expected to achieve facility with the APT language simply by studying this example, but the purpose of each APT command and its general structure should be clear from Figure 5.18. To facilitate the learning process, the objective of this APT program was made identical to that of the block format programs in Case Study 5.1.

The convenience of the computer-type language should be evident from the foregoing example, although this example is really too elementary to display the full contouring power of an NC language such as APT. In the mid-1970s, it appeared that the superiorities of computer-assisted programming in NC languages would cause these languages to dominate and virtually replace the more tedious block formats, but something happened in the early 1970s to reverse the trend. We shall read about this reversal in the sections that follow.

DIRECT NUMERICAL CONTROL (DNC)

A development of the late 1960s was the remote control of NC machines by a mainframe computer on a time-sharing basis. This was called *direct numerical control* or simply DNC. The best way to understand the significance of DNC is to examine the characteristics of the conventional NC systems it replaced.

Conventional NC

The conventional NC approach prior to DNC was to use either a block format to generate a punched paper tape directly or to program in a computer-type language, such as APT, and process this source program off-line on a computer, which translated it into a machine-readable block format program on punched paper tape. The punched paper tape was then removed from the tape punch machine and attached at its ends to form a continuous loop.

The loop was then mounted on the tape reader of the NC machine and read by the machine during execution of the program. The important point to understand here is that the tape was physically processed again through the reader every time a new part was produced or the program re-executed. In real time, the machine would actually execute the NC commands as they were read by the tape reader one instruction at a time. It is no wonder that these early NC "paper" tapes were generally produced on mylar tape instead of paper to permit their repeated usage for long production runs.

Time-Sharing

In the late 1960s, the development of time-sharing on large mainframe computers opened the door to the control of an entire bank of NC machines on a time-shared basis by a single mainframe computer. All of the NC machines were connected on-line to the computer, permitting their programs to be processed in real time *without* the use of punched paper tape. The high speed of the computer permitted it to be interrupted continually by the various NC machines to provide the next sequential instruction for each machine upon demand.

An added advantage of DNC was that entire libraries of NC programs could be maintained on magnetic tape, disk, or bulk storage media, ready for immediate retrieval and loading into active computer memory for execution upon the desired NC machine. If a program change was desired, the DNC computer could make the modification instantly and refile the program on disk. Compare this to the old method of repunching a new paper tape and manually connecting the ends to produce the continuous loop. Even the storage of the old paper tape loops themselves was a nuisance, especially considering their varying lengths.

More current interpretation for the abbreviation DNC is "distributed numerical control," which is more characteristic of the hierarchical systems in use today. Hierarchical control is best explained after the reader understands the next major NC development: CNC.

COMPUTER NUMERICAL CONTROL (CNC)

In the 1970s the term *microcomputer* emerged, and suddenly computers were found to be cheap enough to place a computer on every NC machine in the plant. Thus, each NC machine could have something similar to but better than DNC: the NC machine was not required to time-share its control computer with the other NC machines in the plant. With the new technology, it was possible to do without the data cables that had been so plagued by voltage transients in the factory environment.

The "computer-with-every-machine" approach became known as *computer numerical control,* or *CNC*. Virtually all new numerically controlled machine tools being manufactured today are CNC. Many of the older NC or DNC machines are being updated to CNC. Even conventional, manually operated machinc tools manufactured decades before the development of NC are being upgraded to CNC machines. Several companies market retrofit packages for vertical milling machines, and Figure 5.19 illustrates a Bridgeport vertical milling machine retrofitted with an Aerotech Smart I CNC controller.

A curious twist in the development of numerically controlled machines is that the new CNC machines brought on a revival of interest in the block-oriented formats, especially word address. The reason for this apparent backward step was that with CNC the machinists themselves became able to generate the NC programs using the CNC console by pushing buttons right at the machine. Figure 5.20 illustrates the control panel console

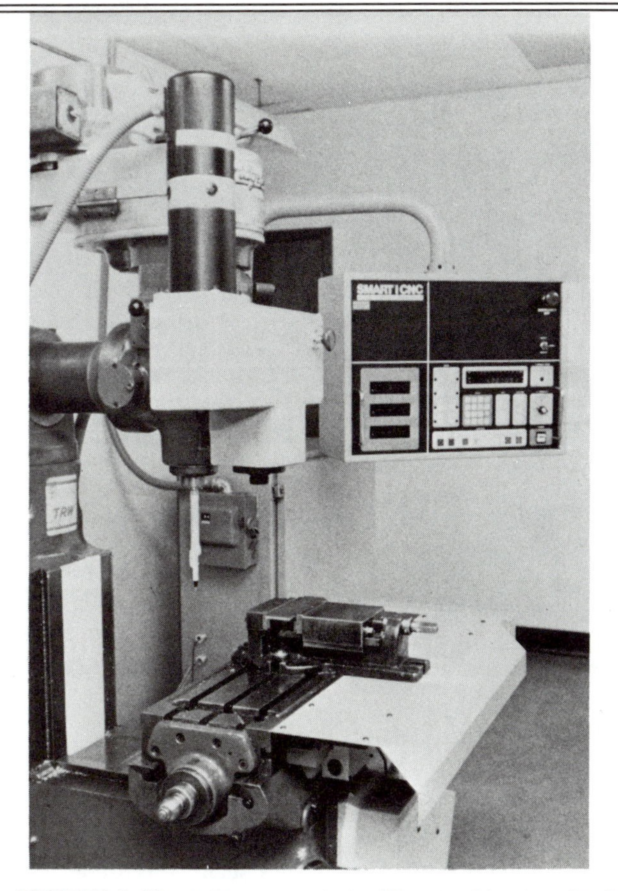

FIGURE 5.19 Brideport vertical milling machine retrofitted with an Aerotech Smart I CNC controller.

for the CNC vertical mill illustrated in Figure 5.19. Its appearance is somewhat similar to the digital read-out panels used by many machinists on conventional, manual machines. The step-by-step programming, characteristic of block-type formats, is similar to the strategy used by the machinist in conventional machine-tool operation. Also, the availability of the computer controller makes debugging feasible right at the machine, not at some remote off-line computing center or engineer's terminal. This new availability of the computer controller right at the machine is generating a new breed of machinist-programmer: The machinist is learning to program, and the programmer is learning to operate the machine. CNC is bringing these two career fields closer together.

Although Figure 5.20 illustrates a particular product, CNC control panels have a great deal in common. The generic CNC panel has the following groups of controls.

1. *Mode selector.* This section of the panel permits the user to elect to run in manual; to execute a program automatically a step at a time, in a complete cycle, or in a continuous series of automatic cycles; to enter a new program; or to edit existing programmed steps.
2. *Control.* This section of the panel permits the user to execute the mode selected. For example, the user can advance a given axis in manual mode; start, clear, or hold a program in automatic mode; enter a new program step in program mode; or display or search for programmed steps in edit mode.

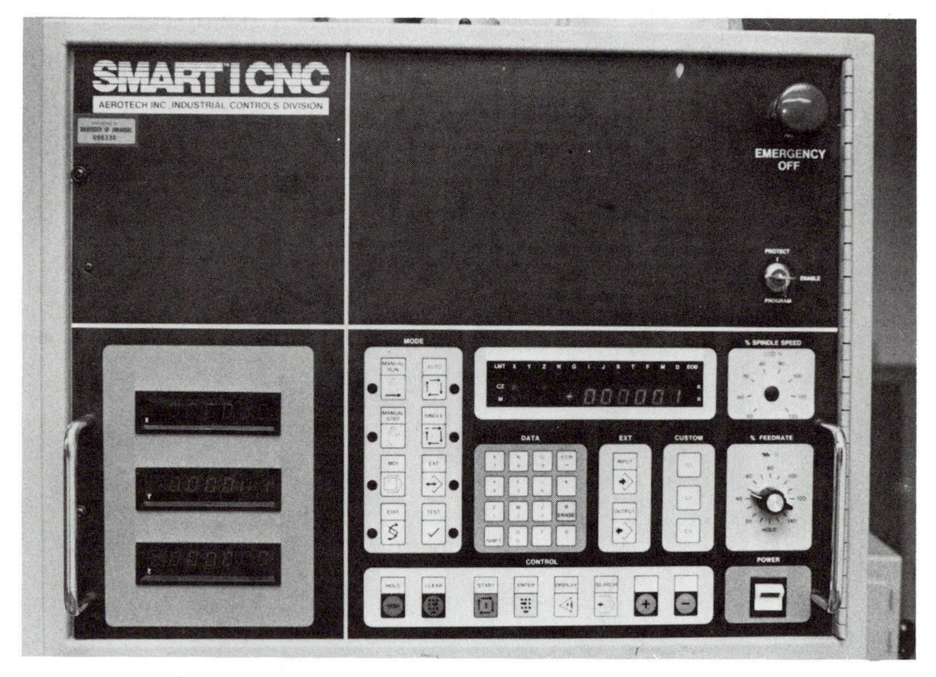

FIGURE 5.20 CNC control panel console.

3. *Keyboard.* A CNC panel must have a numeric keyboard for the entry of coordinate data and miscellaneous values such as feed rate. If the CNC unit uses the word-address NC language (and most of them do), selected alphabetic letters must also be available on the keyboard.
4. *Display.* Finally, the CNC panel must have a means of displaying the current NC command, the current tool position, error codes, and other selected numeric or alphanumeric data. This is usually accomplished by either an LED display or a CRT screen.

Individual CNC models may have cassette readers or additional controls for inputting and outputting programs from an external computer. Another useful control available on most units is a means of varying the spindle speed or feed rate as a percentage of the value programmed. This is particularly useful during program check out.

HIERARCHICAL NUMERICAL CONTROL

It was stated earlier that DNC lives on in a system called hierarchical numerical control. Really, hierarchical numerical control is a system that combines the best features of CNC and DNC. In hierarchical numerical control, each NC machine has its own stand-alone computer controller, but each of these controllers is also connected to a larger central computer that only communicates occasionally with the individual CNC stations on a time-shared basis. Unlike direct numerical control, if the central computer fails in a hierarchical numerical control system, the satellite CNC stations have limited authority to continue operation. The central computer typically monitors production progress on a sampling basis, maintains a library of CNC programs for access by the satellite machine

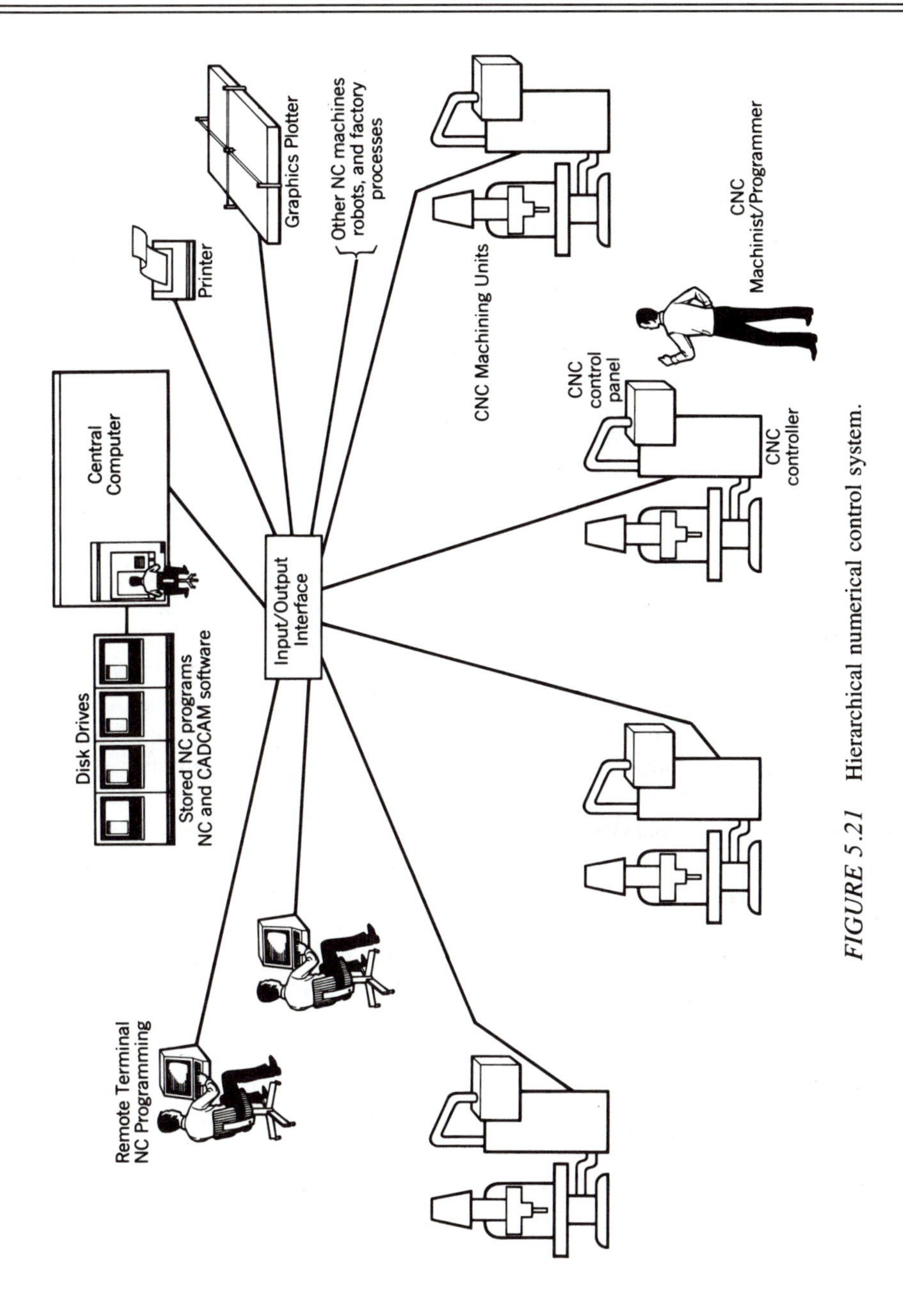

FIGURE 5.21 Hierarchical numerical control system.

stations, and notes and alarms any abnormal conditions. It is even possible to have intermediate supervisory levels among computers in the hierarchy, as will be seen in Chapter 14. Figure 5.21 illustrates the structure of a typical hierarchical numerical control system.

CAD/CAM

The computer can do more for the NC programmer than merely store the programs generated. Note in Figure 5.21 that the disk drives store not only the NC programs but also CAD/CAM software. CAD/CAM represents "computer-assisted design and computer-assisted manufacturing." Another related term is CADAM, or CADCAM, meaning "computer-assisted design and manufacturing." There are many ways that the computer can assist in automated manufacturing, and Chapter 14 will explore further some of these ways. But the computer has a special role in the design phase in the construction of NC programs.

Figure 5.22 illustrates a computer-screen illustration of a product in the design process. It is possible for the NC program to be developed from computer software using the product design as input. This permits the computer to remove the tedium of NC part programming. Even more popular are the less sophisticated systems in which the NC part programmer actually writes the program in an NC language such as APT and then uses the CAD/CAM software to test the program in a simulated machining process executed by the computer. The computer will note and report programming syntax errors before execution and will supply diagnostics. Once a program is syntactically correct, the computer will simulate execution of the program and will construct a drawing of the "machined" part either on the CRT display screen or on a graphics plotter at the election of the programmer. Upon achievement of a satisfactory program, the computer can issue the program to a CNC machining station for execution and can store the program on disk for future production runs. It can also generate three-view drawings, dimension data, and programming for industrial robots that will handle the product from machine to machine during production.

One of the principal advantages of introducing CAD/CAM during the design phase is that complete documentation of the product and process can be generated for addition to the overall information data base. The data base can be accessed to generate production schedules, materials requirements, instructions to operating personnel, quality control checklists, and any other documentation that is both useful to the manufacturing process and is machine-producible.

SUMMARY

Industrial robots were preceded by the automation of machine tools such as milling machines and lathes. The first attempts to automate machine tools employed mechanical cams and linkages—that is, the "hard" automation strategy. But the development of digital computers made possible the reprogrammable automation of machine tools which, in turn, made possible the automation of short production runs. This is the same technological development that later led to the development of the industrial robot. In fact, an industrial robot can be thought of as a variation of a numerically controlled machine tool. Some industrial robots, such as robot welders, perform the same functions as NC machine tools.

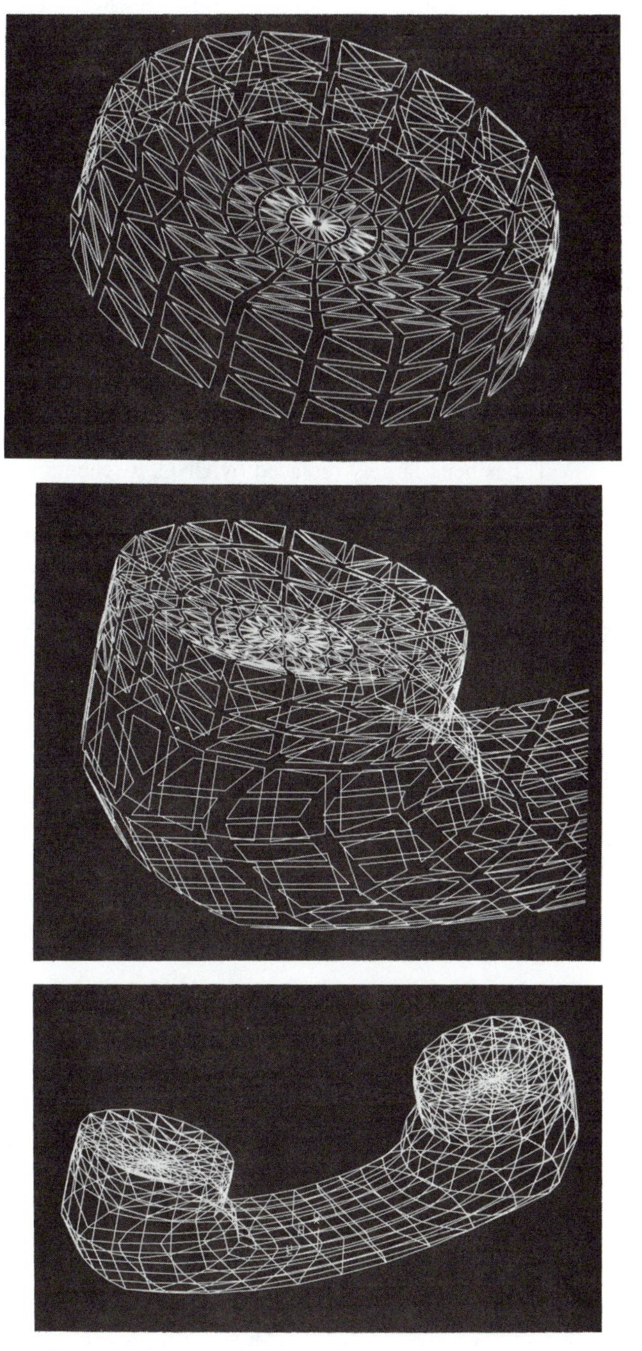

FIGURE 5.22 CADCAM representation of a product in the design phase. (Reprinted by permission of McDonnell Douglas Automation Company, St. Louis, MO.)

Obviously, numerical control brings flexibility to the automatic machine tool, but it does much more than that. Large and complicated workpieces, characterized by aerospace components, are infeasible to produce using hard automation. The convenience of being able to perform a large, complex operation on a single machine and in a single fixturing saves setup costs and jig and fixturing costs, and at the same time it enhances quality.

Machine code for numerical control uses binary-coded decimal (BCD) and monitors read/write errors by means of the parity channel. The axes of motion for each NC machine are predefined, and the tool is commanded to move using either an incremental or an absolute reference. The toolpath can be controlled only at its end points in PTP mode, or the tool can be controlled throughout its motion in contouring mode.

NC machines are programmed by two major methods: block formats and computer languages. Block formats include fixed sequential, tab sequential, block address, and word address. Computer NC languages are typified by the popular NC language APT.

Direct numerical control (DNC) uses a central computer to drive a series of NC machines on-line. DNC has been largely replaced by CNC, a scheme whereby each NC machine has its own control computer. The best features of both DNC and CNC are retained in a system called hierarchical control. Hierarchical control permits simultaneous computer NC programming and CNC console programming at the machining station. A full range of CAD/CAM software support packages are available on disk, accessible on-line to the central host computer. In addition to controlling machine tools, the central computer can control industrial robots, the topic of Chapter 6, which follows.

EXERCISES AND STUDY QUESTIONS

5.1. The NC tape in Figure 5.23 has a parity error. Can you locate it? In this example, one of two types of errors has occurred; what are these possible errors?

5.2. The NC tape in Figure 5.23 is produced in a block-type NC format; what format is it? How can you tell? What is the sequence number of the only complete block in the figure?

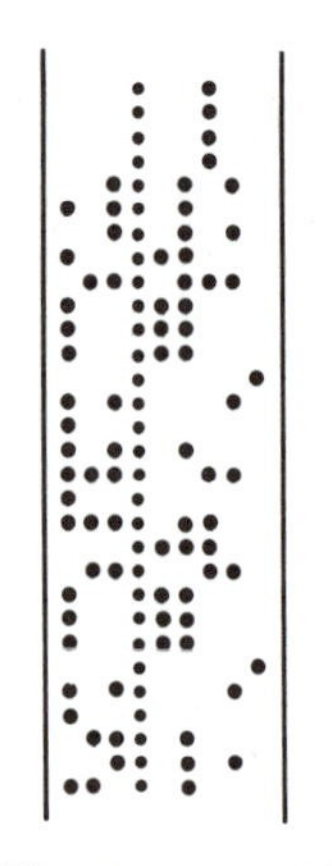

FIGURE 5.23 NC tape with parity error.

5.3. The NC tape in Figure 5.24 is produced in a block-type NC format; what format is it? How can you tell? How many blocks are shown?

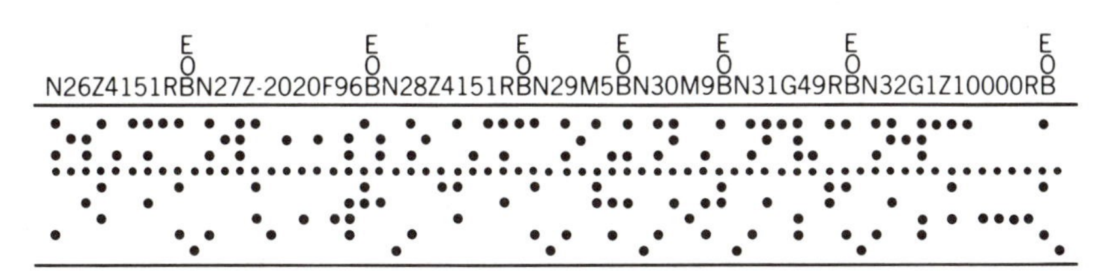

FIGURE 5.24 NC tape in a block-type format.

5.4. The following NC program machines an alphabetic letter in a nameplate clamped to the table; the surface of the table is at $Z = -3800$. The coordinate positions are specified in thousandths of an inch. The depth of cut is two-tenths of an inch.

```
N1 G91 eob
N2 G7 eob
N3 G1 G17 X4000 Y-3000 F100 M3 eob
N4 X-500 Y500 F40 eob
N5 G20 Z-3200 F100 eob
N6 Z-300 F3 eob
N7 G17 X1000 F10 eob
N8 G17 X-1000 Y-1000 F10 eob
N9 G17 X1000 F10 eob
N10 G20 Z1700 F100 eob
N11 G7 M5 eob
N12 M2 eob
```

Which NC format has been used in this example? Is the machine programmed in incremental or absolute? What alphabetic character is machined using this program?

5.5. In Exercise 5.4, what is the thickness of the material? Just before the cut, how close to the material does the tool advance? After the cut, how high above the table does the tool advance? At that point, how high above the material is the tool?

5.6. The following NC program segment is written in absolute mode. Its purpose is to mill a $^{3}/_{8}$-in.-wide slot $^{1}/_{4}$-in. deep in a piece of aluminum. The coordinate words read in thousandths of an inch.

```
00119+03000+04000-0245010065003eob
00217+05500+04000-0245001565003eob
```

Which block-type NC format has been used in this program segment?

5.7. Using incremental mode, rewrite the program segment in Exercise 5.6.

5.8. Using incremental mode, rewrite the program segment in Exercise 5.6 in tab sequential.

5.9. Figure 5.25 dimensions an L-shaped slot to be milled under numerical control. The tool is an end mill with a diameter of 0.200 in., and the depth of cut is 0.150 in. The dimensions show the distances traveled by the tool center point (TCP) under numerical control. What are the overall dimensions of the L-shaped slot?

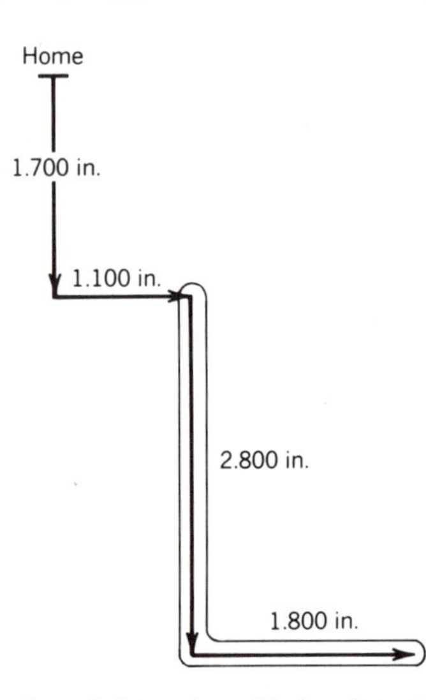

FIGURE 5.25 L-shaped slot to be milled under numerical control.

Why are there discrepancies between the distances traveled by the tool and the dimensions of the L-shaped slot?

5.10. Ignore all N-, G-, F-, S-, and M-functions and write an NC program segment in absolute word-address format to execute the cut described in Figure 5.25, assuming that the beginning point of the tool is at home position 5.500 in. above the surface of the material. (Write only the coordinate-axis-words in this exercise.)

5.11. Using incremental mode and the word address NC format, rewrite the program in Exercise 5.10.

5.12. Why is NC automation preferable to hard automation for small-quantity lots?

5.13. Using the same NC machine material as in Exercise 5.4, what letter would be formed by the following NC program?

```
N1 G91 eob
N2 G7 eob
N3 G1 G17 X4000 Y-3000 F100 M3 eob
N4 X-500 Y500 F40 eob
N5 G20 Z-3200 F100 eob
N6 Z-300 F3 eob
N7 G17 X1000 Y-1000 F10 eob
N8 G20 Z300 F100 eob
N9 G17 Y1000 F40 eob
N10 G20 Z-300 F3 eob
N11 G17 X-1000 Y-1000 F10 eob
N12 G20 Z1700 F100 eob
N13 G7 M5 eob
N14 M2 eob
```

By how much does step N5 cause the machine to clear the top plane of the workpiece?

5.14. Write an NC program similar to the ones in Exercise 5.4 and 5.13 that will (a) make the letter L, (b) make the letter Y, and (c) make the letter M. Notice that the L and M programs are more similar to the Z program, and the Y program is more similar to the X program.

5.15. In Figure 5.11, the hole diameters were omitted and were unnecessary in the design of the NC program to produce this part. How might the hole diameters affect such an NC program?

CHAPTER 6

INDUSTRIAL ROBOTS

The word *robot* epitomizes the public image of industrial automation. This image is only partially correct because: (1) industrial robots are only part of the total automation picture; and (2) the public idea of an industrial robot is highly glorified. Ironically, the public's fascination with robots acted to set back the development of industrial robots in the late 1980s. Schoolchildren and their parents were enticed by television fiction and were eager listeners when promoters began suggesting in the early 1980s that the age of real, industrial, useful robots had arrived. The promoters spoke the truth, but the level of implementation they implied was highly exaggerated. Thousands of unemployed workers sought new careers by going back to school to learn "robotics," expecting a job market for the new skills they would learn. The result was disillusionment and backlash from the excessive optimism and promotion of the early 1980s. This backlash did not expire until the early 1990s, when a more mature technology began to resume sustainable growth.

The objective of this chapter, and indeed this entire book, is to put the industrial robot into perspective as a tool of manufacturing automation, not as the panacea for every problem of lagging industrial productivity. The principles of manufacturing automation apply to industrial robots much as they do to sensors, actuators, controllers, analyzers, drives, and other automation devices and systems studied in this text. We shall also find in this chapter that the robot is itself a system fabricated from these same devices. The enlightened automation engineer is able to discern which industrial problems require the application of a fully integrated industrial robot and which can be served by an assembly of components. In this chapter, we shall begin with a study of the origins of the movement.

Most people think of *Star Wars'* R2D2* when the word "robot" is mentioned. Such a concept is appropriate because the word "robot" was actually coined on the stage, not in the factory. Robots first appeared in New York on October 9, 1922, in a theatre play entitled *R.U.R.* The creator was Czechoslovakian dramatist Karel Capek, and the word "robot" is a derivative of the Czech word *robota,* which means "work" [ref. 105].

In 1956, more than 30 years after Capek's play but before *Star Wars,* a firm named Unimation was formed. Its sole business was robotics. Sixteen years and $12 million later, Unimation turned its first profit in 1972, making real, industrial-grade robots.

*R2D2 is the name of the personable, fictitious robot that starred in *Star Wars,* a popular movie of the 1970s.

Unlike R2D2 or Capek's robots, most real industrial robots are hardly humanoid in appearance. In fact, a more descriptive term for most industrial robots would be "mechanical arm." Figure 6.1 shows a popular model of mechanical arm-type robot—a far cry from R2D2—but much more useful from an industrial standpoint.

The Figure 6.1 robot meets the Robot Institute of America definition of *robot.*

> A robot is a reprogrammable, multi-functional manipulator designed to move material, parts, tools, or specialized devices through variable programmed motions for the performance of a variety of tasks [ref. 106].

Another robot that meets this definition is shown in Figure 6.2. Note that this robot is equipped with a welding head instead of grippers to pick up piece-parts.

A more general definition of robot is provided by Mikell Groover:

> An industrial robot is a general-purpose, programmable machine possessing certain anthropomorphic characteristics [ref. 41].

FIGURE 6.1 Popular model mechanical arm type robot (Reprinted by permission of Unimation-Westinghouse, Inc., Danbury, Conn.)

FIGURE 6.2 Industrial robot equipped with an arc welding head. (Reprinted by permission of Unimation-Westinghouse, Inc., Danbury, Conn.)

Groover's definition does not restrict the concept of robot to manipulator; rather, it leaves open the possibility of other anthropomorphic (humanlike) characteristics, such as judgment, reasoning, and vision.

Not all industrial robots look like mechanical arms. Some robots are enclosed in cubic work envelopes, with the edges of the cubes providing bearing for the robots' movements. This type of robot is sometimes called a gantry-mounted robot.

A key word in the definition of an industrial robot is the word *programmable*. More than any other, this feature of industrial robots, made practical by the advances of inexpensive microchip circuits in the 1970s, has vaulted industrial robots into the workplace. Chapter 15 will discuss these microchips in more detail.

Long before the programmable robots came the mechanically fixed manipulators, whose motions are set by mechanical cams installed in the factory. The Japanese robot industry includes the mechanically fixed manipulators within the definition of robot, but this interpretation is debatable. The disparities in robot definitions can distort statistics that compare robot populations in various countries. The mechanical-cam-type manipulators, whether or not they are called robots, have an important role in factory automation. These manipulators typify the term *hard automation* set forth in Chapter 1 as contrasted with *flexible automation* typified by the programmable industrial robot.

ROBOT GEOMETRY

Since robot configurations vary greatly, some classification of robot geometries is useful before going any further. The industry has settled upon the term *degrees of freedom* to describe the number of ways a robot can move. The form of these movements and the way they are assembled make up the robot *configuration*.

Degrees of Freedom

Every mechanical point on a robot, except in the gripper or tool, at which some form of drive induces motion in a robot part is called a degree of freedom. The motion can be of a pivoting nature or a reciprocal motion as is produced by a pneumatic or hydraulic cylinder. Figure 6.3 displays a robot with six degrees of freedom:

1. Base rotation
2. Shoulder flex
3. Elbow flex
4. Wrist pitch
5. Wrist yaw
6. Wrist roll

Although there are exceptions, in most robots the degrees of freedom are in series. Thus, the first degree of freedom in the robot of Figure 6.3, base rotation, imparts motion to all of the parts of the robot affected by subsequent degrees of freedom. Conversely, the third degree of freedom (elbow flex), for example, has no effect upon the base movement. It follows that the most sophisticated motion in the entire robot is that of the member driven by the highest degree of freedom. Generally speaking, the robot with the most degrees of freedom can produce the most complex movement, but there are other important factors to consider, such as range and quality of motion within a given degree of freedom. This point will become clearer later in this chapter.

The sequence of the various degrees of freedom and their types of motion determine the physical configuration of the robot. Theoretically, there could be a large number of

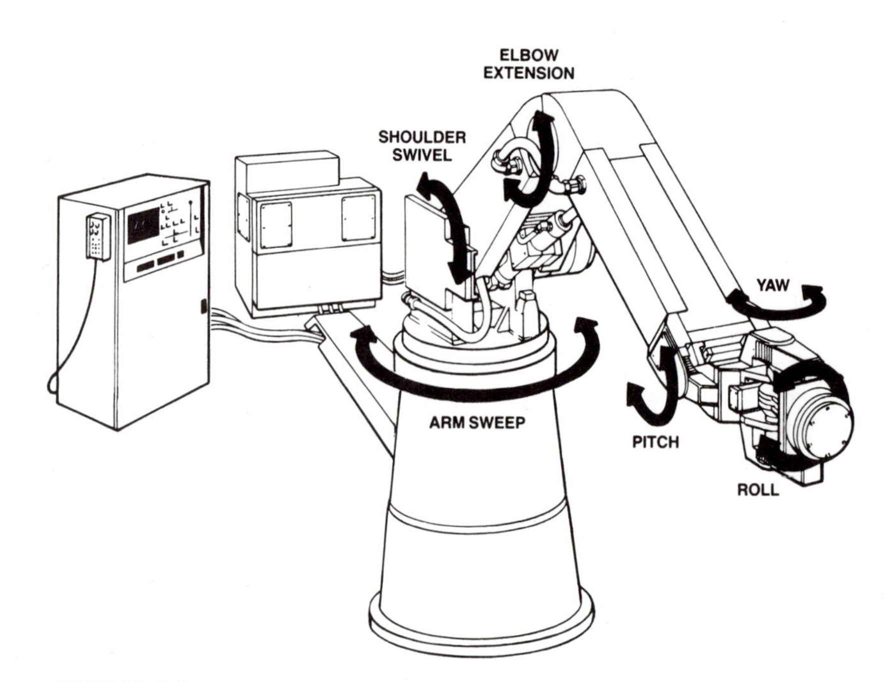

FIGURE 6.3 Industrial robot with six degrees of freedom. (Reprinted by permission of Cincinnati Milacron, Lebanon, Ohio.)

configurations for a robot with six degrees of freedom. From a practical standpoint, however, almost all robots fall into a few popular configuration categories.

Articulating Configurations

Some robots, such as the one in Figure 6.1, actually work like a human arm. Such robots are said to be "articulating." The base rotates in a way similar to a twisting human torso. The shoulder and elbow on most articulating robots pivot on one axis each, perpendicular to the axis of the arm and parallel to the plane upon which the base is mounted. The wrist assembly on articulating robots almost always has pitch but may or may not have yaw and roll. The reader may note that the human hand also is very flexible in the pitch axis but has little yaw or roll. The human wrist normally permits yaw of 60° at most. Virtually no roll at all is permitted by the human wrist, but the forearm and even the shoulder can be used together to achieve about 270° of roll. Thus, a human can screw in a lightbulb or attach a wingnut, but only in a series of twists and regrasps. Most articulating robots have more roll capability than the human arm.

There is a trick that can produce a yaw-type motion in a five-axis articulating robot that has no yaw. For some applications, a 90° roll can be executed and then a pivot can be executed on the *pitch* axis. This method is demonstrated in Figure 6.4. Note that the procedure works best for single-point tools as opposed to grippers. If a robot equipped with a gripper is used to attempt a yaw motion by this procedure, it may be found that the piece-part or tool held by the robot is in the wrong orientation.

A variation of the articulating robot is the horizontal-jointed robot, which simply means that the joint movement in the second, third, and fourth degrees of freedom have their axes arranged *vertically* so that the joint bends are in a horizontal orientation. A horizontal-jointed robot is shown in Figure 6.5. This type of robot may not resemble a human arm as much as the conventional articulating type but is gaining in popularity due to its usefulness in simple handling tasks.

Most practitioners use the name "SCARA" when referring to a horizontal-jointed articulating type robot. The name "SCARA" is so popular that many have forgotten what the letters originally represented (*S*elective *C*ompliance *A*ssembly *R*obot *A*rm). The concept was introduced in the late 1970s as a robot ideally suited for assembly tasks. An assembly task often consists of the following sequence:

1. Pick up a piece-part vertically from a horizontal table.
2. Move the part in a horizontal plane to a point just above another place on the table.
3. Lower the part to the table at the proper point to accomplish the assembly, perhaps including a rotation operation to insert the part into the assembly.

In assembly tasks, sometimes alignment between mating parts is not perfect, and that is where the "selective compliance" feature of the SCARA robot excels. "Compliance" is a robotics term that means that the robot or tooling is capable of adjusting to accommodate misalignment. The horizontal, articulating joints of the SCARA robot make it ideally suited for shifting in the horizontal plane as necessary to make the desired insertion. The vertical insertion motion is also easily accommodated by the SCARA configuration, usually using a pneumatic cylinder. SCARA robots are so popular that they are becoming industry standards and the prices are becoming quite competitive, further increasing their popularity. A popular robot of the SCARA configuration is shown in Figure 6.6.

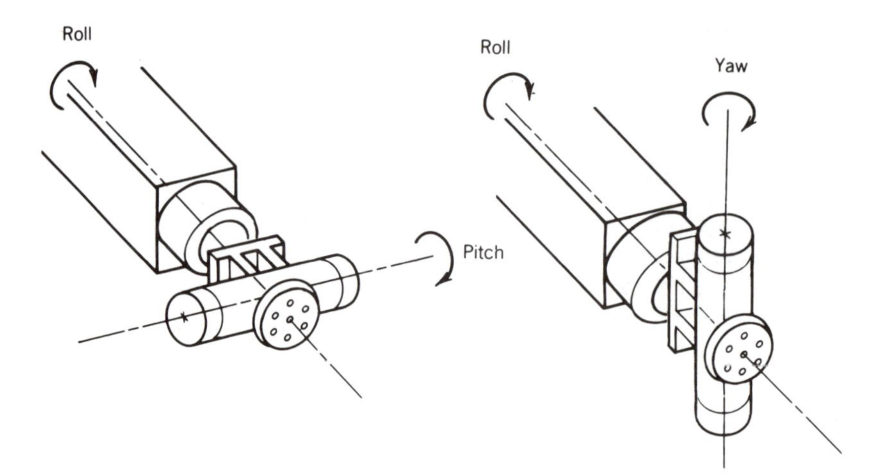

FIGURE 6.4 To obtain yaw motion, roll 90° and then pitch (feasible for certain tool configurations). (*a*) Roll and pitch axes. (*b*) Roll and yaw axes.

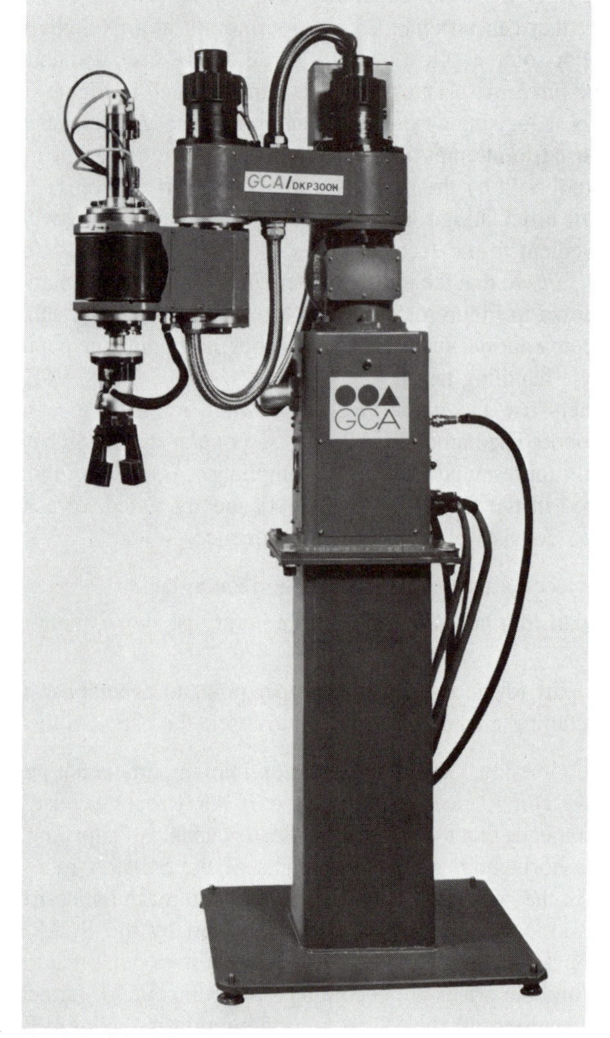

FIGURE 6.5 Horizontal jointed robot. (Courtesy of GCA Corporation, St. Paul, Minn.)

FIGURE 6.6 Articulating robot of the SCARA configuration in the University of Arkansas laboratory. (University of Arkansas Robotics Laboratory photo.)

A variation of the articulating robot is the polyarticulating trunk shown in Figure 6.7. "Trunk" is an apt description of this robot, because it resembles an elephant trunk as it snakes about in a virtually infinite variety of contortions. The advantage of this type of motion is that the robot can extend its trunk into tight workspaces and then orient its tool (usually a paint spray head) into almost any orientation.

Polar Configuration

The polar or "spherical" configuration looks quite different from the articulating configuration but actually is different only in the third axis (third degree of freedom). In the place of an elbow, the polar configuration robot has a pneumatic or hydraulic cylinder that provides extension for the arm. A robot in the polar configuration is seen in Figure 6.8. This type of robot is very popular in the automotive industry.

A typical robot motion is to elevate a workpiece along a vertical path while maintaining the orientation of the workpiece. It will be seen later that for the polar configuration robot, this requires simultaneous, coordinated motion in three axes: shoulder, arm extension, and wrist pitch. The articulating robot has a problem of similar complexity, requiring simultaneous, coordinated motion in shoulder, elbow, and wrist pitch.

Cylindrical Configuration

A robot of the cylindrical configuration has a vertical reciprocating axis for its second degree of freedom, or base extension. This is usually accomplished by a pneumatic or hydraulic cylinder but may be rack and pinion or chain drive in large robots. A cylindrical configuration robot is shown in Figure 6.9. Axes *A, B,* and *C* of this robot generate a

FIGURE 6.7 Polyarticulating trunk-type robot. (Reprinted by permission of Spine Robotics, Molndal, Sweden.)

FIGURE 6.8 Industrial robot of polar configuration. (Reprinted by permission of Unimation-Westinghouse, Inc., Danbury, Conn.)

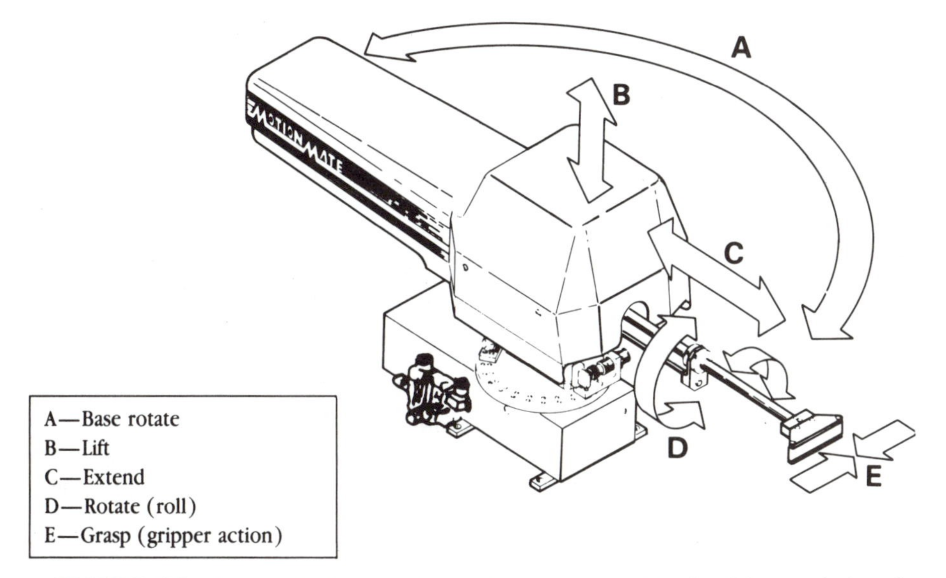

FIGURE 6.9 Industrial robot of cylindrical configuration. (Reprinted by permission of Schrader-Bellows Division of Scovill, Akron, Ohio.)

work envelope in the shape of a cylinder. Cylindrical configuration robots usually have reciprocating motion in their third degree of freedom just as do their cousins of the polar configuration. It is even possible for the cylindrical configuration robots to have *both* base extension and shoulder rotation, giving them an extra degree of freedom between base and wrist. Such robots can have eight degrees of freedom, but this configuration should be considered rare.

The problem of elevating a workpiece or tool in a vertical, straight line is an easy one for the robot of cylindrical configuration. Only one axis of movement is required compared to the three axes of movement dictated by either the polar or articulating configuration. A disadvantage of robots of cylindrical configuration is that they cannot reach around obstacles.

Cartesian Configuration

Machine tool practitioners will feel most at home with robots of the cartesian configuration. This robot is usually gantry mounted but may be mounted on a track on the floor. The first three axes of a Cartesian robot are the familiar x, y, and z axes as in machine tools. The cartesian configuration offers the advantage of rigidity made possible by a boxlike frame to support the robot. Fewer parts on the robot are extended on a cantilevered posture, and even these are only the lighter members close to the tooling. Thus, closer tolerances can be maintained by small Cartesian robots, and for very large robots Cartesian configuration becomes imperative. Figure 6.10 shows a large Cartesian robot.

Work Envelope

With knowledge of the degrees of freedom of the robot and the physical configuration of the assembly of these degrees of freedom, the user can use geometry and trigonometry to determine the position of the robot *tool center point* (TCP) as a function of the positions of the various robot axes. Figure 6.11 shows the computation of the position of the TCP as a function of the axis positions of an articulating robot. Note that each of the angles A, B, and C are measured from a reference parallel to the x axis. Angles A and B are in a counterclockwise direction (positive angles), but angle C is in a clockwise direction and is thus a negative angle. The sine of a negative angle is also negative, so the contribution of the term $(+\ c \sin C)$ to the formula for Y is actually negative. An examination of the diagram of the robot in Figure 6.11 verifies that a negative contribution to the formula for Y is indeed appropriate for the wrist angle shown.

The extreme positions of the robot axes describe a boundary for the region in which the robot operates. This boundary encloses the *work envelope*. Figure 6.12 displays an example of a work envelope. Work envelope for a robot is an extremely important characteristic that should be considered carefully in every robot acquisition. The size of the work envelope obviously determines the limits of reach and is important from an applications point of view. But also important is the viewpoint of safety. A powerful robot placed in a room that is smaller in some dimension than its work envelope will soon wreck its environment, despite the pride and professional care exercised by its programmers. Even with a hypothetically ideal robot that never malfunctions, there are so many variables to consider in programming, and it is so easy for the user to make an error in some detail, that an accident is inevitable if space for the work envelope is not provided properly.

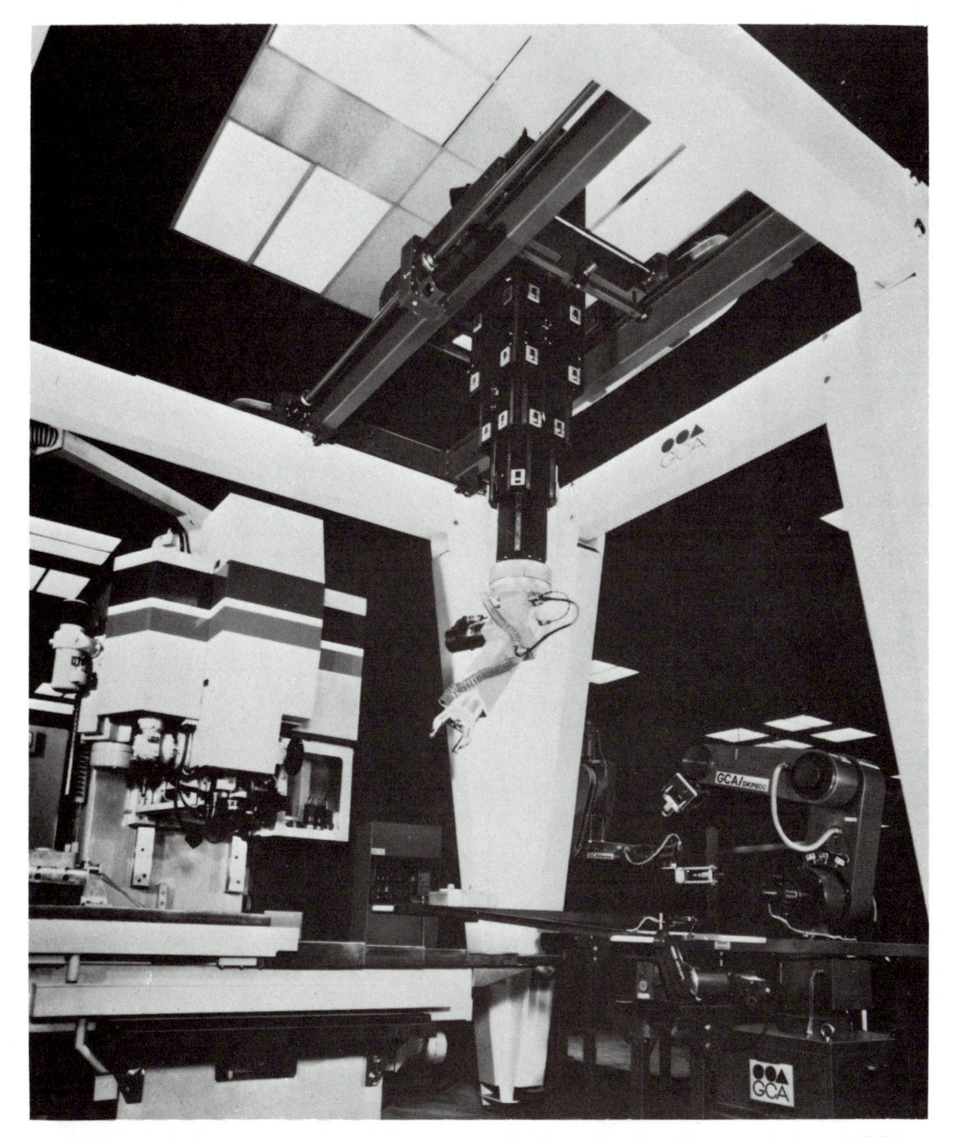

FIGURE 6.10 Large Cartesian "gantry-type" robot. (Reprinted by permission of GCA Corporation, St. Paul, Minn.)

Mobile Robots

The preceding section described robot work envelopes, which are appropriate for most industrial robots because these robots are mounted either on a fixed base or upon a carriage of limited travel on a fixed base. But some robots can walk on legs or move about on wheels. The robot pictured in Figure 6.13 is capable of walking as fast as a human at a brisk walk. It can lift a one-ton payload and can walk with nearly a half-ton payload. In Figure 6.13 the robot not only was capable of climbing into the back of the pickup truck, but it also lifted the back end of the pickup and walked a short distance, turning the truck 90° in the process.

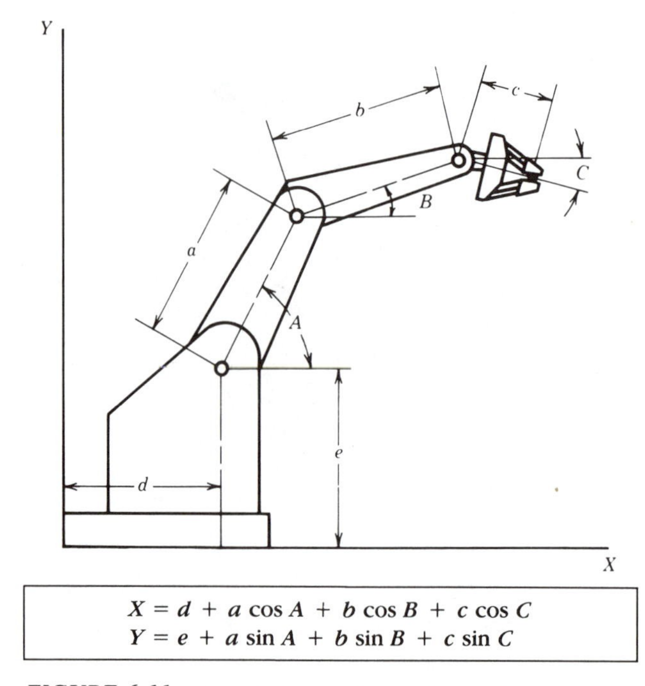

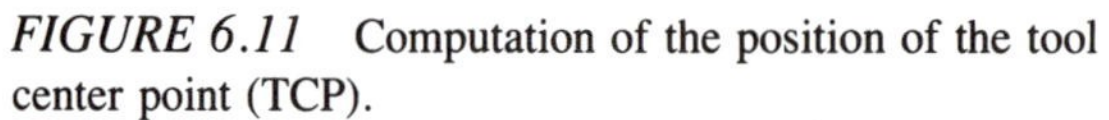

$$X = d + a \cos A + b \cos B + c \cos C$$
$$Y = e + a \sin A + b \sin B + c \sin C$$

FIGURE 6.11 Computation of the position of the tool center point (TCP).

Basic Range and Floor Space Drawings

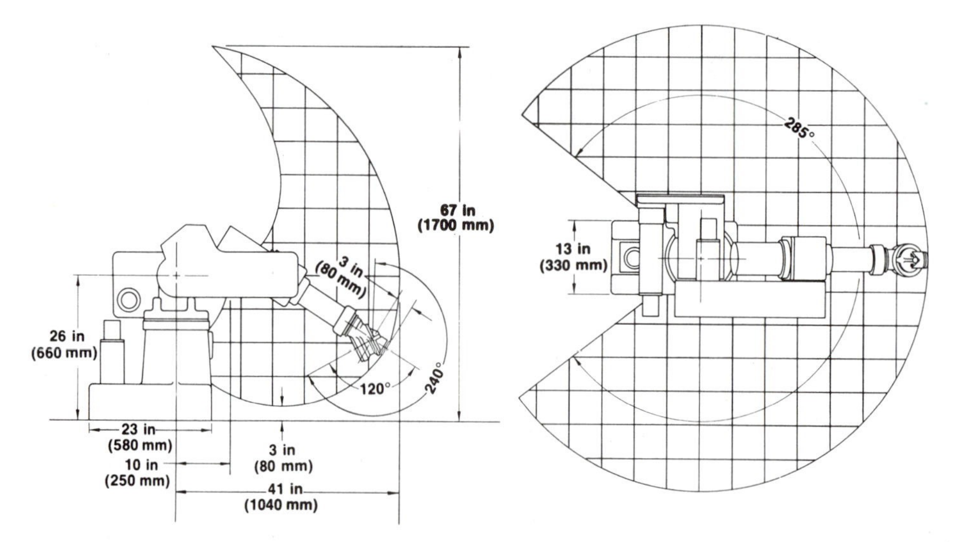

FIGURE 6.12 Example work envelope for an industrial robot. (Reprinted by permission of Cincinnati Milacron, Lebanon, Ohio.)

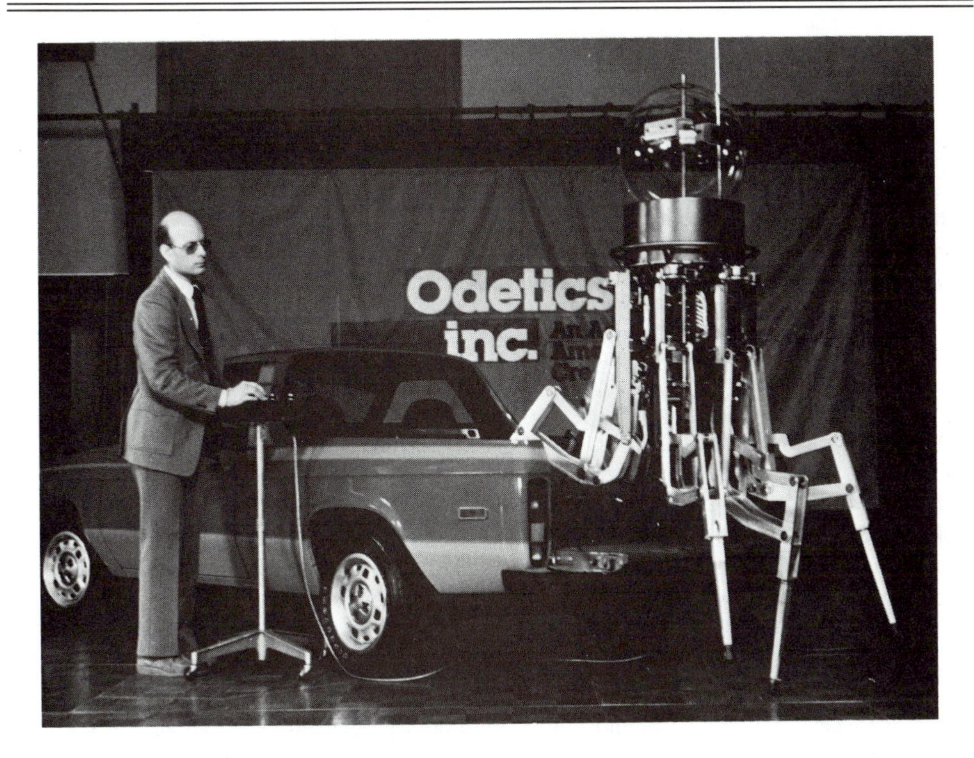

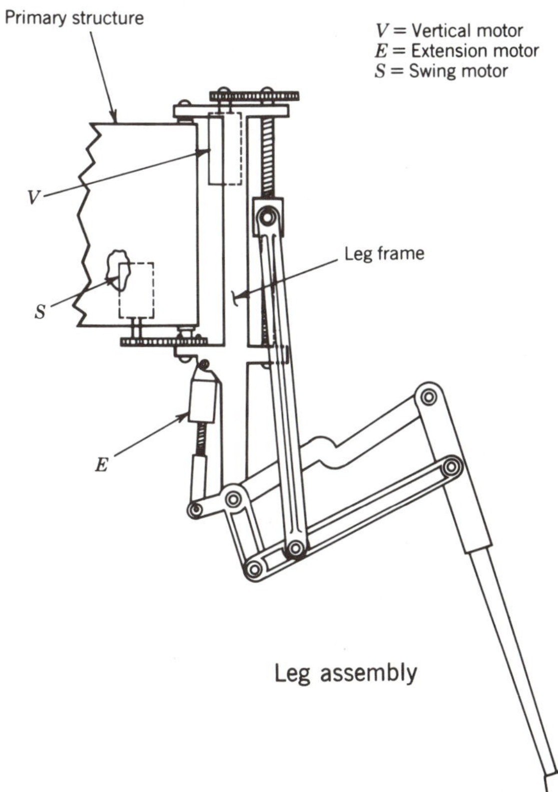

FIGURE 6.13 A walking robot that can lift a ton and can walk away with a 900-pound load. (Photo courtesy of Odetics, Inc., Anaheim, Calif.)

Walking robots are not just curiosities or gimmicks for the entertainment world. Many industries have special problems particularly suited for the capabilities of a walking robot. Some examples are underground mining, space exploration, nuclear accident cleanup, demolition, fire damage inspection, sea floor exploration, and sentry duty. The walking robots are more complex than wheeled models, but they can negotiate a more difficult and uneven surface upon which to travel. The wheeled models have their strong points, too, when used as automatic material handling systems and as interoffice mail delivery systems. To emphasize the usefulness, not the bizarre appearance of the robot in Figure 6.13, its developers (Odetics, Incorporated) dubbed the machine a "functionoid" [ref. 93].

ROBOT DRIVES

The most distinguishing feature used to describe an industrial robot is its power source. The power source usually determines the range of the robot's performance characteristics and in turn the feasibility of various applications, although there is considerable overlap between types. The four principal power sources are now compared.

Hydraulic

From a physical standpoint, the most powerful robots are generally the hydraulic models. Hydraulic robots are able to deliver large forces directly to the robot joints and to the gripper or tool center point. Offsetting this advantage is cost, which is usually higher for hydraulic models than for electric or pneumatic models of equivalent capability. Hydraulic models also require a pump and reservoir for the hydraulic fluid in addition to fittings and valves, all designed for high pressure. Figure 6.14 illustrates a hydraulic robot that features a payload of 225 pounds.

An important application of hydraulic robots is in spray-paint operations. Due to flammability considerations, it may be necessary to employ an explosion-proof robot in paint-spray areas, which require equipment to meet National Fire Protection Association (NFPA) standards for Class I, Division 1 flammable atmospheres. Such standards are almost impossible to meet except by hydraulic- or pneumatic-powered robots.

Popular in the early years, the hydraulic drives are declining in importance among the major robot drives. Most early robots were used in the automobile industry, and the primary application area was spot welding. Many of the spot welding robots are hydraulic. Also, the handling of heavy forgings and die castings called for hydraulic models, typified by Unimation's "first family of robots."

Pneumatic

Some of the least expensive and most practical robots for ordinary pick-and-place operations or for machine loading and unloading are the pneumatic models. The availability of shop air at approximately 90 psi is an obvious advantage. Most factories have compressed air piped throughout their production areas, and this can be conveniently tapped to power a pneumatic robot. The robot in Figure 6.15 is a pneumatic model with a payload of five pounds. The configuration for this robot is cylindrical.

Pneumatic robots usually operate at mechanically fixed endpoints for each axis. Figure 6.16 reveals that the mechanical stop is no different from those employed on pneumatic actuating cylinders extensively used in automation long before the advent of robots. The

FIGURE 6.14 Hydraulic robot with a payload capacity of over 225 pounds including tooling. (Reprinted by courtesy of Cincinnati Milacron, Lebanon, Ohio.)

pneumatic robot is really an assembly of several such cylinders, each one representing an axis of motion.

With motion in each axis controlled only at the end points, the reader may be wondering what can be programmable about a pneumatic robot. But remember that timing and sequence are also important, resulting in an infinite variation of possible programmed setups for the pneumatic robot, even without touching a wrench. By further adjustment of the mechanical stops, even more variety can be achieved. Still, a carefully controlled, continuously varying path is impossible to achieve with the typical pneumatic robot.

It should be mentioned here that there is one type of pneumatic robot, certainly atypical in design, which achieves a continuous, controlled motion through the use of a concept known as *differential dithering*. Differential dithering applies a series of short pulses of compressed air that can act upon the robot member in either direction, causing it to follow a continuous path under control. Figure 6.17 illustrates a commercially available pneumatic robot that uses the concept of differential dithering to produce a low-cost, continuous, and controlled motion.

One of the principal advantages of pneumatic robots is their modular construction and their use of standard, commercially available components. This is true of other robots but is especially true of the pneumatic models. This feature opens up the possibility of a firm deciding to build its own robots, sometimes at considerable cost savings. Some

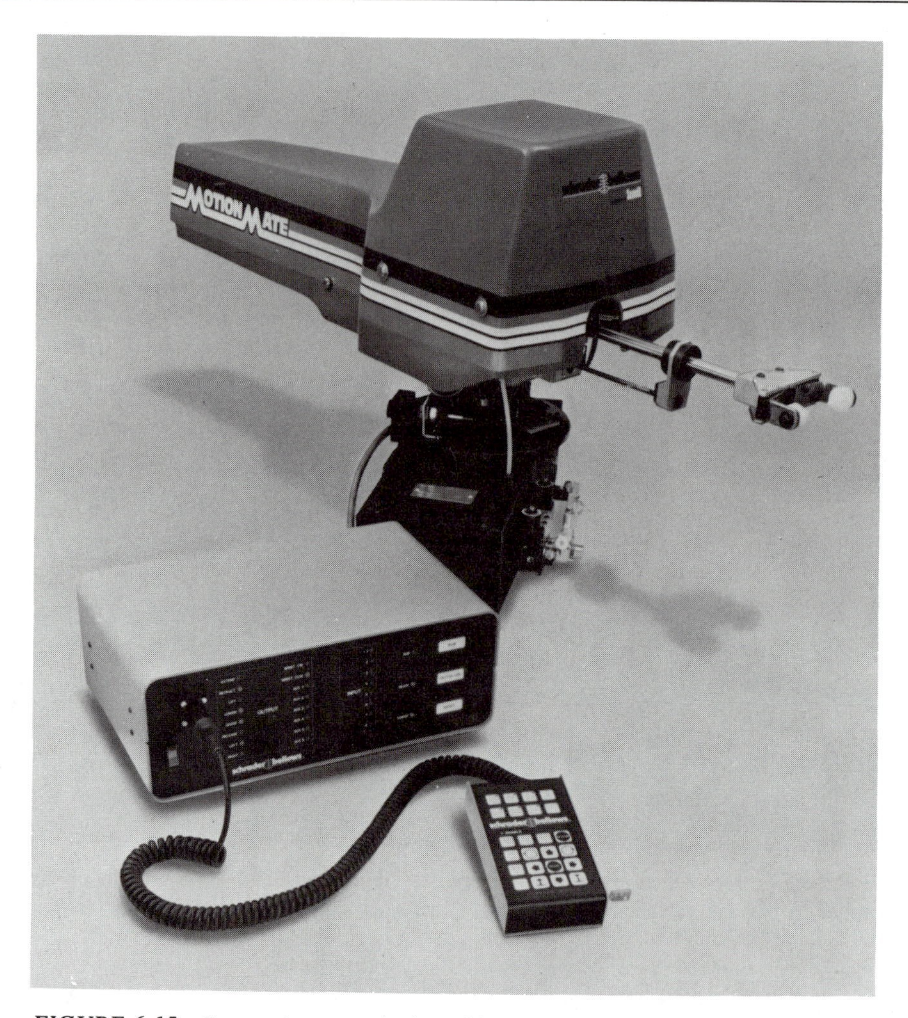

FIGURE 6.15 Pneumatic-powered robot with a payload capacity of 5 pounds including gripper tooling. (Reprinted by permission of Schrader-Bellows Division of Scovill, Akron, Ohio.)

component suppliers emphasize the "build your own" concept in marketing their products. Any firm that decides to embark upon a "build your own" strategy to save hardware costs should also remember to add in the engineering and component procurement costs in addition to the hardware costs.

Electric

Electric robots are popular for precision jobs because they can be closely controlled and can be taught to follow complicated paths of motion. One can argue that many hydraulic models can have the same features, but sophisticated motion control is more typical of the all-electric models.

Electric robots can be divided into two groups according to the types of electric motors that drive each of the axes of motion. One type uses stepper motors (see Chapter 2),

FIGURE 6.16 Mechanical stops for axis limits of a pneumatic robot. (Reprinted by permission of Schrader-Bellows Division of Scovill, Akron, Ohio.)

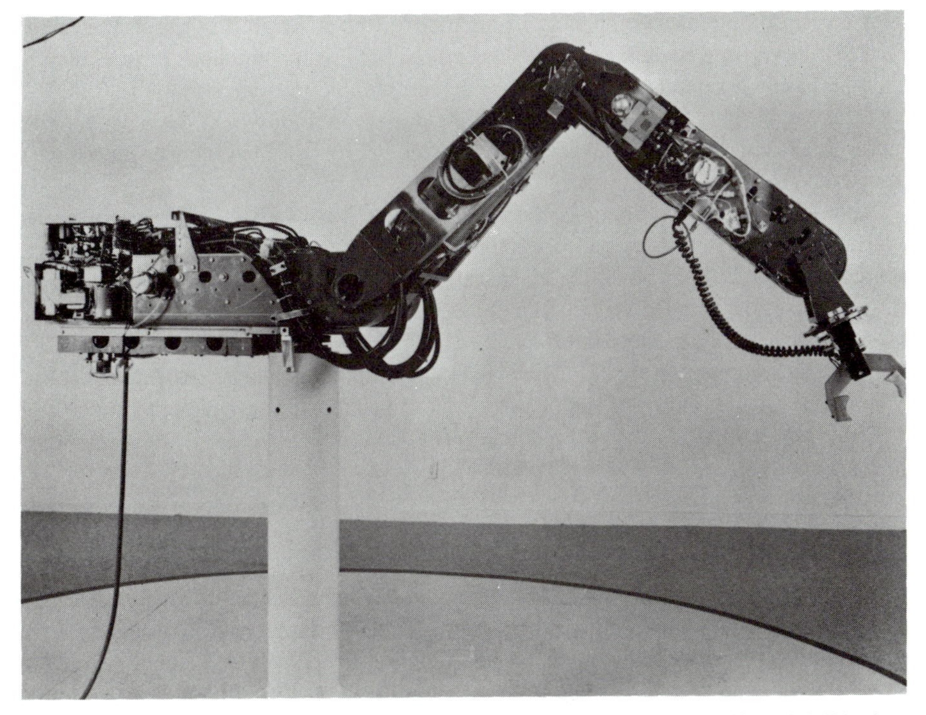

FIGURE 6.17 Pneumatic robot capable of continuous path control by differential dithering. (Reprinted by permission of International Robomation/Intelligence, Carlsbad, Calif.)

which are driven a precise angular displacement for every discrete voltage pulse issued by the control computer interface. The stepper motor movements can be very precise, provided the torque load does not exceed the motor's design limits. Because of this inherent accuracy, the stepper-motor-type robot is sometimes of the open-loop type—that is, the control computer computes the number of pulses required for the desired movement and dispatches the command to the robot without checking whether the robot actually completes the motion commanded. Unfortunately, the robot does not always accomplish the commanded motion because it may encounter an obstacle or for some other reason experience slippage in its mechanical linkage from its drive motors to its mechanical members. When this occurs, the open-loop robot unfortunately "loses its way," and its control computer no longer knows the position of the robot's components. When this occurs, the robot may continue into future cycles with a permanent position error that can make its operation completely useless or even destructive. Fortunately, this predicament has remedies, which we shall see later.

The other species of electric robot is the dc servo-driven type. These robots invariably incorporate feedback loops from the driven components back to the driver. Thus, the control system continuously monitors the positions of the robot components, compares these positions with the positions desired by the controller, and notes any differences or error conditions. DC current is applied to each motor to correct error conditions until the error goes to zero.

It should be noted that feedback loops can be incorporated into the stepper-motor-type robot also. Optical encoders can be used to monitor the actual angular displacement of the driven component. This information is returned to the control computer, which is programmed to take action to correct any error conditions.

Because it can be used either open- or closed-loop, one might expect the stepper-motor design to dominate completely its alternative, the dc servo, in the construction of robots. However, the advantage of the dc servo is that it is a continuous device, thereby making possible a smoother and continuously controllable movement.

Of the two basic types of electric robots, the dc servo-motor type is the most popular. Figures 6.18 and 6.19 show examples of the dc servo type and stepper-motor type, respectively.

The electrically driven robot is the most popular drive today for general-purpose, commercially available industrial robots. The increasing attention toward applications in automatic assembly have made the electrically driven robot, with its superior accuracy and quickness, the favored mode. The Japanese have taken the lead in promoting the electric models, as the United States had emphasized hydraulic models in the early days of robotics development [ref. 17]. Japan's emphasis upon the electric models is natural, as the application of industrial robots in Japanese manufacturing has been primarily in automatic assembly rather than automobile manufacturing.

Mechanical Gear and Cam

For completeness, we will include the manipulators driven by mechanical gears and cams. These types of manipulators are hard-programmed at the factory and do not meet the "programmable" specification found in most definitions of "industrial robot." For that matter, they are found to be generally electrically driven if traced back to the original power source. But the power is delivered to the components by complicated mechanical linkages. Two principal advantages of the mechanical cam-and-gear-drive manipulators are low cost and speed.

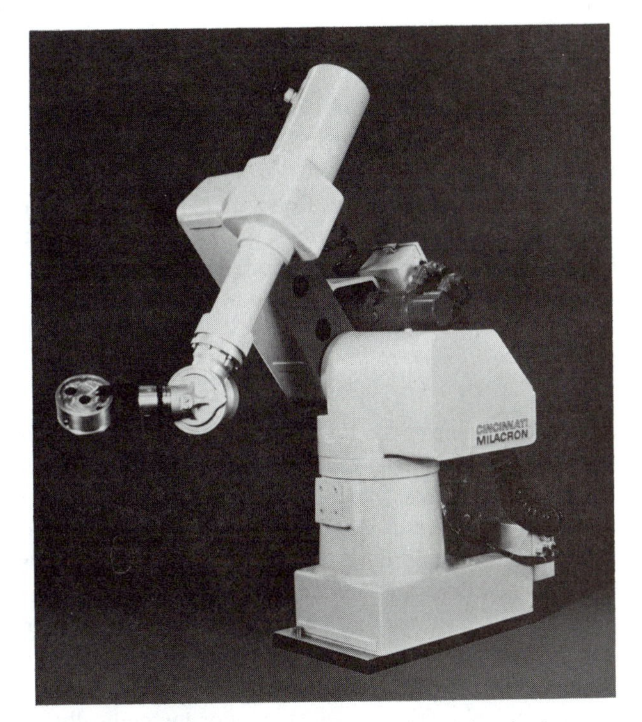

FIGURE 6.18 Example robot driven by electric dc-servo motors. (Reprinted by permission of Cincinnati Milacron, Lebanon, Ohio.)

FIGURE 6.19 Example robot driven by electric stepper motors. (Reprinted by permission of Microbot, Mountain View, Calif.)

MOTION CONTROL

The type of drive used may be the most obvious distinguishing feature of the robot, but just as important, although more subtle, is the degree of control possible over the robot motion. This control is affected by the choice of robot drives, as was seen earlier, but it is not completely determined by the drive. Robot users need to know how much motion control they need for their various applications, because the degree of motion control greatly affects the cost of the robot. Three categories of motion control will now be explained in order of sophistication.

Axis Limit

The least sophisticated and therefore the lowest cost mode of robot motion control is axis limit. This type of control is sometimes called "two-position control" because each robot axis typically has two extreme points. In our description of the pneumatic-powered robots, we saw that the extreme points are usually mechanically adjustable stops.

Users of axis-limit robots should have little or no preference for component motion velocities, as these considerations are beyond their control, although a small degree of speed control can be exercised by varying the power source. In addition, pauses can be programmed between robot motions to permit some selectivity in the speed of the overall cycle. The typical application for axis-limit robots is in machine loading and unloading. Axis-limit robots are invariably either pneumatically or hydraulically powered.

Point-to-Point

Somewhat more versatile than axis-limit control is point-to-point (PTP) control. In this mode, the user can select any point in space in the robot work envelope and move directly to that point. The path and speed of movement en route to the destination point are generally both uncontrollable. Even if speed is controllable, the robot is point-to-point unless the path en route is also controllable. PTP control is good for component insertion, hole drilling, spot welding, and crude assembly applications. Machine loading and unloading operations to or from a pallet or tray also require PTP motion control.

Point-to-point motion should not be confused with straight-line motion. In general, even a simple straight-line motion between two points cannot be accomplished by a PTP robot. One exception to this generality is a straight-line vertical lift by a robot of cylindrical configuration. But straight-line movements are not an easy task for robots, especially the articulating robots. Simultaneous, controlled movement in more than one axis is *always* required to achieve straight-line movement using a fully articulating robot. To counteract this disadvantage, robot manufacturers have developed computer software routines that handle the mathematical mixing of axis voltages, pulses, or valve openings to achieve straight-line movement upon command.

Contouring

As in NC machine tools, the most sophisticated class of robot motion is the full contouring class. Contouring describes motion in which the entire path is continuously or nearly continuously controlled. When the drive is by stepper motor, the control is not quite continuous but still can be classified as an approximation of contouring if there is a feedback loop to the controller and if in addition the controller is capable of varying the *rate* of pulses delivered to the stepper-motor drive.

The difference between PTP motion and continuous path contouring is difficult to distinguish, especially by an observer who did not actually program the robot. By detailed and meticulous programming, the programmer can set up a PTP robot to move in a seemingly continuous contour. Sometimes controller software can provide subroutines to relieve the programmer of specifying the myriad points required to simulate the curvilinear motion with tiny straight-line motions. But programming is not the only problem. The execution time of a PTP robot is dependent upon the number of points specified and thus can become too slow to be effective. Continuous contouring motion provides the ability to control not only the *position* of the robot tool but also the *velocity* of tool motion in each axis controlled by contouring. Contouring motion control is essential for most spray painting, finishing, gluing, and arc welding operations by robots.

Line Tracking

One of the most complex of contouring motions is called "line tracking"—that is, performing an operation while following alongside a continuously moving conveyor. Line tracking is merely another application of contouring, not a separate class of motion. However, the complexity of line tracking demands intricate programming of the robot controller, especially for robots whose bases are fixed to the floor (true of most robots). Some robots designed specifically for line tracking have a horizontal traverse on a track for the first degree of freedom. The traverse can be adjusted to match the speed of the conveyor, giving such robots a distinct advantage over the fixed-base robots with respect to the task of programming. However, fixed-base robots may be equipped with factory-supplied computer software, which, upon demand, conveniently add the line-tracking feature to the user's program. Therefore, the trade-off between the two types is cost of hardware versus cost of software.

Line tracking has obvious advantages. The product being processed can be transported on a *continuous* conveyor instead of an intermittent one. Continuous conveyors are much simpler mechanically and thus are less expensive and more reliable. With more reliable conveyor operation and quicker repair time, the continuous conveyor keeps idle time at a minimum, maximizing both production machine usage and robot usage.

Particularly suited to line tracking is the robot application of spray painting. Spray painting generally is applied to all sides of a piece-part as it is transported by a continuous overhead chain. The robot must be directed at all sides of the part, a feat that can be accomplished most conveniently if the part moves continuously past the work station and if the robot has line-tracking ability. Line tracking is also convenient if multiple operations must be performed by the robot on each part.

ROBOT TOOLING

Although this chapter frequently refers to robot grippers, most robots do not come equipped with such devices. Remembering that programmability and versatility are hallmarks of the modern industrial robot, the reader will see that the robot manufacturers' strategy is to leave the choice of the end-of-arm tooling to the user. Indeed, many robots are programmed to use a variety of tools or grippers in a single setup, automatically selecting and changing tools according to a prescribed sequence.

Grippers

Robot gripper selection is a critical design decision to be made by the automation engineer. This decision can be as important to the success of the application as the selection of the robot. The engineer must be mindful of the conditions of use of the robot gripper and remember that the workplace environment may not be the same as that of the laboratory in which the robot application was tested. Heat, for instance, can cause the grippers to expand, burn, or melt, depending upon the materials used. Abrasive workpieces can cause wear, especially considering the thousands of repetitive operations the robot may be called upon to execute. Perhaps most critical of all is the question of what may happen to the robot gripper if the workpiece is not properly aligned and a collision occurs. Collisions also frequently occur when the robot is programmed improperly.

Grippers come in a wide variety of configurations and are often designed by the customer to fit a particular application. Most grippers close on the part to be picked up, but a large number of them insert their fingers inside the part and then open to grip the part. Many grippers are fashioned to work effectively either way so that the choice is up to the programmer. Figure 6.20 illustrates a variety of grippers.

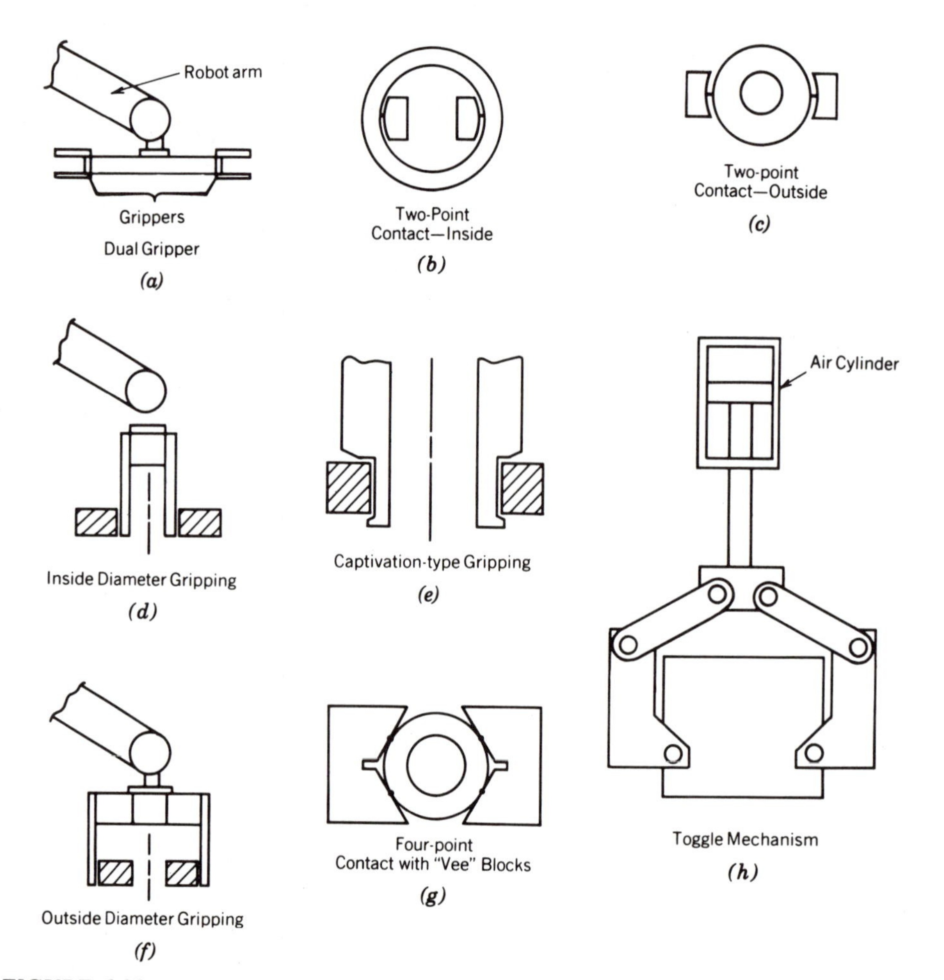

FIGURE 6.20 End-of-arm tooling for industrial robots. (Reprinted by permission of ASEA, Inc., White Plains, N.Y.)

Some applications require a variety of grippers in the same setup, so some type of facility is needed to enable the robot to change grippers or to change from a gripper to a special-purpose end-effector. Many robots are equipped with such tooling, and Figure 6.21 [ref. 2] illustrates a tool changer designed by Keith H. Clark of NASA's Marshall Space Flight Center. Note that both the general-purpose end-effector and the special-purpose end-effector are grippers. However, the general-purpose end-effector has notched fingers for gripping a mating post on the special-purpose end-effector. The long probe on the top of the special-purpose end-effector is for multiple electrical connections that are made in the female connector cone on the general-purpose connector.

For many applications, a double-handed gripper is more efficient than a single-handed gripper. This is typically true of machine loading and unloading, as the robot is able to both unload and load a given station without moving between stations. This saves much time that would otherwise be wasted in repetitious motion in the principal axis (normally base rotation). Base rotation is typically the slowest axis on the robot.

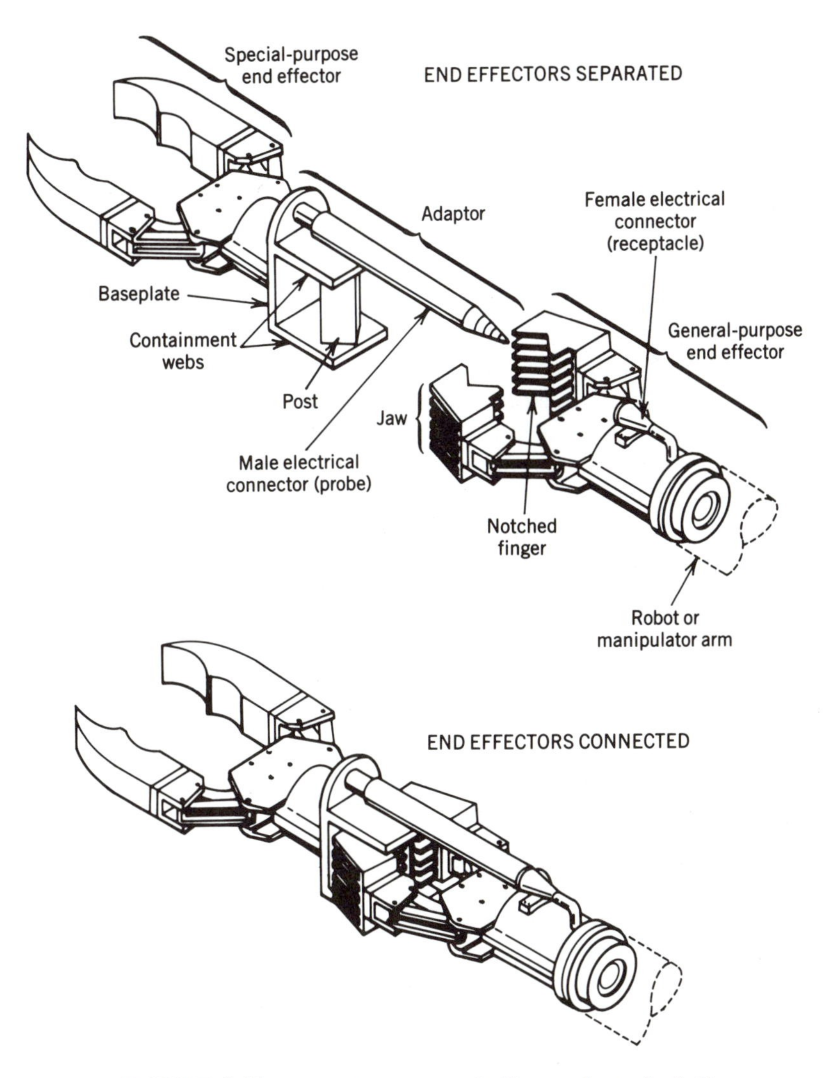

FIGURE 6.21 Special-purpose end-effector adaptor [ref. 2].

Appliances

Besides the common grippers used for piece-parts handling, tool heads of various types can be attached to the end of the robot arm. The wide variety of these end-of-arm tools sets the robot apart from ordinary material-handling devices.

Welding heads are the most common type of robot tool, excluding grippers. Spot welders are the most common, but arc-welding robots are growing in importance, as we will see in the discussion of applications. Spray-painting heads, mentioned earlier in this chapter, are an important type of robot tool. Related to spray painting heads are glue applicators. Both of these tools are useful because of the precision and repeatability of the robot. An unusual tool is a dispenser for electric cable. This tool is used in the programmed assembly of electrical wire harnesses for aircraft and other large equipment.

Part-Compliant Tooling

A tiny misalignment of a piece-part or robot tool can result in complete failure of the process and perhaps damage to the product or to the robot hand. A trick to avoid or at least to blunt the effect of misalignment problems is to mount the gripper with a flexible connection that allows the robot tool to "give a little" when it encounters the object to be picked up. The industry calls this approach *part-compliant tooling*.

One sophisticated type of part-compliant tooling is called *remote center compliance* (RCC). The concept is illustrated in Figure 6.22. In Figure 6.22*a*, the robot is attempting to insert a pin into a hole, but there is lateral misalignment. The chamfer helps, but with rigid tooling the pin still may not enter the hole. A less rigid tooling still might not deliver the desired result because the lateral component of force at the chamfer will tend to rotate the pin about its flexible center of compliance. The assembly in Figure 6.22*b* projects the center of compliance to the leading end of the shaft, which is a remote center of compliance point. The important consequence of this projection is that the pin shifts *laterally* instead of rotating about its top end.

Figure 6.22*c* presents a correct lateral alignment, but an error in angular alignment. Note that this time the pin makes contact with the hole at two places. Lateral components of force on the pin are parallel and opposite each other, but these parallel, opposite forces do not act in the same line because the lateral force from the left acts on a point higher on the pin than the lateral force from the right. This causes a moment to act on the pin, and the RCC assembly shown in Figure 6.22*d* permits insertion to occur. The physical appearance of RCC devices is illustrated in Figure 6.23.

An even simpler strategy that works in some cases is to use rubber or nylon in the construction of the fingers themselves so that a soft, compliant touch not unlike that of human fingers is used to pick up the object. Any such flexible pickup method can be considered part-compliant tooling. The principle is so simple and practical that it would seem to be an obvious solution to gripper alignment problems, but a surprisingly large number of robot applications fail because the user or engineer does not think to try this strategy.

PROGRAMMING

It is worth repeating that a key feature of robots is their capability for being reprogrammed for different tasks. This is the feature that was missing from the mechanical manipulators that were seen before the advent of the industrial robot in the 1970s. Chapter 7 will

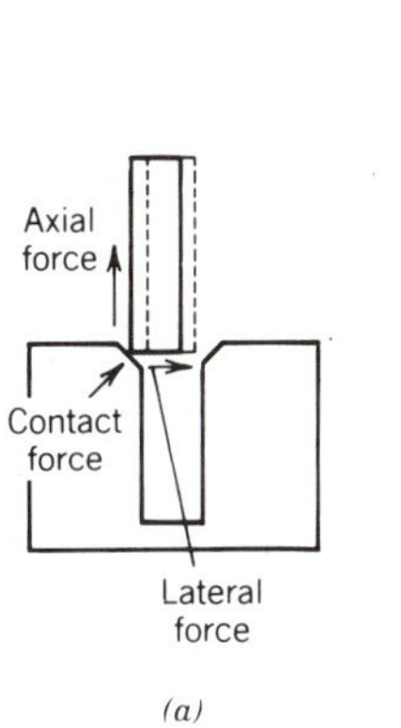

(a)

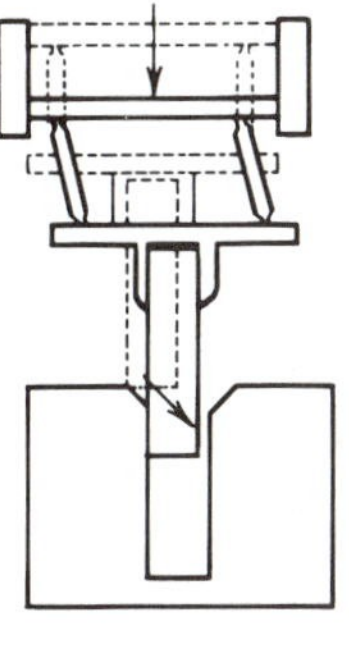

(b)

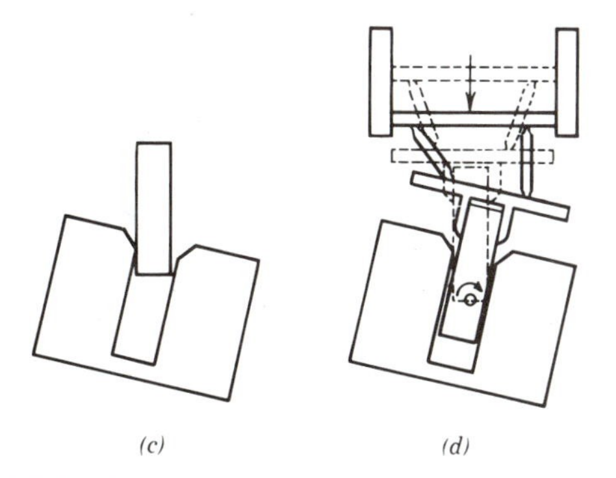

(c) *(d)*

FIGURE 6.22 Remote center compliance (RCC) tooling to deal with misalignment problems.

FIGURE 6.23 Variety of commercially available RCC devices to facilitate close tolerance assembly operations using industrial robots. (Reprinted by permission of Lord Corporation, Mechanical Group, Erie, Pa.)

explain robot programming in more detail, but broad classifications of programming will be identified here.

Some robots have been programmed by manually inserting pegs in a drum to trip switches according to a desired sequence. Others are programmed using a dialect of some popular general-purpose computer language, such as BASIC. One such dialect is the ARMBASIC robot language, aptly named by Microbot, Inc., its creator and copyright owner [ref. 66]. Another BASIC-based robot language is AML, made popular by IBM Corporation's version of the SCARA robot. Another robot computer language is VAL, created by Unimation, Inc., for its PUMA series of sophisticated robots [ref. 112].

The majority of robots have some form of handheld controller called a *teach pendant* (Figure 6.24). The operator manually moves the robot through the desired motions by pressing buttons on the handheld controller. The pace usually is much slower than the production pace, thereby permitting the operator to program the operation carefully. Later, the program is reset to production speed.

Robot reach-pendant controls are very much like the familiar handheld pendant controls for overhead cranes. But the robot pendant control is much more powerful because it has additional controls for commanding the robot to *memorize* points along the path of motion. Also, robot controls can set timers to synchronize the operation, command the sensing of external inputs, and dispatch output signals to peripheral process equipment.

Moving up to a higher plane of sophistication are the robots that are taught by manually seizing the end of the robot arm and actually pushing it through the series of operations in a dry run of the real production process. The robot can be commanded to remember the path during the teach phase, and then commanded to repeat the performance indefinitely. This type of programming is especially good for spray painting and welding applications. Operators skilled in conventional spray painting or welding can teach the robot their skills while simulating an actual manual performance of the job. The robot

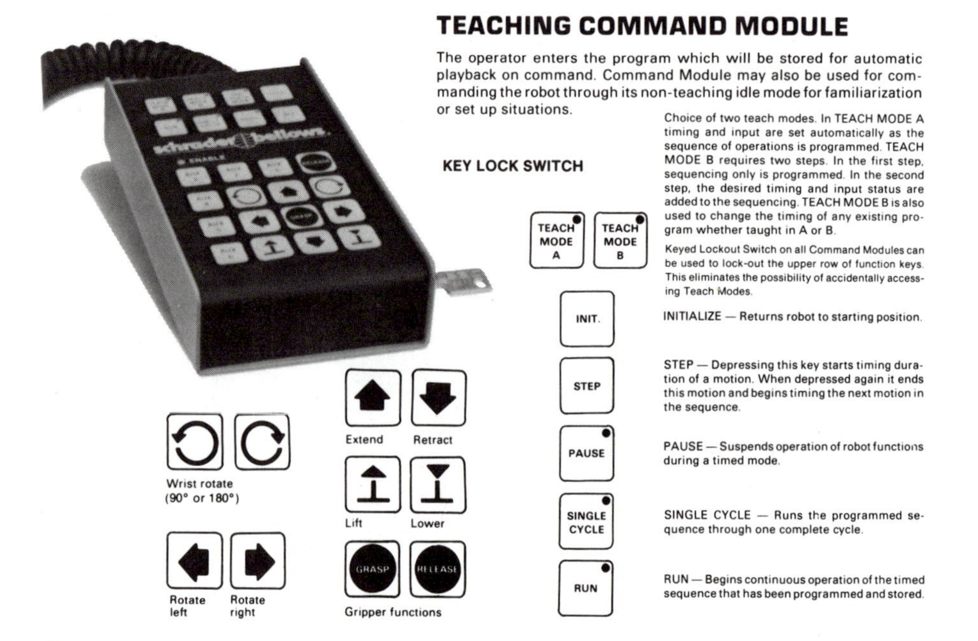

FIGURE 6.24 "Teach pendant" control for industrial robot. (Reprinted by permission of Schrader-Bellows Division of Scovill, Akron, Ohio.)

then merely mimics the actions of its teacher. Of course, the skill acquired by the robot applies strictly to the given application taught, not to spray painting or welding in general.

A variation is to employ a mechanism that mechanically simulates the robot: a robot training arm, sometimes called a *dummy robot*. Compared with the real robot, the training arm (Figure 6.25) is generally lighter and easier to manipulate by the skilled operator charged with the task of teaching the robot. Spray painting is an ideal application for such mechanisms because the skilled operator must feel as though he or she is actually holding a paint spray gun while teaching the robot. The comparatively light training arm can give that kind of feel to the operator. The training arm transmits its path to the control computer during the teach mode. In turn, the control computer drives the real robot through the same path of motion in the run mode.

This text has avoided carefully the use of the terms *leadthrough programming* and *walkthrough programming*. Both of these terms are frequently heard among robotics professionals, but their meanings have often been interchanged. Some say that leadthrough refers to teach-pendant control and that walkthrough refers to the manual dry-run mode employed for spray-painting operations. But other respected professionals interpret the terms vice versa. The reader is advised to use the terms *leadthrough* and *walkthrough* carefully and to insist upon a clarification as to whether a teach pendant is or is not used.

The most sophisticated robots are usually programmed using combinations of the teach modes discussed here. Thus, fixed locations can be taught using the teach pendant, and complex paths like arcs and contours can be programmed using computer software such as Unimation's VAL language. The need for these dual modes of robot programming should become more evident in Chapter 7, which discusses robot programming in detail.

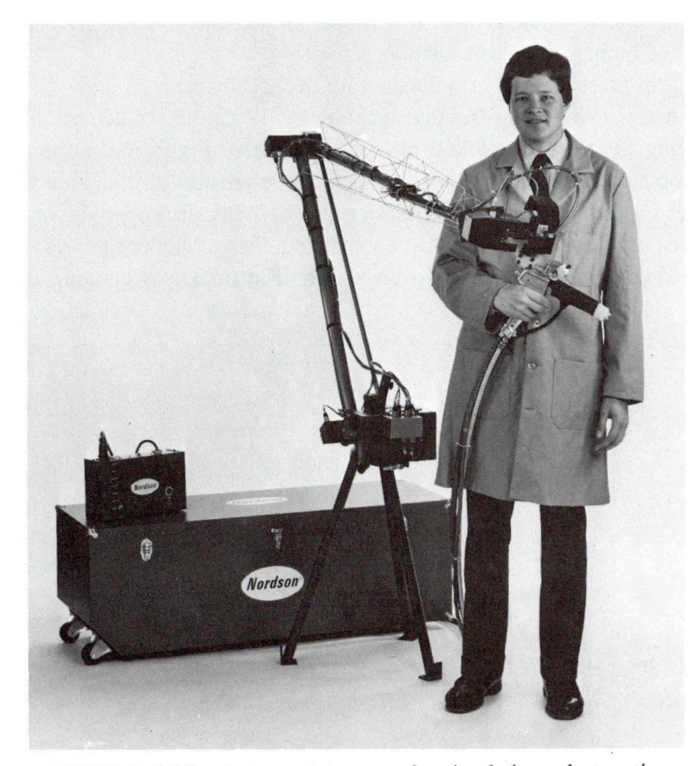

FIGURE 6.25 Robot training arm for simulating robot motion during the teach mode. (Reprinted by permission of Nordson Corporation, Amherst, Ohio.)

SENSING CAPABILITY

Mechanical manipulation has many applications, but used alone its success is highly dependent upon the positioning and orientation of the workpiece. Furthermore, the blind repetition of mechanical sequences can be disastrous if something goes wrong. A robot that can "see" or "feel" its payload and certain aspects of its surroundings has greatly increased utility over its more crude relatives, which operate simply as deaf, dumb, and blind manipulators.

Sensing capability on a robot can have widely ranging degrees of sophistication in addition to a variety of sensing media. For instance, optical sense capability can vary from a simple photoelectric cell to a complex, three-dimensional vision system. Various sensing categories will now be described, beginning with the simplest and most practical and proceeding to the most advanced systems available.

Gripper Pressure Sense

The most elementary sense capability on a robot is probably the ability of the gripper to detect grip force between its fingertips. In its simplest form, the grip sensor consists merely of a limit switch that trips when a given preset grip pressure is reached. Such a limit switch is a practical safeguard against overclosure of the gripper in case of either program error or payload dimensional variability. But the advantages go beyond this safeguard because the limit switch can be used to standardize the grip pressure in a gauging operation. Thus, a robot can be used to gauge thickness by simply closing its gripper upon an object. This feature of a robot is inexpensive to apply, and accuracy surpassing that of human fingers is easy to achieve.

Sometimes grip pressure and grip closure are in fact the same operation. This can be achieved, for instance, by using "tendon technology"—that is, the actuation of the axes by cables leading to motors mounted on the robot base. Figure 6.26 shows how one supplier (Microbot) positions the limit switch to sense tension in the cable that controls gripper closure. Figure 6.27 charts the relationship between grip opening versus number of stepper pulses on the left and grip force versus number of stepper pulses on the right, illustrating the dual role of the grip closure cable. The thickness gauging capability of

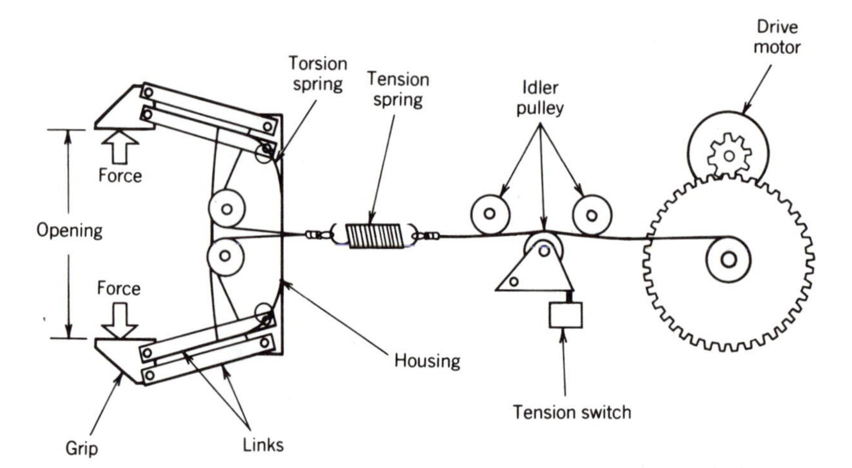

FIGURE 6.26 "Tendon technology": A cable in the robot arm controls both grip closure and grip pressure. Note the position of the limit switch to detect cable tension. (Reprinted by permission of Microbot, Mountain View, Calif.)

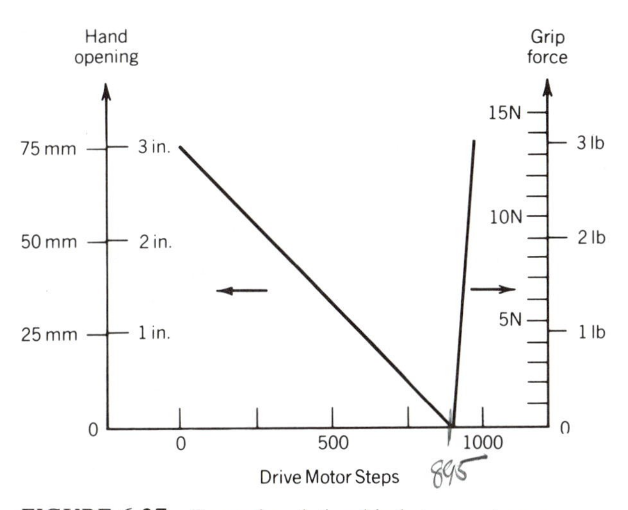

FIGURE 6.27 Example relationship between stepper motor pulses and grip closure and grip force using tendon technology. (Reprinted by permission of Microbot, Mountain View, Calif.)

this device is rated at a tolerance of $\pm 1/16$ in., but a Microbot Minimover 5 has been observed achieving a much higher resolution than the rating specifies.

An example will now be used to illustrate the calculation of stepper-motor pulses to close upon and grip an object of known dimension with specified force.

EXAMPLE 6.1
Determining Robot Gripper Motor Pulses

A robot gripper is actuated by a stepper motor using tendon technology in accordance with Figure 6.27. Suppose the gripper position is open $2\frac{1}{2}$ in. and the command is to close on an object with a force of 20 oz. The object is $1\frac{3}{8}$ in. wide. How many pulses should be issued to the motor?

Solution

The graph in Figure 6.27 intersects the *x* axis at approximately 895 steps, and full travel from gripper wide open to gripper closed and pinched at a force of three pounds is approximately 976 steps. To close on an object $1\frac{3}{8}$ in. wide from a position of open $2\frac{1}{2}$ in., the motor would need the following number of pulses:

$$\text{steps to close} = (2\tfrac{1}{2} - 1\tfrac{3}{8}\ \text{in.}) \times \frac{895\ \text{steps}}{3\ \text{in.}} = 336\ \text{steps}$$

To continue closing to grip with a force of 20 oz would require

$$\begin{aligned}\text{steps to grip} &= 20\ \text{oz} \times \frac{1\ \text{lb}}{16\ \text{oz}} \times \frac{(976 - 895)\ \text{steps}}{3\ \text{lb}}\\ &= 34\ \text{steps}\end{aligned}$$

The total number of pulses required would be the sum:

$$\begin{aligned}\text{total steps} &= \text{steps to close} + \text{steps to grip}\\ &= 336\ \text{steps} + 34\ \text{steps}\\ &= 370\ \text{steps}\end{aligned}$$

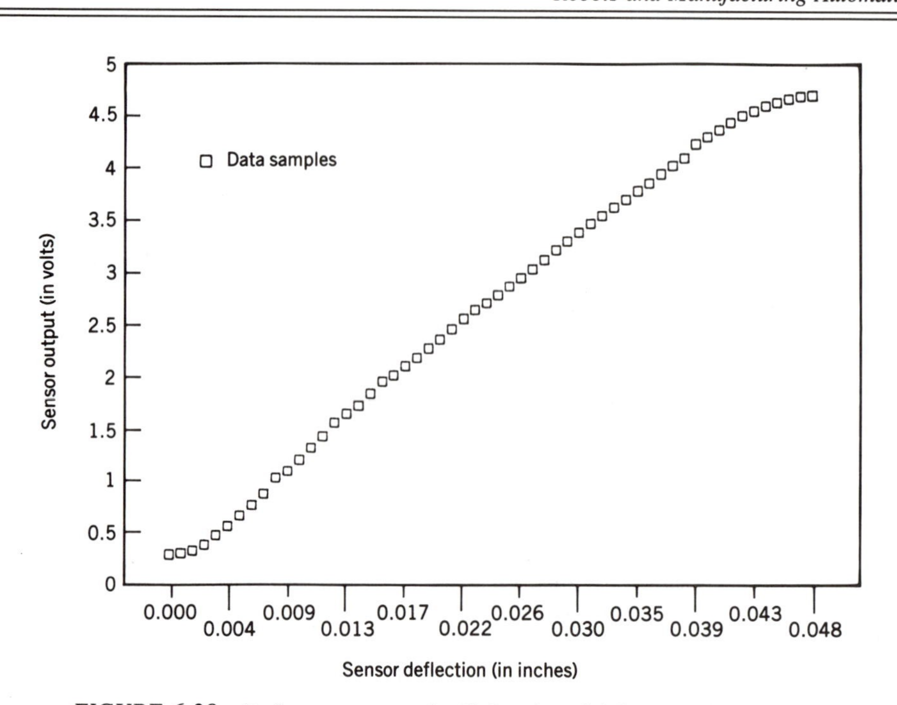

FIGURE 6.28 Performance curve for University of Arkansas Electro-Optic Sensor [ref. 54].

Electro-Optic Force Sensor

Another way to sense gripper closure force is by an ingenious device invented by Marcus Langston at the University of Arkansas Center for Robotics and Automation. The University of Arkansas Electro-Optic Sensor* relies on a flexible gripper contact material through which one or more holes have been drilled to admit an LED light beam for photoelectric sensing at the opposite end of the hole. The force of the gripper closing on an object causes a deformation in the cross-sectional area of the hole in the flexible gripper material, which results in a reduction in the light detected by the photoelectric sensor. This simple, low-cost device is capable of a proportional response to varying gripper pressures with surprising accuracy (see Figure 6.28).

EXAMPLE 6.2
Electro-Optic Force Sense Application

A robot gripper as in Figure 6.20*d* equipped with an electro-optic force sensor is used to vertically pick up a durable symmetrical object weighing four pounds. The coefficient of friction between the object and the gripper fingers is 0.60. Once contact is made, the gripper finger (sensor) deflection is approximately proportional at five thousandths inch deflection per pound of force. The response characteristics of the force sensor are shown in Figure 6.28.

*U.S. Patent No. 4,814,562, March 21, 1989.

The robot gripper is controlled by a dc-servo motor with feedback via the force sensor. At what voltage should the controller halt the closure of the gripper? Allow a margin of error of 25 percent to assure that the object will not slip in the robot fingers.

Solution

Figure 6.20*d* reveals that this gripper is of the inside-diameter type. It is reasonable to assume that equal gripper forces will be exerted on both sides of the inside diameter of the object, and, since the object is symmetrical, the friction force required on both sides will be the same. Thus the weight will be distributed equally so that each gripper finger will exert a closure force sufficient to pick up a weight approximately one-half the weight of the entire object.

Let

$$f_C = 0.60 = \text{coefficient of friction} = \text{ratio } \frac{F_V}{F_N}$$

where F_V = vertical frictional force (parallel to the contact surface) and

F_N = gripper force (normal to contact surface)

$F_V = 1.25\ (\tfrac{1}{2} \times \text{weight of the object})$

Note: The factor 1.25 allows a 25% margin of error

$= 1.25 \times \tfrac{1}{2} \times 4\ \text{lb}$

$= 2.50\ \text{lb}$

$$F_N = \frac{F_V}{f_C} = \frac{2.5}{.6} = 4.17\ \text{lb}$$

D = sensor deflection in inches

$= 0.005\ \text{in./lb} \times F_N$

$= 0.005\ \text{in./lb} \times 4.17\ \text{lb}$

$= 0.021\ \text{in.}$

ΔV = sensor voltage (reference to zero deflection point)

≈ 2.5 volts [from Figure 6.28]

Since the object is "durable," for additional safety from the hazard of dropping the object, the designer might select a trigger voltage of 3.0 volts.

Optical Presence Sensing

Once a robot has grip pressure sense capability, the next step is to add some sort of presence-sensing mechanism, usually a photoelectric cell. A natural place to mount such a sensor is again on the gripper, to detect that an object is ready to be picked up, but this is by no means the only possible mounting place. In fact, the pick-up point is usually located away from the waiting gripper. In such cases, the only physical connection between the sensor and the robot may be an electrical cable.

Photoelectric sensors also can be placed at various points around the robot work envelope to act as a safety device to stop the robot in the event of an unexpected intrusion into the workspace. In fact, at least one leading manufacturer of robots, Prab, has a policy of including with each new robot a chain-link barrier to enclose the work envelope. The barrier is equipped with an electrically interlocked gate; if the gate is opened, the robot stops.

Photoelectric sensors are the most popular devices for presence sensing, but some others deserve mention. Infrared devices have the advantage of not reading ambient light as a false signal. Radio frequency devices are also a possibility, but these systems must be tuned to trigger at the right time for the right object. Radio frequency devices are affected by the size and conductivity of the object to be sensed, and large variations in objects encountered can cause problems.

Robot Vision

The most significant area of sensing capability for robots is the field known as robot vision, a subject to be covered in an entire chapter (Chapter 8) in this book. Formerly regarded as an interesting subcategory of robotics, machine vision has overtaken the field as a central development with far-reaching applications. Vision systems were once thought to be very expensive and beyond the state of the art for most applications. However, experience has shown that in many situations the industrial vision application requires only a portion of the capability of the human eye. Using clever algorithms, vision systems are often able to extract the essential information from even a crude image. Chapter 8 of this text will examine some of the algorithms that make machine vision a practical solution for many robotics applications.

Tactile Sensing

The human sense of touch is a marvelous phenomenon, and scientists have a real challenge in developing robot fingers that can actually "feel" the difference between various textures and surface shapes. But William D. Hillis and John Hollerback at MIT have developed a robot fingertip that can "feel" the difference between screws, washers, and cotter pins. Paul Kinnucan [ref. 49] reports that Hillis and Hollerback used L'eggs Extra Sheer Pantyhose to provide a matrix of 256 (16×16) pressure-sensitive switches to act as data collectors for the computer. The pantyhose material was sandwiched between a piece of conductive silicone rubber and a flexible printed circuit board. The layers were normally separated by the pantyhose, but at pressure points contact was made through the pantyhose mesh, providing a matrix of sensed points. The pattern of the contacts was analyzed by the computer to distinguish objects held by the robot. The pantyhose experiment was a forerunner to modern tactile-sensing systems using electrical contact grids. Helpful for this purpose are piezoelectric materials—that is, materials that emit electric signals when deformed. Figure 6.29 illustrates a tactile sensing grid using the piezoelectric material polyvinylidene fluoride (PVF_2).

Voice Communication

The general public is fascinated with the idea of communicating with robots by voice, and this type of sensory capability is currently under development. There are two modes to such communication, human to robot and robot to human. The former is much more difficult than the latter. Let us deal with the simpler mode first.

There are two general approaches to designing a capability for a robot to speak. The first approach stores segments of prerecorded human voice for recall by a computer program. Although the voice itself is human, the sequence and timing of the voice segments are controlled by a computer. This is the mode used by telephone directory assistance, which uses real human voices with prerecorded digits placed together by computer. Thus, the system can speak a telephone number digit sequence that a human has never spoken or prerecorded, although the digits themselves are prerecorded human voices. The identifying characteristics of these systems is that the prerecorded segments are whole words, not just sounds.

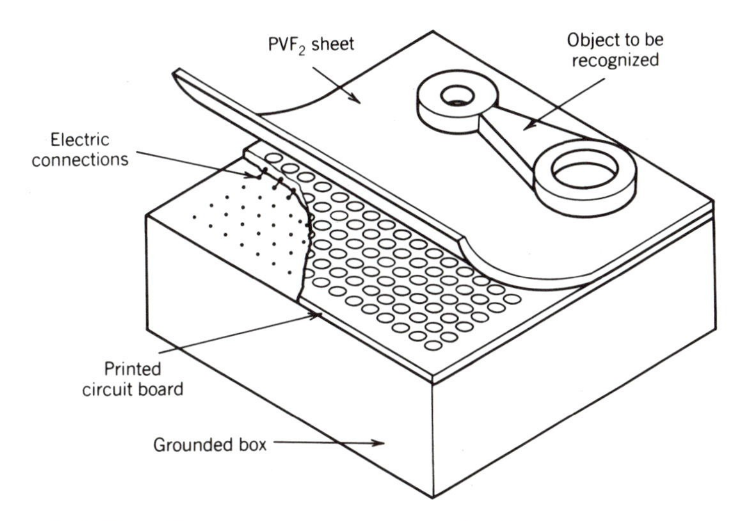

FIGURE 6.29 Electrical contact grid used in the design of a tactile sensor. The polyvinylidene fluoride (PVF_2) sheet has piezoelectric qualities. (Reprinted courtesy of P. Dario and the Society of Manufacturing Engineers, Dearborn, Mich.)

The second and more difficult method of imparting speech capability to a robot is to synthesize speech by assembling elemental synthesized sounds called "phonemes." The 26 letters of the English alphabet actually represent a great deal more sounds because there are "long" and "short" vowels, diphthongs, and combinations of consonants, such as "th," that represent sounds entirely different from the sounds of their component consonants taken alone. Appendix D lists 64 phoneme codes that can be used to synthesize virtually every spoken word in the English language. The codes are in hexadecimal, a computer code that is explained in Chapter 15 of this text. Appendix C also shows how the hexadecimal code is used to construct samples of speech to be spoken by a robot.

It is one thing to teach a robot to speak, but it is an entirely different and more difficult task to teach a robot to *understand* human speech. Building a human-to-robot speech sensory capability is generally beyond the current state of the art except for highly experimental and expensive laboratory equipment. The reason we humans recognize each other's voices is that each person's speech has unique sounds, and these unique sounds demand that a robot programmed to understand speech must be prepared to accept a variety of sounds to represent the same words. This is a difficult problem, and when it is solved, the robot will not only understand speech, but also it will likely be able to recognize who is speaking to it, just as we humans do.

One way we humans make sense out of the variety of sounds we may hear in human speech is context. We can understand a great deal because we know fairly well what to expect from a person's speech. Even if we understand the words perfectly, understanding is often lost without context. The classic example of the importance of context is in the following sentence:

TIME FLIES LIKE AN ARROW.

Three radically different meanings can be attached to this sentence, depending on which of the first three words are interpreted to be the sentence verb. Humans are capable of grasping the meaning if context is supplied. To demonstrate, the three meanings to the sentence will be used in context:

1. Time flies like an arrow, especially when you're having fun!
2. Time flies like an arrow, but fruit flies like a banana. You are probably familiar with the common fruit fly, but you probably have never seen a "time fly." I can tell you that they are really fond of arrows.
3. Time flies like an arrow. You must be very quick whenever you time a fly because flies fly so fast.

Teaching a robot to understand context and be capable of grasping the meaning of any sentence is beyond the capability of current industrial robotics technology.

PERFORMANCE SPECIFICATIONS

The major features that distinguish highly capable, expensive robots from their cheaper alternatives have already been described. But there are some general performance specifications that can make a great deal of difference when selecting a robot for a given industrial application.

Payload

Payload capacity is an obviously important specification but is not as straightforward to determine as it might seem. Most robots can hold a much heavier weight than they can swing about at maximum speed. Also, the shape of the object held and its surface conditions affect the ability of the robot to handle it efficiently. Payload capacity at arm positions close to the base obviously tend to be higher than capacities at full arm extension.

Some robot manufacturers specify two payload capacities: normal and maximum, static and rated, or static and dynamic. The potential robot user should check carefully to ascertain exactly under what conditions the robot manufacturer is determining the rated payload, especially if only one figure is specified.

Repeatability

At this point, the reader should take care to be sure of the difference between "accuracy" and "repeatability." Accuracy is the ability to go to a prescribed point in space defined in terms of x-y-z or some other coordinate system. Machine tools are concerned with accuracy. Repeatability is the ability to return to the same spot again and again after that point has already been taught. For industrial robots, repeatability is a more important consideration than accuracy because the robot is usually taught with a teach pendant the first time. The important test is whether the robot can continue to perform the procedure as taught without slipping off target. Automation engineers are sometimes disappointed to discover that the positioning accuracy of an industrial robot is generally quite inferior to that of numerically controlled machine tools. Robot grippers and tools are generally extended upon a much more flimsy framework than is possible for machine tools.

The tightest specification for repeatability can be held by small pneumatic robots, whose axis positions are stopped mechanically. Some of these robots are rated at ± 0.001 in. or tighter. The big standard hydraulic robots, typical in the automotive industry, are rated at ± 0.050 in. Spray painting and welding robots may have rated positional tolerances as high as ± 0.125 in. or even higher.

There are some tricks that can be used to resolve problems of insufficient repeatability. Figure 6.30 shows a nose cone probe device that is inserted by the robot into a pilot hole

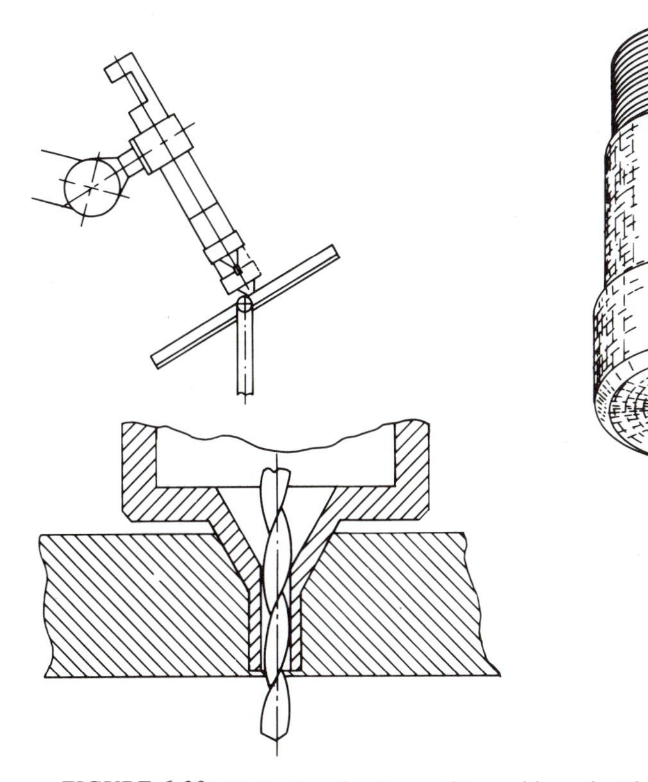

FIGURE 6.30 Probe-type locator used to position robot drill in a drill template, thus improving repeatability. (Reprinted courtesy of T. Molander and the Society of Manufacturing Engineers, Dearborn, Mich.)

just before the operating cycle to assure positioning within a tolerance specified by the customer. In this case, the probe insertion is done at the moment when positioning is most critical.

Another strategy, especially useful for open-loop, stepper-motor-controlled robots, is to go to a home position periodically at some point in the cycle to insert a probe and rezero the axis registers, thus reestablishing home position. The user has the option of programming the rezeroing operation to be executed every cycle or to wait several cycles, depending upon the accuracy desired and the deterioration experienced each cycle.

An interesting comparison is the positioning capability of a robot versus that of the average human. Humans can usually position more accurately than most large industrial robots if they try intently enough. But in a repetitive work assignment a human cannot be relied upon to position intently every time. This explains the superior repeatability of robots over humans. Once programmed, the robot will quickly and consistently go to the same point in every cycle. Although humans may be more *accurate,* they are not equipped to operate with as much *repeatability*.

Speed

Speed is another characteristic that may disappoint some potential robot users. Pick-and-place cycles used in machine loading and unloading are typically rated at two to three seconds for small pneumatic axis-limit robots. Some of these robots can achieve one-second cycles, and cam-operated mechanical manipulators can be even faster. A typical speed for a large, servo-controlled, hydraulic robot is in the range of 50 in./sec.

In many applications, a robot can be slower than its human competitor. Still, use of a robot may be justified on the basis of its higher productivity. This may sound like a contradiction, but it is just another case of the hare and tortoise. The human is the faster "hare," but the robot "tortoise" keeps on working right through breaks, rest periods, lunch hours, and even the night.

ROBOT UTILIZATION AND JUSTIFICATION

By now, the reader should be in a position to enumerate several advantages to applying robots in the workplace. Rather than thinking of robots as being superior to human workers, it is useful to consider the human worker superior and deserving of more meaningful work than is usually assigned to robots. Robots are effective at boring, repetitive jobs that require little or no intelligence or judgment. Robots are also good for extremely fatiguing, hot jobs or for jobs that must be performed in toxic or otherwise dangerous environments.

Besides the thousands of robot applications in boring or dangerous jobs that are unfit for the superior human, there are other jobs for which the robot has features that are superior to those of humans. Robots have higher repeatability on intricate, repetitious tasks. Figure 6.31 shows an intricate grinding operation. Each fine edge must be ground to a precise tolerance. To achieve an effective grind, the tool must be held at a precise angle to the work. It is difficult or impossible for a human to hold the same tool angle while indexing slightly to advance to the next fin. The precise angles and motions can be programmed easily on the robot, in this case a Unimate PUMA.

Jobs that require handling heavy workpieces may be either humanly impossible or extremely fatiguing, thereby impossible for a human to sustain for a full workshift. For these jobs, robots may be more than desirable—they may be essential. Many hydraulic models, and some electric ones, have superior lifting strength when compared to humans.

Many robot applications present multiple advantages for robots. Die casting is a good example. The work is hot and dangerous, and the workpieces to be handled are often heavy and fatiguing. The same can be said of general foundry work. Welding operations combine the necessity for precision with the problems of exposure to hazards. Robots are excellent candidates for both spot-welding and arc-welding jobs. Like welding, spray painting and other spray-finishing operations present hazards but require precision. It may not seem that spray painting is precision work, but left to the discretion of the human worker some units are bound to receive more thorough coverage than others, resulting in variation among the product units. An optimum spray-coverage scheme can be exactly duplicated indefinitely by the robot.

Labor Resistance

Visions of assembly lines operated entirely by robots conjures up fears in the minds of workers who believe they may be replaced by robots on the job. Such assembly lines are a reality in some rare factories, and it is certainly possible for a robot to replace a given worker on a given job. But to see the robots as a cause of general unemployment is to look at the problem exactly backwards. Properly applied, the robot can be used to increase productivity, making the firm more competitive and thus *preserve* jobs. This

FIGURE 6.31 Robot deburrs cooling fins in a heat sink finishing operation, maintaining precise tool angle while indexing slightly for each fin, a difficult job for a human. (Reprinted by permission of Unimation-Westinghouse, Inc., Danbury, Conn.)

rationale can be seen when the automobile industry is examined as a case in point. It would be foolhardy for automobile manufacturers in the United States to abandon robots and automation in order to preserve jobs. Obviously, other automobile-producing countries, notably Japan, would continue producing automobiles as efficiently as possible—using robots and automation. Market forces would soon eliminate the less efficient, nonautomated firms. In the words of James A. Baker, "Henry Ford didn't invent the automobile; he *automated* its production and in doing so created millions of jobs and the backbone of the American economy" [ref. 12].

Economic Justification

Robots range widely in cost from $5,000 on the low end to over $150,000 on the high end, using the early 1990s as a reference point. They are comparable in cost to many machine tools, except for the most expensive NC machine tools, which are higher.

Neither machine tools nor robots are difficult to justify when their roles are vital to the feasibility of the process. For instance, some form of robot may be essential when assembling radioactive components in a product.

Economic justification of a robot that is not vital to the process requires more careful consideration of cost factors. If a robot is being considered as a potential replacement for a human operator, the tendency is to capitalize the annual cost of the human operator and compare it with the initial cost of the robot. But this is a gross oversimplification of the problem. To begin with, a human operator generally works 40 hours per week or less. Overtime is costly, and even then it has its limitations. No worker can work three shifts per day, seven days per week. A robot cannot work continuously either, but a limited amount of maintenance or repair can keep a robot operating three shifts per day, seven days per week. Even with eight hours per week robot downtime, the robot can achieve the equivalent of four human operators, if both human and robot operate at the same pace.

The precision of the robot and its ability to repeat a task continually results in cost savings that may be greater than the savings in direct labor costs. The obvious benefit of robot precision is its effect upon product quality and the accompanying reduction in scrap and rework costs. Less obvious is the savings in paint, glue, and other materials applied by the robot. The human operator can apply too little paint (a quality problem) or apply too much paint (both a quality and a material waste problem). The human tendency favors the application of too much paint, the more serious of the two errors.

More subtle economic factors are the support costs for human operators, such as vacation and sick leave, safety equipment, restrooms, cafeterias, parking lots, and even heating and lights. Reportedly, a plant exists in Japan in which essentially all of the operators are robots. The plant operates in darkness, except when a malfunction occurs, in which case a computer controller illuminates only the portion of the plant appropriate for the needs of the maintenance technician. Upon elimination of the problem, the computer controller extinguishes the lights one by one as the maintenance worker makes his or her way out of the work area.

Robots have economic drawbacks, too, that may be more subtle than their benefits. As with any new equipment purchase, a robot represents a substantial initial cash outlay in return for a future annual cost savings. Cash-flow considerations may make a robot investment difficult even if the expected rate of return is high. Obsolescence can also be a factor but not so much so now that robots are reprogrammable. As was pointed out in Chapter 1, flexible automation equipment including robots are not as obsolescence-prone as special-purpose fixed automation equipment.

Although economics is the common denominator of all robot decisions, some of the decision criteria may be so difficult to quantify that they should be considered separately. Emerson [ref. 35] recommends the consideration of three fundamental facets of robot project feasibility: operational, technical, and economic. Some managers insist that projects be justified upon more than an economic foundation. Even when not required, a sound justification of the operational and technical feasibilities will help to sell the project to management. Managers and decision makers do not want a robot project to go sour by startup problems caused by operational or technical factors such as:

Operational

- Worker resistance to robots
- Production scheduling infeasibilities
- Interruption of work flow during installation and checkout
- Need for parallel manual production during robot installation and checkout
- Inability to use manual backup to recover from a robot breakdown
- Delivery slippages in robot equipment ordered for installation
- Operator training
- Maintenance scheduling
- Spare parts logistics
- Worker safety

Technical

- Equipment incompatibilities
- Robot reliability inconsistent (either high or low) with reliability of equipment it serves
- Insufficient repeatability of robot motion
- Piece-part orientation problems
- Piece-part dimensional variations
- Piece-parts too fragile
- Problems in interlocking robot with process equipment
- Work envelope conflicts

Manufacturing automation engineers who can address each of these operational and technical factors when presenting robot projects to management will engender a feeling of confidence that they have done their "homework" and understand the consequences of a decision to purchase and install a robot.

SUMMARY

Real industrial-type robots bear little resemblance to the glamorized, fictional image born of the media, principally the movies. But the real thing has achieved some glamor of its own, mainly due to the technological advances of microchip circuits. The key feature of industrial robots is their programmability or, more precisely, their *re*programmability, made possible by their microcomputer controllers.

Robots exercise their degrees of freedom in many different ways, depending on their configuration. The axes of motion may be pivotal, cylindrical, or various combinations of each. The maximum extension of the axes of motion result in the boundaries of the work envelope, an important system criterion. The principal power sources driving the axes of robot motion are hydraulic, pneumatic, and electric. Degree of control over the robot motion is a principal determinant of robot capability and likewise its cost.

Industrial robots usually do not come equipped with grippers or other tooling. The wide variety of tools that can be attached to the end of a robot arm is one of the keys to the robot's great versatility. Another key to this versatility is the ease with which robots are programmed using teach pendants, computer languages, or even mimicry of a skilled human operator.

The usefulness of an industrial robot in the workplace can be enhanced by sensors that provide inputs to the robot's controller. Optical and tactile sensors are the most practical at this time, but voice recognition is a possibility for future industrial-grade robots.

When buying a robot, the engineer needs to consider payload, work envelope, repeatability, accuracy, and speed. Robots can replace human workers on undesirable, fatiguing, or unsafe jobs while at the same time preserving other workers' jobs by making the entire operation more productive. In making economic comparisons, the analyst should consider the robot's ability to work three shifts per day if production volumes warrant, but robot downtime and maintenance cost also must be considered. Economic feasibility is not the sole criterion; technical and operational factors also must be considered.

EXERCISES AND STUDY QUESTIONS

6.1. What key feature distinguishes industrial robots from their mechanical manipulator predecessors?

6.2. Suppose your company has a mechanical arm with four degrees of freedom as follows:
   (a) Wrist pitch
   (b) Wrist yaw
   (c) Wrist roll
   (d) Gripper closure
   Explain how you might add additional equipment to convert this mechanical arm to an eight-axis robot.

6.3. When a robot is mounted upon an ordinary x-y table, it has how many additional degrees of freedom?

6.4. What is line tracking? Explain its advantages and disadvantages.

6.5. Which of the following robots are capable of lifting an object vertically while moving in only one axis?
   (a) Articulating robot (Figure 6.2)
   (b) Articulating robot (Figure 6.3)
   (c) Horizontal jointed-arm robot (Figure 6.5)
   (d) SCARA robot (Figure 6.6)
   (e) Polyarticulating robot (Figure 6.7)
   (f) Polar configuration robot (Figure 6.8)
   (g) Cylindrical configuration (Figure 6.9)
   (h) Cartesian configuration (Figure 6.10)

6.6. Name at least two robot languages and describe some of their features.

6.7. Some robotics professionals use the terms *walkthrough* and *leadthrough* when referring to robot programming. What meanings are associated with these terms?

6.8. What is a *teach pendant?* To which mode of robot teaching does this device apply?

6.9. What is *tendon technology?*

6.10. Compare the terms *optical presence sensing* and *robot vision*.

6.11. What has pantyhose to do with tactile sensing by robots?

6.12. What factors must be considered when making an educated comparison of robot payloads?

6.13. Generally speaking, which is more accurate, a human or a robot? Which has greater repeatability?

6.14. What three facets of project feasibility should be considered when justifying a new robot project? Name some factors for each.

6.15. How many human workers are required to operate a single work station 24 hours a day, 365 days per year? Show calculations to justify your answers.

6.16. A robot using tendon technology closes the gripper upon the object and applies grip force according to the following relationship:

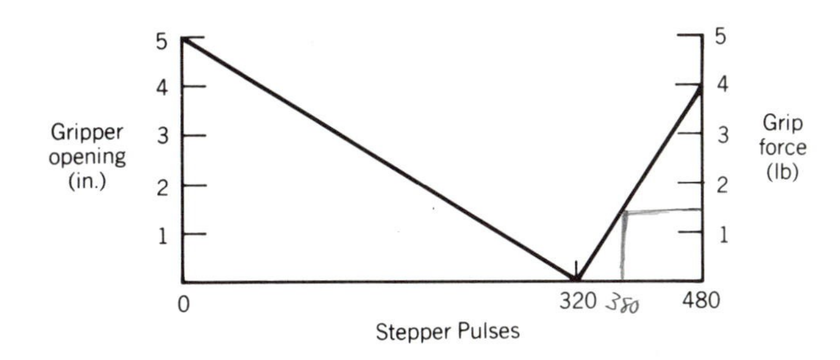

If the gripper starts at a full open position, calculate the number of pulses required to be issued by the robot controller to cause the gripper to close upon a 3-in. block and grip it with a force of 24 oz before picking it up.

6.17. A major airframe manufacturer has a sheet metal operation for which some work stations are operated by humans and some by robots. The human stations experience a 10 percent product quality rejection rate in this operation while the robot stations have virtually no rejects at all. Conservatively speaking and considering downtime, the robot station produces 50 parts per hour at an in-process value of \$5 per piece-part. What is the annual savings in product quality cost if the robot works a standard 8-hour shift per day? What is the quality savings for a round-the-clock operation? State any assumptions made in reaching your conclusions.

6.18. A two-point, contact-inside gripper (see Figure 6.20) has fingers of $\frac{1}{2}$-in. thickness that fully close to result in a 1-in.-thick hand. The gripper also can be used as an outside diameter gripper and has a maximum gripper opening of 3 in. between the jaws. The gripper is driven by a stepper motor using tendon technology for which one step closes or opens the gripper $\frac{1}{64}$ in. Additional steps applied against a rigid resistance increases the force of the grip by $\frac{1}{2}$ oz per step to a limit of 5 lb. Draw a pulse versus grip diagram for this gripper using the format of Figure 6.27.

6.19. Which of the following cylinders can the gripper of Exercise 6.18 pick up?
(a) a cylinder with an outside diameter of $3\frac{1}{2}$ in. and an inside diameter of $\frac{3}{4}$ in.
(b) a cylinder with an outside diameter of 2 in. and an inside diameter of $1\frac{1}{2}$ in.
(c) a cylinder with an outside diameter of 6 in. and an inside diameter of $3\frac{1}{2}$ in.
(d) a cylinder with an outside diameter of $1\frac{1}{2}$ in. and an inside diameter of $\frac{1}{2}$ in.

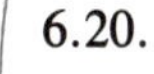

6.20. How many steps must be issued to the driver of the gripper in Exercise 6.18 to cause it to open from a fully closed position and grip the 2-in. inside diameter of a cylinder with a 1-lb force?

6.21. Eggs are to be picked up by a robot gripper equipped with a force sensor that responds as in Figure 6.28. The gripper must close with at least 3 oz of force but with not more than 8 oz of force, or else some of the eggs will be broken. An ounce of gripper pressure deflects the sensor by 0.004 in. At what sensor voltage output should the robot controller cease to squeeze the egg it is intended to pick up?

CHAPTER 7

TEACHING ROBOTS TO DO WORK

In Chapter 6, we learned that today's industrial robot is a much more versatile tool than the mechanical manipulators it is replacing. But this new versatility carries along with it a responsibility for its users: to teach the robot the operation or group of operations desired. As robot versatility varies, so necessarily does the complexity of its teaching—rather, learning. But the same technology that has made the versatility possible is also capable of making the teaching easy, powerful, and user-friendly, and the most successful robot manufacturers are those who have concentrated on this human aspect.

THE ROBOT'S WORLD AND THE REAL WORLD

To get a grasp upon the scope of the problem of teaching a robot, stop reading and take a moment to conduct the following experiment. Picture yourself as standing between two bulletin boards. With your left hand, use your thumb and forefinger to simulate reaching for a straight pin stuck in the first bulletin board one foot to the right of your right shoulder and slightly in front of it. You may turn your head to watch your left hand make the reach, but keep your body trunk facing forward as nearly as possible. Now move the imaginary pin in a straight line to a point as far as you can reach to the left of your left shoulder and simulate sticking it into the second bulletin board at that point. If you moved in a straight line, your hand should have passed very close to your face and neck en route. Repeat the motion and observe carefully the position and action of your elbow, shoulder, and wrist. The motion is quite complicated, the joints take on peculiar positions, and their flexures are even more extraordinary as the joints twist, turn, accelerate, and decelerate, even when the velocity of the imaginary pin is held constant. Such is the complexity of the articulating arm and also our ability to take for granted the way we learned to draw a straight line in kindergarten.

Complicated joint motion is a problem we did not have in Chapter 5, when we studied the programming of CNC machines. Most machine tools are designed to deal directly with the familiar real world of *x*, *y*, and *z* axes. But this is not true of articulating robots nor of their cousins of polar or cylindrical configuration. This has led robot engineers to refer to the robot's world as described by "joint coordinates" (Figure 7.1) compared to Cartesian *x*-*y*-*z* coordinates, which are sometimes called "world coordinates" (Figure 7.2).

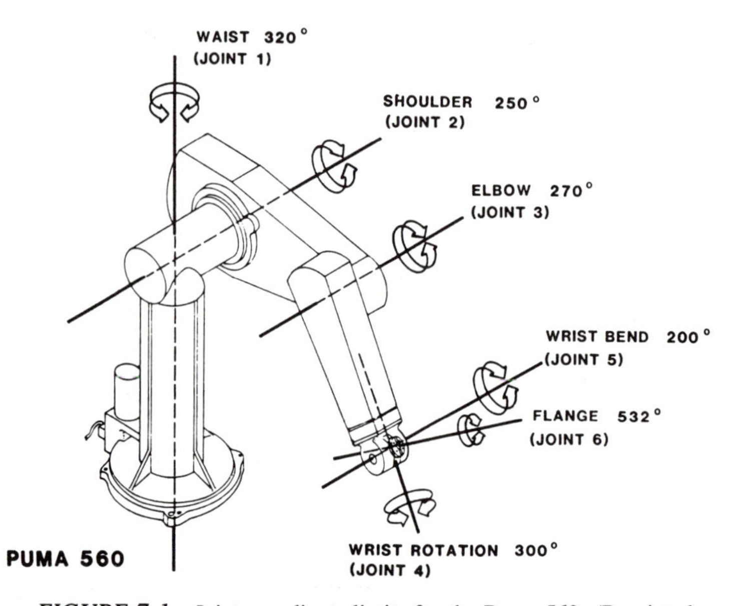

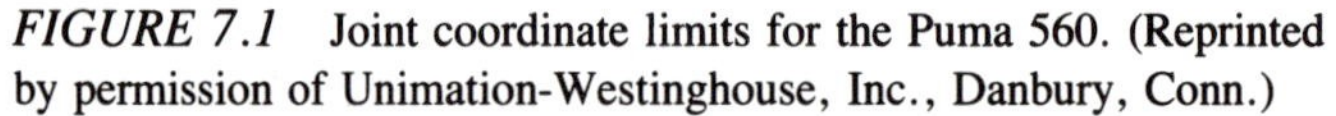
FIGURE 7.1 Joint coordinate limits for the Puma 560. (Reprinted by permission of Unimation-Westinghouse, Inc., Danbury, Conn.)

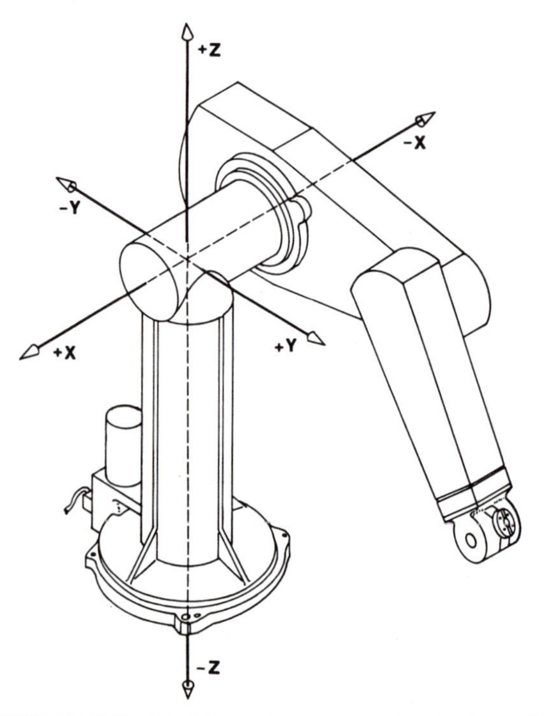

FIGURE 7.2 World coordinate system for the Puma 560. (Reprinted by permission of Unimation-Westinghouse, Inc., Danbury, Conn.)

Thus, every point in the conventional Cartesian space of the robot's work envelope can be described by a set of positions of the robot's axes of motion—that is, angles in the case of articulating robots.

But there is yet another world—the world of the robot *tool*. Because the robot wrist can have various configurations, the robot gripper, tool, or other end-effector can be pointed in various directions even though the tool center point (TCP) is in exactly the same position. Thus, it can be recognized that a single position in the real world (world coordinates) can be represented by a family of positions in the robot's world (joint coordinates). But it may be desirable to program motion from the tool's reference, not the robot's and not the real world's reference. Picture the problem of backing a pneumatic screwdriver out of a recessed screw hole. It is neither the position of the robot nor the orientation of the real world (the work environment) that is important; it is the orientation of the tool. The "back-out" motion of the screwdriver must be parallel to its centerline. This is usually interpreted to be negative motion along the *z* axis in a system called "tool coordinates" (Figure 7.3). Other tool motions might call for movements in the tool *z* plane, instead of along the tool z axis.

It should be remembered that it is possible to calculate every joint movement and move everywhere in the real-world envelope or in the tool-world envelope by commanding the robot in the joint coordinate system alone: possible, yes, but practical, no. In Chapter 6, the complexity of using trigonometric functions of the joint angles and their corresponding joint lengths to perform the necessary transformation to the world *x* axis alone was shown in Figure 6.11. Similar calculations are necessary for the world *y* and *z* transformations, not to speak of the *compound* transformation necessary to relate the *tool* coordinate system to the real world. The problem becomes even greater if the robot is

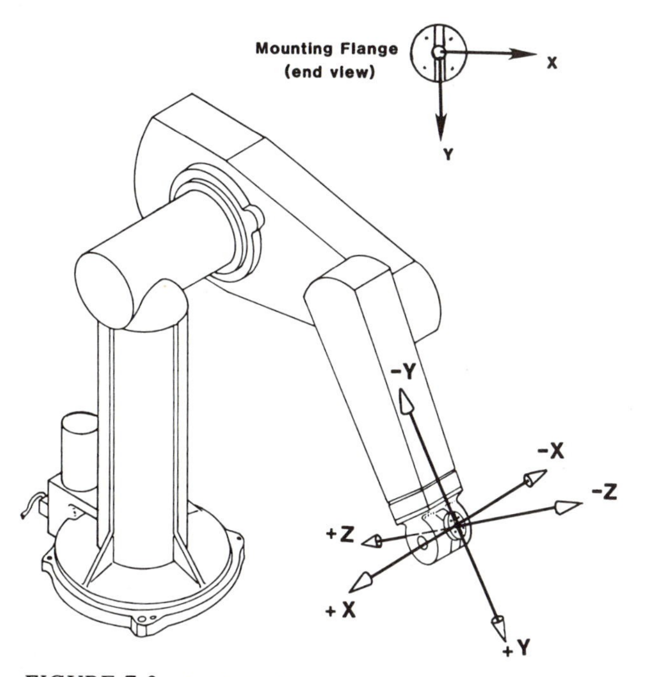

FIGURE 7.3 Tool coordinate system for the Puma 560. (Reprinted by permission of Unimation-Westinghouse, Inc., Danbury, Conn.)

capable of continuous path motion, making the task one of calculating not only point-to-point transformations but axis velocity transformations as well.

To summarize, the first requisite of a powerful, user-oriented robot programming system is a built-in axis transformation capability performed by the robot control computer, preferably in real time. If the computer is not fast enough to perform the necessary transformation computations in real time, two programming passes may be necessary, the first to establish positions in the operator's world coordinates, and the second to reset the motions in the robot's joint coordinates. Real-time transformation capability is especially important to sophisticated, continuous-path robots. Practicality of programming demands that these sophisticated robots have switches on the teach controls (whether on a teach pendant, keyboard, or console) to enable the programmer to switch conveniently from joint mode to world mode to tool mode.

PROGRAMMING METHODS

Reviewing the principal programming methods, the most basic robots are at least in part programmed mechanically—that is, with a wrench. In addition to the mechanical setting of stops, many of the basic robots are taught with teach pendants, and most industrial robots in general have a teach pendant to facilitate programming. Some continuous path robots, especially the paint-spray models and some welding models, are taught by physically moving the robot tool or a facsimile of it through the desired motions. Finally, there is console keyboard programming in a robot computer language; this is the most tedious but most powerful teaching method of all.

There is not much to be said in a college textbook about the setting of mechanical axis limits or the teaching of robots by physically moving the robot tool through the desired continuous path. Therefore, these two programming methods will receive little attention in this chapter. Teach-pendant control deserves some examination, however, and there is even more to be studied in the subject of keyboard programming using robot computer languages.

Teach-Pendant Programming

The programming of even the simple, axis-limit robots can be facilitated by the use of a teach pendant. In the axis-limit models, the function of the teach pendant is twofold: (1) to sequence the operations in the cycle, and (2) to specify the timing between operations. Figure 7.4 displays the teach-pendant control for a popular axis-limit robot.

Perhaps the most noticeable aspect of Figure 7.4 is the availability of two teaching modes: A and B. Teach mode A is for teaching both the sequence of operations *and* their timing. If the operator is capable of thinking quickly enough to program in real time—that is, anticipate every move and key it in at exactly the pace desired for the robot motion—teach mode A is the only mode needed to program the robot. But most of us need time to think through each step while we program, so we need an additional mode for timing the learned sequence: teach mode B. The need for teach mode B becomes obvious to any beginning programmer as soon as he or she has an opportunity to watch the robot stepping through his or her first program, inevitably pausing at various points in the program that the operator had embarrassingly overlooked. Having already taught the *sequence* in teach mode A, the operator can concentrate on *timing* in teach mode B, using the STEP key to cue each successive programmed step in real time. Besides the regular teaching modes, there is a practice mode or experimental mode in which the operator may move the robot through various sequences to observe its operation without

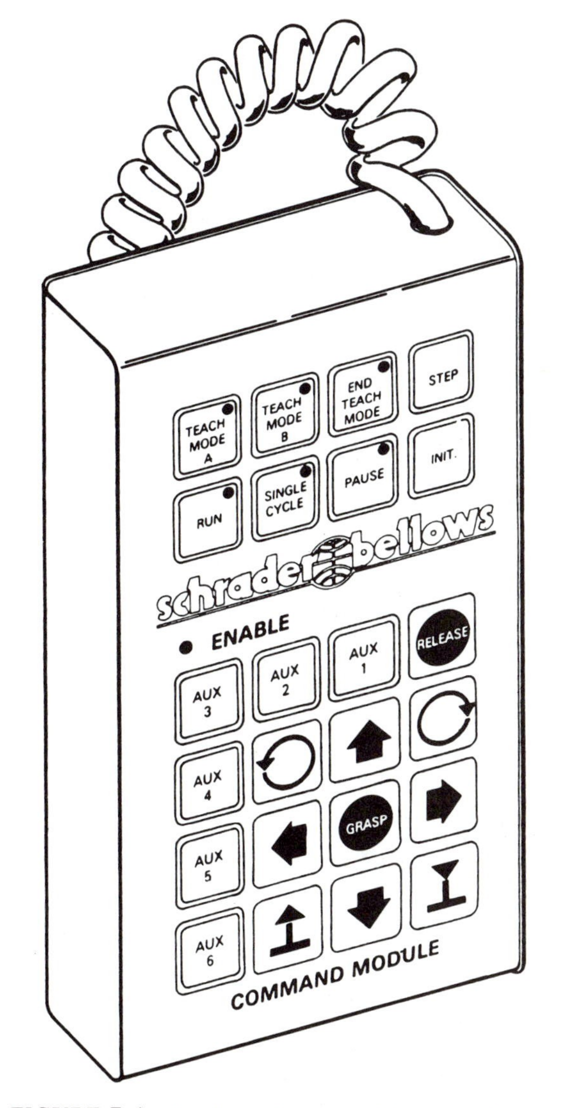

FIGURE 7.4 Teach pendant control for a popular axis-limit robot. (Reprinted by permission of Schrader-Bellows Division of Scovill, Akron, Ohio.)

specifying that the robot memorize the steps that have been taken. This is recognized as the familiar MANUAL RUN mode studied in Chapter 5 for CNC machines, and in Figure 7.4 it is accomplished for the industrial robot whenever the ENABLE indicator is illuminated.

The actual programming is accomplished by the array of keys in the lower right portion of Figure 7.4. These keys usually are labeled with symbols instead of words to make them more user-friendly. Despite attention to the human factors of keyboard design, however, some confusion is likely to occur when first programming with a teach pendant. One of the problems is identifying right versus left, whether the keyboard uses the words *right* and *left* or representative arrows. The problem is in deciding whether the definition reference is the robot's or the human operator's. In the pendant of Figure 7.4, the convention is to take the robot's viewpoint, but this can seem backwards to the operator, who usually faces the robot while programming. The problem can be alleviated if the operator will stand somewhat behind the robot and "look over its shoulder" while programming.

The AUX keys on the teach pendant of Figure 7.4 are for cueing external functions controlled by the robot's control computer. Desired external functions must be prewired to the robot controller's output terminal, and then the operator can cue these functions at the appropriate point in the robot cycle by pressing the AUX key for that function while programming in teach mode A. Any or all AUX functions may be cued in any program as often as desired, and the robot controller will remember the sequence and timing of these functions in the same way it remembers its own physical movements. Thus, the robot controller can be used as a control computer or programmable controller for the entire work station, not just the robot.

Analogous to the AUX functions, it is also possible to have prewired inputs to the robot controller to enable the robot to recognize and respond to external conditions that may affect the robot's operation. In the robot controlled by the teach pendant of Figure 7.4, there are no keys to program the responses to these inputs because the response mode is fixed: Any external input signal is interpreted as a universal signal that the robot should stop. This interpretation is useful for such external inputs as:

1. Limit switches on in-feed material hoppers to indicate when a hopper is empty.
2. Photoelectric beam switches to detect piece-part misalignment or misorientation.
3. Limit switches or photoelectric beam screen patterns to detect the intrusion of a human operator or other obstruction into the robot's path that could cause an accident.
4. Limit switches that detect jamming in the material in-feeders or discharge path.
5. Electrical sensors that monitor the robot arm's mechanical status or axis adjustments.

The reader should be able to envision ways that this list could be extended.

Although it is obviously valuable to be able to stop the robot upon command of an external input, it is also useful for some external inputs to cue other robot actions. The classic example is machine loading and unloading. The robot must wait while the punch press, die-casting machine, injection-molding machine, or other processing machine executes its own cycle. The beginning of that cycle can be triggered either by the robot (via an AUX function) or by a sensor input to the processing machine controller, if it has one. But after the machine cycle is completed, how is the robot to know to reenter the machine and remove the processed piece-part? One solution to this problem is to time the machine cycle during setup and program a pause in the robot cycle sufficient to accommodate the machine cycle. This works, but there are several disadvantages to this approach. To begin with, machine process times may not be exactly uniform. This necessitates that the robot be programmed with a long enough pause to permit the longest of machine cycles to reach completion. Otherwise, a collision may occur between the robot and the machine it feeds! But such a long pause will inevitably waste valuable total cycle time in all but the longest of machine cycles. Also, a sensor-type trigger that signals the successful completion of the machine cycle provides some confidence that there has been not only a sufficient time period to complete the machine cycle, but that the cycle has been completed *successfully* and the point of operation is "all clear" for part removal. Remember "Murphy's Law" and picture the possibility of a press punch sticking in the die upon closure; this is no time for a mindless robot to attempt blindly to extend its gripper into the die area.

As the robot system moves up the scale in its sopl
external cues that would be useful inputs to the robo
stop" cues listed earlier might be converted to a
programmed to handle these exigencies. If the in-f
be preprogrammed to reload the hopper from a ne
might trigger a robot regrasp and repositioning in
If a human operator has intruded into the robo
preprogrammed to pause, issue an audible war
workspace is cleared. This type of sophisticati
programming alone.

In summary, the teach-pendant method of pro
to teach a robot to do a simple task. It is especially effective
pick-and-place operation when both the pick location and the place location are
addition, auxiliary external outputs can be programmed into the robot cycle, but the programmed response to external inputs becomes awkward because each input logically triggers a different sequence of robot actions. Also, with a teach pendant alone, it can be difficult to program geometric paths such as circles or even straight lines. Finally, teach-pendant programming does not accommodate the powerful computer techniques of subroutines and iterative programming.

Keyboard Programming

The preceding section described the convenient features of teach-pendant programming but suggested that the method leaves some things to be desired. As the reader has probably surmised, keyboard programming is the answer to these shortcomings. But before going further, it should be understood that teach-pendant control, or a keyboard facsimile of it, for physical manipulation of the robot axes is fundamental to a practical, efficient *keyboard* programming system.

This is not to say that robots do not exist that are commanded solely by keyboard computer commands. This is the realm of the robotics hobbyist, and several interesting models are available. Figure 7.5 displays a popular hobbyist robot along with its programming keyboard that is mounted right on the robot, making the robot, control computer, and programming self-contained in one compact, mobile unit. On any mobile robot it is awkward to have a separate control computer, which must be connected to the robot by either an umbilical cord or telemetry.

The programming language of the robot pictured in Figure 7.5 is microprocessor hexadecimal code, to be covered further in Chapter 15. Hexadecimal assembly language is a highly efficient code from the standpoint of computer memory, hardware, and software required but is a detailed and tedious language. But if the hobbyist is willing to devote the time and effort, a sophisticated program can be developed with a minimum of capital cost and with the most compact of control systems.

To program an industrial robot easily and quickly to perform a variety of useful tasks, a powerful, easy-to-learn computer language is needed, accompanied by a teach pendant for manipulating the robot during the programming and teaching phase. To see the need for a powerful computer language in addition to the teach pendant, consider the robot task of arranging a unit load or "cube" of blocks, as shown in Figure 7.6. It would be possible with a teach pendant to teach the robot, one block at a time, to pick up each and every block and place it in the precise position desired. The feasibility of this method would depend upon the size of the robot controller's memory because every move for every block would have to be memorized by the robot control computer. For the unit load of blocks, the program easily could consist of thousands of memorized robot

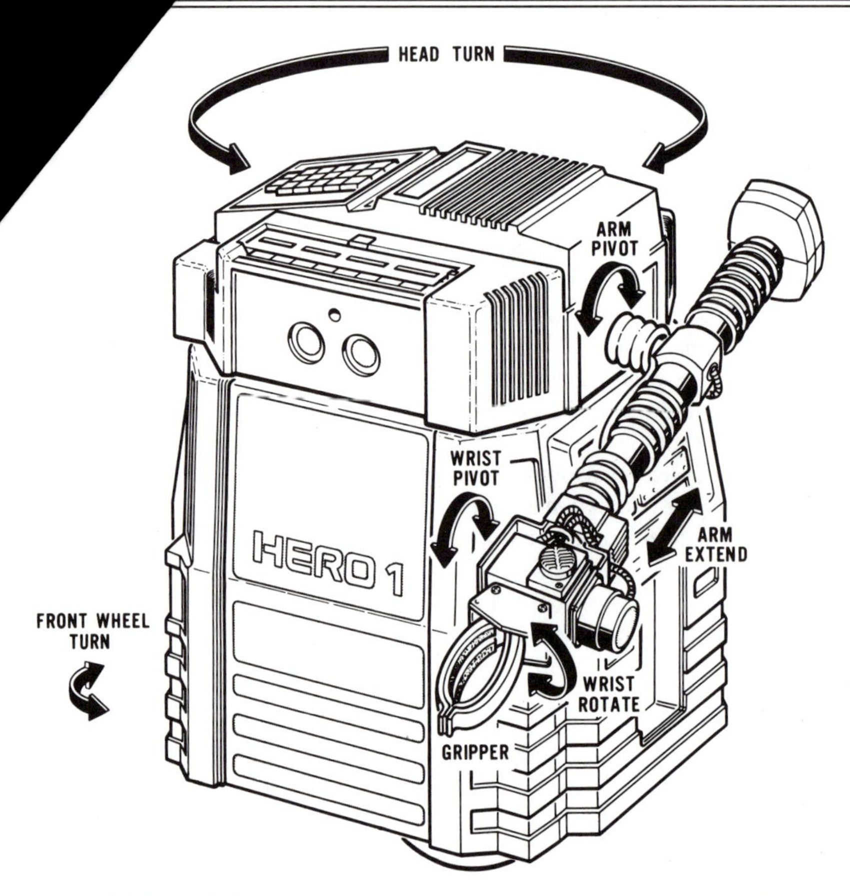

FIGURE 7.5 Self-contained, mobile, hobbyist robot with on-board programming keyboard. (Reprinted by permission of Veritechnology, Electronics Corp., Heathkit Educational Systems, St. Joseph, Mich.)

ments. Then picture the programmer's dismay upon learning that a reconfiguration of the work station requires that the position of the load be shifted slightly! To deal with this problem a convenient feature available in the programming software of many industrial robots is what is referred to as "offset branching" by Diesenroth [ref. 31]. Often a robot is required to do a repetitive operation such as drive a screw, apply a glue pattern, or paint in a prescribed sequence of motions. For a given routine, such as stacking a concrete block, the operations are identical; only the locations are different. With offset branching it is not necessary to teach the robot all of the new subpoint positions for each location in which the routine is desired. Through offset branching the robot maintains consistent relationships among all the relative subpoints during execution of the routine, regardless of the starting point. This feature is also convenient for such repetitive operations as driving screws, applying glue patterns, and painting in a prescribed sequence of motions.

With a combination teach pendant and keyboard computer language, the robot could be taught a few key locations by manipulating the robot arm using the teach pendant in a convenient coordinate system such as world coordinates. Then a keyboard computer program could be constructed that would index upon the learned key locations and perform an iterative process to stack each level in an orderly fashion, taking advantage of known dimensions of the blocks being stacked. The computer even could be programmed to accept different-sized blocks, changing the stacking strategy to optimize the process for

FIGURE 7.6 Unit load of blocks.

each size block. Finally, various subroutines could be constructed to specify robot actions to be taken when confronted with various external contingencies.

Computer languages designed to accomplish these objectives can be easily learned in a day or two. Persons with computer programming background can usually adapt to the robot computer language in a matter of hours. Some of the robot languages are even written as a dialect of some existing, popular, general-purpose computer language such as BASIC. Three widely used robot programming languages, Unimation's VAL® (ref. 112) IBM's AML/2 (ref. 6), and Microbot's ARMBASIC® (ref. 66), have been selected to illustrate the format and usage of robot computer languages in general.

VAL

Many characteristics of the VAL language will be familiar to the reader who has background in a general-purpose computer language such as BASIC. As in BASIC, VAL employs editor software for program preparation on a keyboard. VAL's EDITOR system provides the additional service of advising the programmer when some language (syntax) error is made in constructing a program statement. Of course, errors can still be made in programs without the EDITOR detecting them; such programs follow all of the language rules of VAL but command the robot to do something illogical or at least different from what the user intended.

After the EDIT phase is completed, the VAL program may be EXECUTEd on command from the terminal keyboard; the same is true of BASIC. But at this point, the characteristics of VAL and BASIC begin to diverge. In BASIC, the EDIT phase and the EXECUTE phase are typically separate. But in VAL, it is entirely possible to modify the program using the EDITOR during execution. As soon as the programmer exits the EDIT phase after modifying an existing program that is currently in execution, the robot will commence the modified execution right in the middle of the execution of the previous version of the program. However, it does linger in the old program long enough to complete the program *step* it was executing at the instant the keyboard command to exit EDIT was received.

Another important step in VAL programming is the physical identification of location points using the teach pendant. This teach-pendant phase provides the link between the robot, which must operate in the real physical world, and the VAL program, which is written in terms of symbols such as POINTA and POINTB. This is not to say that the teach-pendant phase has to be used to show the robot all of the locations to which it is programmed to go. Only a few key points have to be shown, such as the location of a machine, the end of a conveyor belt, or the *corner* of a pallet. Thousands of other points to which the robot can be directed can be referenced from these few key points. This is the power of VAL.

Before going on to the program coding, let us examine this physical locating process that uses the teach pendant. Figure 7.7 displays a Unimate Puma teach pendant. In the second row of push buttons can be seen the three mode keys: TOOL, WORLD, and JOINT (keys 9, 10, and 11). These three keys refer to the three sets of coordinate systems explained at the beginning of this chapter in the discussion about the robot's world versus the real world. The three coordinate systems are mutually exclusive states, and the LED indicators just above each push button tell which state the system is in. The six keys arranged vertically at the bottom of the teach pendant (keys 21 to 26) are used to drive the robot physically to the locations desired for recording. The programmer can select whatever coordinate system is most convenient (usually WORLD or TOOL) for moving the robot into position. The robot, of course, lives in the world of JOINT coordinates, but it continually performs coordinate transformations to allow the user any mode selected. When a desired location point is reached, the operator types the command HERE and then types the symbolic name that will be used to refer to that location point in the VAL program. For example,

```
HERE POINTA
```

identifies the current physical position of the robot with the location name POINTA.

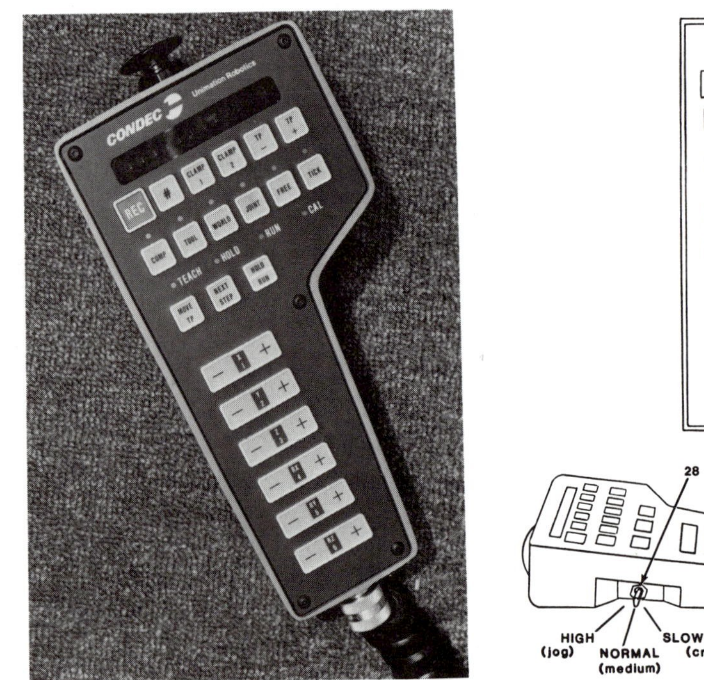

FIGURE 7.7 Teach pendant for Unimate Puma robot. (Reprinted by permission of Unimation-Westinghouse, Inc., Danbury, Conn.)

A location name is any string of letters and numbers and periods except that the first character must be an alphabetic letter and there must be no intervening blanks. Some valid location names are:

```
POINTA
POINT.A
A
MACHINE
PICK.POINT
PLACE.POINT
PALLET.CORNER
PROGRAM.1
PGM.2
```

Now let us turn our attention to the program of VAL instructions, which will refer to the locations identified in the teach-pendant phase.

To begin with, every VAL program must have a *name* so that the program can be called upon to EXECUTE, be stored in a computer file, or be displayed again for review and changes using the EDIT mode. The choice of the name for a program is up to the user, but of course it must be unique so that VAL does not confuse the program with other programs that exist in its memory. In fact, writing a new program by the same name and then storing it in the computer will destroy the older version of the program automatically. For program names, the same rules apply to the selection of alphabetic letters, numbers, and periods as applied in the selection of names for locations, discussed earlier.

The VAL instruction format mimics BASIC, with one instruction per line, and the principal field is the instruction word (or abbreviation) separated from other fields by a space. Variables (arguments) used to specify the instruction further are separated by commas. An optional label precedes the instruction if it is desired later to jump to that instruction from elsewhere in the program. The following example illustrates the VAL command format:

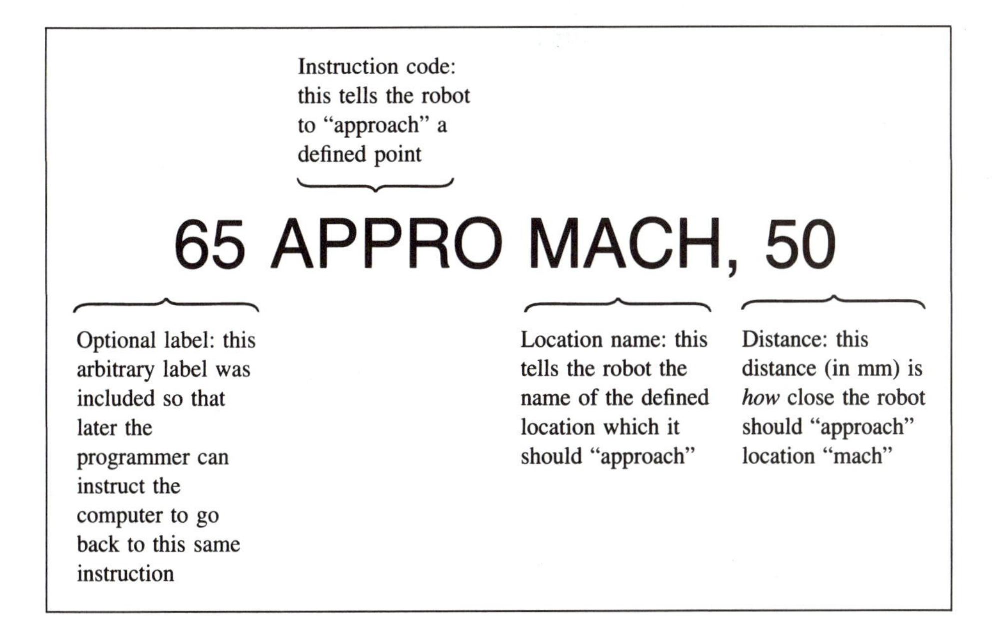

One of the most effective ways of learning a new language—any language—is to begin using it, even without a full understanding of it. We will use that approach here by writing a simple program in VAL. To the right of each step is an explanation of what the program is instructing the robot to do. At the top, we give the program a name suggestive of its purpose, which is simply for the robot to pick up objects from one defined location *A* and deposit them in another location *B*.

program	A.TO.B.	This is the program name.
	OPEN	Open the robot gripper jaws.
	MOVE A	Move the robot gripper to defined position *A*.
	CLOSEI	Close the gripper jaws immediately.
	MOVE B	Move the gripper to defined position *B*.

This is a working VAL program, but it is not a very good one, as we shall see in a moment. But first note the apparent inconsistency in using the command OPEN to open the jaws but the command CLOSEI to close the jaws immediately. Actually, we could have correctly said OPENI to open the jaws immediately. The OPEN instruction without the "I" appended tells the robot to begin opening the hand during the execution of the next instruction. Provided that there is plenty of distance to the point *A*, there will be no harm in allowing the robot to open its hand during the move to *A*. However, it is essential to close the hand upon the object before commencing the MOVEB instruction, or the robot gripper will leave the position without having gripped the object. Hence, the CLOSEI instruction was used to close the hand upon the object immediately—that is, before the MOVEB instruction.

Still, the program A.TO.B. is not a very good one. When commanded to MOVE, the robot takes a direct path using its knowledge of the identified locations and of its own axis movement capability. The path traveled during the MOVE will not necessarily be straight, but it will be quite direct. Picture the robot hand moving directly to a position where an object is to be picked up. Except under unusual circumstances, the object will be struck by the hand instead of being picked up by it. Furthermore, during the move to *B* the hand will move directly also, meaning that if both points *A* and *B* are on the surface of a table, the hand will skid across the table top instead of picking up anything! Finally, the program will end with the robot gripper clenched over location *B*, the point at which it was supposed to deposit the object.

We can certainly improve upon program A.TO.B, so let us write a new one. We could modify the old one using the EDIT program and retaining the name A.TO.B, but let us give the modified version a new name instead, for ease in distinguishing the two versions.

BETTER.A.TO.B.	New program name.
OPEN	
APPRO A,25	Approach location A within 25 mm.
MOVE A	
CLOSEI	
DEPART 25	Back away from A 25 mm.
APPRO B,25	Approach location B within 25 mm.
MOVE B	
OPENI	
DEPART 25	Back away 25 mm.

This program has resolved the problems described earlier for program A.TO.B. The APPRO instruction causes the robot to hover directly over the target position instead of actually going all the way to it. The expression *hover over* is used loosely here because the object to be picked up might be attached to a vertical wall or even be overhead. Figure 7.8 shows various feasible pickup positions that the robot gripper could assume, depending upon the definition of location *A*. Note in Figure 7.8 that no matter what the plane of the pickup surface is, and no matter from what angle the main robot arm is approaching, the gripper jaws themselves have become aligned to approach the object from the correct angle—that is, a direct *z*-axis offset in TOOL coordinates. The reader may want to refer back to Figure 7.3 to recall just what TOOL *z* axis represents. When the programmer uses the APPRO instruction, the VAL software performs the necessary axis transformations *automatically* to assure that the position of the approach will be correctly oriented along the *z* axis of the gripper. This is a convenient instruction, not only for gripping objects to be picked up, but also for drills, automatic screwdrivers, and other tools that might be mounted in the robot hand.

Suppose that the desired approach to the object to be picked is *not* perpendicular to the table, as was shown in Figure 7.8. An effective angle of approach for pickup of some objects might be from the side, not the top. The approach can be from any consistent angle, and the APPRO instruction can still be utilized. VAL's method of interpreting what angle the user wants is to match exactly the tool orientation that was used back in the teach-pendant phase when the position was originally defined with the keyboard command HERE A. In fact, the approach angle for position *A* can be made *different* from the approach angle to position *B* simply by employing a different tool orientation in the teach-pendant phase while positioning *B* with the instruction HERE B. Thereafter

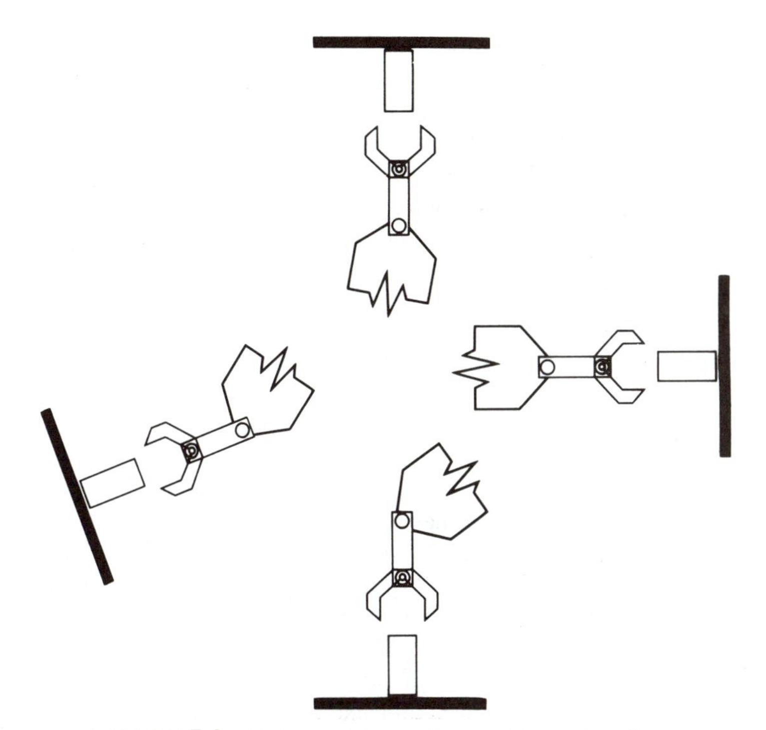

FIGURE 7.8 Various gripper pickup positions depending upon the definition of location *A*.

in all executions of program BETTER.A.TO.B., VAL will maintain the desired gripper orientation for each approach.

Program BETTER.A.TO.B is an improvement, but there are still some potential problems in applying it in a real factory environment. The program's success is dependent upon rather rigid requirements imposed by the robot upon the work station it serves. Every time the robot returns to position *A*, it expects the position to have a workpiece precisely located, ready for the robot to pick up. Then every time it reaches position *B*, the robot expects to have an empty space to accommodate the next workpiece. So either a dependable human operator or a reliable automated system must be installed to work with the robot as it is programmed in program BETTER.A.TO.B or the station will cease to function.

A word about execution of VAL programs is in order here. A VAL program will execute as many times as specified in the EXECUTE command. If no number of times is specified, it will execute only once. In most production settings, one would want a VAL robot program to operate indefinitely until halted or ABORTed. To execute a program indefinitely, the operator merely types any *negative number* for the desired number of executions. Since a negative number of executions makes no sense at all, VAL's convention is to use this as a signal to operate the program continuously. Upon completion of the last command in the program, VAL merely returns control to the first command without skipping a beat.

At any time during execution, the speed of the overall program may be altered by a keyboard command SPEED, abbreviated SP. The integer number typed after the word SPEED is taken to represent a percentage of standard speed, ranging from 0 to 327 percent. This is useful for slowing down the program at first, during checkout, and speeding it up later, even during execution. An ABORT command, or simply A, typed at the keyboard terminal halts the program (at completion of the current step) as soon as the return key causes the typed character A to be entered. Also an emergency power off (EPO) push button for the arm power supply is available for emergency use.

Taking advantage of VAL's indexing capability, computer program loops can be used to cause VAL to index locations slightly at point *A* so that it can unload an entire tray of piece-parts instead of picking up objects at a single point in space each time. Suppose we have a tray of 30 piece-parts arranged as shown in Figure 7.9. Note in the figure that there are five rows and six columns. The tray could be a shipping pallet or a conveyor belt tray, depending upon the size of the piece-parts and the robot to handle them.

To program the robot to unload the tray of 30 parts, we need to know the distances between rows and between columns. If these two distances are constant, our program will be much more efficient. Thus, the importance of order and uniformity in production is made apparent in this example. The actual distances should be known quantities in a planned, orderly, automated system, but if they are not known, the robot teach pendant could be used to obtain a reading. But for our purposes let us assume that the distances are already known to be 30 mm between rows and 40 mm between columns, measured from the center of one piece to the center of the next. Thus, if the robot begins at the front left position and unloads the tray by rows, every time it approaches for another pickup it should shift to the right 40 mm. At the end of each row, the robot should shift back to the left and also to the rear to start a new row. This process should continue until the last piece is removed from the tray. At this point, the robot might type "FINISHED" or "NEW TRAY, PLEASE" on the terminal screen or typewriter and pause a few moments to allow the operator or automated material-handling system to replace the empty tray with another one full of parts. Let us now write a program to accomplish this.

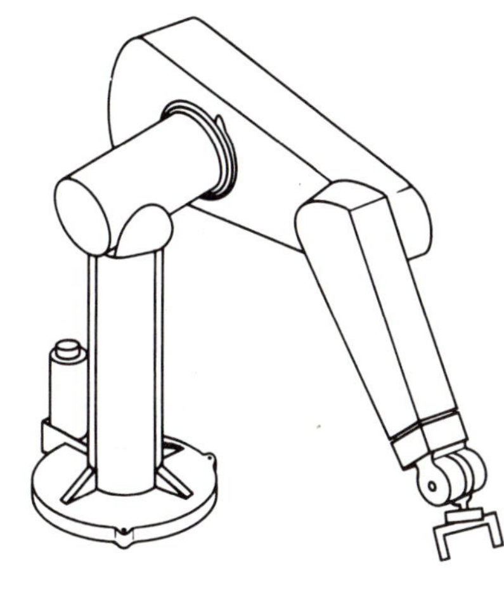

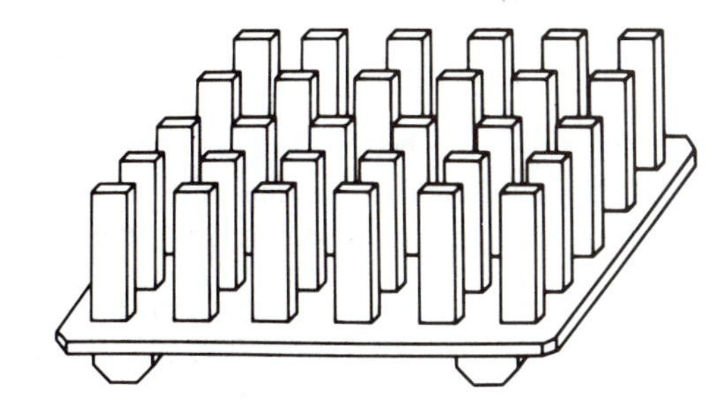

FIGURE 7.9 An entire tray of piece-parts can be loaded or unloaded by the robot using indexed computer loops.

Program	Comment
TRAY.TO.B.	Start with corner position
95 SET A = AA	AA.
SETI NBR = 0	Set an integer variable named NBR = 0. This variable will count pieces up to 30 as they are placed at position *B*.
85 GOSUB TRAY	Go to a subprogram named TRAY, which will handle row and col shifting.
OPEN APPRO A,25 MOVE A CLOSEI DEPART 25 APPRO B,25 MOVE B OPENI DEPART 25	Same as program BETTER.A.TO.B.
SETI NBR = NBR + 1	Add 1 count to the total NBR every time an object is place at *B*.

IF NBR LT 30 THEN 85	If the total NBR is still not 30, jump to command line labeled 85 and go move another piece.
TYPE NEW TRAY, PLEASE	If the total NBR *has* reached 30, the tray is finished; type a message to that effect.
DELAY 10	Wait 10 seconds for a new tray.
GOTO 95	Jump back to command line labeled 95 and start all over again with a new tray.

Since program TRAY.TO.B. had an instruction "GOSUB TRAY," we must now write another program (rather, subroutine) for it to go to. The purpose of this subroutine is to put all in one package the sequence of program steps needed to manipulate the rows and columns and to shift the robot hand over 40 mm as it indexes from *column* to *column* down each *row,* one row at a time. Then, at the end of each row, the robot hand is shifted back to the far left to start a new row. The shift is negative along the world x axis:

$$\text{Shift} = -40 \text{ mm} \times 5 \text{ col} = -200 \text{ mm}$$

We only multiply by 5 instead of 6 because we do not need to shift to pick up the first object in the corner of the tray. In addition to the shift back *along* the row, we must shift to a *new* row that is 30 mm away, this time parallel to the world y axis.

TRAY	Subprogram name
IF NBR EQ 0 THEN 60	If this is the very first piece in the corner of the tray, don't shift at all and jump to command labeled 60.
IF COL EQ 6 THEN 50	If COL is already equal to 6, we have finished an entire row and are ready to shift back to col one on the left and begin a new row. Jump to command labeled 50 to do this.
SHIFT A BY 40	Shift 40 mm along the world x axis. This is a shift right down the row.
SETI COL = COL + 1	Add one to the column count as we index down the row.
RETURN	Go back to the main program to the command right after the GOSUB and pick up an object in the new shifted location.
50 SHIFT A BY −200,30	Shift 40 mm × 5 back along the world x axis and shift 30 mm to a new row.
60 SETI COL = 1	Set an integer named COL to one to signify that we are beginning a new row on the far left column (Col. 1).
RETURN	Go back to the main program to the command right after the GOSUB and pick up an object.

The new program TRAY.TO.B. supplemented by the subroutine TRAY to do the shift manipulations is considerably more sophisticated than our first program A.TO.B. Before going further, let us review some of the features of the new VAL instructions we used in the latest version of our program. The SHIFT instruction causes a simple shift of a known location along the *x, y,* and *z* axes of the world coordinates. Our tray was lined up nicely with the world *x* axis along the rows and the world *y* axis along the columns. Had they been reversed, our two shift instructions would have been

```
SHIFT A BY 0,40

SHIFT A BY 30, -200
```

Since the *x*-axis shift is 0 in the first of the above two statements, we could have omitted it, as follows:

```
SHIFT A BY ,40
```

But note that we still need a comma preceding the 40 to assure that the 40-mm shift will be interpreted correctly as a *y*-axis shift, not an *x*-axis shift. It is permissible to shift *x, y,* and *z* axes all in the same shift, as follows:

```
SHIFT PLACE BY 25,-8,16
```

which signifies:

x axis: shift +25 mm

y axis: shift −8 mm

z axis: shift +16 mm

We were lucky that the tray was lined up with two of the world axes, one way or another, because if it had been at some random angle, a simple SHIFT command would not have been capable of handling the job. VAL can handle this type of situation with a FRAME instruction, but this type of move is beyond the scope of this text.

Another new instruction used in program TRAY.TO.B was the SETI command. This instruction merely allows the programmer to set an integer variable (I stands for *integer*) to a value that will be useful to the logic of the program. The name of the integer is an arbitrary selection of the programmer but must follow the same rules as for program names and location names. As was seen in program TRAY.TO.B., it is possible to add an integer constant (or variable) to an integer variable with a SETI instruction. It is also possible to subtract (−), multiply (*), integer divide (/), and compute an integer remainder (%). The SETI instruction should not be confused with the SET command, which is used to set the world and tool coordinates of a specified location name equal to the world and tool coordinates of another location already stored in memory.

Another instruction used in TRAY.TO.B. was the IF instruction. This instruction permits checking the value of a previously set integer variable and taking some logical action accordingly. In TRAY.TO.B., the IF instruction used the abbreviation LT to represent "is less than." In subroutine TRAY, two IF instructions used the abbreviation EQ to represent "is equal to." All of the permissible abbreviations in IF instructions are:

EQ—is equal to

NE—is not equal to

LT—is less than

GT—is greater than

LE—is less than or equal to

GE—is greater than or equal to

In all IF statements, if the IF clause is true, the program jumps to whatever statement in the program agrees with the label typed after the word THEN. If the jump to another labeled instruction is always to be taken, there is no need to use an IF instruction; use a "GOTO" instruction instead, as was also illustrated in program TRAY.TO.B.

In program TRAY.TO.B., a DELAY instruction caused the robot to wait ten seconds for a new tray of parts to be served by either an operator or an automated system. The DELAY instruction is contrasted with the PAUSE instruction; the latter causes an indefinite delay until the operator types "PROCEED" on the keyboard to signal to the robot to continue. The keyboard entry would be time-consuming and impractical for use every time a new tray is needed. With the use of a DELAY instruction, the ten-second time period must be set long enough to permit ample time for the tray to be set up, but if the delay is too long, production time will be wasted with the DELAY instruction also. Since the time may vary somewhat, it is better to have the delay too long than too short. Still, the solution to the problem is not the best because vital seconds wasted, when the robot is idle but the new tray has already been set up, can spell the difference between economic success or failure for the robot. This suggests the use of an input sensor that *automatically* informs the robot that the new tray is in position. In addition, the robot could send an output signal to trigger the material handling system every time the robot completes a tray and is ready for a new one. Then not only could the arbitrary ten-second delay be eliminated, but the TYPE statement could also be eliminated. The robot may still have to wait for a new tray, but it will be able to proceed *as soon as* the new tray is ready. A new VAL program will now be written, which we shall name BEST because it is as far as we will go with VAL in this introduction.

program BEST

|
.
.
.

first 12 commands are
identical to the first 12
commands in program TRAY.TO.B.

.
.
.
|

IF NBR LT 30 THEN 85	(Same as before.)
SIGNAL 1	Send an output signal on *output* channel number 1 to signify to the automated material handling system that the robot is ready for a new tray.
WAIT 1	Wait for an input signal at *input* channel number 1 that will signify that the new tray is in position and ready for the robot to proceed.

SIGNAL −1	Turn the ouput signal back off as the robot begins work on a new tray.
GO TO 95	Same as in previous program TRAY.TO.B.

An output signal *from* the robot or an input signal *to* the robot can take on only one of two values: high or low. This is to say that such signals are dc logic variables, a subject to be studied further in Chapters 11, 12, and 13. In the VAL language, a SIGNAL instruction can send either a high or low signal on as many available input/output lines as are desired. The arithmetic sign of the I/O line specifies whether high or low is intended. For example,

```
SIGNAL 6,−3,5
```

directs the robot to turn on output channel 6 (high), turn off output channel 3 (low), and turn on output channel 5 (high), all at the same time.

The WAIT instruction can specify only one input channel and causes the robot to hold further execution of the program until the specified input channel achieves the specified state. The specified state is high if a positive channel number is specified and low if a negative number is specified.

For example,

```
WAIT −3
```

directs the robot to wait until input channel 3 is *low* and then proceed with the next instruction. The instruction

```
WAIT 2
```

directs the robot to wait until input channel 2 is *high* and then proceed with the next instruction.

In VAL program BEST, we saw the robot's capability to communicate with the outside world by means of the SIGNAL and WAIT instructions. There are other VAL instructions that permit the robot to react more quickly to external communications. The VAL instruction

```
REACTI −2,DANGER,ALWAYS
```

would cause the robot to immediately interrupt whatever it is doing if input channel 2 becomes low, branching immediately to subroutine DANGER. The purpose might be to halt operations if a sensor detects that a person or object has violated the robot's work envelope, for example. The principal difference between the WAIT instruction and the REACT instruction is as follows. If the programmer specifies the WAIT instruction, the robot will sit idle while it "waits" for the input channel to go high (+) or low (−) as specified by the programmer. But if the REACT instruction is specified, the robot will *continue* executing subsequent VAL instructions until the specified input channel condition is satisfied. The robot remains on alert for the specified input channel condition until the channel condition causes the interruption actually to occur or until the programmer decides to silence the REACT instruction with a corresponding IGNORE instruction. The relationship between REACT and IGNORE is illustrated as follows:

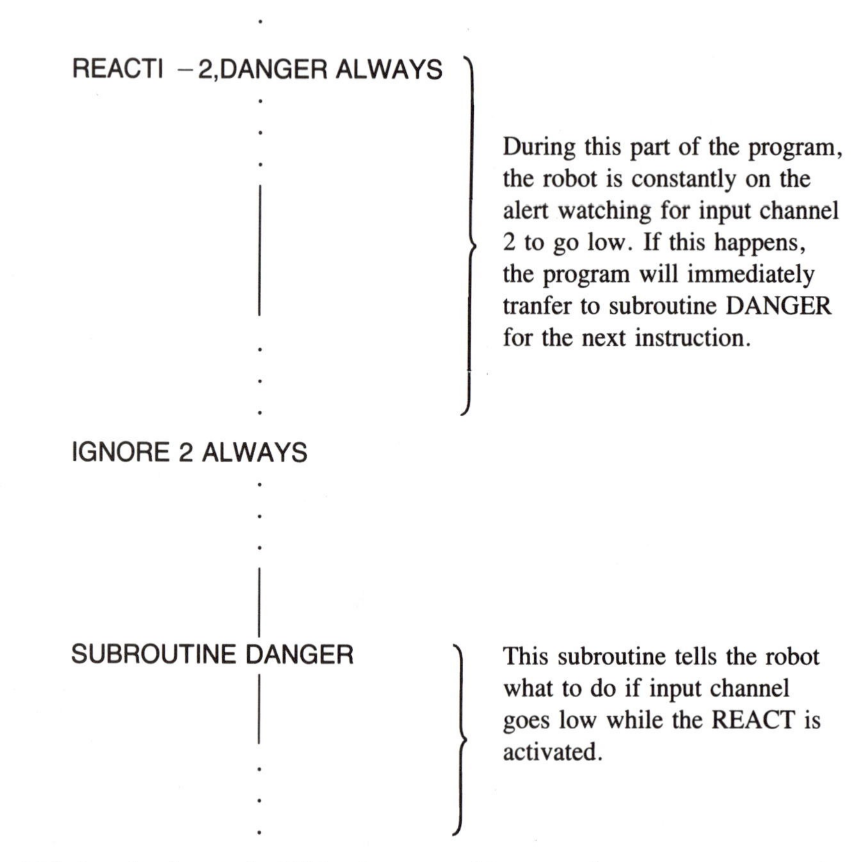

This introduction to the VAL robot control language by no means covers the power of the VAL language software. The idea here is merely to give the reader a glimpse of some of the things that can be achieved by a robot under the control of any powerful robot language in general, and Unimation's VAL language in particular.

AML/2™

Another example programming language for robots is IBM's AML/2. Whereas VAL uses a format familiar to programmers in BASIC, AML/2's format is similar to the popular general-purpose computer language C. One of these similarities is the program modularity found in both AML/2 and C. AML/2 provides for the use of subroutines, analogous to the use of functions in C, to allow the temporary passing of control to a separate segment of code. AML/2 was intended to be a general language for various types of automation equipment, not just for robots. This generality is suggested by the name represented by the initials AML, which stands for "A Manufacturing Language." The language will be briefly described here. A more complete description of the features of the language is available from IBM [ref. 6].

The complete AML/2 package includes an editor, a simulator for safe-testing of programs before execution, and an online debugger. The editor is for keyboard generation and checking of programs for syntax, as was described earlier for VAL. Comments can be interspersed in AML/2 programs to facilitate understanding of the program's construction. There are several formats for comments:

```
/*  ..............comment ..........................  */
/*  ............................comment ................
..........................................(cont) .....
..............(cont) .............  */
##  .............comment ............................
==  ............comment .............................
```

As in the C language, each executable program statement is ended with a semicolon.

AML/2 is used primarily for programming robots of the SCARA and cartesian configurations. Each of these robots has an *x-y-z* world coordinate system. In addition, they may have a rotational axis (*r* axis) along the tool *z* axis. Joint mode is also available in AML/2. The assumed units for the Cartesian axes are as follows:

x axis: millimeters
y axis: millimeters
z axis: millimeters
r axis: degrees (positive is counterclockwise when viewed from above)

The first programming task when writing an AML/2 program is usually to define various robot location points (axis positions) to be used by the program. Location points can be defined using the teach pendant and keyboard together. The procedure will be demonstrated with an example for a location named "POINT": With the teach pendant, position the end-effector in the desired position and axis orientation to be defined as location "POINT." Then, on the keyboard, type:

```
POINT: NEW PT
```

Then type function key

```
ALT F7
```

The *x, y, z,* and *r* coordinates of the robot's current axis positions will automatically be recorded in a location named "POINT." A line in the program representing this action will be printed as

```
POINT: NEW PT(x,y,z,r);
```

where *x, y, z,* and *r* are the appropriate numerical constants corresponding to the axis positions of the robot when the robot is at "POINT."

If the desired axis positions are known or are stored in a program location variable, the robot can be commanded to move to the desired point, using the command

```
PMOVE();
```

The location variable name is placed between the parentheses. Note the similarity between the AML/2 command PMOVE and the VAL command MOVE. The location variable in the PMOVE command can be an arithmetic expression, in which case the axis positions are computed to determine the desired destination point. A practical application of this feature is to add a safe clearance to the z axis on a SCARA robot to permit the robot gripper to be moved to a point hovering just above the desired pickup point. This is what the APPROACH command accomplished in VAL. For example, if a SCARA robot is intended to pick up an object at point A, it could be given the command

```
PMOVE (A + SAFEZ);
```

provided that SAFEZ has been defined to provide a positive z-axis offset.

It is possible to direct a move in only one axis, by using the MOVE command, instead of PMOVE, formatted as in the following example:

```
MOVE(4,90);
```

in which the constant 4 represents the fourth axis (rotation) and the constant 90 represents the absolute position to which axis four is to be rotated (90 degrees). It is often necessary to move in the z axis only, so a special command has been reserved for this purpose:

```
ZMOVE();
```

A variable containing the absolute position of the z axis for the destination point is enclosed in the parentheses. A typical usage would be:

```
ZMOVE(TOPZ);
```

in which the value of the variable TOPZ is zero to represent the limit of upward travel for the z axis.

In Chapter 5 the programming of NC machines included both incremental and absolute modes. AML/2 also permits both incremental and absolute motion commands. To move to a point relative to the current position of the end-effector the following command is used:

```
DPMOVE();
```

The parentheses enclose the name of a set of coordinates to be added to the current position of the end-effector to obtain the set of coordinates for the new point to which the end-effector will move.

The AML/2 gripper commands are

```
GRASP();
```

and

```
RELEASE();
```

which correspond to VAL commands CLOSE and OPEN, respectively. To achieve the approximate equivalent of VAL's OPENI and CLOSEI in AML/2, the following command is used after either of the gripper commands:

```
DELAY();
```

The parentheses enclose a constant that specifies the time in seconds that the robot delays its next action. The user must select the appropriate time to allow the gripper to open or close before commencing the next motion in the program. The AML/2 DELAY command has the same function as the VAL DELAY command.

AML/2 has a standard way of dealing with palletization so that general palletization commands can be used instead of the more fundamental SHIFT commands that were used in the VAL subprogram TRAY described in the previous section. AML/2 assumes that column and row spacing in a pallet are constant, although the row spacing may be different from the column spacing. The spacing is computed by AML/2 by examining the corner points defined in the following palletization command:

```
name: NEW PALLET(LL,LR,UR,PPR,N)
```

where

LL means lower left
LR means lower right
UR means upper right
PPR means parts per row
N means total number of parts on the entire pallet

Let us apply this command to the same example that we used in the VAL programming section using the pallet of 30 parts illustrated in Figure 7.9. From AML/2's perspective the pallet appears as shown in Figure 7.10.

For this example the palletization command would be as follows:

```
TRAY: NEW PALLET(FIRST,RIGHT,LAST,6,30);
```

	25	26	27	28	29	30	UR
	19	20	21	22	23	24	
	13	14	15	16	17	18	
	7	8	9	10	11	12	
LL	1	2	3	4	5	6	LR

FIGURE 7.10 AML/2 robot perspective of 30-item pallet load as in VAL program TRAY (see Figure 7.9).

Where FIRST, RIGHT, and LAST are predefined locations representing the respective points in the palletization command definitions LL, LR, and UR, respectively. Note that AML/2 always looks at a pallet as beginning in the lower left and proceeding to the right. Rows are horizontal and columns are vertical. The last item on the pallet is in the upper right corner. The plane and orientation of the pallet is specified in the "NEW PALLET" command as illustrated above. Therefore, "horizontal" and "vertical" and "left" and "right" are meaningful only with respect to the plane and orientation defined. In the real world, therefore, it would be possible for a given pallet's rows to be vertical and columns to be horizontal.

During palletization operations, AML/2 keeps track of which position on the pallet the robot is acting upon; it does this in a register called the "current part indicator." The following AML/2 commands, illustrated for the pallet named "TRAY," permit operations relevant to the current part indicator:

Command	Description
SETPART(TRAY,NBR);	Sets the current part indicator for the pallet named TRAY to the integer value of the variable NBR.
GETPART(TRAY);	Moves the robot end-effector to the *x-y* position for the current part (part number in the current part indicator) in pallet TRAY. The *z* and *r* axes are unaffected.
NEXTPART(TRAY);	Index the current part indicator to the next part on pallet TRAY.
PREVPART(TRAY);	Sets the current part indicator back one.
LOC=PALLETPT (TRAY,NBR);	Sets the value of variable LOC to the *x-y-z-r* location coordinates of the part indicated by the integer value NBR on pallet TRAY. If NBR is omitted it is assumed to be the value of the current part indicator.
NBR = TESTP(TRAY);	Sets the value of variable NBR to the current part number.
TESTP(TRAY,NBR,NEXT);	Compare the current part number for pallet TRAY with the integer value of the variable NBR. If they are equal, the program will branch to location NEXT; otherwise the program continues with the next sequential instruction.
PAUSE();	Causes the program to stop and wait for a cue from the operator console or from a host control computer. Useful for synchronizing the execution of the program with the process.
LOCA = WHERE(N);	Sets the variable LOCA equal to the position of axis *N* for the current position of the end-effector. *N* may be omitted, and if it is, AML/2 will return all four axis positions to PT variable LOCA. Note the functional similarity between this command and VAL's command HERE.

DISPLAY(); — Used for printing messages on the screen for operator intervention or other purposes. A typical message is an alphabetic character string, in which case the string is enclosed in single quotation marks (') within the parentheses.

Example: DISPLAY ('NEW TRAY, PLEASE'); Note the similarity to VAL's command TYPE.

BRANCH(); — Used to transfer control of the program to a command not next-in-sequence. A command label is named between the parentheses.

Example: BRANCH(START); This command is similar to VAL's GOTO.

This text has attempted to provide an introduction to the format and style of AML/2 and to enable the reader to write rudimentary programs. The language is much more powerful than is suggested here. For instance, many of the MOVE commands can be used for moving to a set of points in sequence, not just to a single point. Built-in interpolation facility permits curvilinear motion, much as is achieved in NC toolpath control. In addition, there are convenience commands for constructing iterative sequences such as are found in the C language. The C programmer will recognize the following such commands:

WHILE. . . . . . . .DO

and

IF. . . . . .THEN. . . . . .ELSE

Now, having completed a brief introduction to AML/2 commands and format, we will write a short program that corresponds in purpose to the VAL program TRAY.TO.B shown in the previous section. Having essentially the same program in both languages for comparison will enable the reader to better understand both VAL and AML/2. Point *B* is the destination point, as was true for VAL program TRAY.TO.B. For this program we shall assume that point *B* is somewhere on the same table (at the same vertical height) as pallet TRAY. As was done in the VAL program we shall assume that ½ in. (25 mm) is a safe approach distance for points on pallet TRAY and for point *B*.

```
/* AML/2 PROGRAM TRAY TO B
DEFINING LOCATIONS WITH TEACH PENDANT:                          */
        B:  NEW PT          ## POINT B
        LL: NEW PT          ## PALLET POINTS
        LR: NEW PT          ## PALLET POINTS
        UR: NEW PT          ## PALLET POINTS

        LASTPT();
/*
SAFE APPROACH DEFINITION                                        */
        SAFEZ: NEW⟨0,0,25,0⟩;
```

```
/*
PALLET DEFINITION:                                                          */

        TRAY: NEW PALLET(LL,LR,UR,6,30);

/*
MAIN PROGRAM:                                                               */

        MAIN: SUBR();
        PALLET(TRAY);                         ## Set up to pick
STRT:   SETPART(TRAY,1);                      ##   first part
        PMOVE(PALLETPT(TRAY) + SAFEZ);        ## Hover over Pt 1
PICK:   PMOVE (PALLETPT(TRAY));               ## Pick up each
        GRASP ();                             ##   part in turn
        DELAY (1);                            ## Allow 1 sec to close
        PMOVE(PALLETPT(TRAY) + SAFEZ);        ## Move up, then over to
        PMOVE(B + SAFEZ);                     ##   hover over Pt B
        PMOVE(B);                             ## Deposit each
        RELEASE();                            ##   part at Pt B
        DELAY(1);                             ## Allow 1 sec to open
        PMOVE(B + SAFEZ);                     ## Move safe distance
        TESTP(TRAY,30,DONE);                  ## Pallet all finished?
        NEXTPART(TRAY);                       ## No, index to next part
        GETPART(TRAY);                        ## Go to next part position
        BRANCH(PICK);                         ##   and repeat sequence
DONE:   DISPLAY('NEW TRAY, PLEASE');          ## Pallet finished
        DELAY(10);                            ## Wait for next pallet
        BRANCH(STRT);                         ## Start over on new pallet
        END;
```

ARMBASIC

Many college and university laboratories are equipped with Microbot Minimover and Teachmover robots, so the ARMBASIC [ref. 60] language may be of classroom benefit to students. But the real objective here is to equip the student with a general understanding of robot language capabilities and limitations as preparation for entry into the *industrial* arena.

The ARMBASIC language is used for open-loop control of robots whose axes are driven by stepper motors. The ARMBASIC outputs to the motors are a series of pulses, as was explained in Chapters 2 and 6. Since the control is open-loop, the control computer does not know the position of the robot arm. However, the control computer is capable of keeping track in memory of the algebraic sum of pulses it has issued to each of the robot joint motors. The control computer keeps these motor pulse counts in a separate register for each motor that drives a robot axis. These pulse count registers may be called *position registers,* but whether the robot actually has moved in accordance with these registers depends upon whether an obstruction has been encountered, the speed of the move has been too fast, or any other cause has made the stepper motors slip. The programmer can rezero the position registers at any time convenient in the program. Such rezeroing can be considered a calibration of the robot if the robot is moved to a known fixed location just prior to the rezeroing.

The approach in ARMBASIC is to structure the robot commands as subroutines of the BASIC language instead of constructing a separate but similar complete language. This means that the ARMBASIC programmer must also know the BASIC language. But what this also means is that users can take advantage of all of the standard features of the BASIC language presently available in their microcomputers for other purposes and merely access specific ARMBASIC routines whenever needed. In addition, all of the EDITOR software available from the user's standard BASIC system continues to be available while programming in ARMBASIC.

ARMBASIC commands or routines are surprisingly few and simple. There are only six commands as follows:

@RESET Initializes the "position registers" to zero—that is, it establishes a "home" reference point for all the stepper motor counts.

@STEP Moves the robot by delivering pulses to stepper motors as specified.
Example: @STEP SP,A,B,C,D,E,F
This command directs the robot to move at a speed specified in the variable *SP*. Variable *A* specifies the stepper motor pulse count to be advanced by the first robot axis (the base), variable *B* the second axis (shoulder), variable *C* the elbow, and so on. Variable *F* is for axis six, the robot gripper. The motion is "incremental," meaning that the steps moved by each axis are relative to axis position just prior to the @STEP command.

@SET Places the robot in a manual mode under control of the keyboard (see Figure 7.11). This is in essence teach mode control in joint mode.

@READ Records the algebraic sum of pulses delivered to each stepper motor since the previous @RESET.
Example: @READ A1,A2,A3,A4,A5,A6
This command records the position registers (pulse counts) of the robot axis stepper motors relative to the most recent @RESET command when those position registers were rezeroed. In this example, the variable A1 records the pulse count for the base, A2 the shoulder, etc., . . . , A6 the gripper. These variables can later be manipulated and pulse counts recalculated to return the robot to this position by means of an @STEP command.

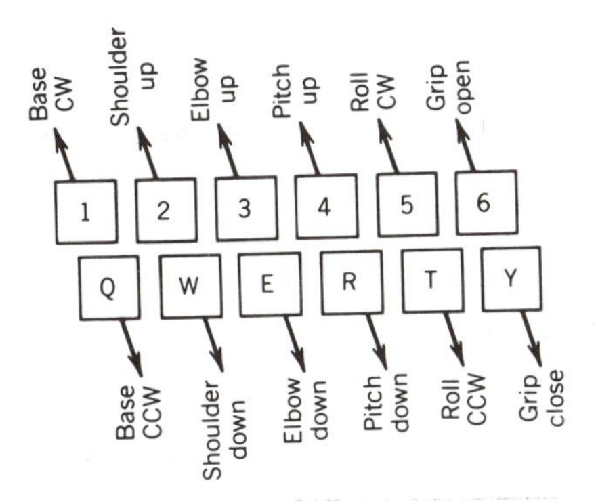

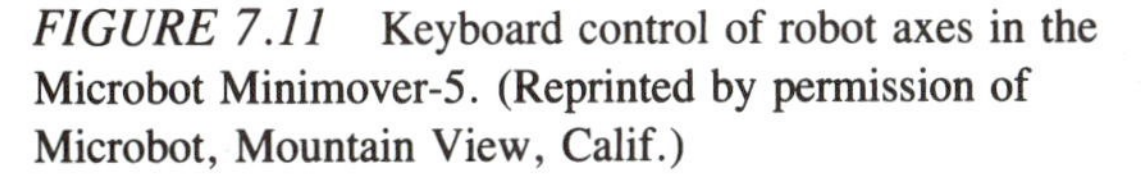

FIGURE 7.11 Keyboard control of robot axes in the Microbot Minimover-5. (Reprinted by permission of Microbot, Mountain View, Calif.)

@CLOSE	Closes the gripper until a sensor detects that the hand has encountered a fixed resistance.
@ARM	Selects the computer port number to be used. This is useful in setup, not programming.

Note that each ARMBASIC command is preceded by an @ sign, which distinguishes the ARMBASIC commands from ordinary BASIC commands. Also note that the language has an @CLOSE command but has no corresponding OPEN command. It is possible to open (or close) the hand merely by sending a series of pulses to the robot motor that opens the gripper. This can be done with either the @STEP or the @SET command. The difference with the @CLOSE command is that driving of the gripper motor will continue until it is *switched* off by a sensor on the robot (a limit switch on the Microbot Minimover 5).

It might seem that ARMBASIC programming would require the programmer to make meticulous calculations involving trigonometric functions of the robot dimensions and the angular displacement rate of each stepper motor pulse, but this is not the case. Because of the manual mode capability provided by the @SET command, the robot can be moved manually in joint mode to the position desired and then an @READ executed. This saves the absolute pulse counts (relative to the last @RESET) for future reference. The next time the programmer wants the robot to go to that same place, an @STEP command is used to specify pulse counts for each axis equal to the *difference* between the saved pulse counts at the source location and the destination location. Figure 7.12 explains the procedure for using these saved location points to make programmed moves. The trajectory of the robot during the programmed moves is smooth because ARMBASIC pulses each stepper motor at a rate that produces a smooth direct motion from point A to point B. The path will not normally be a straight line but will not be as curved as shown in Figure 7.12, which was exaggerated for expository purposes.

The principal lesson to be learned from Figure 7.12 is that the programmer did not have to calculate or even be aware of the actual values for the pulse counts in variables A1 through A6 and B1 through B6. The programmer, by naming the variables, merely assigns computer memory to hold the stepper pulse data. Then, during the manual teach phase, ARMBASIC actually assigns numbers to these variables in computer memory. Finally, remember that this entire procedure is an open-loop system that is dependent upon no slippage in the stepper motor action due to encountering an obstruction or from moving too fast. If slippage has occurred, the computer will have no way of knowing that the robot is not actually where the location data in computer memory says it is.

All of this is best understood by an example. Let us now program a robot in ARMBASIC to accomplish the same pick-and-place operation for which we wrote the VAL program BETTER.A.TO.B. First we need to define variables that will identify the stepper count for all of our robots axes for every point we want to access. By stepper count, we mean the total number of motor steps required for that axis to reach the given point from a home position that has been established with an @RESET command.

Let Ai = stepper count for axis i for the pickup point

Bi = stepper count for axis i for the place point

Ci = stepper count for axis i for the *approach* to point A

Di = stepper count for axis i for the *approach* to point B

Suppose we have a six-axis robot in which the sixth axis is the gripper. This means we have 4 points $\times$ 5 axes or 20 variables for which we must determine the stepper count

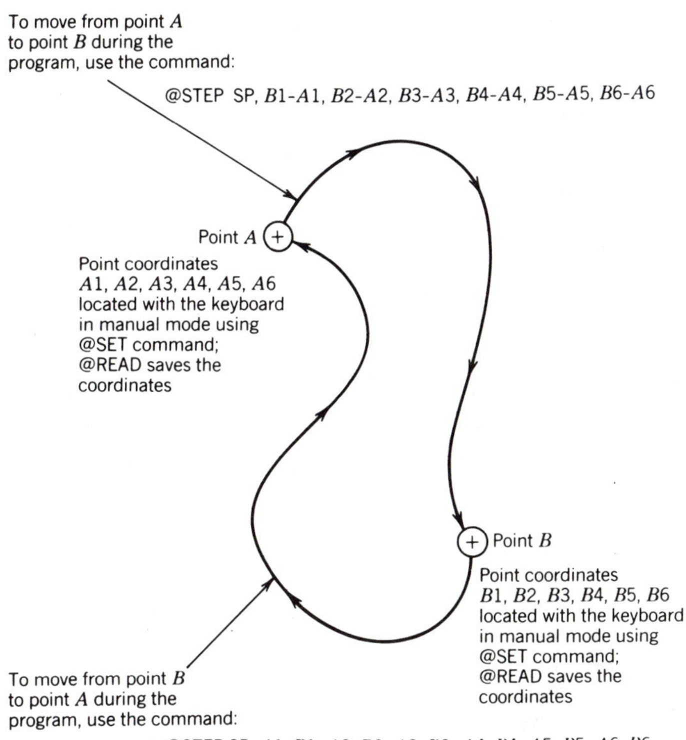

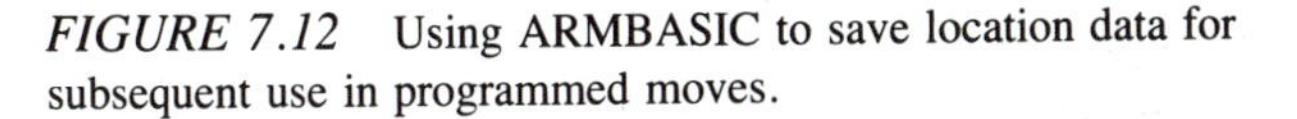

FIGURE 7.12 Using ARMBASIC to save location data for subsequent use in programmed moves.

from the home position. In addition, we must establish the axis 6 (gripper) stepper count difference between an open gripper and a gripper closed upon the object we want to pick up.

The most practical way to enter these 21 variables into the control computer is to move the robot physically to the desired positions using @SET commands recording the stepper counts using an @READ command after each @SET command. This can be recognized as the manual teach phase, analogous to the HERE A, HERE B portion of VAL program BETTER.A.TO.B. A good strategy in accomplishing this is to select one of the points *A*, *B*, *C*, or *D* as the home position. This will reduce the number of variables we need to save by five, from 21 down to 16. Further, it will avoid some arithmetic in the @STEP commands during execution of the program. The approach point *C* with gripper jaws open at least as wide as necessary to pick up the object is selected as a convenient reference for home. It will be the starting and ending point for each execution of our pick-and-place program. In ARMBASIC,

100 @SET	Puts robot in manual mode so operator can move it to point *C* (the approach point for pickup).

Operator uses the keyboard (in joint mode) to position the robot at point *C* with gripper jaws open wide enough (with some margin) to grasp the object to be picked. Operator types 0 to end manual mode.

110 @RESET	Makes point *C* the home position.
120 @SET	Manual mode

Operator uses the keyboard to position the robot at another point to be taught—this time point *A*, then types 0.

130 @CLOSE	Closes the gripper upon the object to be picked up.
140 @READ A1,A2,A3,A4,A5,G	Save the stepper count referenced back to home (point *C*). The stepper count to close the gripper upon the object is in *G*.
150 @SET	Manual mode.

Operator uses the keyboard to position the robot at point *D*, then types 0.

160 @READ D1,D2,D3,D4,D5	Save the stepper count referenced back to *home* (*not* point *A*).
170 @SET	Manual mode.

Operator moves the robot to point *B*, then types 0.

180 @READ B1,B2,B3,B4,B5	Save the stepper count referenced back to home.

The robot has now learned all of its points including point *C*, which was established as the home position. The reader may be wondering why the @READ commands at statements 140, 160, and 180 specified only five instead of six robot axes. But remember that axis 6 (the gripper) maintains the same degree of closure in most of the moves. Thus, it was necessary to save only 16 variables: 5 axes times 3 locations plus one value G to represent the steps for gripper closure. We are now ready for the execution phase, but the first thing the robot must do is release the sample object it picked up during the teach phase and return home. This is accomplished by the following:

190 @STEP 50,0,0,0,0,0,-G	Open the gripper; speed 50.
200 @STEP 50,D1-B1,D2-B2,D3-B3, D4-B4,D5-B5	Back out from position *B* to *D;* speed 50.
210 @STEP 50,-D1,-D2,-D3,-D4,-D5	Go home; speed 50.

The robot is now ready to execute the pick-and-place routine, which is effected by the following ARMBASIC commands:

220 @STEP 50,A1,A2,A3,A4,A5	Move to position *A*.
230 @CLOSE	Close gripper to pick up object.
240 @STEP 50,-A1,-A2,-A3,-A4,-A5	Back out to home point *C*.
250 @STEP 50,D1,D2,D3,D4,D5	Advance to point *D*.
260 @STEP 50,B1-D1,B2-D2,B3-D3, B4-D4,B5-D5	Move object to point *B*.
270 @STEP 50,0,0,0,0,0,-G	Release object.
280 @STEP 50,D1-B1,D2-B2, D3-B3,D4-B4,D5-B5	Back away.
290 @STEP 50,-D1,-D2,-D3,-D4,-D5	Go home.

```
300 GOTO 220        Repeat sequence indefinitely.
310 END             End program.
```

The reader may have noticed that statements 270, 280, and 290 are identical to statements 190, 200, and 210. Indeed, statement 190 could be changed to GOTO 270, and statements 200 and 210 could be eliminated from the program. Either way, as soon as the operator signals that the robot has been taught the last point (point *B*) by typing a zero in statement 170, the robot will go immediately into execution.

What has been shown is merely the essentials of an ARMBASIC program to move through a simple pick-and-place routine similar to BETTER.A.TO.B. shown earlier. To make the program more practical, it would be necessary to:

1. Program prompts to be displayed on the screen directing the operator to locate points in the teach phase and to enter 0 when a new point is located.
2. Program a pause between the manual teach and execution phases, so that the robot will wait until the operator is ready to begin execution.
3. Provide a method of user control over the number of executions.

All of these features and other conveniences could be added using the conventional BASIC programming language.

For instance, a programmed pause could be accomplished by a BASIC subroutine such as:

```
350 FOR N = 1 TO 2000
360 NEXT
370 RETURN
```

The subroutine does nothing but index the variable *N* to pass time. The actual amount of time passed would depend upon the speed of the particular control computer, the limit of the loop index (in this case, 2000), and perhaps the length of the computer program in which the routine is used. In practice, the programmer usually experiments until the most desirable loop index limit is found for a particular application. For sake of discussion and for use in exercises at the end of this chapter, we shall assume that the sample routine illustrated above causes a one-second programmed pause.

To program an external device such as a robot tool to turn on or off, the programmer might use a BASIC statement such as:

```
OUT 250, NN   (to turn the tool on)
```

or

```
OUT 250, FF   (to turn the tool off)
```

where variables *NN* and *FF* represent the appropriate numerical (binary) value to signify on and off, respectively.

SUMMARY

It is the programmability of robots that gives them their great versatility and stimulates the creativity of automation engineers to use them in industrial applications. The programming of industrial robots can range from a simple manipulation of the robot tool through a desired sequence of motions that the robot will later mimic. At the other end of the spectrum is the intricate coding of a compact, on-board microprocessor using the hexadecimal code of the processor.

One of the principal problems in robot programming is to make the translation from the robot's frame of reference to the operator's frame of reference, and vice versa. People think of space as right versus left, forward versus back, and up versus down, but robots think of space in terms of the types of movements of their axes (joints) required to move to a given point in space. Another frame of reference is the tool's. For the most practical and effective teaching and programming of an industrial robot, the programming system must be able to accommodate transformations among the three coordinate systems: joint, world, and tool.

The most powerful and versatile of programming methods use both a teach pendant and a keyboard language. The proprietary languages VAL, of Unimation, Inc., AML/2, of IBM, and ARMBASIC, of Microbot, Inc., were used in this chapter to illustrate the relationship between the teach mode and the keyboard programming mode in those robot systems that use both modes of programming.

EXERCISES AND STUDY QUESTIONS

7.1. Write a VAL program for a robot that will unload a shipping pallet loaded with bronze castings, as shown in Figure 7.13. The castings are spaced in three uniform rows of four castings each. The clearance between the edges of the castings as they sit on the pallet is two inches.

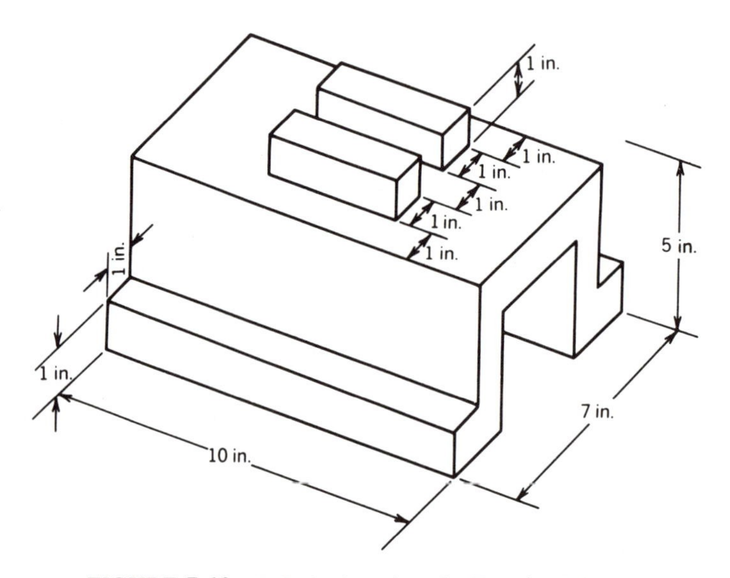

FIGURE 7.13 Palletized castings for Exercise 7.1.

7.2. Would a teach-pendant phase be required to implement the VAL program written for Exercise 7.1? Why or why not?

7.3. Write an ARMBASIC program for a robot to accomplish the task described in Exercise 7.1.

7.4. What points would need to be taught to the robot in a teach-pendant phase to accomplish the ARMBASIC program for Exercise 7.3?

7.5. A robot is to be programmed to apply a glue pattern as shown in Figure 7.14. The glue applicator tool held by the robot is turned on by output signal 1. Extra glue is applied at critical spots (see five spots in Figure 7.14) by pausing half a second at each spot while the glue applicator tool remains on. Write a VAL program to accomplish this robot task.

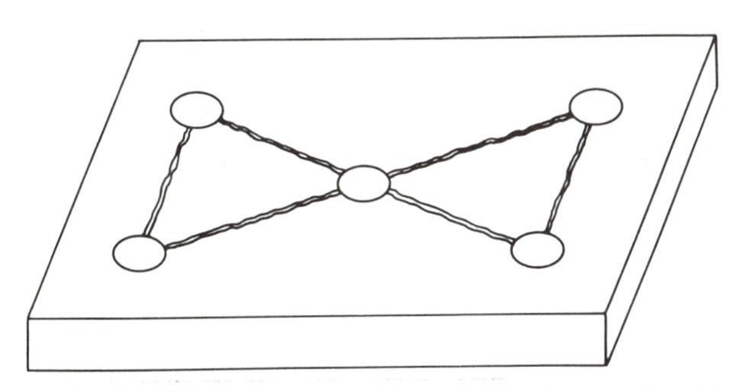

FIGURE 7.14 Glue pattern for application by an industrial robot.

7.6. Write an ARMBASIC routine for the robotic glue application process described in Exercise 7.5. You may assume that control computer output #250 is connected to the glue applicator tool and is turned on by variable *NN* and turned off by variable *FF*.

7.7. Graphite disks similar to hockey pucks are unloaded from a conveyor and fed to a press for processing. Suppose the following conditions for input and output signals are defined for a robot system to handle the graphite disks.

Input	*Output*	*Definition*
1		Graphite disk in position for pickup from conveyor
4		Press ram closed or still in motion
5		Part is present in gripper jaws
	1	Alarm light (no disk in position)
	2	Audible alarm (horn)

Write a VAL program to make use of the I/O signals during the operation. (There are, of course, many correct solutions to this exercise.) Program the horn to blast intermittently at one-second intervals in the event that there is no operator response to the alarm light.

7.8. An industrial robot loads and unloads a forging press in which the parts handled are extremely hot. To prevent the robot gripper from getting too hot, the robot dips its hands in a cooling bath every cycle. Write a VAL program to pick up hot forging blanks, feed the forging press, signal the press ram to actuate, cool the hand, unload the forging press, drop the hot forging in a pallet-hopper, and continue to the next cycle.

7.9. The forging press machine cycle time in Exercise 7.8 is 1.5 sec. The robot requires 3.8 sec to retreat to its cooling bath and return to the ready position to unload the press. To increase the productivity of the overall system, a sensor is installed on the robot gripper to signal input 1 (ON) whenever the gripper exceeds a specified temperature. Program the robot to cool its hand only when necessary.

7.10. In Exercise 7.9, the hot forging, while held in the robot's hand, trips the heat sensor even though the gripper itself may not be too hot. Explain how the robot program can deal with this problem.

7.11. Just for fun, a voice synthesizer is installed on output port 4 on the robot for Exercise 7.9. The synthesizer is programmed to say "Ah-h-h!" upon command of the output port. Insert the appropriate VAL command(s) to cause the robot to say "Ah-h-h!" while it cools its hand.

7.12. A robot has a set of ten defined locations for insertion of electronic components in circuit boards. Another set of ten locations identifies the pickup points for the robot to obtain the components for insertion. Write a VAL program for robot insertion of all ten components upon an initial input signal from the material handling system. At the completion of the tenth insertion, the robot outputs a signal to the material handling system.

7.13. Write an ARMBASIC program for a robot to accomplish the electronic circuit board component insertion task described in Exercise 7.12.

7.14. In AML/2 program TRAY TO B the robot picks up pallet parts one at a time and deposits each at point B. The robot begins in the lower left corner and finishes in the upper right corner of the pallet. Rewrite the program to begin by picking up the part in the upper right corner of the pallet, proceed backwards, and end by picking up the part in the lower left corner of the pallet.

7.15. A stack (Stack A) of uniform objects, each of height 20 cm, is to be restacked one at a time so that in the new stack (Stack B) the objects will be in reverse order. Write an AML/2 program to accomplish this restacking task. The program should define points ABOVEA and ABOVEB, located as follows:

		Robot Axes			
		x	*y*	*z*	*r*
Points:	ABOVEA	−195	400	0	0
	ABOVEB	0	400	0	0

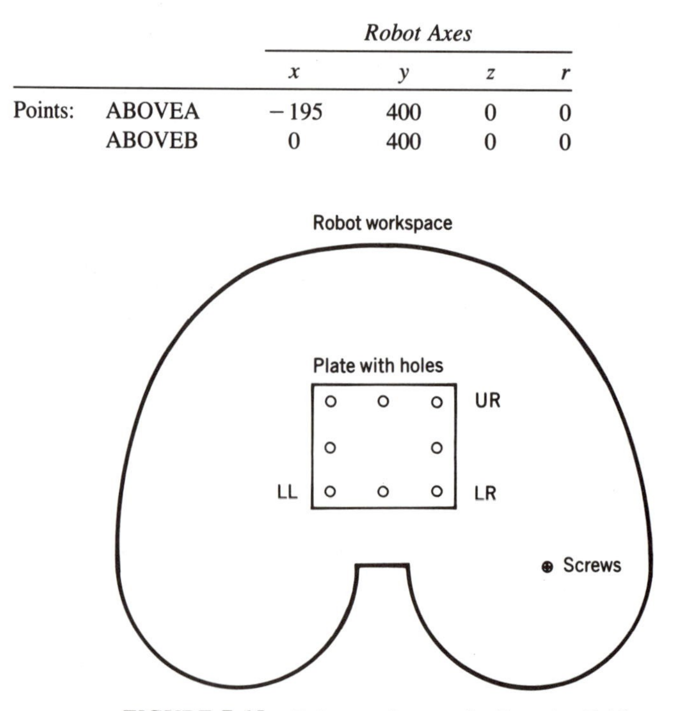

FIGURE 7.15 Robot workspace for Exercise 7.16.

Points ABOVEA and ABOVEB are located 20 mm above the pickup point for the top object in the respective stack when the stack is full. (A stack is full when there are five objects in the stack.)

7.16. This exercise is to use the palletization commands of AML/2 to assemble screws to a square plate with eight screw-holes located uniformly around its perimeter as shown in Figure 7.15. The screws are located in a dispenser at a predefined pickup point named SCREWS. The robot is equipped with a magnetic gripper/screwdriver tool so that it is able to pick up each screw at point SCREWS (without any gripper action), take it to the respective hole in the plate, and apply two clockwise turns to the screw to complete the assembly for that screw.

CHAPTER 8

MACHINE VISION SYSTEMS

This chapter makes its first appearance in this, the revised edition of the book. In the original edition machine vision was relegated to a small subtopic in the chapter on robotics. The author had speculated that practical machine vision systems would not be commonplace for many years. This forecast turned out to be shortsighted, for indeed the growth rate of robotic machine vision systems applications has apparently surpassed that of robotic manipulators. The error in the author's forecast was due to a common misconception of what the term "machine vision" really represents.

Machine vision is often thought of as the capability to sense, store, and reconstruct a graphic image that matches the original as closely as possible. The objective is actually more specific than that. Machine vision systems usually have a specific assignment, such as checking for proper part orientation, identifying parts, searching for specific defects, or checking alignment for assembly. A procedure can be devised to efficiently and effectively accomplish the specific goals of the machine vision requirement, usually without requiring a full implementation of the graphics-reproduction capability of the system. These procedures take advantage of various features of the image and use algorithms to extract the essential data required from the picture.

The focus of this book is upon applications, and the purpose of this chapter is to impart insights into the capabilities of the equipment and software to be adapted to a given problem in manufacturing automation. With this objective in mind we will examine only what we need to know with respect to the technology of the equipment and concentrate upon the ways automation engineers are finding to employ machine vision systems for practical tasks.

IMAGE ACQUISITION

The business of a machine vision system can be divided into two somewhat sequential steps: (1) the gathering of the image into a machine-readable form, and (2) the manipulation and analysis of that image to interpret it and accomplish the goals of the user. We will begin with the first step in the sequence, image acquisition.

Image Scanning

The first step is to gather input data from the original image or object. A camera lens arrays an image on a glass faceplate that consists of a very fine grid of photosensitive material. Each tiny spot on the grid is electrically charged in proportion to the light that impinges upon it. These tiny spots are the source of the most basic picture elements, called pixels. The electrical charge for each pixel is gathered by the system, usually with a scanning electron gun. A typical system uses a raster scan, similar to that employed by a black-and-white television camera that gathers pixels to reproduce an entire picture on a CRT screen, as shown in Figure 8.1.

The quality of the image reproduced is dependent upon the ability of the scanning system to distinguish between two closely separated points, a characteristic called "resolution," not unlike the resolution of industrial robots, as studied in Chapter 6. The horizontal resolution of the scan depicted in Figure 8.1 is the width of the screen divided by 640. The vertical resolution is the height of the screen divided by 480.

Another determinant of image quality is contrast. This is the ability of the scanning system to detect shades of difference from pixel to pixel. The lowest quality machine vision system uses a simple two-state sensor that registers each pixel as being either black or white. Despite its low quality, surprisingly good images have been achieved by systems that are capable of producing only an array of black dots. High resolution "dot-matrix" printers achieved wide acceptance in the decade of the 1980s and are still in use in the 1990s, although they are being replaced by higher quality laser printing systems.

Moving up the quality scale from simple bistate contrast systems are systems that are able to detect and reproduce four, six, or eight shades of gray from black to white for each pixel. An ordinary computer printer can be used to demonstrate this concept also

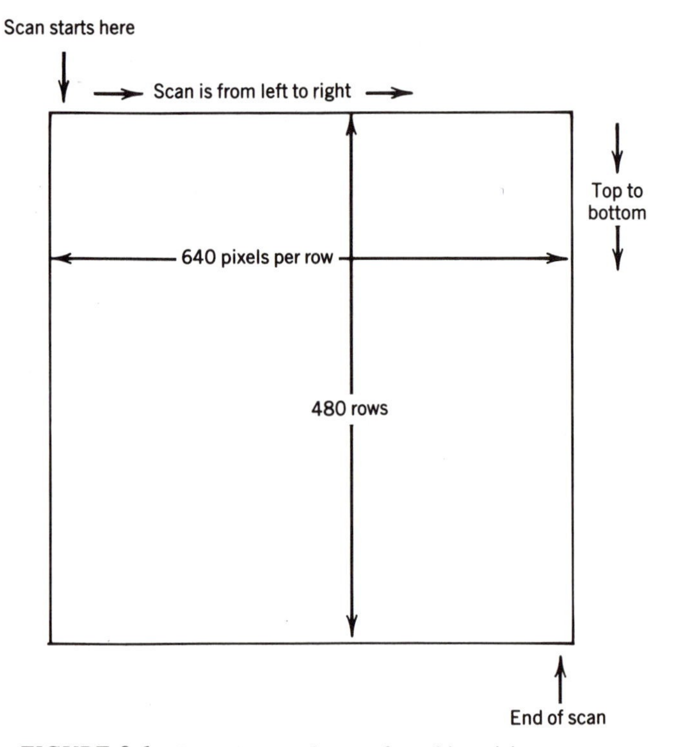

FIGURE 8.1 Raster scan scheme of machine vision systems.

```
MMMMMMMMMM   BBBBBBBBBB   IIIIIIIIII
MMMMMMMMMM   BBBBBBBBBB   IIIIIIIIII
MMMMMMMMMM   BBBBBBBBBB   IIIIIIIIII
MMMMMMMMMM   BBBBBBBBBB   IIIIIIIIII
MMMMMMMMMM   BBBBBBBBBB   IIIIIIIIII
MMMMMMMMMM   BBBBBBBBBB   IIIIIIIIII
```

FIGURE 8.2 Comparison of print densities of various alphabetic letters.

by taking advantage of the varying print densities of the alphabetic letters, numerals, and special characters within the character set capability of the machine. For example, the letter "M" is a denser character than the letter "B" or "I" and thus can be used to represent a pixel of a darker shade of gray in a multilevel contrast system, as is illustrated in Figure 8.2.

In the early days of computing, images were often printed using ordinary print characters as the pixel element. The varying densities of the print characters were used to achieve a gray scale of impressive quality. The pictures were best viewed from a distance considering that the resolution was 10 pixels/in. horizontal and 6 pixels/in. vertical. Photoreduction improved the resolution and produced surprising results. The most frequent application was wall calendars, especially for college dorm rooms.

For the objective of realistic representation of the original subject, a high range of gray scale is ideal. But, curiously, for the industrial applications of machine vision systems, a simple black-and-white pixel system may be superior, not to speak of its lower cost. The objective of a machine vision system is usually to discern a target feature, not to make a pretty picture. The automation engineer should design each application system with the objectives of the application in mind. If the objective is to find the position of a known part, detection of the edge of that part is of paramount importance. It matters not whether the machine is able to determine whether the part is gray or black at the extreme points of its edges; what does matter is that the machine is able to locate exactly those extreme points.

High resolution, as with gray scale, can also have its drawbacks, and matching of system to application is important here, too. When an automation engineer selects an expensive, high-resolution vision system for an application that does not need it, he or she may find that the system "can't see the forest for the trees." High resolution is analogous to magnification, as both enhance detail. But the question the automation engineer must ask is whether such detail is needed. Figure 8.3 contrasts two views of a spark plug at different levels of magnification. If the objective is for a machine to measure the gap between the electrodes, the surface imperfections revealed in the higher magnified version may detract from the process rather than enhance it.

Lighting

Lighting of the subject should be considered an important element of the image acquisition system, and it is mentioned separately here because lighting plays such an important part in the success of the application. Anyone who has studied the moon with a telescope or even with binoculars knows that the terrain features are greatly enhanced by the angle of incidence from the dominant light source, the sun. Thus, mountains and canyons are much more visible during waxing and waning portions of the cycle, when viewed near the sunset or sunrise line, than at full moon. Machine vision systems, too, can be made more effective by cleverly taking advantage of unusual lighting effects.

FIGURE 8.3 Too much magnification may detract from rather than enhance the image.

Experience with machine vision systems usually leads the designer to choose dedicated lighting systems rather than relying upon ambient light. Dedicated light may consist of a point source that enhances sharp features or takes advantage of sharp shadows. For other applications the object may be surrounded with multiple source light to eliminate the presence of shadows that may give rise to dark pixels that can be falsely interpreted. Sometimes a silhouette is desired so that edges can be detected; in this case backlighting may be desirable. But if the object has an important feature to be detected on its face, such as a hole or a groove, backlighting will be counterproductive.

Another lighting problem is glare. Black, shiny surfaces can be sensed as white owing to reflection of the light source. The human eye more easily distinguishes between glare and white background than does a machine vision sensor. Although glare from a black surface can be brilliantly white, the human eye (and brain) can recognize the context of the scene and ignore glare much as the human ear (and brain) can interpret speech from context. Because humans take this capability for granted, they sometimes fail to adequately deal with the problem of glare when designing a machine vision system.

Digitization

The sensor that gathers the image for a machine vision system is typically an analog device. However, the computer version of the image must be digital. A central problem of automated manufacturing is the gathering of analog signals and converting them to digital approximations for storage and analysis by computer. Analog to digital conversion will be studied in more detail in Chapter 14, but in this chapter on machine vision we will examine a major application of the concept.

Once again, we return to the idea of gray scale when we refer to digitization. The process of digitization converts an analog signal, which can take on an infinitely variable

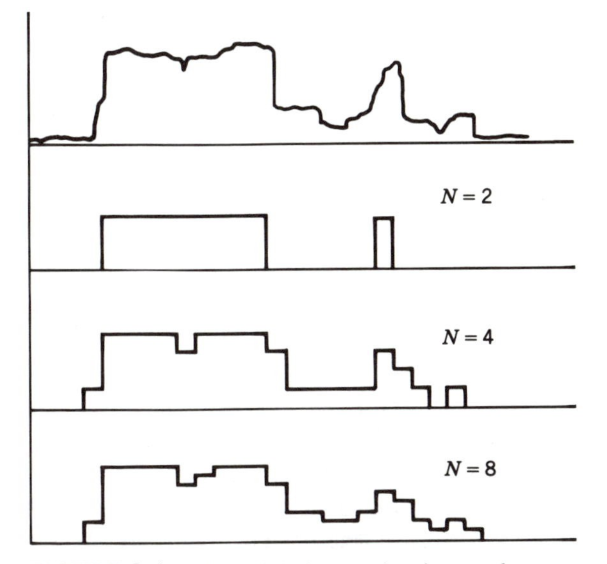

FIGURE 8.4 Three digital approximations to the same continuous analog signal representing light intensity in the scan of an image. The value of N is the number of binary digits used to represent the digital values.

value, to an integer from 1 to N, where N represents the degree of gray scale recognizable by the system. Invariably, N is a power of two because machine vision systems are designed to store digitized image data in binary registers. Typical values are 2, 4, 8, 64, and 256, which represent binary registers of length 1, 2, 3, 6, and 8, respectively. The first level ($N = 2$) contains no gray-scale capability at all, because its two levels represent absolute black and absolute white. Still the $N = 2$ system remains popular because of its low cost, its low storage requirements for the digitized image, and its appropriateness for industrial applications of machine vision systems. In fact, we will see later in image analysis techniques that simplifying an image by reducing it back to an $N = 2$ image in a technology called "thresholding" can actually enhance its usefulness.

Figure 8.4 compares the digitization of a 32-pixel scan using $N = 2$, 4, and 8, as might be encountered in a raster scan of a machine vision image. Note that more detail is lost in the $N = 2$ version, but this version consumes only 33 percent as much computer storage space as does the $N = 8$ version. One strategy is to retain the gray-level contrast data gathered in the image-acquisition phase in order to permit experimentation with various thresholds, or transitions between black and white, in the image analysis phase.

Having scanned and stored the image in the machine vision system computer, we turn our attention to the second major task of the machine vision system—the analysis of the image to accomplish the system's objectives.

IMAGE ANALYSIS TECHNIQUES

Windowing

Windowing is a means of concentrating vision system analysis on a small field of view, thereby conserving computer resources of run time and storage. Windowing is an essential first step for virtually every robotic vision system analysis, and though we usually do not think about it, our human vision systems also are constantly performing a windowing process. Although largely taken for granted, the human vision system has an amazing facility for looking directly at a small field while maintaining a much larger field within the range of peripheral vision. The human brain quietly monitors the peripheral field and ignores it unless unexpected motion or something out of place interrupts the attention span and brings the disturbing influence to the full attention of the observer. Machine vision systems are greatly inferior to human vision systems in this regard, but the use of windowing enhances the capability of the system to emulate the work of a human concentrating upon a small detail.

The most practical applications of windowing employ fixed windows, that is, the window is always set up in the same place within the image. This usually means that some sort of fixturing must be used to identically position every workpiece so that consistency of the window subject is maintained. The principles of positioning and orientation are discussed in Chapters 1 and 3 are seen now to be important for machine vision system effectiveness also.

More sophisticated machine vision systems are able to employ adaptive windowing, in which the system is able to select the appropriate window out of context. In such systems a search of the entire image detects known landmarks that identify the position and orientation of the subject workpiece. The landmarks can then be used by the system to find the window area of interest and proceed as in a fixed window scheme. The advantage is obvious; orientation and positioning become unnecessary, resulting in dramatic savings in production costs. Some robots, equipped with adaptive windowing capability, have experienced success with "bin-picking," the selection and picking up of workpieces piled randomly in a bin, an easy task for humans, but a very difficult one for a robot.

Whether its position is fixed or adaptive, the window can be adjusted in size for the purpose of the application. A window can be compressed to a single pixel if desired to pinpoint a color or gray level with a single sample. However, because of the possibility of spurious readings due to irregularities in the object or lighting or variations in the sensor operation, it is usually more practical to open a window to sample a reasonable number of pixels and compute an average gray level even for a point measurement.

A lesson can be learned from the need to use "windowing" in a machine vision system. An ever-present trade-off in machine vision systems is speed versus sophistication of the analysis. If multiple objectives are present and a very smart system is required, perhaps even utilizing forms of artificial intelligence, there will be some sacrifice in speed. Perhaps some of the objectives can be sacrificed in a compromise to make the analysis fast enough to be economically feasible. When no such compromise is possible, windowing may offer a way out of the dilemma.

Once a window of interest has been identified, a variety of analyses can enhance the target features or identify and describe them in detail. The most basic of these analyses is thresholding.

Thresholding

This chapter has already described thresholding as reducing an image to binary black or white pixels. Thresholding is the oldest, simplest, and still one of the most effective methods of image analysis. One reason it was so popular in the early days of machine vision is that thresholding is capable of screening out the operational variability of the sensing system.

Any binary imaging system is usually designated as thresholding, though the system is not necessarily limited to a single threshold. Consider the problem of capturing the image of a gray snake on a black-and-white checked tile floor. Black and white pixels could both be assigned a value of 1 (white) and only the pixels in the gray region between two thresholds could be transformed to a value 0 (black). This would convert the image to a black snake on a white floor. Perhaps robots do not need to worry about snakes on the floor, but in an industrial setting two well-chosen thresholds could be used to select a band of gray level and thus pick up only a specifically desired object, perhaps of a different color from the undesired objects or the background. Because of this utility, thresholding systems are usually designed to accept two thresholds. If single-level thresholding is desired, one of the two thresholds can be adjusted completely to the right or left extreme of the brightness scale. It is difficult to conceive of an industrial application of a thresholding scheme that provides more than two thresholds in a system that produces only a binary image.

Finding exactly the right threshold for a binary vision system can be the key to success in an automation application. This problem is trickier than it may seem. Selection of a threshold exactly halfway between extreme light and dark can result in a totally dark picture, if both light and dark are below midrange. The machine vision system needs to somehow employ some intelligence in the selection of an appropriate threshold, whether that intelligence is supplied by a human or by reasoning algorithms programmed into the machine.

Histogramming

A popular method of selecting a threshold is by "histogramming." A frequency histogram is constructed for the pixel counts at each level of gray accommodated by the system. If the image is simply a dark object on a light background, the histogram will be bimodal, as in Figure 8.5. The shape of the object is irrelevant as is illustrated by Figure 8.6. Histogramming can also be used to select a given shape, if its gray-level is different from other shapes and if its pixel count is approximately known relative to the pixel count of other shapes in the image.

Shape Identification

The selection of shape is stretching the limits of capability for histogramming and thresholding techniques. There are other clever ways to identify a desired shape or to perceive its orientation, and a variety of algorithms have these purposes. We use the term *algorithms* because in this area of image analysis, mathematical or computer procedures, often iterative in nature, are used to process the data to arrive at some conclusion about an object's shape or orientation in the image.

One method of finding the orientation of a known shape is to scan a series of straight lines across a binary image looking for characteristic "run-lengths" of black or white pixels. Suppose, for instance, that it is desired to properly orient gaskets, as shown in Figure 8.7. The task is to align the minor axis of symmetry (y-y in the top diagram) with the direction of travel of the conveyor.

A simple approach to this problem takes advantage of the computational speed of the system computer to quickly strike a variety of scan lines across the image at various, perhaps random, angles. This type of "scanning" should not be confused with the original raster scan of the entire image during the image capture. Rather, this scan is an imaginary one in which the control computer is examining the existing pixel matrix and searching for runs of either black or white pixels or a specific combination of both. For the gasket in Figure 8.7 the system is attempting to find the major axis (x-x) by the following series of runs:

white-black-white-black-white-black-white-black-white

The two white runs at the series extremes represent the background. Besides these beginning and ending white runs, there are three white runs in the series—one for each of the two small holes and one longer run for the larger center hole. If a gross orientation is all that is desired, an industrial robot equipped with a vacuum gripper can immediately proceed to pick up and reorient the gasket as soon as the first conforming run series is detected. Finer alignment can be made by repeated scans in near proximity to the first successful conforming scan, in an attempt to maximize the pixel length of the interior white run. The maximum run length would indicate that the diameter, rather than a chord, has been identified for the inner hole.

Perhaps even simpler than looking for the series having the correct number of black–white run transitions would be a search for the longest straight line pixel run of the following pattern:

black-white-black-white-black-white-black

Here the interior transitions are ignored and the objective of the search is to find the first and the last black pixel and maximize their separation. Once this maximum is found, the major axis of the gasket will have been identified. A variation of this procedure would work even if the gasket had no holes.

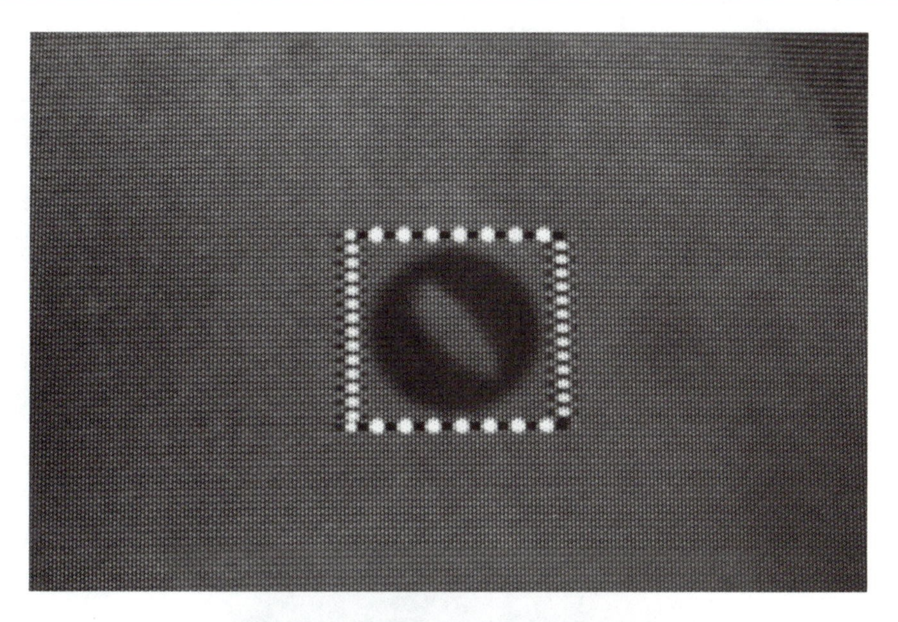

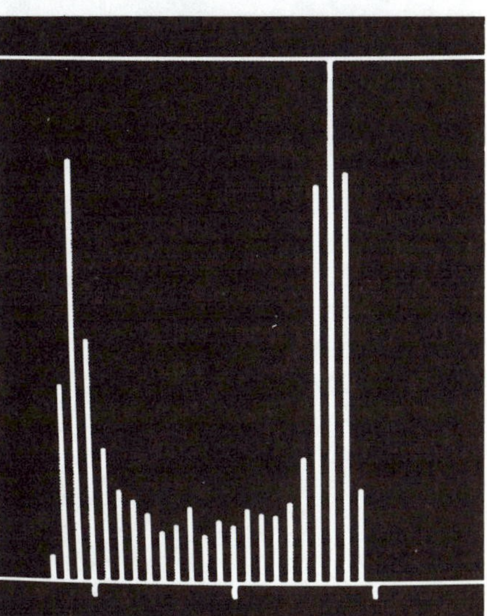

FIGURE 8.5 Bimodal histogram of a dark object on a light background. (University of Arkansas Robotics Laboratory photo)

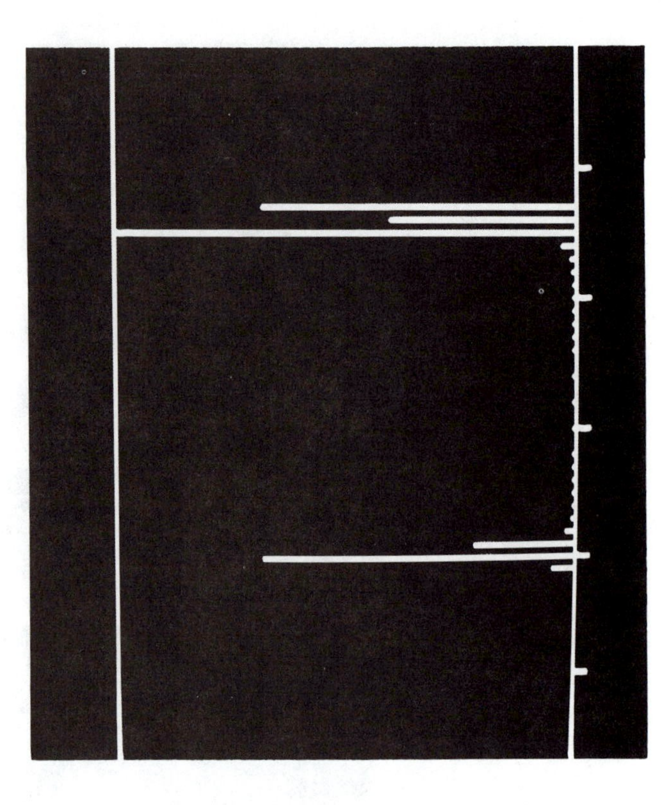

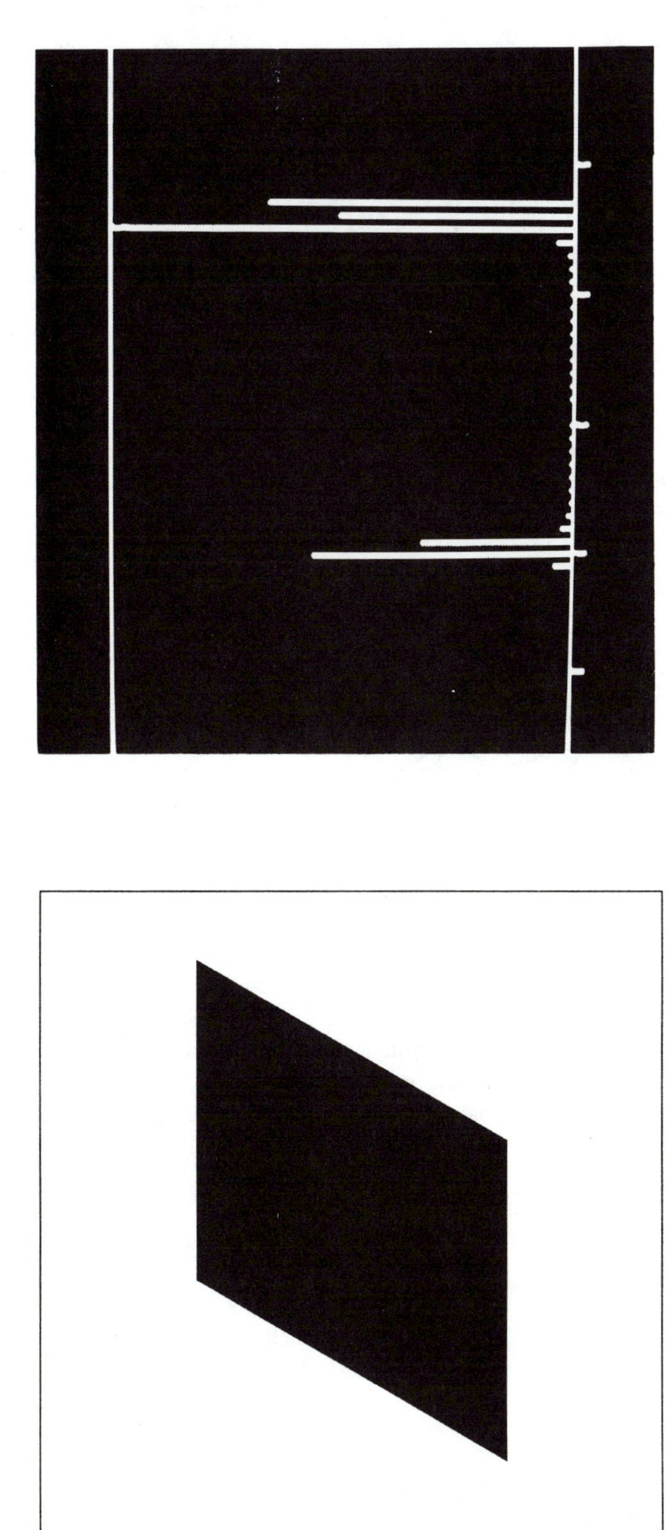

FIGURE 8.6 The shape of an object is irrelevant in the array of its histogram of pixel counts at each gray level. The two dissimilar objects shown have similar histograms. (University of Arkansas Robotics Laboratory photo)

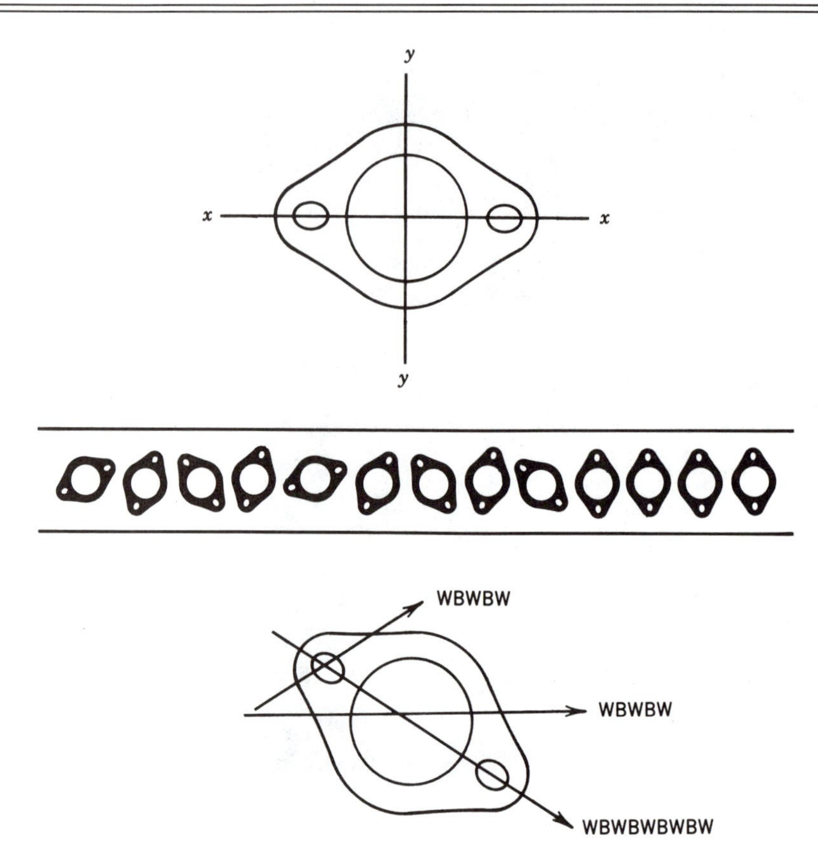

FIGURE 8.7 Using a shape identification algorithm to select and properly orient gaskets on a conveyor line.

Within this application, possibilities exist for more intelligent searches than a random procedure. Note that the shape of the gasket shown in Figure 8.7 has a local maximum on the minor axis of symmetry in addition to the global maximum on the major axis. One can see how the principles of optimization techniques can be applied to machine vision systems analysis in a manner similar to the way in which these principles are applied to the field of operations research. Industrial engineers are well-founded in principles of operations research and are finding new applications for their skills in the field of machine vision.

Template Matching

A binary pixel image in the machine vision system memory enables comparisons, pixel-by-pixel, with a template matrix that has the desired pattern. The method will be illustrated with an example.

EXAMPLE 8.1

Consider the 171-pixel matrix in Figure 8.8 that contains six familiar patterns—six letters of the alphabet. The task is to locate the letter *F* using a 3 × 5 pixel template for comparison. In nested iterations throughout the matrix each 3 × 5 set of pixels can be examined for degree of match with the template. As suspected, the letter *E* is a close match, attaining a score of 13 out of a possible 15. The letter *F*, of course, rates a perfect 15. In real applications the search can consume

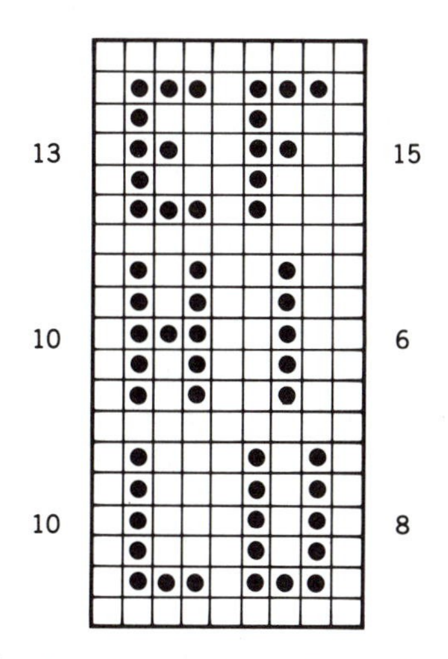

FIGURE 8.8 Template matching. The numbers show the frequency of correspondence between pixels in the test area and pixels in the template for the alphabetic letter F. Since the template is 3×5, a perfect score is 15.

a large amount of computer time as the number of trial areas expands exponentially. Even in the small example illustrated, consider how many comparisons would have to be made to check all of the feasible positions in all of the feasible orientations for the 3 × 5 template superimposed upon the 171-pixel image screen. Then multiply this task by all of the target templates within the template library.

The enormous computation task of an exhaustive search for a template match gives a person new appreciation for the pattern-recognition capability of humans, which, though we take it for granted, is phenomenal compared to machine vision speeds and accuracy. To deal with this problem, Mehrotra and Grisky [ref. 63] have devised a strategy for cutting down the search space by employing a data-driven scheme that attempts to make sensible, closed line drawings from the given data. Working with the data without the benefit of a template, the researchers have succeeded in constructing polygons that bear some semblance of the object's shape, as can be seen in Figure 8.9. These polygonal

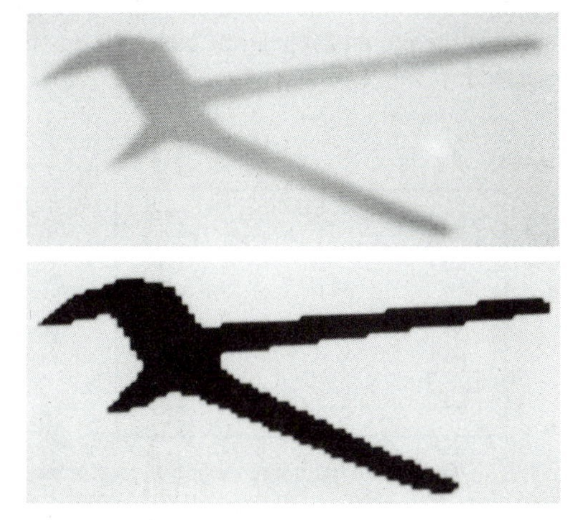

FIGURE 8.9 Polygonal approximation. (Photo courtesy of Center for Robotics and Manufacturing, University of Kentucky)

approximations are then identified with template classes to drive the system toward a more efficient search for an appropriate template to confirm the match. This approach is particularly appropriate when the contents of the image are not known and the objective is to find a match with any of a variety of templates stored within the machine vision system's database.

In template matching it is sometimes better to take advantage of one peculiar feature instead of trying to find a match for the entire object. Consider the design of a metal stamping as shown in Figure 8.10, in which the stamping has an unusual and easily recognizable feature in the crosses positioned at each of the four corners. The crosses are ideal for identification and also serve to ascertain the object's orientation because their axes are parallel with the sides of the stamping. Even if the axes were not so aligned, if they were of a fixed and known angle with the sides, the machine vision system could be programmed to figure out the orientation of the workpiece. In such an application there is no reason that the angles would not be known, as they would become fixed in the original design of the stamping die set.

Earlier in this chapter we considered the technique of "adaptive windowing," and there are obvious similarities between adaptive windowing and the method discussed here. Bolles and Cain [ref. 18] have given identity to the strategy of concentrating upon a peculiar feature by calling it the "local-feature-focus method." A primary advantage of this method is that it makes feasible the identification of a partially occluded object. A haphazard stack of stampings of the type shown in Figure 8.10 would be difficult to deal with under the best of circumstances, but taking advantage of a local feature might move the machine vision task into the realm of feasibility.

The reader may be having reservations about the concept of local-feature-focus because it may seem that workpieces do not normally have such convenient and unique features. The stamping design in Figure 8.10 for instance may appear to be contrived to facilitate the method. The response to this doubt is to acknowledge the strategy of overtly contriving the design to facilitate the identification, orientation, handling, and assembly of the workpiece. Is this not one of the basic principles of manufacturing automation we examined in Chapter 1?

Edge Detection

Edge detection is a procedure that uses binary logic to guide a search through an image, one pixel at a time, as it first finds an edge and then follows it completely around the object until it repeats itself. At the point of repeat the conclusion is reached that the entire object has been circumscribed.

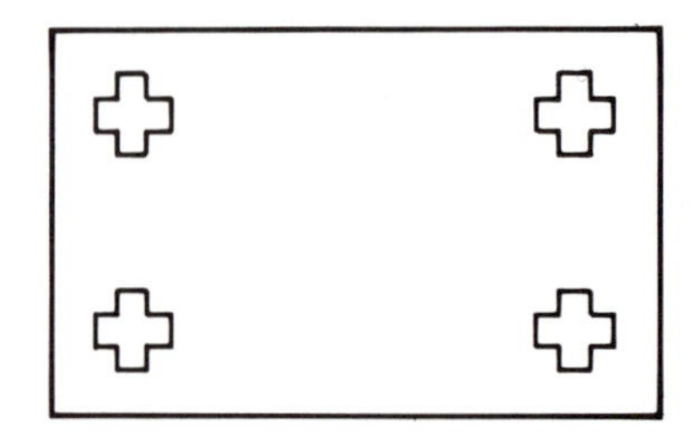

FIGURE 8.10 This metal stamping design incorporates an easily recognizable feature that facilitates identification, location, and orientation of the workpiece.

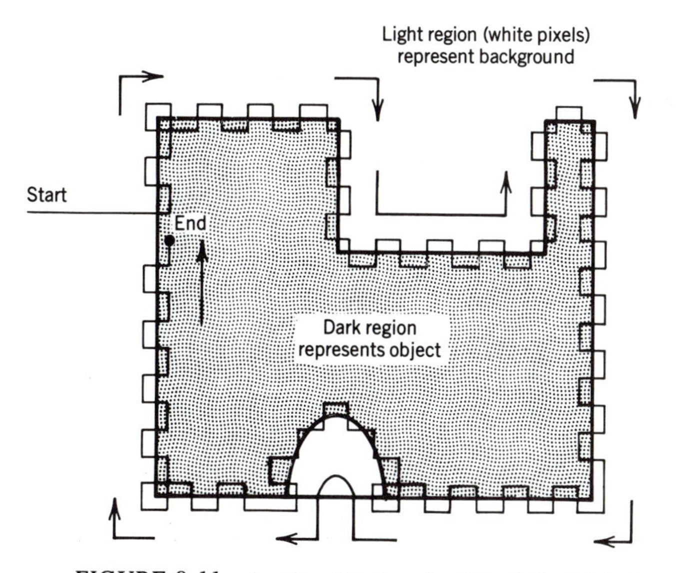

FIGURE 8.11 An edge detection algorithm finds and then, by continually crossing it, follows the edge completely around the object.

An example edge-detection procedure is described in Figure 8.11, displaying a binary image in which the object is dark (binary 1) and the background is light (binary 0). A similar procedure (with opposite logic) works as well for a light object on a dark background. The system begins on a straight path across the background, examining each pixel it encounters and determining whether it is light or dark. As long as it encounters light pixels it continues on a straight path. As soon as the system encounters a dark pixel, it knows that it has crossed the boundary into the region within the object, so it turns in an attempt to maintain the edge. From that point forward the system keeps turning so that by continually crossing the boundary it maintains contact with it. Note that the edge-detection system is able to detect both interior corners and exterior corners and in a crude fashion can even follow a curve, although the computer image will be squared off into a series of tiny square corners.

The construction of a practical procedure for edge detection is in truth a design problem, because there are many different ways the job can be done. As a design case study, Example 8.2 will demonstrate the construction of a logic flow diagram for the edge detection procedure described in Figure 8.11.

EXAMPLE 8.2 (DESIGN CASE STUDY)

Design a simple logic flow diagram suitable for developing a computer program for executing an edge-detection algorithm as sketched in Figure 8.11.

Solution

One design for this edge detection procedure appears in Figure 8.12.

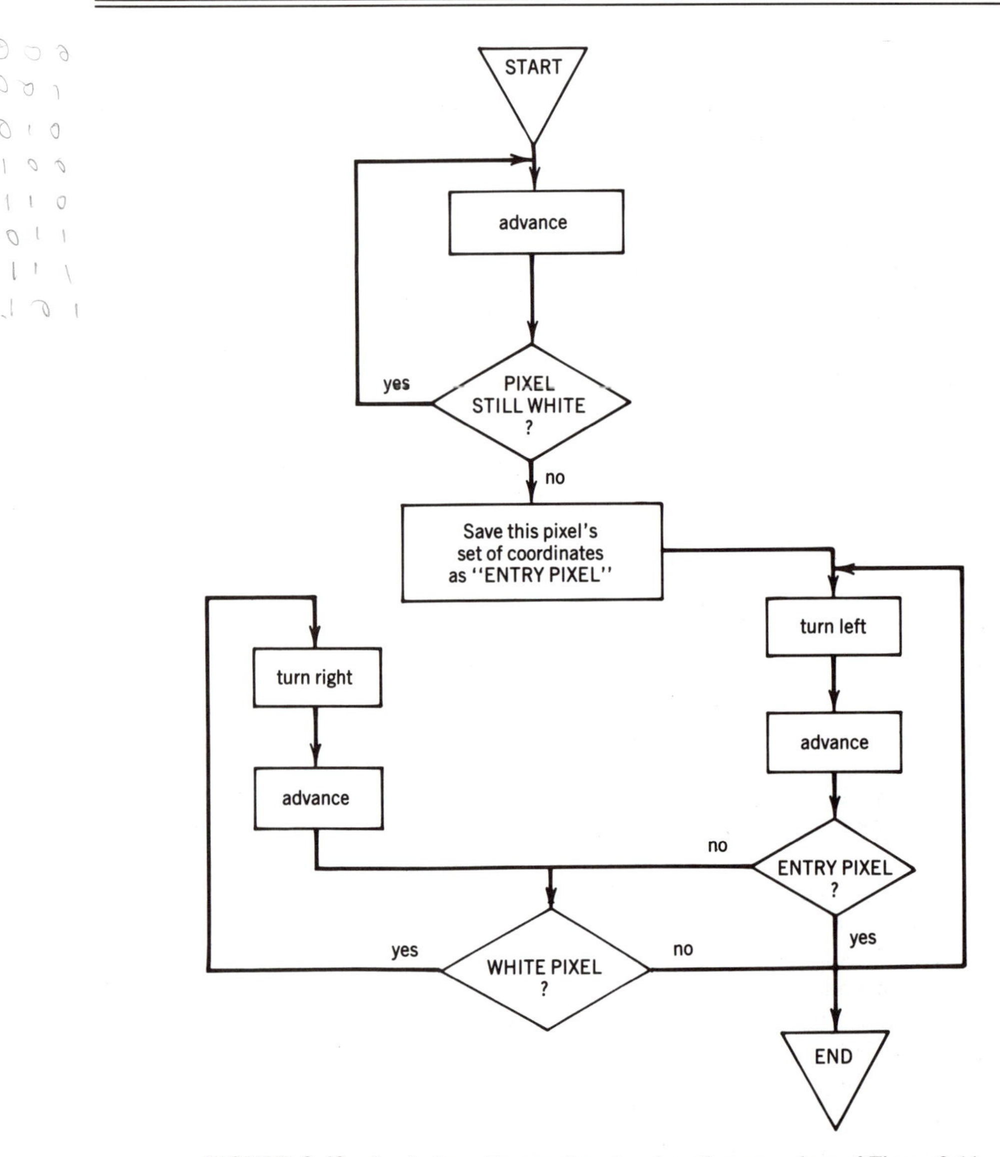

FIGURE 8.12 Logic flow diagram for edge detection procedure of Figure 8.11.

One problem with the edge-detection procedure is that if it ever makes a wrong turn, it can get lost in a tight loop, constantly turning left if it is lost inside the object, and constantly turning right if it is outside. An algorithm can watchdog this situation, and if it is ever discovered that pixels are being repeated in a tight loop, the system can go back to the pixel repeated the most and know that this is where the circle pattern began. At that point the system can reverse the turn taken in that pixel and be back on track. Such an algorithm is especially useful when negotiating curves, for which a clear decision regarding a single pixel may not be obvious.

The logic flow design shown in Figure 8.12 is unable to deal with possible wrong turns that can result in the system being caught in a tight loop either within or outside the edge of the object. One of the exercises at the end of this chapter is a design problem in which the student is asked to add this sophistication to the design of the logic flow diagram shown in Figure 8.12.

Roberts Cross-Operator

An algorithmic method of edge detection makes use of the Roberts cross-operator [ref. 102]. The algorithm computes the square root of the sum of the squares of the adjacent diagonal differences between pixel gray-scale values. This process is illustrated in Figure 8.13.

The idea is to identify a transition (edge) by finding points at which the diagonal differences are greatest. The Roberts cross-operator is calculated for all pixels in the image for which the adjacent diagonals to the right and down exist. This is to say that the operator is undefined for the bottom row of pixels and the right column of pixels. An example will illustrate the calculation.

EXAMPLE 8.3
Calculation of Roberts Cross-Operator

Calculate the Roberts cross-operator for the following pixel matrix:

3	1	2
2	3	3
0	2	3

Solution

1	$\sqrt{5}$	—
3	1	—
—	—	—

The reader can verify the results for each cell by applying the formula shown in Figure 8.13. The Roberts cross-operator exists for only four cells in this nine-cell matrix. The matrix was intentionally kept small to illustrate the calculation, but the results do not

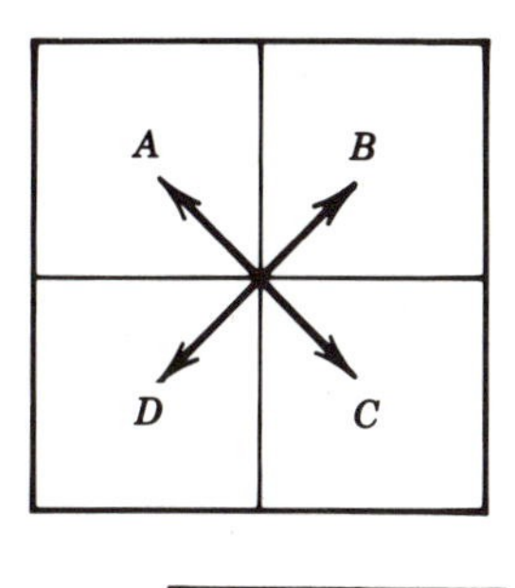

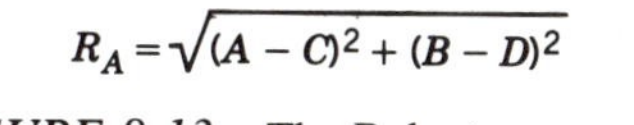

$$R_A = \sqrt{(A - C)^2 + (B - D)^2}$$

FIGURE 8.13 The Roberts cross-operator.

illustrate the capability of the Roberts cross-operator to detect an edge. A more comprehensive example with an idealized image will now be used to better illustrate the edge-finding ability of the Roberts cross-operator analysis.

EXAMPLE 8.4
Using the Roberts Cross-Operator Matrix to Find an Edge

On a 16 × 9 pixel ideal white field let us place a carefully selected ideal black image—an isosceles triangle with base exactly equal to 10 pixels and height exactly 5 pixels, as shown in Figure 8.14*a*. Let us use a gray-scale range of 2^6 so that ideally all white pixels will be gray scale 63 and all black pixels gray scale 0. By carefully placing the triangle in our ideal image, all pixels will either be wholly within or wholly outside the triangle except for those pixels on the diagonal sides of the triangle. Since these pixels are exactly half-in and half-out in our idealized model, we will assign each of them a gray scale of exactly 31. The resulting pixel matrix is shown in Figure 8.14*b*. Figure 8.14*b* gives us a precise model for calculating and testing the Roberts cross-operator.

Solution

We now calculate the Roberts cross-operator for each feasible pixel in the matrix. The result is Figure 8.14*c*. Note that the bottom row and right column are empty, as the Roberts cross operator is not defined for these border cells. Figures 8.15*a*, 8.15*b*, and 8.15*c* utilize carefully selected thresholds to illustrate the ability of the Roberts cross-operator analysis to detect the outline of the triangle in Figure 8.14*a*. Note that the selection of an inappropriate threshold in Figure 8.15*c* gave a rather misleading result. Since threshold selection can be critical to the success of a Roberts cross-operator analysis, it may be necessary to calculate limits for the operator for a given imaging system in advance. Such calculations follow directly from the formula and are reserved as exercises at the end of this chapter.

Edge-detection capability opens up a host of machine vision applications. One of these is dimensional gauging. If the opposite edges of an object can be detected, then the distance between these edges can be gauged by the vision system's control computer, whether the dimension be x,y or a computed dimension at an angle between the axes. Edge detection can also be used in the process of "skeletonization," in which all white pixels are reduced to a one-pixel-wide outline on the periphery of the white field. A similar process with dark pixel areas can produce familiar line drawings. We now examine a small sample of practical industrial applications of machine vision systems.

INDUSTRIAL APPLICATIONS

From what has been studied thus far in this chapter, it should be apparent that the typical "machine vision system" is not a set of artificial eyes to permit a robot to see where it is going and navigate about its workplace while dealing with a changing environment. It is true that some robots are equipped with object avoidance systems, but such systems bear little resemblance to machine vision systems, which are performing a wide variety of manufacturing operations from inspection to measurement and component assembly. In this section we will examine some of the current application areas to suggest situations that are favorable to the introduction of a machine vision system.

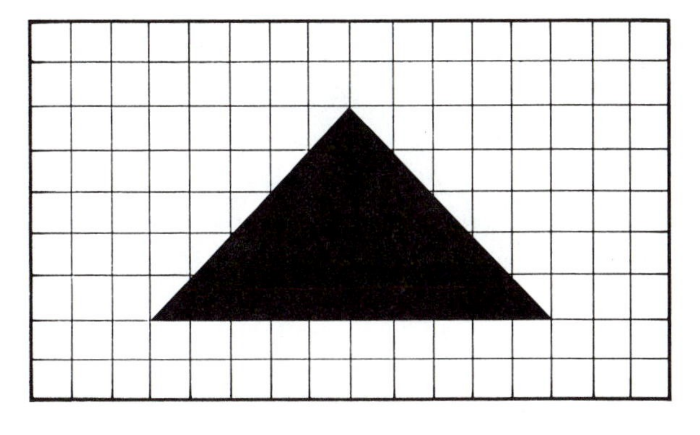

(a)

~	~	~	~	~	~	~	~	~	~	~	~	~	~	~	~
~	~	~	~	~	~	63	63	63	63	~	~	~	~	~	~
~	~	~	~	~	63	63	31	31	63	63	~	~	~	~	~
~	~	~	~	63	63	31	0	0	31	63	63	~	~	~	~
~	~	~	63	63	31	0	0	0	0	31	63	63	~	~	~
~	~	~	63	31	0	0	0	0	0	0	31	63	~	~	~
~	~	63	31	0	0	0	0	0	0	0	0	31	63	~	~
~	~	~	63	63	63	63	63	63	63	63	63	63	~	~	~
~	~	~	~	~	~	~	~	~	~	~	~	~	~	~	~

(b)

~	~	~	~	~	~	0	0	0	0	~	~	~	~	0	u
~	~	~	~	~	0	32	45	32	0	0	~	~	~	0	u
~	~	~	~	0	32	63	44	63	32	0	0	~	~	0	u
~	~	~	0	32	63	31	0	31	63	32	0	0	0	0	u
~	~	0	32	63	31	0	0	0	31	63	32	0	0	0	u
~	0	32	63	31	0	0	0	0	0	31	63	32	0	0	u
~	0	32	71	89	89	89	89	89	89	89	71	32	0	0	u
~	0	0	0	0	0	0	0	0	0	0	0	0	0	0	u
u	u	u	u	u	u	u	u	u	u	u	u	u	u	u	u

(c)

FIGURE 8.14 An ideal image is used to demonstrate the computation of the Roberts cross-operator and the ability of the operator to reveal an edge.

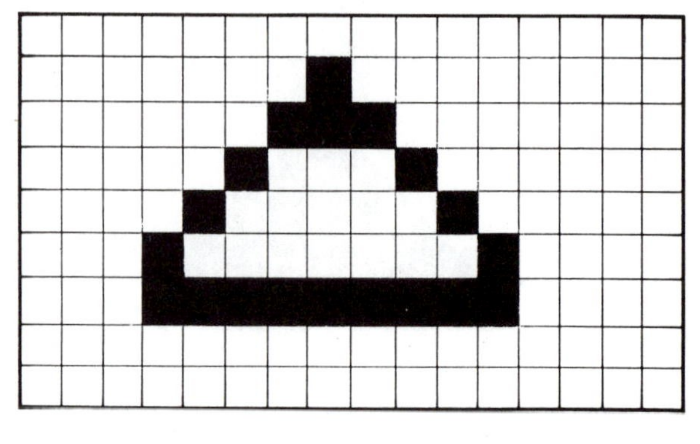

(a)

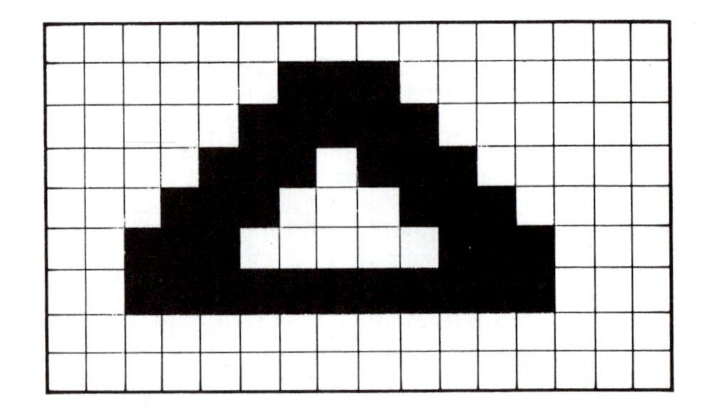

(b)

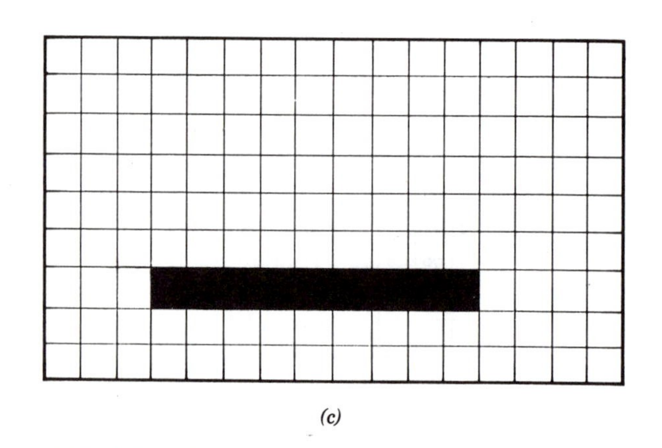

(c)

FIGURE 8.15 Edge detection capability of the Roberts cross-operator is shown at various threshold selections. Compare with ideal image in Figure 8.14(a).

Inspection

It is natural to think of a machine vision system as suitable for inspection tasks, but such suitability greatly depends upon the type of inspection task to be accomplished. The author of this text was once asked by a faucet manufacturer whether a robot could take over the final inspection process for chrome faucets, examining each for surface qualities and defects of widely varying types. The question can be quickly recognized as tongue-in-cheek because present-day robots and vision systems have meager capability for judgmental decisions regarding the overall quality of a product presented for final inspection. In such tasks humans still excel. On the other hand, if there are a few specific defects that are highly recognizable in a high-contrast two-dimensional image, a machine vision system might be very effective. A good example would be checking for missing parts, broken pieces, or parts assembled upside down. A candidate application for machine vision systems has usually been performed for many years by a human inspection system, but the human inspection system has somehow failed to achieve its objective.

Why does a human inspection system sometimes fail to do the job? The detailed repetition of some inspection tasks is simply beyond the capability of humans. Anyone who has used a word processor can appreciate the ability of a machine to perform a completely effective and exhaustive "SEARCH" for a given pattern of characters within a lengthy document. A human searcher has difficulty focusing upon such a repetitive task for more than a few minutes, but a machine search is incredibly fast and accurate. Even if a human were capable of doing such a highly repetitive operation, the job would no doubt be intolerably boring. It is also conceivable that an inspection task could be hazardous to a human, though such motivations have not been seen frequently to lead to actual machine vision system applications. Human boredom and propensity for errors in repetitive inspection tasks are more dominant reasons for seeking a machine vision system.

Electronics Manufacturing

The progressive miniaturization of components and circuits in the electronics industry has generated demands for new manufacturing methods. Locating insertion points in assembly operations and inspecting printed circuits are among the applications of machine vision systems to electronics manufacturing. The accuracy and resolution of the machine vision system working in tandem with a highly accurate robot with similar resolving power enables the robotic system to perform high-speed electronic component through-board insertion or surface mount component placement at speeds that would be impossible for humans.

Figure 8.16 reveals two tiny reference points on the face of a printed wiring board; these points are called "fiducial points."* These points are used for location with sufficient precision to perform robotic component insertion or surface mount placement. Thus, even if the mechanical fixturing of the board is of insufficient precision for subsequent robotic operations, the fiducial points provide the precise reference upon which critical operations can be based. Figure 8.17 is actual size for a 100-pin gull-wing bumpered quad flat pack (BQFP), a typical integrated circuit package circa the early 1990s. Figure 8.18 is a close-up revealing the power of the machine vision system to resolve the 25-mil spacing of the tiny pins around the edge. In Figure 8.19 we see a robotic flexible assembly system for electronics manufacturing. The machine vision camera is mounted directly on the robot, an IBM model 7565 SCARA.

*Or "fiduciary points."

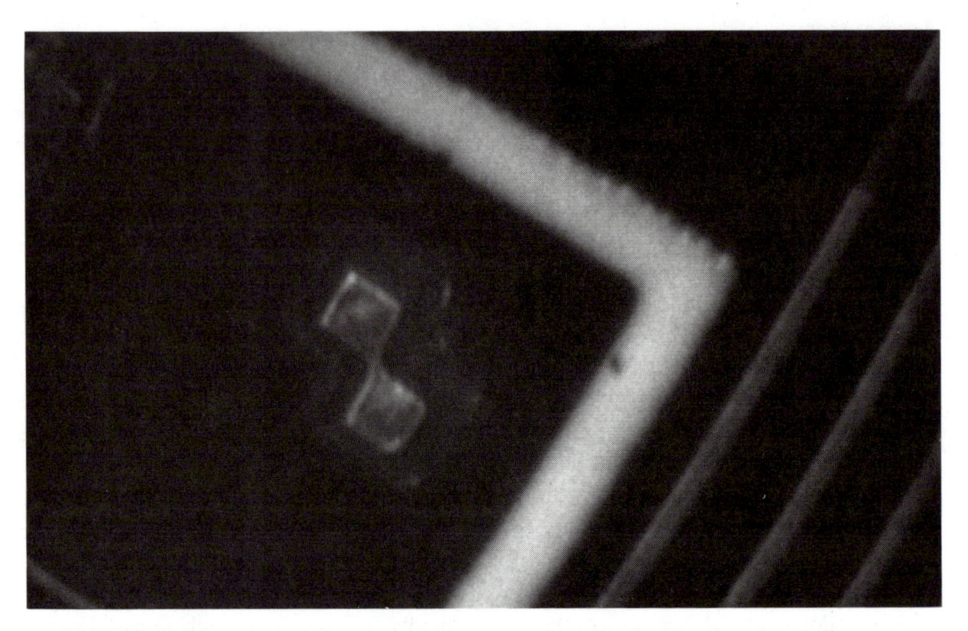

FIGURE 8.16 A machine vision system view of a printed wiring board revealing two fiducial points at a spacing of 25 mils. (University of Arkansas Robotics Laboratory photo)

As machine vision systems become more widely accepted, and there is ample evidence that they will, it is believed that machine vision systems keying from fiducial points will supplant hard fixturing in many cases. Some BQFPs in actual use have a 15-mil lead spacing and contain up to 304 leads.

Investigation of a creative implementation of fiducial points is reported by Witriol and Cowling at Louisiana Tech University [ref. 121]. The researchers have used luminescent fiduciary "codes" on components to be located. The machine vision system scans the work under ultraviolet illumination along with normal illumination. A second image, scanned in the presence of normal illumination only, is subtracted from the first image. The result is a high-quality, high-contrast image of the fiduciary codes without the clutter of other details in the work being processed. Since the fiduciary codes are intentionally applied, different codes could be used to identify different features or components. Also, a code could be placed upon a critical spot close to the TCP on the end-effector of a robot arm, and a machine vision system would thus be able to track the position of the robot with respect to the fiduciary point—in real time, according to Witriol and Cowling.

FIGURE 8.17 Actual size for a 100-pin gull-wing bumpered quad flat pack (BQFP), a typical integrated circuit package of the early 1990s. (University of Arkansas Robotics Laboratory photo)

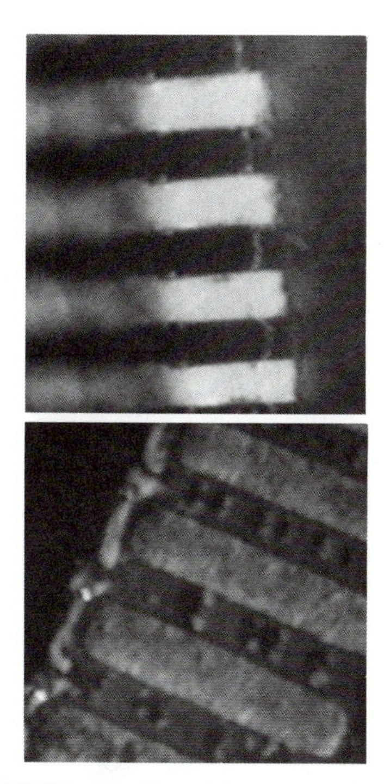

FIGURE 8.18 Machine vision system close-up of 100-pin BQFP, resolving the 25 mil (.025 in) spacing of the leads. (University of Arkansas Robotics Laboratory photo)

FIGURE 8.19 Robotic flexible assembly system for electronics manufacturing. The robot has automatic gripper change capability and can deal with dual in-lines (DIPs), J-lead plastic leaded chip carriers (PLCCs), and bumpered quad flat packs (BQFPs). (University of Arkansas Robotics Laboratory photo)

Apparel Manufacturing

The application of vision systems to electronics manufacturing is not surprising but other, less likely industries are also finding utility in robotic vision systems. In the manufacturing of clothing the piece-parts to be manipulated present several challenges because they are irregular in shape, nonrigid, and difficult to pick up. The mating and folding of various pieces before stitching is particularly difficult for a robot, but vision systems are assisting in this task. Edge detection is an important task in identifying key breakpoints that are used to guide folding and start points for seams. A vacuum table is usually necessary to hold the cloth workpiece in place. A light reflective surface is sometimes used on the vacuum table to provide contrast with the darker workpiece [ref. 16].

Agricultural Applications

Incredible as it may seem, robot vision systems have been found to be of sufficient accuracy and, more importantly, speed to economically sort coffee beans one by one based upon color of the individual beans. The beans are individually picked up using a vacuum-type system and presented one at a time to the vision system. The controller then diverts the beans according to color in a highly synchronized operation. Research by Dr. Earnest W. Fant at University of Arkansas is carrying the idea a step further—to individually sort peanuts and even to inspect grains of *rice!*

SUMMARY

Machine vision is coming to the forefront of developments in robots and manufacturing automation. As this chapter shows, machine vision projects often have more specific objectives than perceived by the layman.

Machine vision functions can be divided into two categories: image acquisition and image interpretation. The acquisition phase has two principal parameters: resolution and contrast. Lighting is an important factor to be considered also and can be selected to enhance the image features of interest. The digitization of image data is a classic example of analog to digital conversion that has application to other areas of manufacturing automation.

The interpretation and analysis phase is where specific objectives are being pursued with a variety of specialized techniques. Windowing is a means of concentrating the analysis upon a small field to conserve computer resources, especially run time. Thresholding is one of the most effective methods of image analysis. Related to thresholding is histogramming, which can be used to select appropriate light intensity levels for thresholding. Shape identification is accomplished by a variety of clever techniques including run-length searches, template matching, polygonal approximation, and local-feature-focus. Edge-detection algorithms can be devised using logic flow diagrams. A mathematical transformation upon the pixel matrix using the Roberts cross-operator also assists in identifying edges. Edge-detection schemes are useful in constructing line drawings or for skeletonization.

As might be expected, inspection tasks make up a principal application of machine vision systems, but these tasks are quite feature specific and usually involve hazardous operations or repetition too monotonous for humans. In judgmental inspection operations humans still excel. Electronics manufacturing is a significant application area for machine vision, especially inspection, fiduciary point identification, and component insertion.

Apparel manufacturing and agricultural applications including real-time sorting of coffee beans, peanuts, and rice make up some of the more challenging applications of machine vision.

EXERCISES AND STUDY QUESTIONS

8.1. For an image in which the pixel matrix rows are designated by subscript i, the columns by subscript j, and the pixel gray scale intensity by g_{ij}, write a generalized equation for the Roberts cross-operator R_{ij} for a matrix of n rows and m columns.

8.2. Write a computer routine in BASIC, FORTRAN, or C that develops the Roberts cross-operator for a 318 × 480-pixel image. How many operator values will be generated by the program?

8.3. Calculate the number of Roberts cross-operators that can be developed for an 8 × 10-in. (8 in. vert, 10 in. horiz) image that has the following resolution.

Horizontal: 100 pixels/in.
Vertical: 60 pixels/in.

8.4. Calculate the theoretical maximum for the Roberts cross-operator for an image for which the pixel gray scale is represented by a three-digit binary number.

8.5. Refer to Figure 8.8 and calculate the number of unique ways the *entire* 3 × 5-pixel template can be superimposed on the 171-pixel image. Consider all positions and all orientations (vertical-up, vertical-down, horizontal-right, horizontal-left), but exclude all diagonal orientations because of incompatible spacings between pixels.

8.6. The three matrices appearing below are *identical*. Perform the binary thresholding operation indicated for each to receive a message from the author's home school. If 15 ≤ pixel intensity ≤ 35, pixel → dark; otherwise pixel → light.

20	15	31	7	35	24	33	8	58	49	60	7	45	54	51
28	5	36	9	26	10	31	11	46	12	6	4	50	3	61
30	43	27	14	28	3	27	14	55	9	45	1	49	36	46
34	2	29	2	30	7	29	10	50	11	51	13	64	1	53
29	16	21	5	32	25	34	8	59	47	57	6	52	48	60

If 20 ≤ pixel intensity ≤ 60, pixel → dark; otherwise pixel → light.

20	15	31	7	35	24	33	8	58	49	60	7	45	54	51
28	5	36	9	26	10	31	11	46	12	6	4	50	3	61
30	43	27	14	28	3	27	14	55	9	45	1	49	36	46
34	2	29	2	30	7	29	10	50	11	51	13	64	1	53
29	16	21	5	32	25	34	8	59	47	57	6	52	48	60

If 45 ≤ pixel intensity, pixel → dark; otherwise pixel → light.

20	15	31	7	35	24	33	8	58	49	60	7	45	54	51
28	5	36	9	26	10	31	11	46	12	6	4	50	3	61
30	43	27	14	28	3	27	14	55	9	45	1	49	36	46
34	2	29	2	30	7	29	10	50	11	51	13	64	1	53
29	16	21	5	32	25	34	8	59	47	57	6	52	48	60

8.7. Construct a pixel matrix of the type illustrated in Exercise 8.6 above, which carries a message of your choice that can be deciphered by selecting different binary thresholds for each word embedded in the matrix.

8.8. **(Design Problem)** Design a logic algorithm that automatically corrects the edge-detection algorithm shown in the design solution for Example 8.2 in the event that the system makes an isolated error by erroneously turning right, instead of left, as soon as it makes a white-to-black pixel transition or erroneously turning left, instead of right, as soon as it makes a black-to-white pixel transition. Draw a logic diagram for the edge-detection algorithm that incorporates your added error-correction routine.

8.9. The cliché, "a picture is worth a thousand words," may be conservative. Perform calculations to estimate the number of computer "words" (bytes) required to represent a single graphic image of typical size, resolution, and contrast. Justify your estimate.

CHAPTER 9

ROBOT IMPLEMENTATION

During the writing of this textbook the author had an opportunity to witness the proof-testing of a robotics application for a major U.S. manufacturer, the Singer Company's Motor Products Division, Clarksville, Arkansas, plant. Company representatives had gathered to witness the test because this robot was the first ever to be installed in this plant, and besides, robot projects always attract more than their share of attention. All was in readiness, and the test began. A few seconds into the test and CRASH: a sequencing error caused a collision between the robot and its simulated machine test stand, which literally fell over and was caught in the arms of an observing engineer. Embarrassment? Perhaps a little. But failure? No. The key word in the foregoing story is *simulated*. The test was a mock-up of the real thing, and simulated machines had been substituted for real factory equipment. The engineer at this plant had the foresight and experience to know that automation projects are tricky undertakings and unforeseen problems are to be expected. What could have meant disaster for the automation project and ridicule from labor and management alike resulted instead in a revision of the robot sequence, a resetup of the test stand, and a retry of the application. Although many intermediate setbacks were observed in the continuing automation of this highly successful company, the ultimate result was success in automation. We would all do well to follow the example of this company's automation engineers, and it is for this objective that this chapter was written.

Assuming a basic knowledge of industrial robots from Chapter 6 and of robot programming from Chapter 7, this chapter addresses the problem of planning, selecting, developing, testing, and installing a successful robot application. The principles and pitfalls of robot system implementation covered in this chapter are probably more important than the technical details covered elsewhere in this text. Although this chapter is about robots, the reader should recognize that many of the principles cited apply to the implementation of any automated system for manufacturing.

NOT SO FAST

Having learned what robots are, some of their capabilities, and the basics of how to program them, it is natural to want to apply this new knowledge to applications within our own factories or facilities. The most important consideration here is caution. The temptation is to leap into a project by buying a robot and "doing something, even if it's wrong." Many companies have done just that, and sure enough, it did turn out to be wrong.

The problem of implementation of robots and other manufacturing automation projects are almost always underestimated. Automation projects have an extraordinary susceptibility to Murphy's Law. Planning ahead for those inevitable pitfalls can permit steps to be taken to minimize their impact, especially if the problems can be discovered in the automation development laboratory, not on the factory floor. Careful planning also can moderate management's expectations into goals that have reasonable prospect of attainment.

Five phases can generally be identified in the implementation of a robotics application, despite the wide variation among robots in size and type. These phases and the associations among them are shown in Figure 9.1, which is intended to show a somewhat sequential relationship. The primary flow of activity is downward through phases, as is indicated by the heavier arrows, but at each level there are the alternatives of returning to a previous phase or of aborting the project.

At first, most companies try to skip one or more of the implementation phases shown in the diagram, but if they do, they invariably find themselves on a path back up the diagram. The resultant chaos has represented a significant setback for robotics in those companies and in others as the news spreads. Some case studies will illustrate.

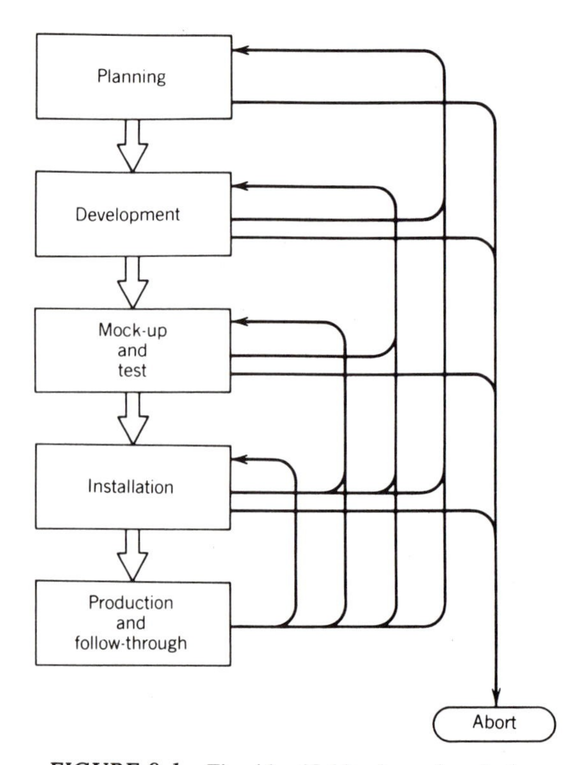

FIGURE 9.1 Five identifiable phases in robotics application implementation.

EXAMPLE 9.1 (CASE STUDY)

At a Ford Motor plant in St. Louis, snags in 200 production-line robots delayed the 1986 introduction of the Aerostar minivan. Then, the discovery that the same robots had been skipping many key welds led to the recall three months later of some 30,000 of the vehicles* [ref. 17].

EXAMPLE 9.2 (CASE STUDY)

A Campbell Soup plant in Napoleon, Ohio, was outfitted with a $215,000 system designed to lift 50-lb cases of soup. But anytime it encountered defective cases, the machine would drop them, causing thousands of dollars in damage. Eventually the robot was donated to a local university and replaced by three humans. Says Warren Helmer, the company's manager of engineering research and development: "Campbell's was ready for robots, but robots weren't ready for Campbell's"* [ref. 17].

It is virtually impossible to skip one of the five key phases successfully and achieve a satisfactory new robotics implementation. Each of these phases will now be examined.

PLANNING

Even the planning phase carries a prerequisite—education—but in this chapter we are going to assume a basic knowledge of robot types, capabilities, and control methods. If only one person in the plant has any knowledge of robotics, it would be wise to hold some basic robotics orientation seminars in-house to broaden the exposure of other engineers, technicians, and operating personnel to the potential of robotics applications in manufacturing. This increases the chances of finding a good *first* application.

Do not forget to include key labor representatives in the preliminary orientation seminars. Considering some of the benefits to be gained by automation and robotics, it is entirely possible that labor representatives will take a positive attitude toward certain robotics projects. And you can be sure that if labor is made to feel that their input on robotics projects is *not* welcome, they will most certainly become the project's adversary. Labor representatives are likely to be the best source for identifying those projects that workers themselves would choose to automate. While the labor viewpoint is just one among several, sometimes opposing viewpoints, it certainly is a viewpoint not to be overlooked or ignored.

Isolate Potential Application

The importance of the first application cannot be overemphasized. More than for any other type of manufacturing productivity improvement project, the success or failure of the first attempt to install a robot in the plant will be well known by all personnel. Do not be surprised if the word even reaches the general media. Robot stories, whether they have good or bad endings, make hot news items.

Using a basic knowledge of robotics, you can usually identify a good candidate for implementation, if one in fact exists in the operation under study. The planning phase

*© 1987 *Time* magazine, reprinted with permission.

now can proceed with this candidate, recognizing that later phases of implementation may precipitate a decision to scrap this candidate in favor of a better one.

Identify Objectives (Benefits)

Figure 9.2 displays a block diagram for the planning phase; the first step after isolation of a potential application is to define objectives in terms of benefits to be achieved. Such

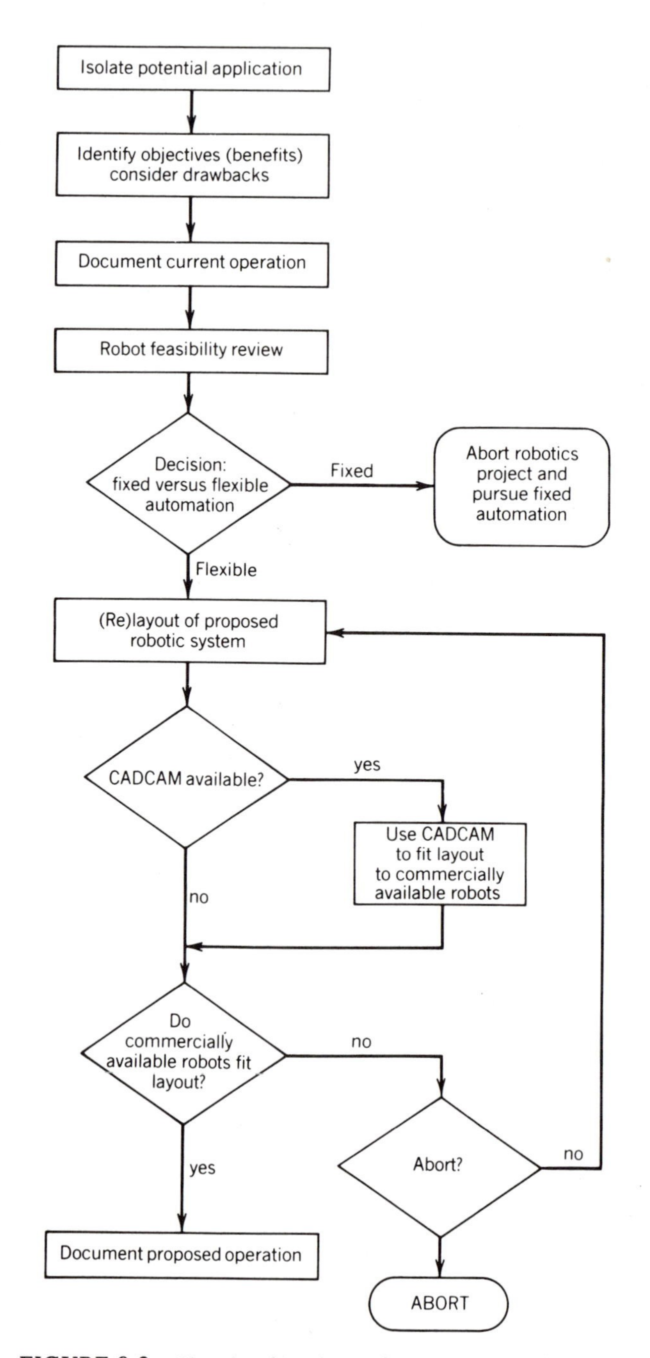

FIGURE 9.2 The planning phase of a robotics implementation.

advice may seem to be both obvious and trite, but it is easy to forget why a robotics project is being undertaken. Sometimes the reason is that "the boss wants to try a robot out somewhere—anywhere—in the plant." This is a valid, but not a very good reason; but if it is the real reason, and it often is, this reason should be openly stated and understood by all parties. There is a rationale for trying a new technology in the interest of education and experience, but it is better to gain experience on a successful and productive project, not on one that remains purely experimental.

A more robust objective would be to increase productivity, reduce labor costs, or reduce cycle time on a critical operation. Another good objective would be to eliminate an undesirable job or to assist in the more undesirable aspects of that job. Safety is a valid criterion, and the project may be initiated solely to protect workers from a dangerous job or from exposure to hazardous agents or conditions. Even if the robotics project increases costs and reduces productivity, it may be justifiable by the criterion of worker safety and health. Product quality might be another criterion, or a combination of objectives might be stated. In this early planning phase, it is risky and probably a waste of time to attempt to set quantitative goals, such as rate of return on investment. These will come later.

Consider Drawbacks

While naming objectives, a recognition of potential drawbacks to the project should be acknowledged at this point. The robot will likely have a significant impact upon one or more workers' jobs. This impact can add up to any of the following:

1. Good for the company, but bad for employees
2. Bad for both company and employees
3. Bad for one employee, but good for most employees
4. Good for the company and good for all employees

It is a noble objective to shoot for number 4 above if possible, but some robot projects may be worthwhile even if some workers are temporarily displaced. What is imperative is that the project planners consider worker impact and know where they stand about it relative to the particular project in question. Specifically, if the robot will be doing something that is currently done by a human worker, the project planners need to include in their thinking a plan for where the replaced human worker will be after the project is on-stream. You can bet that the subject worker will be doing some thinking about the future on his or her own, and if the worker's questions are given no answers—or worse, conflicting answers—a major obstacle to the success of the project can easily develop.

An honest look needs to be taken at the project impact upon production schedules and maintenance. Sometimes a project can be expected to lower production costs slightly in the long run but simply not be worth the disruption during setup and testing. Also the questions of model changes, product discontinuations, and potential process changes need to be considered. If these issues are addressed early on, it is possible that a more flexible and realistic robot system will be installed.

While considering production and maintenance issues, it is useful to construct worst-case scenarios and perceive plans for how the system could deal with these situations. Consider the following questions:

1. What would happen to the system if sudden power outage should occur?
2. Will the new system "burn the bridges" for returning to a manual system at a later date?

3. Will it be feasible to maintain production manually while the system is down for repair or maintenance?
4. Suppose the quality of vendor-supplied piece-parts being handled by the robot deteriorates. Will the system be able to handle these variations or even be able to recognize them when they do occur?
5. Can the robot be utilized in another system if its present application ceases to exist for whatever reasons?

Concerning robot motion, be sure to consider the possibilities of errors of both commission and omission:

Error of commission: Suppose the robot makes a wrong move, either the wrong motion at the right time or the right motion at the wrong time. What will be the result?

Error of omission: Suppose the robot fails to move when it is supposed to move. What will be the result?

Either error can affect production schedules, quality, damage to equipment, and even personnel safety. Both errors can be dealt with in the design of the system if they are anticipated.

From a quality standpoint, robots usually mean more product consistency, but will the resultant quality be consistently *better* or consistently *worse*? Process consistency is generally a prerequisite to quality, but robots sometimes handle products roughly and can mar finishes or crush fragile components.

Safety

Safety was mentioned earlier as a benefit to be gained by installing a robot, but the robot can be a hazard, too. Although the incidence rate is low, there have been cases of severe injury by robots, even a few fatalities. There is ample reason to emphasize safety in planning for robotics applications, even more than for other capital equipment projects. The public recognizes that some risks are necessary to accomplish the tasks of production, but the public has little tolerance for robots that injure people. There is some hostility to the idea that robots should be doing the work that people do anyway, and when a robot injures someone, that hostility is exacerbated. Further aggravating the problem is the familiar theme used by the entertainment industry depicting robots as warriors or sinister instruments of technology used by space aliens or forces of evil.

The initial planning phase is the time to consider what guards, interlocks, alarm routines, emergency power-offs (EPOs), and other safety measures may be needed to make the prospective application safe. Most of these measures apply to the production phase of the application after all programming and testing have been completed. The production phase, however, is not the most dangerous stage in the development of an application. Even though most of the time of a robot's operational life is spent in the production phase, most accidents occur during programming, setup, and subsequent maintenance.

During the programming of an industrial robot, the best protection against injuries is a well-trained technician and the availability of EPOs. One rule that is often applied is that no one is permitted in the work envelope while robot arm power is on. The intent of this rule is good, but unfortunately it is often impractical to follow during the programming phase. A rule that is ignored is worse than no rule at all. Certainly, the programmer should have immediate control of arm power via the teach pendant while programming. Only persons actively engaged in the programming and setup of the robot should be permitted within the work envelope of the robot.

For the production phase, some type of work envelope security is usually desirable to keep personnel out of the work area of the robot during operation. In Chapter 6 the method of optical presence sensing using photoelectric sensors was presented as a safety device for protecting against personnel intrusion into the work envelope. Infrared barrier devices can also be used for this purpose by interlocking them to the robot's control system. Any intrusion through the invisible barrier immediately stops the robot.

An even simpler method is to use a fixed perimeter guard or fence. Access gates are necessary for maintenance, but these can be equipped with limit switches to immediately interrupt the action of the robot in the event the gate is opened during robot operation.

Besides maintenance there are other reasons personnel may need to have access to the robot work envelope during operation. Sometimes a human worker and a robot share a work station in which each has a role. One example would be in which the human operator presents piece-parts to the robot for pickup. In such an operation it would seem that worker and robot would each need access to the same physical location. However, a sliding feed mechanism (see Figure 9.3) can serve to transfer parts to the robot from a separate human work station outside the robot's work envelope and at the same time transfer completed parts from the robot's work station back to the human work station, as is seen in the setups for some punch presses and other hazardous equipment. The human operator can perform the unload, load, and slide operations while the robot is busy with its own cycle.

It is possible to stop the unsafe intrusion of a robot into a protected area by the construction of heavy posts or other rigid obstructions. This method of dealing with safety should be considered a drastic one. Unfortunately, the post may seriously damage the robot, or worse, it might result in a worker being pinned between the robot and the post.

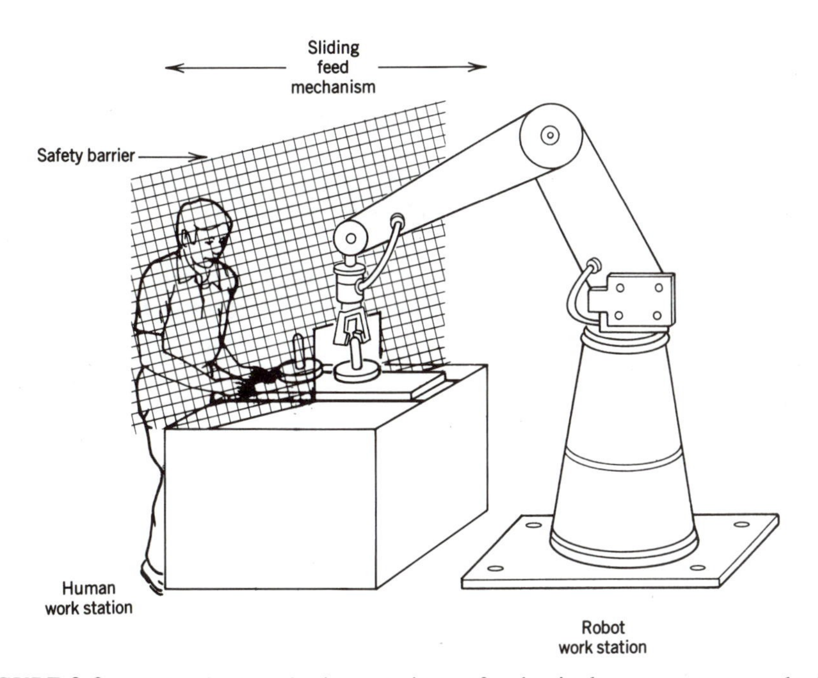

FIGURE 9.3 Sliding feed mechanism permits a safety barrier between operator and robot.

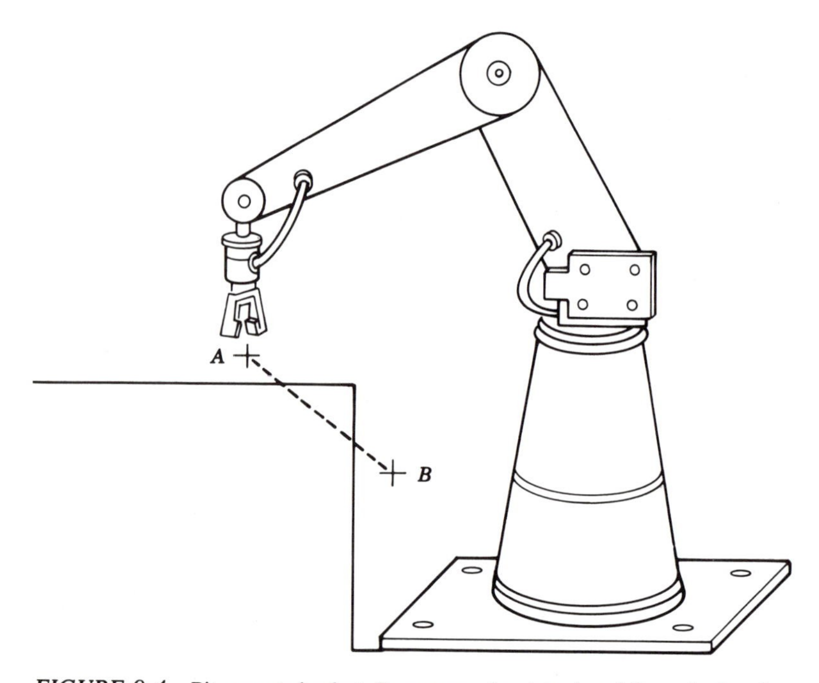

FIGURE 9.4 Pit-mounted robot. Programmed points *A* and *B* are in the clear, but the controller selects a path between the points that results in a collision with the floor.

Even the area overhead may need protection from robot collision. If the robot's reach extends into the rafters or other overhead structure of the building, some control may be needed to prevent damage to the building, the robot, or both. One choice is to lower the robot into a pit, but there are problems with this solution to the problem. If the *x-y* area of the pit is smaller than the area of the robot's work envelope, the robot may collide with the floor. Extreme care would be required in the programming of such a robot to prevent an intersection between the robot's path and part of the floor. Even if the programmer selected points that were in the clear, the robot controller might select a path between points that results in a collision. This concept is illustrated in Figure 9.4.

The alternative is to make the *x-y* area of the pit large enough to contain the *x-y* area of the entire work envelope of the robot. This solution can also be awkward, making the robot difficult to reach by personnel or the material handling system.

Yet another solution to the problem of protecting against a robot collision with the ceiling or superstructure is to provide a photoelectric sensor to signal the controller to arrest the robot's motion if the robot arm is lifted too high.

Document Current Operation

If the robotics project has survived the planning phase to this point, it is time to document carefully the existing operation to be modified. These data will be valuable both for the detailed design of the proposed automated system and for its eventual evaluation. The documentation should include layouts, machine-cycle times, manual operations times, maintenance histories, and production and quality performance records. In addition, the documentation should include product design data such as dimensions, tolerances, and other quality standards. Most of these data will be on file already and may even be computer-accessible.

The close scrutiny of operations required to commit their details to written documents usually results in discovery of some opportunities for improvement even before the robot is installed. This is an argument for embarking upon a robot implementation project even if it results in a decision not to use a robot.

Robot Feasibility Review

Upon completion of the documentation for the existing configuration of the candidate operation, it is time to examine specifications for commercially available robots to determine whether available robots can feasibly be applied to this type of job. The emphasis at this point is upon technical feasibility, not economic feasibility. Things to look for are rated payloads, general speed, repeatability, and motion control. Some ballpark economics are also in order as a prelude to the next decision that is to be made: the decision between fixed versus flexible automation.

Fixed versus Flexible Automation

At this point a hard look must be taken at the alternatives to robots. Even though a robotics application that looks satisfactory is beginning to take shape, this does not mean that an even better solution does not exist. The more conventional designs for automation employ fixed or hard automation. The robot with its general-purpose design and reprogrammable control system is the more flexible choice, but it is sometimes the more expensive choice. Hard automation can also be expensive, more so than are robots if the equipment is highly specialized.

If the production run is not very long, the hard automation route is especially expensive. Hard automation loses much of its value at the end of a production run and may even become worthless. An industrial robot, however, is more likely to hold some value because of its utility in other potential applications. Remember from Chapter 6 that robots are usually best for moderately high production, not for extremely high production.

Speed may be an important criterion in the decision between fixed and flexible automation. Fixed automation is usually faster than are industrial robots. Besides the matter of sheer speed, there is the question of concurrent function. Although robots can be equipped with double grippers and their controllers can signal multiple functions to be performed by other machines, robots themselves are generally single-function devices. Fixed automation is often best when entire banks of simultaneous operation are placed under the direct control of a single automated process controller.

This text is principally about robots and flexible automation methods, so the remainder of this chapter will assume that at this stage of planning a project decision is made to pursue the application of an industrial robot instead of the installation of fixed automation equipment. But the principles and procedures of a rational implementation of an automation project are nearly identical whether the choice is to use a robot or hard automation.

Proposed System Layout

Having decided between an industrial robot and fixed automation, the next step in the implementation process is to make drawings of the proposed automatic system, including the robot, conveyers, parts feeders, and process machines that will interact with the robot. This step is appropriate whether the robot end-effector is to be equipped with grippers, welding heads, paint spray heads, or any other tooling.

Earlier in the process, the current or manual operation was documented with layout drawings and production data. The earlier drawings can be a useful point of departure for the proposed system layout, but the physical arrangement of the process will likely change when the robot is installed. The changes may be minor to accommodate the work envelope of the robot or may involve a whole new approach to the process. In this planning phase, bold new approaches to the process should be given opportunity to surface and be evaluated.

In the layout of the proposed processes, do not forget the lessons learned from Chapter 4, and consider the possible need for buffer storage. Even when the proposed robot is reported to be extremely reliable, will it be reliable in this application? Will the interface from conveyor to robot and process machine to robot have the same high degree of reliability? If there is no provision for buffer storage, the robotic system workcell may have considerable downtime and fail to achieve minimum economic production levels.

As discussed in Chapter 6, speed is not the prime virtue of an industrial robot, and an automated system that simply replaces a human with a robot in a workcell of the same physical configuration may be missing some real opportunities for improving efficiency. Consider the following actual case in point. In a manual process, it was no problem for the human operator to reach over a small guardrail intended to hold workpieces in position on a conveyor. For the robot, however, this type of reach was more time-consuming and necessitated an eccentric design for the double gripper. It was discovered that the guardrail could be removed for a short distance along the conveyor right at the pickup point without affecting the guardrail function but greatly facilitating the robot function. This seemingly tiny change greatly enhanced the effectiveness of the new robot system.

The layout of a proposed robot system is an iterative process in which a trial layout is compared with the capabilities and work envelopes of commercially available robots. Then if the layout proves to be an impractical fit, a new layout is prepared in the next iteration. The iterative nature of the process is illustrated in Figure 9.2, which summarizes the entire planning phase of a robot implementation. When preparing layouts, remember to include vertical elevations as well as an overall plan view. The robot will be working in a three-dimensional environment, and all dimensions must be compatible. Project managers, pressing to get the robot system on-line, may not see the value in painstaking three-view drawings of the proposed setup, but successful robotics and automation engineers are finding such efforts to be time well spent.

One time-saver in the iterative process of relaying out of the proposed robot system and comparing it with commercially available robot hardware is to make use of a computer-assisted design package, such as McAuto's PLACE system. Figure 9.5 illustrates the "wire-model" method of computer-assisted design for the purpose of comparing robot work envelopes and studying potential machine interferences. Some of these packages have preprogrammed software to represent various commercially available robots in the display. The automation engineer merely names the desired robot for display, and the wire figure of that robot is immediately placed in the base image of the workcell.

To complete the documentation of the proposed robotic system, it is a good idea to attempt to quantify payloads, operation times, and robot arm paths. These details can be altered to conform to particular robot hardware at a later date, but target performance specifications for the process, particularly operating times, can be very useful in the development phase that follows.

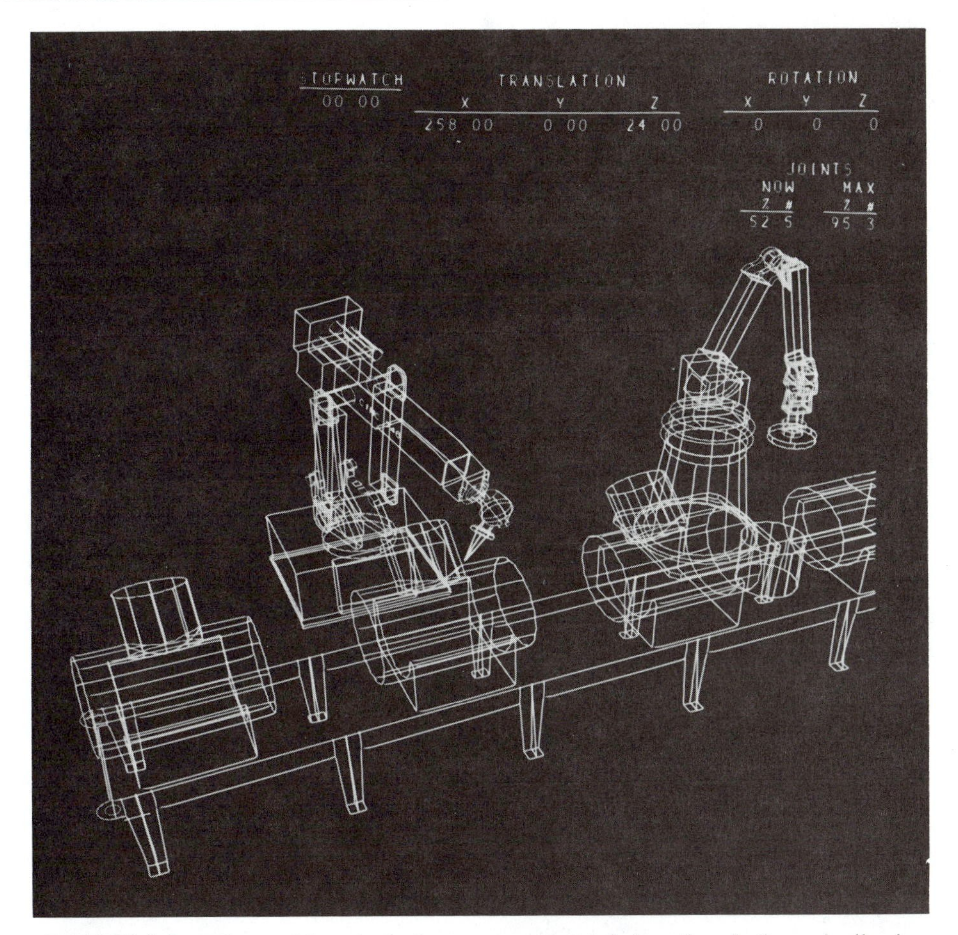

FIGURE 9.5 Wire-model method of computer-assisted design of a robotic workcell using PLACE. (Reprinted by permission of McDonnell Douglas Automation Company, St. Louis, Mo.)

RTM Analysis

The need for quantifying operation times has led to the development of the RTM (robot time and motion) method of analysis developed at Purdue University [ref. 78]. Most industrial engineers are familiar with MTM (Methods Time Measurement), the popular system for analyzing human work using standard time data. If precision measurement and standardization of individual work elements is viable for human tasks, which are subject to many causes of variation, an analogous system for robots should be even more effective and practical.

Table 9.1 lists the basic ten RTM elements, many of which are embellished by variations depending upon the circumstances of the operation or motion. The scheme of the RTM system is to have complete element tables on file for each robot model that is to be analyzed by the RTM system. In addition, computational formulas are developed for each robot model, using statistical regression techniques, and the regression constants are stored for each robot model to be analyzed. Both of these computer files of standard data are then accessed by a processor called the "RTM Analyzer."

TABLE 9.1 RTM Symbols and Elements

Element No.	*Symbol*	*Definition of Element*	*Element Parameters*
1	Rn	*n-segment reach:* Move unloaded manipulator along a path comprised of *n* segments	Displacement (linear or angular) and velocity or Path geometry and velocity
2	Mn	*n-segment move:* Move object along path comprised of *n* segments	
3	ORn	*n-segment orientation:* Move manipulator mainly to reorient	
4	SEi	*stop on position error*	Error bound
4.1	SE1	Bring the manipulator to rest immediately without waiting to null out joints errors	
4.2	SE2	Bring the manipulator to rest within a specified position error tolerance	
5	SFi	*Stop on force or moment*	Force, torque and touch values
5.1	SF1	Stop the manipulator when force conditions are met	
5.2	SF2	Stop the manipulator when torque conditions are met	
5.3	SF3	Stop the manipulator when either torque or force conditions are met	
5.4	SF4	Stop the manipulator when touch conditions are met	
6	VI	Vision operation	Time function
7	GRi	*Grasp an object*	
7.1	GR1	Simple grasp of object by closing fingers	Distance to close/open fingers
7.2	GR2	Grasp object while centering hand over it	
7.3	GR3	Grasp object by closing one finger at a time	
8	RE	Release object by opening fingers	
9	T	Process time delay when the robot is part of the process	Time function
10	D	Time delay when the robot is waiting for a process completion.	Time function

To use RTM, the analyst lays out a procedure for the robot to accomplish a required task. The procedure, not unlike computer programming, is documented in a format in which the RTM elements are mnemonics analogous to computer op codes. Constants are specified in fields that follow the RTM element field, with the constants separated by commas, as in the modifier fields following op codes in computer programming. Comments are optionally added to explain each line of RTM code, the same as comments are used in computer programming. Table 9.2 shows a sample listing of RTM code for a manual insertion task performed by a Cincinnati Milacron T3 industrial robot.

The RTM Analyzer accepts the "program" of RTM code and performs an evaluation and interpretation procedure somewhat analogous to the compilation of a computer program. During this "compilation" the element tables and computational performance models for the given robot are accessed and applied to the specific constants supplied by the user in each line of code. The result is a complete evaluation of the specified method, including a listing of the elemental times and the overall time totals in milliseconds. Using stopwatch-measured time as a standard for comparison, experiments with RTM have shown a close correspondence with reality. Nof and Lechtman [ref. 78] report a record of predictive accuracy in the range of −2 percent to +3 percent.

For all of the precision shown by RTM and its advantage over MTM in not being subject to the random variation of human performance, there is a major drawback to the method. Each robot has its own set of performance characteristics, and though these characteristics are very precise and nonrandom, they still must be assembled and stored in the computer for reference by the RTM Analyzer. For example, for a single RTM subelement (R1) an entire matrix of times has been developed for a single robot make and model at Purdue University. This two-dimensional matrix accommodates variations in move distances and move speeds in centimeters per second. Imagine the size of the database required to support RTM research using all the RTM elements and subelements, accomodating all move distances, considering all the variations of move speeds on all available makes and models of industrial robots! The pioneering effort to lay the groundwork for this enormous research task has been the objective of the research laboratories of the Department of Industrial Engineering at Purdue University. Though the research required to develop RTM standard data represents an ideal opportunity for students to practice work measurement and statistical skills, it is questionable whether universities should be expected to undertake such an immense effort without substantial support. The robot manufacturers might be assumed to take responsibility for this task for the convenience of their customers in analyzing and justifying potential applications using their robots. However, for the most part, the robot manufacturers have not answered this challenge. This raises the question whether RTM will ever achieve universal industry acceptance as a practical and reliable tool for analysis of robot applications. It is possible that RTM was a vision of the 1980s that might experience a rebirth of interest in the 1990s with the advent of the mass storage media described in Chapter 15.

A final admonition for the entire planning phase of a robotics implementation is to keep it simple. The admonition also applies to the development phase that follows. It is so easy to be overly optimistic about the powers of the machine and launch into a project that will not work. Certainly if the application will be the first robot installed in the plant, do not attempt to redefine the state of the art, particularly with regard to sophisticated sensing systems and artificial intelligence. A modest success is much better than an immodest failure. Use of the simplest robot that will accomplish the objective, and while looking for the simple solution, keep open the possibility of using no robot at all.

TABLE 9.2 RTM Input Data for Insertion Task by the T^3

Statement		*Comments*
T^3		Robot type
Insertion Task		Task title
*Conditions		
. . . .		Input of condition signals, if any were used
*		
[REPEAT 1 TO 26 SEVEN TIMES]		Repetition command could be stated here
1 R1	5, 7.5	Reach 7.5 cm at 5 cm/sec to start position above base
2 GR1		Grasp base
3 M1	5, 5.0	Raise base
4 M1	25, 48.0	Move to above assembly position
5 M2	25, 12.5, 5, 5.0	Bring base down in two segments movement
6 RE		Release it
7 R1	25, 23.0	Rise
8 R1	25, 55.0	Reach fixture
9 D 2		Wait 2 sec (for continuation)
[IF SIGNAL. NE. VALUE) GO TO END]		Conditional branching could be used
10 R1	5, 15.0	Bring fingers above peg 1
11 GR1		Grasp peg 1
12 M1	5, 15.0	Raise peg 1
13 M1	25, 55.0	Move it to above base
14 M2	25, 11.0, 5, 9.0	Insert peg in two steps
15 RE		Release peg 1
16 R1	25, 22.0	Rise
17 R1	25, 50.0	Reach above peg 2
18 D 2		Wait 2 sec
19 R1	5, 12.5	Bring fingers above peg 2
20 GR1		Grasp peg 2
21 M1	5, 12.5	Raise it
22 M1	25, 50.0	Move it above base
23 M2	25, 10.0, 5, 5.0	Insert it in two steps into peg 1
24 RE		Release peg 2
25 R1	25, 17.5	Rise
26 R1	25, 60.0	Reach start point

DEVELOPMENT

The demarcation between planning and development is not clear, but a general sign is the appearance of prototype hardware. Welding robots, spray-painting robots, and virtually all categories of robot application require some related hardware development, but it is especially true of the robots that will manipulate piece-parts. The most important areas of hardware development are the following three robot interfaces:

1. Robot interface with product, especially piece-parts
2. Robot interface with handling equipment
3. Robot interface with processing machines

The biggest problem is usually the gripper design to handle piece-parts. Robot manufacturers are probably wise to market "handless robots," leaving the end-of-arm tooling development to the user. The users know their own products, how these products have been handled using the manual system, previous attempts at automated handling that have not worked out, and the possibilities for design change of the products themselves. Considering the importance of gripper design, we shall return to it in a moment. However, first let us review the prerequisite process stability analysis and some other contributors to the equipment/product interface, such as jigs and fixtures and piece-part design.

Process Stability Analysis

A key to the design of a robot gripper system is the uniformity of the piece-parts that the gripper will pick up. Product quality suddenly can take on new meaning as dimensional tolerances become critical to the robot's ability to pick up the parts. The key specification is not the nominal dimension; it is the tolerance. Besides dimensional uniformity, there is the problem of positional uniformity, both in orientation and in point in space. If existing conveyors and fixtures are to be used, what is the positional capability of this equipment?

An analysis of process stability may consist of a review of the documentation prepared for the existing operation, but it will usually involve more than this. It is worthwhile in such an analysis to examine worst-case product rejects and worst-case misorientations. Consider such questions as:

1. How often will the robot encounter piece-parts that are presented upside down?
2. Will the robot ever encounter missing piece-parts?
3. If the robot will handle assemblies, is it possible that some of these assemblies will be incomplete or have misaligned parts?
4. Will piece-parts ever have unexpected burrs or mold flash?

The analysis of process stability may cause the developers to return to the planning stage, but process stability problems can perhaps be overcome by part redesign, jig and fixtures design, or by the employment of parts presentation equipment.

Part Redesign

Chapter 1 introduced the fundamental concepts for designing products for robots and manufacturing automation. Some adherence to these basic principles is prerequisite to a successful program of automated production, but at this phase of the robotics implementation process it is wise to return to these principles and determine whether additional changes to the product design will be advantageous to the robotic system. Particular attention should be given to the mating of parts to each other or even the mating of parts to the machines that will perform operations upon them. For instance, it might have been obvious to the original product designers to provide taper pins for the mating of two halves of a product housing. But the same designers might have overlooked the tapering of a shaft that is to be inserted into a hole in an assembly machine to facilitate the production process. Indeed, the original designers may have omitted the shaft entirely because it served no final purpose in the product. An in-process purpose in an operation such as robotic assembly may make product redesign very worthwhile.

Jig and Fixture Design

The advent of robotics has promoted the development of special-purpose pallets and trays that maintain both position and orientation for parts and assemblies to be handled by robots. The development of foam molding techniques has made such special-purpose

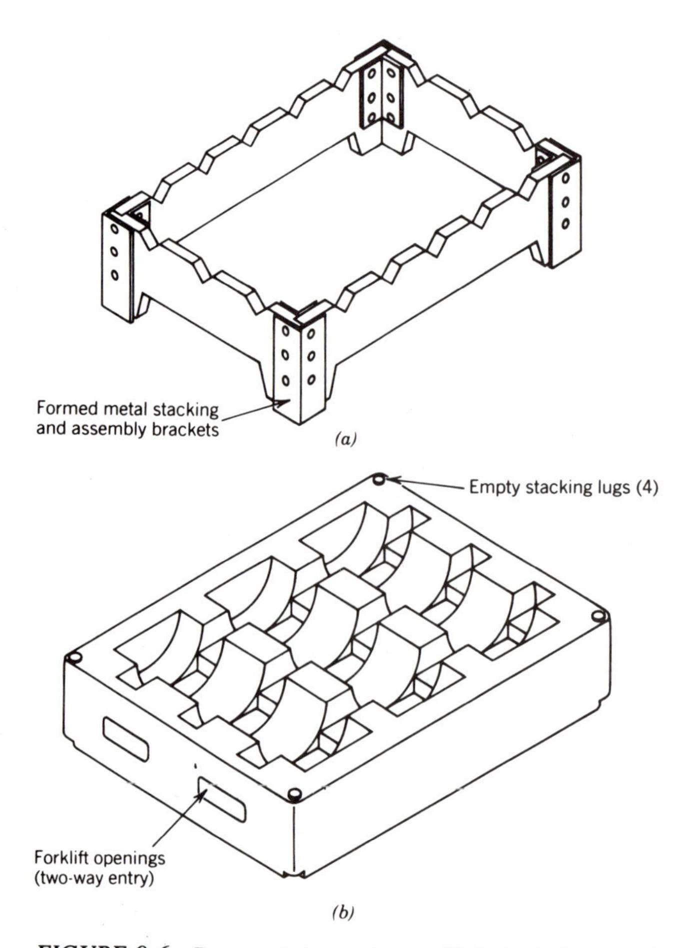

FIGURE 9.6 Representative custom-molded plastic frame and tray type figure facilitate parts orientation for robot handling. (a) Stacking frame. (b) Separator—tray pallet. (Reprinted by permission of Thermodynamics Corporation, Broken Arrow, Okla.)

fixtures effective and economically worthwhile. Figure 9.6 illustrates pallet- and tray-type fixtures useful for robot loading and unloading applications.

Unfortunately, special-purpose trays and pallets can be quite expensive to procure during the development phase of a robotics implementation. The answer is usually to fashion fixtures manually from alternative materials such as corrugated cardboard or wood during the development phase.

Gripper Design

We now return our attention to the design of the robot gripper itself. Grippers are often subjected to a great deal of punishment and are notorious for wear. This malady suggests the use of hardened tool steel for gripper jaws, but such jaws are not recommended for the development and mock-up phases. Gripper designs typically evolve through several modifications during robotics application development. Therefore, a soft metal is often used for experimentation, and aluminum is the most frequent choice.

Gripper design can make or break a robotics project, and its importance cannot be overemphasized. The product design and application will dictate the type and complexity of the gripper. Ordinary machine loading and unloading employs perhaps the simplest gripper designs. But if the application is handling forgings, the complexity increases, as will be seen in Chapter 10 in the discussion of representative robotics applications. The most complicated gripper designs are those used in automatic assembly. It is not unusual for assembly grippers to have multiaxis movement, spring-actuated intermediate steps, and on-board subtooling.

In the design of a robot gripper, do not forget to consider the possibility of using a double-handed gripper. Double grippers can be especially productive for machine loading and unloading using a strategy whereby the robot handles both a finished and an unfinished workpiece at the same time. Thus, every time the robot goes to the machine it both unloads *and* loads it. The advantage to this strategy is that the robot can move parts to other points while the machine is in its run cycle instead of sitting idle at the machine awaiting completion of the cycle. Such systems will be analyzed in detail in Chapter 10.

MOCK-UP AND TEST

Up to this point in the robotics implementation, a robot still has not been purchased. The purchase can perhaps be delayed even further. It is acknowledged that the choice of robot hardware was quite firm by the end of the planning phase. Some commitment to that choice was required in the development of the gripper and associated fixture hardware. A real robot will be needed for the mock-up and test phase, but there are ways to experiment without purchasing, as will be seen in a moment. First, a test stand must be set up.

Test Stand

Most industrial robots interact with other process equipment, some of which may be massive, very expensive, or too critical to the existing process to permit it to serve a role in a robot test setup. Only a portion of each process machine is generally necessary to prove the robotics setup. The reasonable course of action, then, is to build a model or merely a skeleton of each machine, duplicating only the salient features. Figure 9.7 illustrates a skeletal model of an automatic machine for winding electric motor armatures at the Singer Company. The machine was one of two machines to be loaded and unloaded

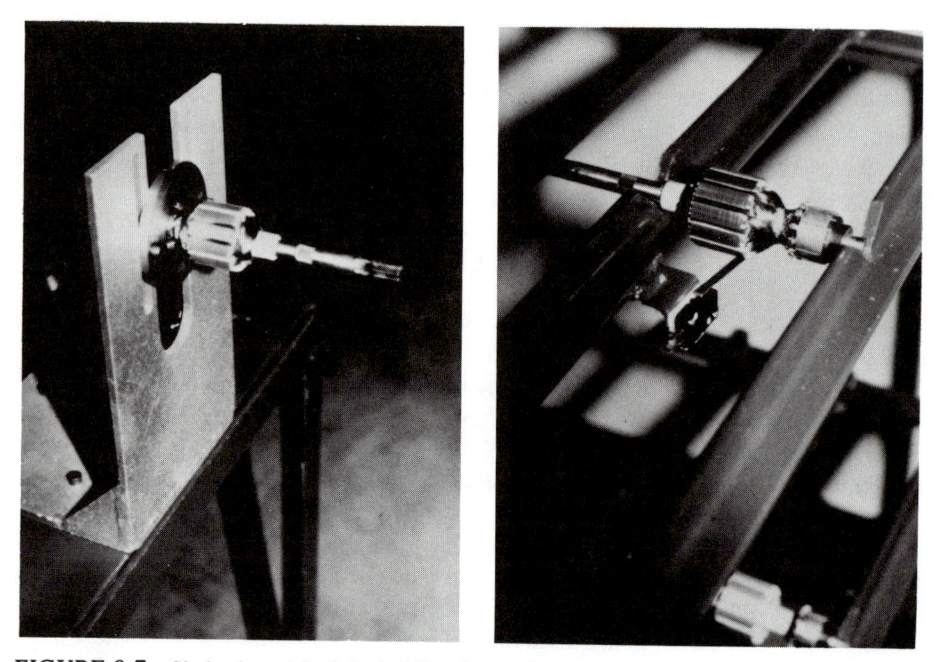

FIGURE 9.7 Skeletal model of the holding fixture for an automatic machine for winding electric motor armatures. The skeletal model was used during application testing of an industrial robot for loading and unloading the machine. Also shown is a similar skeletal model of the conveyor pickup point.

by an industrial robot, but neither could be spared from the production line during the robot-application testing phase. The only feature of interest on the winding machine during the test was the *fixture* that would receive the unwound armature shafts. Therefore, as can be seen in Figure 9.7, a simple frame for holding this fixture is all that is needed in the robot test stand.

Sensors and Actuators

The industrial robot rarely works alone, free from any interaction with related processes. Limit switches, proximity switches, photoelectric cells, and other process sensors must be devised as inputs to the robot so that the robot can synchronize its movements with the process or products it serves. Conversely, the robot may send to the process an output that signals some function to commence. Such inputs and outputs apply not only to robots whose jobs are to handle and manipulate workpieces. Spray-painting robots, spot-welding robots, and assembly robots also have such needs for input and output. These I/O elements are generally conceived earlier, in the application planning period, but choice of devices is often still underway during the system mock-up. The subject of industrial logic analysis and control of these robot inputs and outputs, usually by programmable controllers, will be the subjects of Chapters 11, 12, and 13. The function of the desired inputs and outputs can usually be tested by programmable controllers prior to acquisition of the robot itself.

Robot Test

The point is finally reached at which a real robot must be added to the test stand for experimentation and prooftesting. However, actual purchase of the robot may not be necessary. If the company already has in operation a robot of the type desired in the potential application, it can perhaps be temporarily set up in the test stand for experimentation. To do this, however, one must consider the potential schedule disruption of seizing a working robot from the production line.

Some universities and private robot laboratories have "robotics centers" where industry can try out potential robotics applications without purchasing the hardware. Such centers generally have some kind of cooperative terms requiring industrial users to contribute to a foundation so that the university or laboratory can acquire robots, maintain them, and sustain the operations of the robotics center. Such an arrangement can serve to minimize the investment risk of the industrial robot user while representing a significant contribution to the education of young engineers and technicians. Exposure to real industrial robots and application hardware is so important to college students because such equipment has only recently become available. Industrial concerns seeking to reap the benefits of a robotic laboratory arrangement with a college or university should also consider the tax merits of contributing funds to a college or university. United States tax legislation of the early 1980s made possible tax credits for contributions to colleges, universities, and other nonprofit institutions when the funds are earmarked for research. It is important to understand the distinction here: tax credits are far superior to tax deductions.

When using an institutional robotics center for mock-up and test of a robotics application, it will usually be necessary for the potential user firm to supply the test-stand equipment peculiar to its own operation. This will generally consist of product samples, product-specific jigs, fixtures, robot grippers, and skeletal models of the process equipment. The institution will generally supply the robot itself, its controller, and a programmable controller for the overall system. The institution also may provide lasers and other sophisticated tools for alignment, measurement, and evaluation. Student technicians may develop gripper designs from product samples.

When setting up a robot test stand at a robotics center, it is sometimes useful to convert proposed layout drawings to full scale. Full-scale layouts are useful for laying a pattern on the floor or bench where a test stand is to be set up. This was the approach taken in the motor armature winding application for the Singer Company described earlier in this chapter. This robotics application was tested at the University of Arkansas Center for Robotics and Automation.

Another avenue may be available to the potential industrial robot user, still without buying the robot. Some robot suppliers have been known to lend an industrial robot to a potential purchaser for a period of perhaps 60 days. During this period, the potential robot application can be fully tested and evaluated. If the application is successful, the robot can be purchased; otherwise, it can be returned.

Experimentation

Upon obtaining access to the real robot, either acquired or to be acquired, some preliminary experimentation can be performed in manual mode, if the robot is so equipped. Using the teach pendant, the operator or technician can direct the robot through its intended cycle in manual mode without teaching the robot. This can be done as a dry run or with the actual product in the grippers, or paint in the spray gun, or with welding current on, depending upon the application. Most firms will want the initial experimentation and familiarization phase with the robot to be done without supplementary process equipment models in place in the test stand.

The next step in the experimentation is to program the robot through a slow-motion execution of its intended cycle. This would correspond to teach mode A, as described in Chapter 7. Continuous-path robots may be programmed using a slow-motion specification in the program. While operating in slow motion, the entire robot system, including inputs and outputs from and to the process, can be checked for conformance to plan.

After the slow-motion test, the robot can be reprogrammed in teach mode B—that is, speeded up to real time. This is the first full-scale, real-time test of the robot application in the test stand, and its success is still not certain. Personnel should still be ready for the unexpected to happen, such as was described at the beginning of this chapter. Personnel may also need to be available to supply product pieces manually on the upstream side of the robot during the simulation and remove finished production on the downstream side.

One additional point must be emphasized on the subject of experimentation. Work-envelope hazards are at their highest levels during the experimentation process. Numerous engineers, technicians, operating personnel, and simply spectators usually gather to witness and assist in the experimentation of a new robot. At this point, work envelopes are subject to change as the setup is revised. Envelope boundaries are usually not clearly delineated with barriers during the experimentation period. Technicians need to enter the work envelope often to adjust valves, mechanical stops, grippers, or workplace fixtures. Persons who operate the robot power and controller during these periods should be certain that the envelope is clear during robot operation. Use of verbal signals such as "CLEAR!" can facilitate safety, but the operator should assure that any such signals to clear the robot work envelope have in fact been followed before actually initiating robot motion.

Repeatability Evaluation

Once the full-scale, real-time application is operational, it should be run continuously for a number of cycles to check for repeatability. This should be done in the test stand

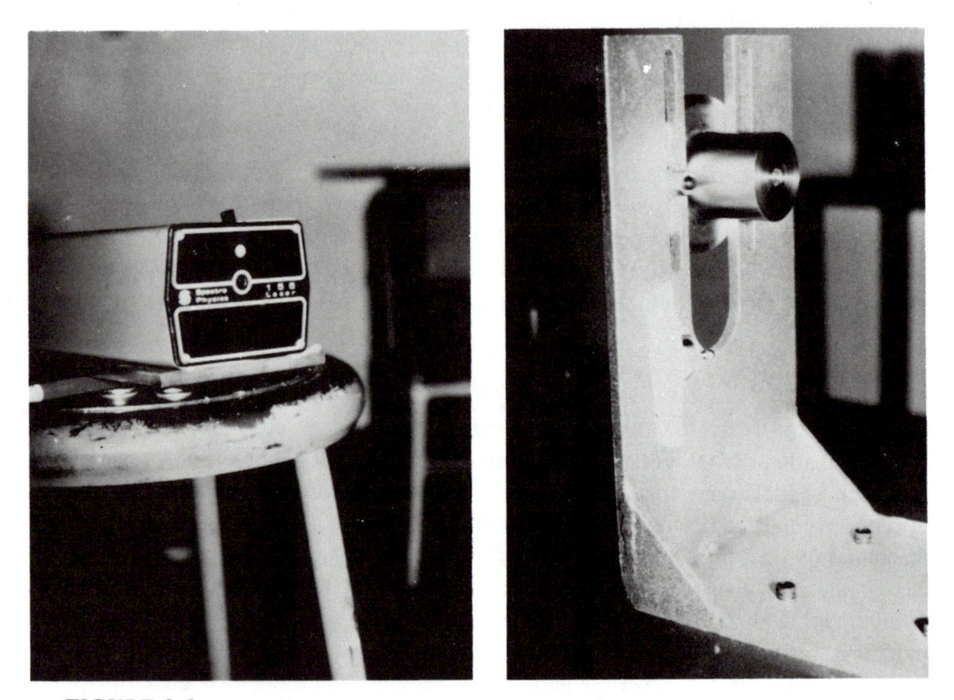

FIGURE 9.8 Laser is used to verify alignment of skeletal model in robot test stand.

before converting the application over to continuous production operation. Sometimes it is found that a robot system can operate nicely for a few iterations of its cycle at full speed, but then problems can develop. During this continuous test, alignment of the equipment should be checked frequently and the repeatability of the robot monitored.

A low-power laser is a useful alignment tool for robot-application test stands and for checking the repeatability of the robot itself. In the Singer electric motor winding application described earlier, tolerance for misalignment was low, and a laser (Figure 9.8) was used to check for slight shifts in position during the testing. A simple dentist's mirror was taped to the machine model frame and reflected the laser beam to a spot on the wall of the laboratory to provide a multiplying effect. The principle is illustrated in Figure 9.9. The tiny spot on the wall made by the highly collimated light of the laser is simply marked with a pencil to indicate original position (point A). At any time during test, the laser spot can be checked against the pencil spot. The laser selected for this purpose was of low power (0.95 milliwatt) so that personnel were not endangered even if they walked in front of the beam, provided they did not look directly into the beam. A case study will now be used to illustrate the multiplying effect of the laser when used as an alignment tool.

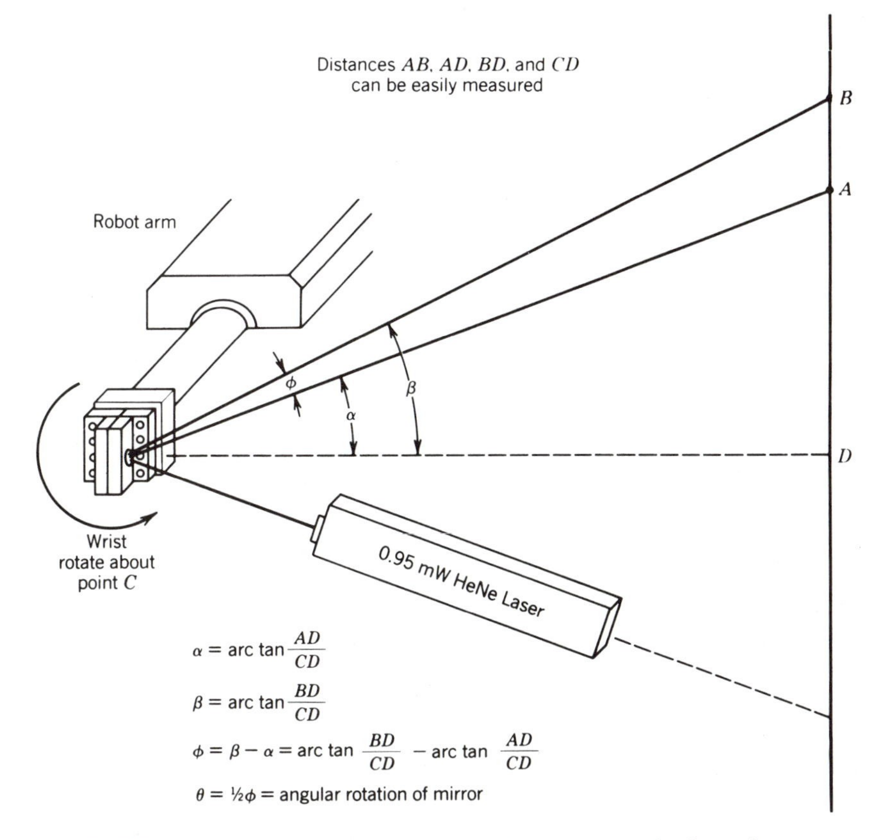

FIGURE 9.9 Laser beam permits a multiplying effect to check angular misalignments of robot or test stand equipment. Points A and B are tiny spots of laser light reflected from the dentist's mirror, which is mounted on the robot. The tiny spots of light on the wall can be as much as 40 ft away from the robot; the farther away, the greater the multiplying effect.

EXAMPLE 9.3
Robot Alignment Check Using a Laser Optical Lever

A robot test stand is set up in a large room. The robot gripper is $3\frac{1}{2}$ ft above the floor. A low-power laser beam is aimed at a dentist's mirror rigidly secured to the gripper and is reflected to a spot on the wall 40 ft away. At the original position of the robot, the laser makes a point of light 6 ft up the wall measured from the floor. After 100 cycles of robot operation, including 200 rotations (180°) of robot axis 5 (wrist rotation), the robot is returned to the original position in its cycle. At this point, the laser spot has moved up the wall to a position 6 in higher than it was before the test. Assuming that the mirror has remained fixed to the robot and the orientation of the laser source has not been disturbed, what has been the shift in angular rotation of the wrist?

Solution

Referring to Figure 9.9, distances are as follows:

$$AD = 6 \text{ ft} - 3\tfrac{1}{2} \text{ ft} = 2\tfrac{1}{2} \text{ ft} = 30 \text{ in.}$$

$$BD = 30 \text{ in.} + 6 \text{ in.} = 36 \text{ in.}$$

$$CD = 40 \text{ ft} = 40 \times 12 \text{ in.} = 480 \text{ in.}$$

$$\phi = \beta - \alpha = \text{arc tan}\,\frac{BD}{CD} - \text{arc tan}\,\frac{AD}{CD}$$

$$= \text{arc tan}\,\frac{36 \text{ in.}}{480 \text{ in.}} - \text{arc tan}\,\frac{30 \text{ in.}}{480 \text{ in.}}$$

$$= 4.29° - 3.58° = .71°$$

$$\theta = \text{angle of rotation of the mirror}$$

$$= \tfrac{1}{2}\phi = \tfrac{1}{2}(0.71°) = 0.36°$$

Tweaking

The term *tweaking* was apparently coined by engineering students working late hours perfecting their robotics laboratory projects on the night before final presentations. Tweaking is the use of trial-and-error methods to make tiny adjustments to the robot's programmed cycle in an attempt to squeeze the highest possible productivity level from a workcell. Some engineers disparage such procedures in favor of analytical optimization techniques. But in the industrial world, tweaking, by whatever word you might want to call it, is the normal method of solving a problem. Industry, just as the student on the night before a project's due date, often does not have time to research the theory or perfect the analytical formulae, especially for slight adjustments to cycle that can improve productivity. When operating at or near optimality, trial and error or tweaking may be the fastest and most efficient method of fine-tuning the process.

INSTALLATION

Testing, adjusting, evaluating, and tweaking have been completed, and the robot workcell is ready for installation on the production floor. If the layout, mock-up and test were performed correctly, the actual installation should be straightforward. It needs to be that way to prevent line delays and to minimize the distraction of employees during setup. Robot installations always get a healthy share of attention on the factory floor.

Conventional project scheduling tools can be useful in scheduling, controlling, and managing the robot installation. Such tools include PERT and critical path methods (CPM), which of course apply to any project-type installation, not just robot installations. But the typical urgency associated with robot installations enhances the value of project management tools to expedite the installation. Table 9.3 shows a critical path activity schedule that was actually used in the installation of a Cincinnati Milacron HT^3 Industrial Robot at the laboratories of the University of Arkansas. The associated precedence network diagram is shown in Figure 9.10. The actual critical-path computations are beyond the scope of this text and are the subject of another course. A suggested reference on this subject is Moder, Phillips, and Davis [ref. 69].

A final caution in the robot installation phase is to use care to see that main utility shutoffs are not inadvertently located within the work envelope of the robot. In one facility, a wall-mounted telephone was installed inside the safety perimeter fence enclosing

TABLE 9.3 Robot Installation Schedule

University of Arkansas
CINCINNATI MILACRON HT^3 INSTALLATION
CRITICAL PATH ACTIVITY SCHEDULE—July 13

Activity	*Time Required*	*Scheduled Start*	*Scheduled Completion*	*"Drop Dead" Date*	*Schedule Slack*
A. Area layout	3 days	Yesterday	July 15	July 15	None
B. Uncrate, check contents	3 days	Yesterday	July 15	Sept. 5	33 days
C. Clear floor area	3 days	Yesterday	July 15	July 15	None
D. Remove air conditioner duct	2 days	July 15	July 19	July 19	None
E. Electrical installation (460 V–3ϕ)	8 weeks	July 19	Sept. 14	Sept. 14	None
F. Water supply and cooling tower	3 weeks	July 19	August 9	Sept. 14	5 weeks
G. Foundation, anchor studs (incl. 2-week cure), class A ground	4 weeks	July 15	August 12	Sept. 6	16 days
H. Procure hoist	2 weeks	Yesterday	July 26	Sept. 5	29 days
I. Robot placement	1 day	August 12	August 15	Sept. 7	16 days
J. Hook up water and electrical	3 days	Sept. 14	Sept. 19	Sept 19	None
K. Perimeter fence	1 week	August 15	August 22	Sept. 15	17 days
L. Fence interlock	2 days	August 22	August 24	Sept. 16	17 days
M. Lighting	3 days	July 15	July 20	Sept. 19	42 days
N. Signs, seating, display	2 weeks	August 15	August 29	Sept. 21	16 days
O. Start-up test	2 days	Sept. 19	Sept. 21	Sept. 21	None

Note: "Days" means working days. Saturdays, Sundays, and Monday, September 5 (Labor Day) are excluded.

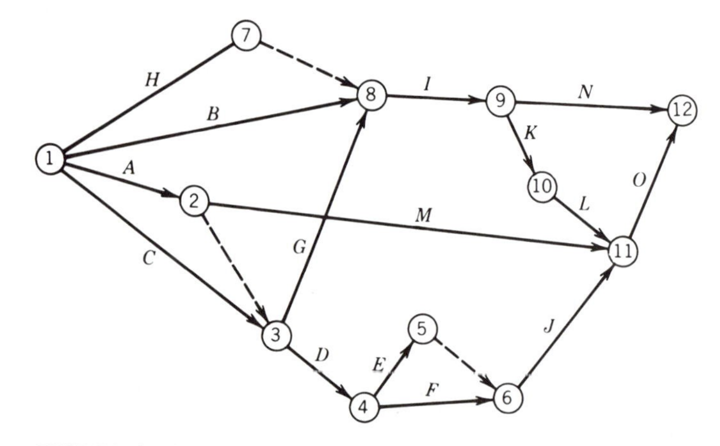

FIGURE 9.10 Precedence network (PERT) diagram for robot installation at University of Arkansas.

an industrial robot work area. The telephone was outside the robot work envelope, and the idea did have merit because it afforded the maintenance technician easy access to communication with the manufacturer's field service engineer during troubleshooting, but the scheme had its drawbacks. On occasion, the telephone would ring during robot operation, tempting personnel to enter the restricted area to answer the phone.

Preliminary Production

The physical installation is only part of the overall installation. The new robotics workcell should be operated temporarily on a pilot basis to prove the process. This may seem to be a duplication of the mock-up test, but remember that the pilot operation is with real processing machines and real product. "Pilot operation" implies production at a reduced volume or rate. This may be a luxury that cannot be afforded if the process equipment is critical and loaded to full capacity. If a pilot operation is not economically feasible, at least a parallel operation is desirable even though it may have to operate at full speed. By "parallel" we mean the operation of separate, new automatic and old manual workcells *simultaneously* until the newer automatic cell is proven. In some applications, parallel operation simply will be impossible, and full-scale stand-alone production will be imperative. In such cases, a great deal of importance must be placed upon the previous phase of the robot workcell implementation: mock-up and test.

PRODUCTION

Once the new robotics application is in full operation, it still should not be ignored. Continuing evaluation of the application is necessary to identify problems that need correction, and not to be overlooked is the need for evaluation data for decision-making when other potential robotics applications arise. Hard questions that need answering after production is well underway are the following:

1. Were original objectives for this application reached?

2. If original objectives were not reached, was the application still successful from an economic standpoint?
3. Is the quality of the product distinguishable from that of the product from the former process? If so, is it better or worse?
4. Is the system uptime proving the attainment of reliability and maintainability objectives?

MANAGEMENT AND WORKER COMMITMENT

Conspicuously absent in this chapter has been reference to the securing of management approval and worker support of the proposed robotics project. There was a warning early in this chapter against automatic management approval—the cases where the boss wants a robot, *period*. Lasting success of a robotics project requires technically and economically sound projects that can be clearly justified on these terms.

If the robotics project is the *first* robot to be installed in the plant, it is wise for management approval to be secured at each of the five major phases of implementation. In this fashion, management can approve only a piece of the implementation at a time. This in turn places a responsibility upon project personnel to prove that the project remains viable phase by phase as the project progresses. Capital investment criteria such as rate of return should be utilized to evaluate each new expenditure of the firm's resources.

Worker acceptance may be more tricky than management acceptance, but such acceptance is facilitated by genuinely seeking worker ideas and inputs to the process. There is a marked difference between seeking worker acceptance "so that workers won't obstruct the project" and sincerely seeking worker inputs in order to make the project more successful. Workers generally know more about the processes they operate than do the engineers and technicians who are attempting to implement the robotics application to facilitate those processes. If the knowledge and experience of these workers can be utilized to modify, improve, limit, or change the focus of a robotics project, workers can take pride in claiming co-authorship of the robotics project if it is successful. The best part of this strategy is that the robotics project approached in this way probably will be *technically* more successful as well as more successful from a worker acceptance standpoint.

Worker resistance to robotics projects is often well justified. If someone is going to lose a job to satisfy the ego of a manager, technician, or engineer who happens to want to see a robot installed in the plant regardless of its economic viability to the company, workers have every reason to question the rationale for the project. This is a question of *ethics* in automation, a topic that shall be discussed further in Chapter 17.

The resistance to robotics projects encountered from operating supervisors can be surprising. Supervisors are a part of management's team, but when it comes to robotics projects, supervisors may exhibit reactions more akin to labor's because the supervisor may feel just as threatened as labor does. A conversion to automation may mean that employees will be displaced, perhaps reducing the number of personnel supervised and in turn the supervisor's status or importance. The robotics and automation equipment may also be more technical than before, and the supervisor may feel unqualified to handle the new responsibilities. As was credited earlier to labor, supervisors also have a great deal of knowledge and experience with the processes they supervise, and their contributions to robotics projects within their areas can be invaluable.

TRAINING

In the early stages of production with a new robotics workcell, the technicians who installed the robot will usually remain responsible for robot operation, program modification, and maintenance. In-house training during this period can be used to transfer knowledge in operations and maintenance to personnel who normally operate and supervise the processes in the area in which the robot has been installed. Sometimes robot manufacturers will have schools for training operating and maintenance personnel on their particular equipment. This is especially true for maintenance personnel. One word of caution is in order on this point. Robot manufacturers should not be expected to provide basic training to customer maintenance personnel in such subjects as basic hydraulics and electronics. Some robot users send new employees having no maintenance training or experience to robot school, expecting them to become experts in a few days. The adverse consequences that these firms suffer are to be expected.

Robot programming training often will be necessary, especially if the robot has a keyboard programming mode. Keyboard programming is better done by a proficient technician or engineer than by operating and maintenance personnel. The robot manufacturer's customer training program should be accessed early in the acquisition process if project personnel are not already proficient in robot programming prior to the planning phase of the robot implementation project.

SUMMARY

Enthusiasm for technology and fascination with industrial robots cause many managers and manufacturing engineers to leap prematurely into robot applications. It is easy to overestimate the capability of the robot, especially in dealing with unexpected product or process variations.

Five phases in the systematic implementation of a robotics application can be identified: planning, development, mock-up, installation, and production. The process is somewhat iterative, with the possibility of returning to earlier phases at each step of the process. Aborting the project or switching to a fixed automation solution is also an alternative. Even if the project eventually is aborted, the phases of implementation of the robotics project generate valuable data and contribute to the efficiency and stability of the production process. Sometimes these side benefits are more important than the direct benefit provided by the industrial robot.

Since the actual acquisition of an industrial robot remains an option throughout most of the implementation process, the planning, development, and testing phases often can be accomplished in outside laboratories or in university centers for robotics and automation. Such centers can be ideal for experimentation and can provide necessary instrumentation for monitoring cycle times and process repeatability. For example, a low-power laser is a useful tool for monitoring robot alignment over a series of process cycles. Sophisticated timers, programmable logic controllers, and student researchers are other useful resources provided by such centers.

Whether the experimentation and implementation of a new ideas for robotics takes place in-house or in an off-site laboratory, both management and workers need to be knowledgeable of the objectives and progress of the project. This interaction should consist of more than a simple appraisal of what is going on; ideas and participation in the development process should be solicited actively. The workers who operate the process to be automated and their first-line supervisors have knowledge and experience with the product and process that cannot be duplicated by the technicians and engineers directly charged with developing the robotics application. This participation by the worker and line supervision in the implementation process also will pay off in the production phase when operating personnel must be trained to run and maintain the new robotics workcell.

EXERCISES AND STUDY QUESTIONS

9.1. Name the five major phases of a robotics project implementation.

9.2. What is the recommended prerequisite to the five major phases of robotic implementation?

9.3. At what point in the implementation of a robotic project should key labor representatives be included in the implementation?

9.4. Why is the success of the first robotics implementation more important than for other projects in automation?

9.5. In which phase of robot implementation does safety begin to become a consideration?

9.6. Describe two ways of protecting against unsafe personnel intrusion into a robot's work envelope during robot operation.

9.7. Describe a method of preventing the necessity of personnel intrusion into a robot's work envelope when the worker must share a work station with a robot.

9.8. Why would a robot be installed in a pit? Describe disadvantages of this strategy.

9.9. At which phase of the robotics project implementation does it become necessary to purchase the robot?

9.10. What principal drawbacks to a robotics project should be considered in the planning phase?

9.11. What is to be gained by documenting the old manual operation prior to installation of a new robot?

9.12. How do robots compare with fixed (hard) automation in terms of speed?

9.13. What is the potential advantage for supplying buffer storage in the layout of a proposed robotic system workcell?

9.14. What are three principal interfaces between the robot and its environment?

9.15. Why do robot manufacturers leave the end-of-arm tooling development to the user?

9.16. What is process stability analysis and why is it critical to the implementation of a robotic project?

9.17. What two classes of product uniformity are essential to the implementation of robots?

9.18. Describe several possible defects or misorientations that might render a robotics workcell useless.

9.19. Why is aluminum often used in experimental gripper design for robots?

9.20. What robot applications employ the most complicated gripper designs?

9.21. How can the use of double-handed grippers decrease the robot system cycle time in machine loading and unloading?

9.22. What is the primary use for a laser during robot workcell implementation?

9.23. A fixed laser beam projects a spot on the wall reflected from a small mirror placed upon the body of a robot. The mirror on the robot is 50 ft away from the wall. If the base of the robot is rotated clockwise to a position where the reflected laser beam is perpendicular to the wall, the spot on the wall is located 4 ft away from the original spot. Upon robot execution of a series of "base rotate" operations and then a return to the original position, the spot is found to have moved horizontally 12 in., increasing the reflected angle. By how much angular displacement has the original base robot position deteriorated, assuming laser, mirror, and other equipment have remained securely mounted?

9.24. What role does *tweaking* play in robot implementation?

9.25. What are PERT and CPM, and what is their relationship to robot implementation?

9.26. What are two alternatives to a pilot operation of a robot workcell?

9.27. Why do operations supervisors often resist robot implementations within their departments?

CHAPTER 10

INDUSTRIAL APPLICATIONS OF ROBOTS

This chapter is a review of existing robot applications by type, offering readers opportunities to generate ideas applicable to their operations and to confirm the feasibility of their own ideas.

WELDING

Welding is the first application addressed, because welding is the number-one application of industrial robots as of the early 1990s. The majority of these applications are in *spot* welding in such heavy assembly-line industries as automobile and truck manufacturing, but arc-welding robots are on the increase.

When performed manually, both spot welding and arc welding are subject to personnel safety hazards. In addition, welding is undesirable to workers because of the protective equipment that must be worn, especially for arc welding. When robots are not used, heavy loads, especially heavy-duty spot-welding equipment, may be handled by personnel. Not the least important motivation to use robots in welding operations is the quality and product uniformity attainable, for both spot welding and arc welding.

Figure 10.1 illustrates the key application of spot welding on an automobile body line. Since the line moves continuously, automatic line-tracking capability is usually required for spot-welding robots on assembly lines.

One example of an arc-welding application is in the welding of structural members of ship hulls. Arc welding inside a ship's hull is often extremely cramped and can be very dangerous because of toxic fumes and gases liberated in the confined space. Add to this danger the sheer discomfort of attempting to weld in a tiny space. Figure 10.2 illustrates the use of an arc-welding robot for this ideal application.

A large determinant of the success of a robot arc-welding operation over a manual operation is the improvement in "arc-on-time." Manual-welding operations often have very low arc-on time percentages (20–30 percent) because the remaining 70–80 percent of the time is consumed in adjusting helmet, respirator, or other personal protective equipment that would not be needed if a robot were used instead. Sometimes one skilled human welder can operate and control several robot arc welding systems, making possible production levels two to four times that of a single welder working without the benefit of robots. The following example illustrates how such benefits can be compounded.

FIGURE 10.1 Spot welding robots on an automobile body assembly line. (Reprinted by permission of Unimation-Westinghouse, Inc., Danbury, Conn.)

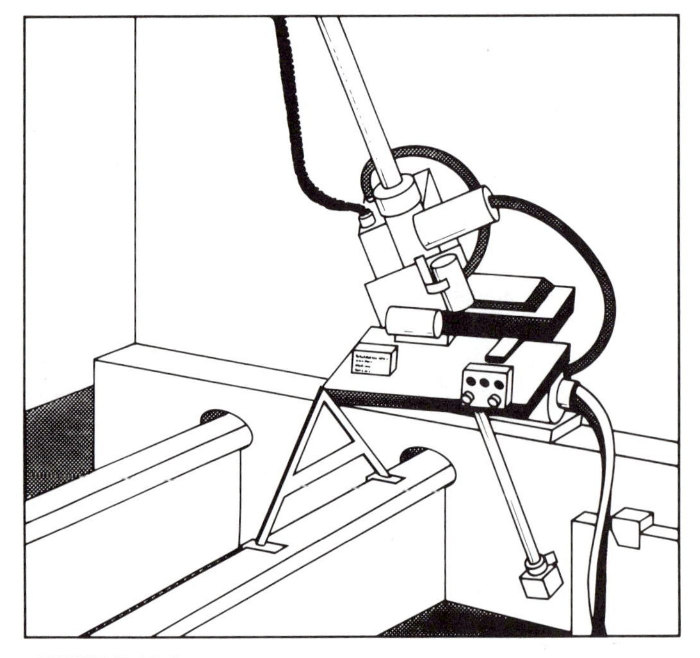

FIGURE 10.2 An arc welding robot welds the inner bottoms of ship hulls. (Reprinted by permission of Unimation-Westinghouse, Inc., Danbury, Conn.)

EXAMPLE 10.1
Arc Welding of Bridge Trusses

The fabrication of bridge trusses from standard structural steel components is a major arc-welding operation. For this study, compare an average arc-on time for manual welding of 30 percent with an assumed arc-on time for a robot welder of 80 percent. Further assume that the robot welder is 40 percent faster while laying the weld and that one welding operator can tend simultaneously four robot welders. How many manual welders would be required to accomplish a productivity level the same as one welder supervising this system of four arc-welding robots?

Solution

$$\begin{aligned}\text{productivity (robotic system)} &= 4 \times \frac{80\%}{30\%} \\ &\quad \times 1.4 \text{ productivity (manual welder)} \\ &\approx 15 \text{ manual welders}\end{aligned}$$

The dramatic productivity comparison observed in Example 10.1 is not an exaggeration. In fact, the productivity improvement can be even greater for arc-welding operations involving "weave patterns" for the weld path.

Weaving is a type of arc-welding path in which the welder lays a higher quality weld by oscillating the welding tip orthogonal to the path of the weld. In other words, the welder lays a zigzag path in the way a special sewing machine makes a zigzag stitch for very much the same purpose. Weaving requires considerable skill on the part of the human arc welder, and even a skilled welder hardly can maintain a uniform weave over a long path. Even when the skill of the welder makes a quality weave feasible, the speed of the process is seriously reduced by the requirement to oscillate the tip. This is where the arc-welding robot excels, because the robot can be programmed to oscillate the welding head many times faster than is humanly possible. This author has observed a production TIG welding operation (tungsten-inert-gas) being performed using a weave pattern in which a single robot was achieving a production rate eight to ten times faster than the human welder it replaced. Besides the faster speed being achieved by the robot, the superior quality of the weld was immediately visible.

MACHINE LOADING

Machine loading and unloading by industrial robots offer some of the same key advantages as does robot welding: safety and relief from handling heavy loads. The job of punch press operator historically has been one of the most dangerous factory jobs because of the risk of amputations while feeding the press by hand. The risk has been greatly reduced in recent years due to:

1. Increased use of robots and automated press feeding equipment.
2. Increased attention to industrial safety standards, especially OSHA standards for safeguarding the point of operation of presses and other dangerous machines.

Both influences have implications for the field of robotics, the first directly and the second indirectly, in that serious attention to machine safety standards is driving the movement toward machine-loading automation. One example of robots loading and unloading machines was seen in Chapter 6 (Figure 6.7), where a robot was shown loading and unloading a machine tool. We will now consider some additional examples and will analyze the interaction between the robot and the machines it loads.

Multiple Robot and Multiple Machine Loading

The plan view diagram in Figure 10.3 illustrates a machine loading/unloading application in which four robots are used to feed four milling machines and a conveyor is used to transport workpieces. The product is truck windshield wiper motor housings, and the production rate is over 28,000 per day (two-shift operation) with a cycle time of eight seconds. The milling machines are themselves of an automatic indexing design, resulting in a highly automated operation. The line is flexible enough to handle several slightly different models in model batches without refixturing. The primary objective of this application was to eliminate the production slowdowns caused by parts hanging up in the manual machine-load operation. The high operating speeds and small clearances made manual loading tricky. But the robots delivered the desired speed and accuracy to eliminate the workpiece hang-ups. A side benefit was the reduction of scrap. The composite economic benefits were expected to produce a one-year payback period. The following case study illustrates the calculation of multiple-robot/multiple-machine system productivity.

EXAMPLE 10.2 (CASE STUDY)
Robot Machine Loading System Productivity

The robot machine loading/unloading system diagrammed in Figure 10.3 has an eight-second cycle time and a daily two-shift production level of over 28,000 workpieces. Do the four Kingsbury milling machines perform sequential operations upon each workpiece in series, or do all four milling machines perform the complete machining cycle in parallel with each other—that is, is each part processed completely by only one machine, not all four?

Solution

Series Operation

$$\begin{aligned}\text{production rate/day} &= 1 \text{ part/8 sec} \times 60 \text{ sec/min} \times 60 \text{ min/hr} \\ &\quad \times 16 \text{ hr/day} \\ &= 7200 \text{ parts per day (two shifts)}\end{aligned}$$

Parallel Operation

$$\begin{aligned}\text{production rate/day} &= 1 \text{ part/8 sec} \times 60 \text{ sec/min} \times 60 \text{ min/hr} \\ &\quad \times 16 \text{ hr/day} \times 4 \text{ machines} \\ &= 28{,}800 \text{ parts/day (two shifts)}\end{aligned}$$

Since the daily production level was quoted to be "over 28,000," it can be concluded that the system operates on a parallel basis whereby each milling machine completes an entire workpiece while the other three also each complete an entire workpiece.

For a second example in machine loading, we turn to a press-loading operation, a prime application for industrial robots. Stauffer [ref. 101] reports a robot-loading application for a 150-ton, straight-side press used for blanking triangular-shaped reinforcing gussets for oven doors. Most blanking operations are from sheet or strip stock, but the

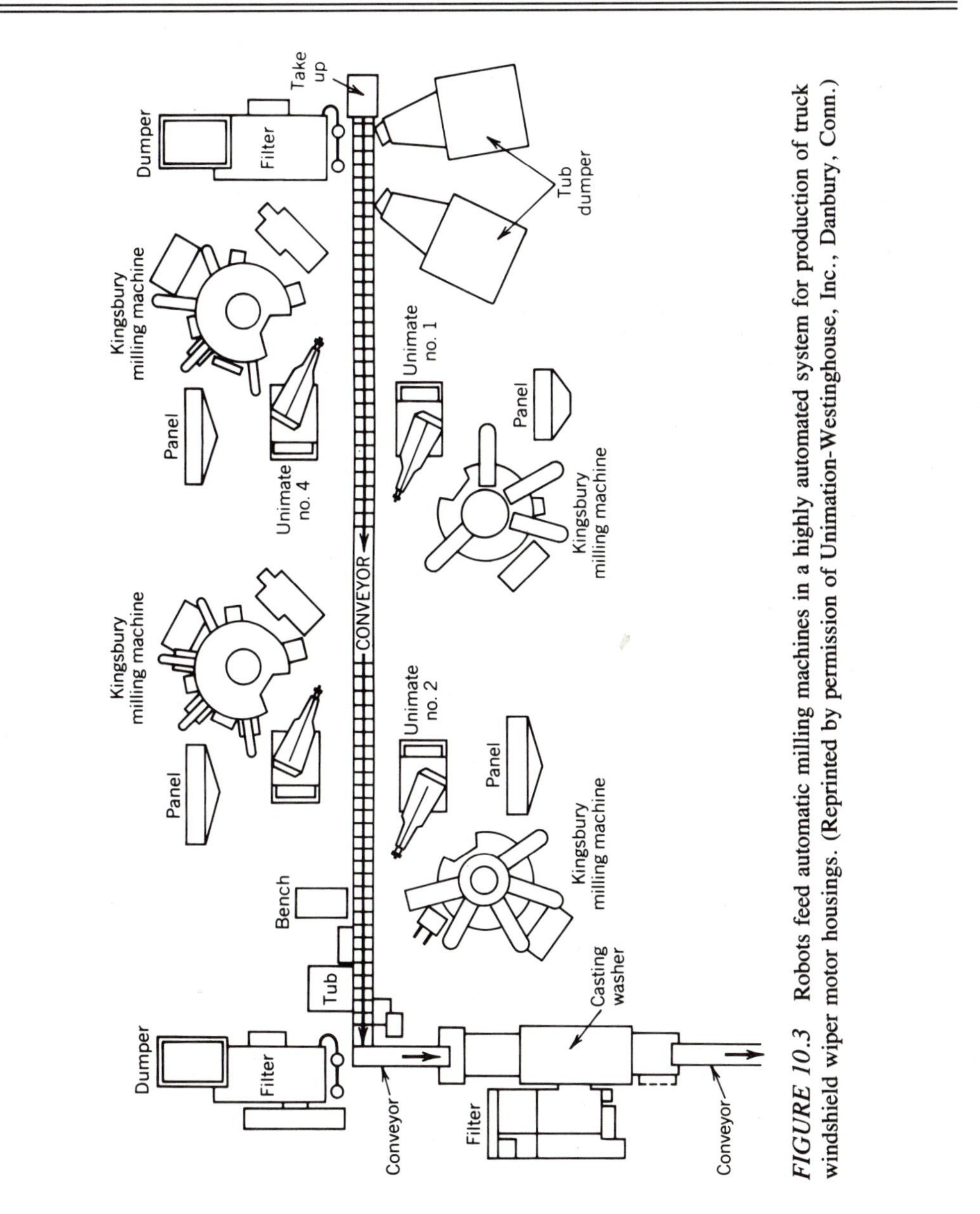

FIGURE 10.3 Robots feed automatic milling machines in a highly automated system for production of truck windshield wiper motor housings. (Reprinted by permission of Unimation-Westinghouse, Inc., Danbury, Conn.)

gussets in this operation were blanked from scrap pieces from a prior operation. For this application, it was desirable to maintain flexibility so that the press could be converted back to manual operation or even switched back and forth between manual and robotic operation depending upon the model number and part description. Two of the characteristics of this operation were frequent die changes and low-volume production lots for each model and part number. This characteristic favored the use of programmable robots over their alternative, hard automation, the latter often being selected for press loading and unloading in long production runs.

The oven-door gusset-blanking operation was no small workcell, and a forklift truck was used to manipulate die sets during re-setups at production-lot changeovers. But the presence of the large robot prevented access to the press by the forklift truck. The solution to this problem was to place the robot on casters with a home position plate provided for consistently locking the robot in place at every new setup with no loss in position accuracy from production lot to production lot.

The robot for this application, a hydraulic model, was chosen with an objective of maintaining high repeatability while keeping costs as low as possible. The robot was of polar (spherical) configuration and had four nonservo, axis-limit, programmable axes. Recalling the concepts learned in Chapter 6, we know that this robot would have mechanical stops on each axis and would have excellent dimensional repeatability at its axis limits but would have no positional control at intermediate points between stops. This characteristic presented a problem when the robot was to pick up another scrap blank from an infeed stack as shown in Figure 10.4. As the height of the stack decreases, the radial distance from the center of rotation of axis 2 changes slightly, as is illustrated in exaggerated form in Figure 10.4. The slight change in position is inconsequential to the pick-up function, but later when the robot places the blank into the press die, the workpiece will be in an incorrect position for accurate positioning in the die. The solution to this problem was to stack the blanks in a curve to match the arc traveled by the robot hand in its extended position. This solution required a slight modification to the magnetic fanning device for presenting the blanks for pickup. The cost of modifying the fanning device was easily justified by the cost avoidance of not going to a more expensive robot with more sophisticated axis motion control.

The actual pickup system for this application used suction cup pickup devices on the robot gripper. A key advantage to this type of gripper is that the rubber suction cups afford a small amount of compliance to tolerate a minor misalignment when the workpiece enters the press and positions itself against the locating pins in the die. Rubber suction cups are popular press-feeding tools even for manual press-feeding operations, but they are even more advantageous for use by robots because of this position compliance feature.

Figure 10.5 exhibits an integrated manufacturing system consisting of three robots and a variety of machine tools served by a single conveyor. The product is a copper fuser roll for Xerox Corporation duplicating machines. Stainless-steel caps are brazed into each end of each fuser roll, which is then processed through several stages of machining. The line has a capacity of 100,000 assemblies per year.

Double-handed grippers on each robot are a key to the success of this robot application. One hand holds the unfinished part, while the other hand unloads the finished part from the machine. Then the unfinished part is loaded into the machine by the other hand, saving the robot an extra trip back to the conveyor while the machine tool waits. Reduced machine idle time during unload/load cycles increases each machine's productivity and in turn the productivity of the entire line. The following two examples will demonstrate the computations to compare robot loading of machines using one-handed grippers with those using two-handed grippers.

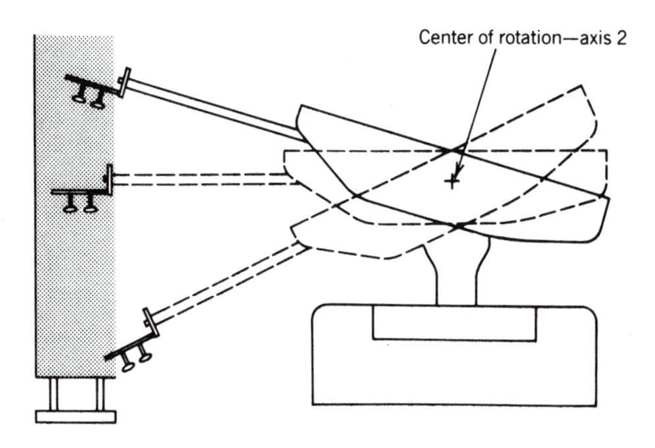

FIGURE 10.4 Diagram to exaggerate the positioning problem at the pickup point when using a nonservo, axis-limit robot of the polar (spherical) configuration.

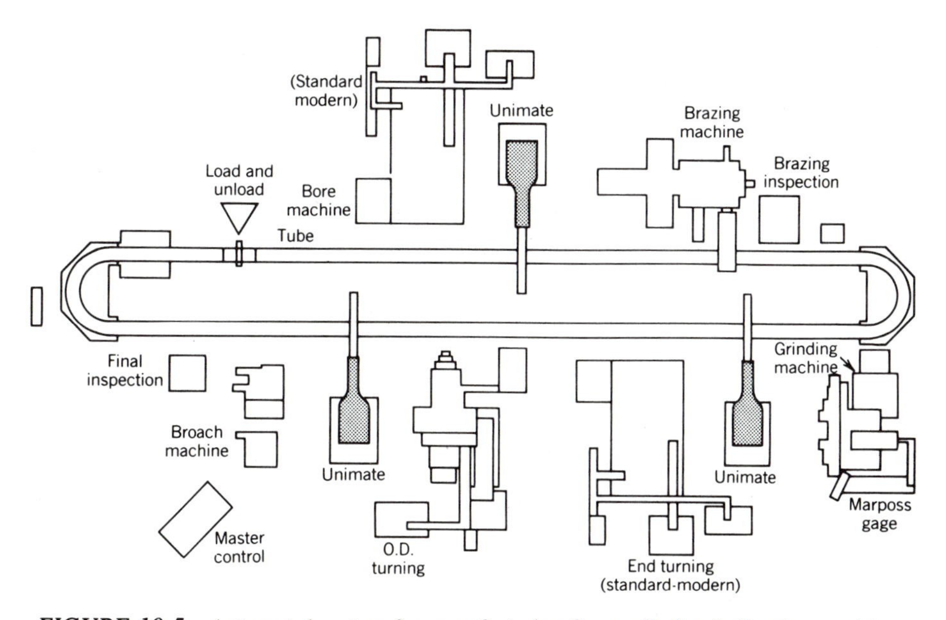

FIGURE 10.5 Automated system for manufacturing fuser rolls for duplicating machines. (Reprinted by permission of Unimation-Westinghouse, Inc., Danbury, Conn.)

EXAMPLE 10.3
Milling Machine Loading and Unloading

A robot loads and unloads an end-turning machine from a central conveyor. The following average robot operation times apply:

Pick up part from conveyor (including average wait time for part to arrive in pickup position)	2.6 sec
Move robot hand from conveyor to machine	1.7 sec
Load part into machine and back hand away from machine so the machine can start	1.1 sec
Unload part from machine	0.8 sec
Move robot hand from machine to conveyor	1.7 sec
Deposit part onto conveyor	0.3 sec

Suppose the turning machine operation cycle requires 24 seconds. Assuming an average 20 percent system downtime for maintenance, clearance of malfunctions, and other causes, what is the daily eight-hour shift production level, (a) assuming a one-handed robot gripper, and (b) assuming a two-handed robot gripper?

Solution

For One-Handed Gripper

Assume that the robot has no other duties and waits at the machine in order to unload it as soon as the machine cycle is complete. Arbitrarily selecting the beginning of the machine operation cycle as the beginning of the system cycle, the typical operation sequence is:

Machine operation cycle	24.0 sec
Unload machine	0.8
Move to conveyor	1.7
Deposit finished part on conveyor	0.3
Pick up new part	2.6
Move to machine	1.7
Load into machine	1.1
Total	32.2 sec

The production then for an 80 percent efficient eight-hour shift would be:

$$\begin{aligned} \text{production} &= 1 \text{ unit/32.2 sec} \times 60 \text{ sec/min} \times 60 \text{ min/hr} \times 8 \text{ hr/shift} \times 0.80 \\ &= 715 \text{ units/shift} \end{aligned}$$

For Two-Handed Gripper

The appropriate operation sequence would be:

Machine operation cycle	24.0 sec
Unload machine	0.8
Load into machine	1.1
Total	25.9 sec

Note in the case of the two-handed gripper that the robot operations of move to the conveyor, deposit part, pick up the new part, and move to the machine could all be performed by the robot *during* the machine operation cycle and are therefore omitted from the timed sequence. The improved production level could be computed as:

$$\begin{aligned} \text{production} &= 1 \text{ unit/25.9 sec} \times 60 \text{ sec/min} \times 60 \text{ min/hr} \times 8 \text{ hr/shift} \times 0.8 \\ &= 890 \text{ units/shift} \end{aligned}$$

Therefore, the double-handed gripper makes possible a per-shift production increase of

$$890 - 715 = 175 \text{ units}, \qquad \text{or } \frac{175}{715} \approx 24.5\%$$

without purchasing any additional robot or machine equipment.

EXAMPLE 10.4
Double-Handed Robot Loading and Unloading of a Fast Production Press

The previous example was for robot loading/unloading of a machine operation, for which the machine operation time (24 seconds) was considerably larger than the robot handling time portion of the overall cycle. Now let us see how the picture changes if the machine operation is much faster, as in a mechanical press operation, for example. Let us say that the press cycle time is 0.9 seconds and compare the one-handed gripper operation with the two-handed gripper operation as we did in the previous example.

With One-Handed Gripper

The corresponding operation sequence would be (using the same robot operation times as were used in Example 10.3):

Press operation cycle	0.9 sec
Unload press	0.8
Move to conveyor	1.7
Deposit finished part onto conveyor	0.3
Pick up new part	2.6
Move to press	1.7
Load into press die	1.1
Total	9.1 sec

The eight-hour shift production, using the same 80 percent efficiency assumption, would be:

$$\begin{aligned} \text{production} &= 1 \text{ unit/9.1 sec} \times 60 \text{ sec/min} \times 60 \text{ min/hr} \times 8 \text{ hr/shift} \times 0.8 \\ &= 2531 \text{ units/shift} \end{aligned}$$

For the Two-Handed Gripper

When the two-handed gripper is used with the fast-cycling mechanical press, the robot is not able to accomplish as much of its handling cycle during the machine cycle as was possible with the slower machining cycle of Example 10.3. Therefore the press cycle time will be performed during the robot handling cycle, instead of the reverse, as was true of Example 10.3. To tabulate the cycle sequence, we will begin with the robot operation "unload the press," as the press cycle time will be absorbed and will not affect the overall cycle time:

Unload the press	0.8 sec
Load into press die	1.1
Move to conveyor	1.7
Deposit finished part onto conveyor	0.3

Pick up new part	2.6
Move to press	1.7
Total	8.2 sec

The shift production would be computed as:

$$\text{production} = 1 \text{ unit/8.2 sec} \times 60 \text{ sec/min} \times 60 \text{ min/hr} \times 8 \text{ hr/shift} \times 0.8$$
$$= 2809 \text{ units/shift}$$

The per-shift production improvement afforded by the double-handled gripper is:

$$2809 - 2531 = 278 \text{ units or } \frac{278}{2531} \approx 11\%$$

but is not nearly as dramatic as when used with the longer machine cycle operation of Example 10.3.

Sequential Machine Loading

The analysis of robot machine-loading applications becomes a bit more complicated when a single robot has the task of feeding several machines in an organized sequence. If the automation engineer has timed and planned the operation carefully, the robot can be programmed to anticipate cycle completions at appropriate stations and move to these stations in advance to shorten machine idle time while waiting for the robot.

FIGURE 10.6 Robot loads/unloads drilling and boring machines for castings for truck differential assemblies. (Reprinted by permission of Prab Robots, Inc., Kalamazoo, Mich.)

In an actual robot application, the robot pictured in Figure 10.6 is sequentially loading and unloading three drilling and boring machines to process 20-lb castings to be used in truck differential assemblies. Figure 10.7 illustrates the production cell layout for this application. Station *A* represents the pickup point; stations *B*, *C*, and *D* represent the three machines; and station *E* is the discharge conveyor for completed parts. The robot gripper as well as the three drilling and boring machines are designed to handle two parts at a time for higher production.

An obvious objective in this application was to eliminate the laborious loading and unloading of the 20-lb parts. Before the workcell was automated, machine operators handled approximately 12 tons of workpieces per shift. The automation of this workcell relieved operators of the drudgery of handling the heavy workpieces and permitted them to concentrate upon inspection and quality control. Besides the savings in direct labor, the productivity of the workcell increased nearly 60 percent.

In this real example the machining stations were equipped to handle two parts at a time, and accordingly a double-handed gripper on the robot was essential to enable the robot to service the machine sequence with any degree of efficiency. Were it not for the fact that the machines could handle two parts at a time, however, the double-handed gripper on the robot would have had little value. This appears to be a contradiction of the principles illustrated in Examples 10.3 and 10.4 in which the double-handed gripper was seen to markedly improve the performance of the system. The difference here is that

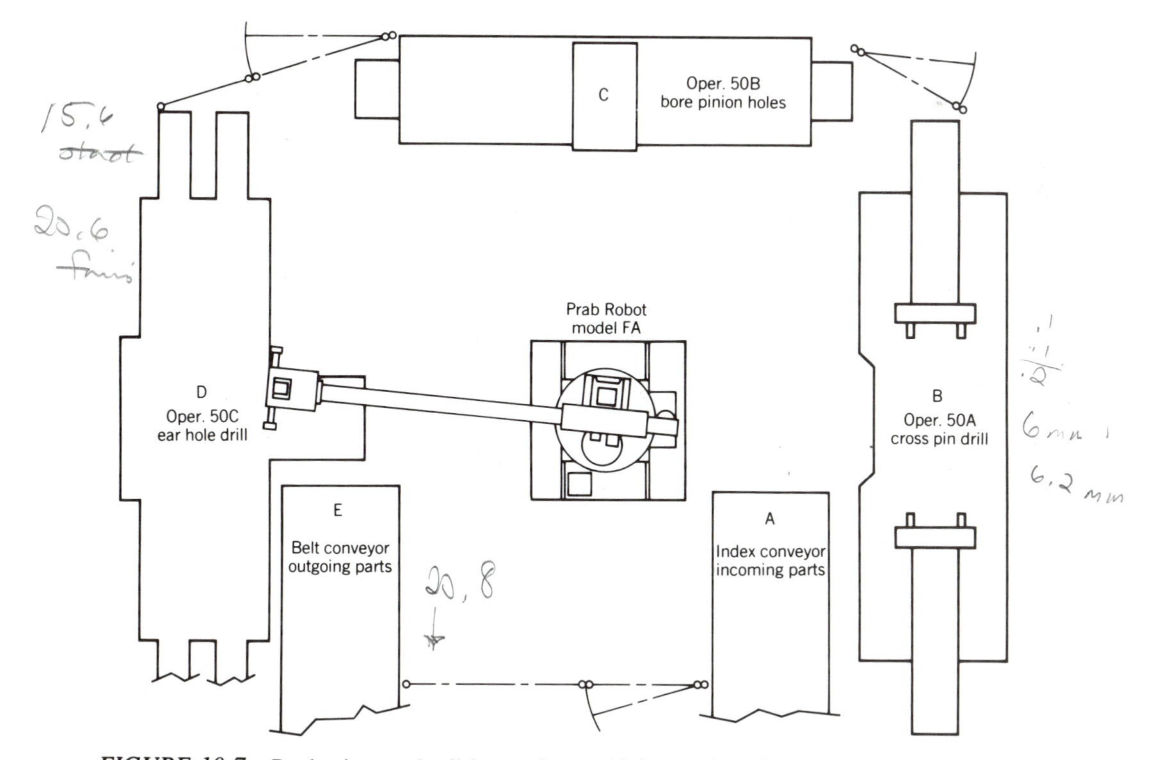

FIGURE 10.7 Production workcell layout for machining castings for truck differential assemblies. The letters denote the sequence of workpiece positions. (Reprinted by permission of Prab Robots, Inc., Kalamazoo, Mich.)

the robot is being used to sequentially load a series of machines instead of simply loading and unloading a single machine.

A subtle reality to the sequential machine loading problem is the constraint that, once the system is loaded and operating, the robot must follow a rigid sequence of steps to feasibly service all machines in the sequence. This rigid sequence must be held regardless of the relationships between the machine cycle times, the time required for the robot to move from station to station, and the load/unload times. This rigid sequence is a methodical retrogression from the last machine in the sequence to the first, consisting at each station (station i) of the following steps:

1. Unload machine at station i.
2. Move to station $i + 1$.
3. Load machine at station $i + 1$.
4. Move back (retrogress) to station $i - 1$.
5. Unload machine at station $i - 1$.
6. Move to station i.
7. Reload station i.

As the robot retrogresses through the system, steps 5, 6, and 7 of the above sequence for machine i becomes steps 1, 2, and 3 for station $i - 1$. The process continues through the machine sequence backwards until the first machine is loaded by a new workpiece from the incoming parts conveyor. At that point the robot (or human operator for that matter) *always* must proceed directly to the last machine in the sequence to unload it and begin the retrogression anew. This rigid sequence must be held regardless of the comparative cycle times of the machines in the sequence. It might seem that the robot should proceed to whatever machine needs unloading as soon as possible, but there is no reason to unload a machine if there is no place to put the finished part while waiting for the next machine in the sequence to accept it. Similarly, if the previous machine in the sequence is not ready to supply a part to the next machine in the sequence, there is no reason for the robot to unload the machine to make ready for it. Therefore, after the system is loaded and operating, the most efficient strategy for the robot is to proceed in an orderly fashion from the last station to the first in the unloading and reloading process. Whether the robot is required to wait for a machine to finish its cycle, and at which station it is required to do its waiting, is a function of the relationships between machine cycle times and the time required for the robot handling operations. These same relationships may instead dictate that one or more of the machines must sit idle and wait for the robot to complete its handling operations. Regardless of whether machines must wait or the robot must wait, the sequence of steps taken by the robot follows the rigid retrogression series in the loading and unloading of the machines in the sequence.

A curious result of the necessity that the robot follow the rigid retrogression sequence in the loading and unloading of a series of machines in which there is no storage between stations is that a double-handled gripper provides no advantage in efficiency. Contrast this with the usual machine loading application in which a double-handed gripper on the robot can significantly reduce handling time and robot moves between stations. (See Examples 10.3 and 10.4.)

The analysis of robot loading/unloading of a sequence of several machines is one of the more interesting and involved studies of the productivity of robot systems. Robotics and manufacturing automation engineers should be well grounded in such analyses, and the following two case studies will illustrate analysis methods as well as to point to shortcuts for streamlining the computations.

EXAMPLE 10.5 (CASE STUDY)
Analysis of Production Rate: Robot Sequential Machine Unloading

Suppose in the truck differential machining operation of Figure 10.7 that the following operation times apply:

Machine 1 (station B) time = 9.1 min

Machine 2 (station C) time = 9.0 min

Machine 3 (station D) time = 5.0 min

Gripper pickup time (all stations) = 0.1 min

Gripper release time (all stations) = 0.1 min

The robot's move times are shown in Table 10.1. The problem is to find the system's cycle time and production rate.

TABLE 10.1 Move Times (minutes)

	To				
From	*A*	*B*	*C*	*D*	*E*
A	—	0.3	0.6	0.9	1.2
B	0.3	—	0.3	0.6	0.9
C	0.6	0.3	—	0.3	0.6
D	0.9	0.6	0.3	—	0.3
E	1.2	0.9	0.6	0.3	—

Note: In this case the robot cannot move past A in the clockwise direction, nor past E in a counterclockwise direction.

Solution

To analyze the case and to determine system productivity, we begin by constructing Table 10.2 to show each step taken in the cycle. The system's production rate can be determined when the cycle begins to repeat itself.

Remember, a machine cannot be unloaded until it has finished its machining cycle on the truck differential. The lines labeled by * in Table 10.2 are the points in the cycle in which station D is unloaded, which represents the arbitrary end points in our cycle. After the system becomes fully loaded, the difference in clock times between successive "unload D" operations represents the system cycle time. For example:

Clock time when unloading workpiece #2 from machine D	35.4
Clock time when unloading workpiece #1 from machine D	−24.7
System clock time	10.7 min

Notice that the system cycle time could have been obtained from any two identical operations *after* the system reached its normal operating procedure. For example, we could have chosen the operation "move to D":

Clock time—move to D when machines hold workpieces #4, #3, and #2	33.2
Clock time—move to D when machines hold workpieces #3, #2, and #1	−22.7
System cycle time	10.7 min

The production rate is the reciprocal of the system cycle, so:

$$\text{production rate} = 1 \text{ unit}/10.7 \text{ min} \times 60 \text{ min/hr}$$
$$= 5.6 \text{ units/hr}$$

TABLE 10.2 Cycle Analysis: Robot Sequential Machine Loading

Clock Time	Robot Position	Stations Loaded B	C	D	Operation	Time Required	Scheduled Completion
0	*A*				Pick up at *A*	0.1	0.1
0.1	*A*				Move to *B*	0.3	0.4
0.4	*B*				Load machine *B*	0.1	0.5
			Workpiece				
0.5	*B*	#1			Machine at *B*	9.1	9.6
9.6	*B*	#1			Unload *B*	0.1	9.7
9.7	*B*				Move to *C*	0.3	10.0
10.0	*C*				Load *C*	0.1	10.1
10.1	*C*		#1		Machine at *C*	9.0	19.1
10.1	*C*		#1		Move to *A*	0.6	10.7
10.7	*A*		#1		Pick up at *A*	0.1	10.8
10.8	*A*		#1		Move to *B*	0.3	11.1
11.1	*B*		#1		Load *B*	0.1	11.2
11.2	*B*	#2	#1		Machine at *B*	9.1	20.3
11.2	*B*	#2	#1		Move to *C*	0.3	11.5
19.1	*C*	#2	#1		Unload *C*	0.1	19.2
19.2	*C*	#2			Move to *D*	0.3	19.5
19.5	*D*	#2			Load *D*	0.1	19.6
19.6	*D*	#2		#1	Machine at *D*	5.0	24.6
19.6	*D*	#2		#1	Move to *B*	0.6	20.2
20.3	*B*	#2		#1	Unload *B*	0.1	20.4
20.4	*B*			#1	Move to *C*	0.3	20.7
20.7	*C*			#1	Load *C*	0.1	20.8
20.8	*C*		#2	#1	Machine at *C*	9.0	29.8
20.8	*C*		#2	#1	Move to *A*	0.6	21.4
21.4	*A*		#2	#1	Pick up at *A*	0.1	21.5
21.5	*A*		#2	#1	Move to *B*	0.3	21.8
21.8	*B*		#2	#1	Load *B*	0.1	21.9
21.9	*B*	#3	#2	#1	Machine at *B*	9.1	31.0
21.9	*B*	#3	#2	#1	Move to *D*	0.6	22.5
24.6	*D*	#3	#2	#1	Unload *D*	0.1	24.7*
24.7	*D*	#3	#2		Move to *E*	0.3	25.0
25.0	*E*	#3	#2		Drop at *E*	0.1	25.1
25.1	*E*	#3	#2		Move to *C*	0.6	25.7
29.8	*C*	#3	#2		Unload *C*	0.1	29.9
29.9	*C*	#3			Move to *D*	0.3	30.2
30.2	*D*	#3			Load *D*	0.1	30.3
30.3	*D*	#3		#2	Machine at *D*	5.0	35.3
30.3	*D*	#3		#2	Move to *B*	0.6	30.9
31.0	*B*	#3		#2	Unload *B*	0.1	31.1
31.1	*B*			#2	Move to *C*	0.3	31.4
31.4	*C*			#2	Load *C*	0.1	31.5
31.5	*C*		#3	#2	Machine at *C*	9.0	40.5

(Columns Clock Time, Robot Position and Stations Loaded B, C, D are under *Status **Before** Operation*.)

TABLE 10.2 (Continued)

*Status **Before** Operation*							
		Stations Loaded					
Clock Time	*Robot Position*	*B*	*C*	*D*	*Operation*	*Time Required*	*Scheduled Completion*
31.5	*C*		#3	#2	Move to *A*	0.6	32.1
32.1	*A*		#3	#2	Pick up at *A*	0.1	32.2
32.2	*A*		#3	#2	Move to *B*	0.3	32.5
32.5	*B*		#3	#2	Load *B*	0.1	32.6
32.6	*B*	#4	#3	#2	Machine at *B*	9.1	41.7
32.6	*B*	#4	#3	#2	Move to *D*	0.6	33.2
35.3	*B*	#4	#3	#2	Unload *D*	0.1	35.4*

Now let us consider a similar case having the same setup and the same robot move times but having shorter machining times.

EXAMPLE 10.6. (CASE STUDY)
Analysis of Production Rate: Revised Machining Times for Robot Sequential Machine Loading Application

In this example, we will have machine times as follows:

Machine 1 (station B) time = 2.0 min

Machine 2 (station C) time = 1.0 min

Machine 3 (station D) time = 2.6 min

Find the system cycle time and production rate.

Solution

The cycle analysis is detailed in Table 10.3.

As in Case Study 10.5, to determine system cycle time, we find the difference in clock times after the system has reached its normal operating procedure:

Clock time—when unloading workpiece #2 from machine *D*	12.8
Clock time—when unloading workpiece #1 from machine *D*	−8.4
	4.4 min

System cycle time

$$\text{production rate} = 1 \text{ unit}/4.4 \text{ min} \times 60 \text{ min/hr} = 13.6 \text{ units/hr}$$

The machine operation times for the two examples considered in Case Studies 10.5 and 10.6 were chosen very carefully to illustrate a point. In Case Study 10.5, the machining time affected the system cycle time because the robot had to wait for one of the machines during the cycle. In Case Study 10.6, however, each of the three machine cycle times was too small to cause the robot to wait. Therefore, in Case Study 10.6 the system cycle time was limited to the time required for the robot to complete all the steps in its sequence *without regard to the machining operations*. This can be verified for Case Study 10.6 adding all robot times in the cycle, as shown in Table 10.4.

TABLE 10.3 Cycle Analysis: Revised Robot Machine Loading Application

*Status **Before** Operation*							
		Stations Loaded					
Clock Time	*Robot Position*	*B*	*C*	*D*	*Operation*	*Time Required*	*Scheduled Completion*
0	*A*				Pick up at *A*	0.1	0.1
0.1	*A*				Move to *B*	0.3	0.4
0.4	*B*				Load machine *B*	0.1	0.5
			Workpiece				
0.5	*B*	#1			Machine at *B*	2.0	2.5
2.5	*B*	#1			Unload *B*	0.1	2.6
2.6	*B*				Move to *C*	0.3	2.9
2.9	*C*				Load *C*	0.1	3.0
3.0	*C*		#1		Machine at *C*	1.0	4.0
3.0	*C*		#1		Move to *A*	0.6	3.6
3.6	*A*		#1		Pick up at *A*	0.1	3.7
3.7	*A*		#1		Move to *B*	0.3	4.0
4.0	*B*		#1		Load *B*	0.1	4.1
4.1	*B*	#2	#1		Machine at *B*	2.0	6.1
4.1	*B*	#2	#1		Move to *C*	0.3	4.4
4.4	*C*	#2	#1		Unload *C*	0.1	4.5
4.5	*C*	#2			Move to *D*	0.3	4.8
4.8	*D*	#2			Load *D*	0.1	4.9
4.9	*D*	#2		#1	Machine at *D*	2.6	7.5
4.9	*D*	#2		#1	Move to *B*	0.6	5.5
6.1	*B*	#2		#1	Unload *B*	0.1	6.2
6.2	*B*			#1	Move to *C*	0.3	6.5
6.5	*C*			#1	Load *C*	0.1	6.6
6.6	*C*		#2	#1	Machine at *C*	1.0	7.6
6.6	*C*		#2	#1	Move to *A*	0.6	7.2
7.2	*A*		#2	#1	Pick up at *A*	0.1	7.3
7.3	*A*		#2	#1	Move to *B*	0.3	7.6
7.6	*B*		#2	#1	Load *B*	0.1	7.7
7.7	*B*	#3	#2	#1	Machine at *B*	2.0	9.7
7.7	*B*	#3	#2	#1	Move to *D*	0.6	8.3
8.3	*D*	#3	#2	#1	Unload *D*	0.1	8.4
8.4	*D*	#3	#2		Move to *E*	0.3	8.7
8.7	*E*	#3	#2		Drop at *E*	0.1	8.8
8.8	*E*	#3	#2		Move to *C*	0.6	9.4
9.4	*C*	#3	#2		Unload *C*	0.1	9.5
9.5	*C*	#3			Move to *D*	0.3	9.8
9.8	*D*	#3			Load *D*	0.1	9.9
9.9	*D*	#3		#2	Machine at *D*	2.6	12.5
9.9	*D*	#3		#2	Move to *B*	0.6	10.5
10.5	*B*	#3		#2	Unload *B*	0.1	10.6
10.6	*B*			#2	Move to *C*	0.3	10.9
10.9	*C*			#2	Load *C*	0.1	11.0
11.0	*C*		#3	#2	Machine at *C*	1.0	12.0

TABLE 10.3 (continued)

	*Status **Before** Operation*						
		Stations Loaded					
Clock Time	*Robot position*	*B*	*C*	*D*	*Operation*	*Time Required*	*Scheduled Completion*
11.0	*C*		#3	#2	Move to *A*	0.6	11.6
11.6	*A*		#3	#2	Pick up at *A*	0.1	11.7
11.7	*A*		#3	#2	Move to *B*	0.3	12.0
12.0	*B*		#3	#2	Load *B*	0.1	12.1
12.1	*B*	#4	#3	#2	Machine at *B*	2.0	14.1
12.1	*B*	#4	#3	#2	Move to *D*	0.6	12.7
12.7	*B*	#4	#3	#2	Unload *D*	0.1	12.8

Note that Table 10.4 leads to the same cycle time computed for Case Study 10.6. This can greatly simplify the computation of cycle time and production rate for robot machining-loading operations because Table 10.4 is obviously simpler than Table 10.3. However, what Table 10.4 does not reveal is the maximum station machining time allowable to make calculation of the system cycle time dependent only upon the robot times. This too can be calculated, however, as will be seen in the development that follows.

The total robot cycle time calculated in Table 10.3 can be divided into two major segments:

Unload a given machine
.
.
.
.
Reload that same machine } Cycle segment α

All other robot operations } Cycle segment β

TABLE 10.4 Tabulation of Robot Cycle Time: Robot Machine-Loading Application

Time Required	*Operation*
0.1	Unload *D*
0.3	Move to *E*
0.1	Drop at *E*
0.6	Move to *C*
0.1	Unload *C*
0.3	Move to *D*
0.1	Load *D*
0.6	Move to *B*
0.1	Unload *B*
0.3	Move to *C*
0.1	Load *C*
0.6	Move to *A*
0.1	Pick up at *A*
0.3	Move to *B*
0.1	Load *B*
0.6	Move to *D*
4.4 min	

TABLE 10.5 Computation of Cycle Segments α and β

Time Required	*Operation*	
0.1	Unload D	Cycle segment α = 1.6 min
0.3	Move to E	
0.1	Drop at E	
0.6	Move to C	
0.1	Unload C	
0.3	Move to D	
0.1	Load D	
0.6	Move to B	Cycle segment β = 2.8 min
0.1	Unload B	
0.3	Move to C	
0.1	Load C	
0.6	Move to A	
0.1	Pick up at A	
0.3	Move to B	
0.1	Load B	
0.6	Move to D	
Total 4.4 min		

In Table 10.5 we arbitrarily select machine station D to compute cycle segments α and β, but regardless of which machine station is selected for computation, α and β remain constant in this example. The reader can verify this by selecting machine station C or B and achieve the same results as shown in Table 10.5 in which machine station D was selected for computation of cycle segments α and β.

In cases in which the robot portion of the cycle is constant from machine to machine, the recognition and computation of robot cycle segments α and β greatly simplify the computations for total cycle time and system production rates for sequential machine loading applications of robots. The procedure is as follows:

1. Calculate robot cycle segments α and β.
2. Determine the maximum machine time M among all machine stations.
3. (a) If $M \leq \beta$, total system cycle time $T = \alpha + \beta$.
   (b) If $M > \beta$, total system cycle time $T = \alpha + M$.

Verifying these results for both Case Studies 10.5 and 10.6:

For Case Study 10.5:

$$\alpha = 1.6 \text{ min}$$
$$\beta = 2.8 \text{ min}$$
$$M = 9.1 \text{ min}$$

Since $M > \beta$, $T = \alpha + M = 1.6 + 9.1 = 10.7$ min/cycle

For Case Study 10.6:

$$\alpha = 1.6 \text{ min}$$
$$\beta = 2.8 \text{ min}$$
$$M = 2.6 \text{ min}$$

Since $M \leq \beta$, $T = \alpha + \beta = 1.6 + 2.8 = 4.4$ min/cycle

Note that both results agree exactly with the computations made earlier in Case Studies 10.5 and 10.6 but that the lengthy cycle analyses represented by Tables 10.2 and 10.3 were avoided.

Forging and Die Casting

When piece-parts to be handled are hot, there is even greater motivation for using robots in machine loading and unloading. Thus, robots are finding acceptance in forging, die casting, and other hot-parts-handling operations. Anyone who has watched a drop forging operation and experienced the environmental heat and noise knows that this is a prime candidate for the application of robots.

Figure 10.8 diagrams the loading and unloading of forging presses by two robots in each of three workcells. The first robot in each workcell picks up the hot forging blank from the induction furnace and loads it into the forging press. The second robot unloads the forging press and places the hot forging into a secondary trimming operation. Note that all work at the three stations is performed inside a protected area.

Although robot hands are more durable and heat-resistant than human hands, robot hands can get too hot, too. However, the robot can be programmed to dip its hand into a cooling bath at appropriate intervals. This was treated as an exercise in robot programming in Chapter 7.

Another problem in using robots in forging applications is that the forging operation changes the shape of the piece-part being handled. This may necessitate the design of a special gripper that can accommodate both shapes. Figure 10.9 illustrates a robot gripper that has extension springs to accommodate the preforged billet of narrow cross section and later the forged, flattened billet of wider cross section. Also note the long, flattened

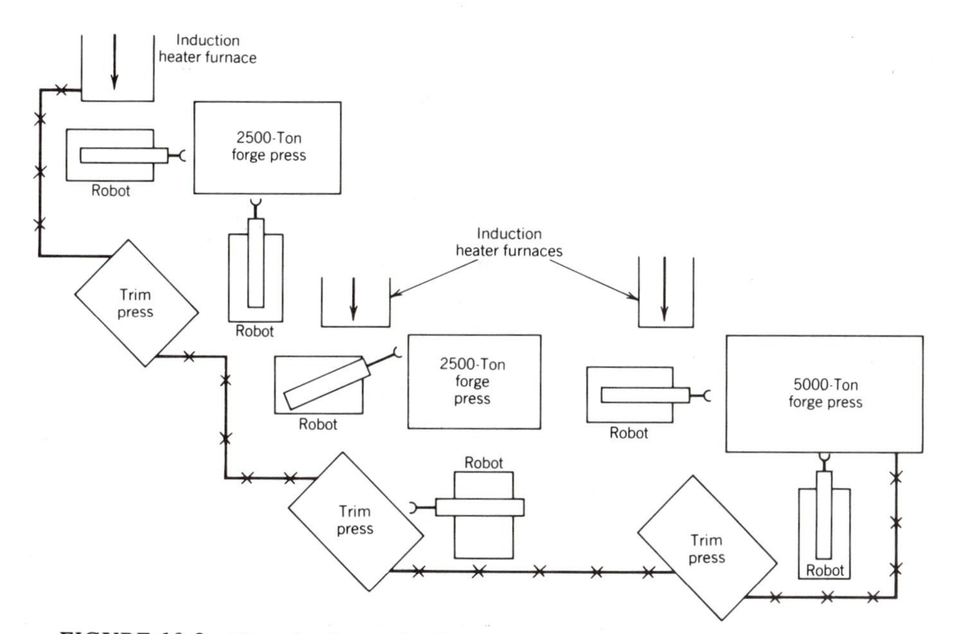

FIGURE 10.8 Three forging workcells served by six industrial robots. (Reprinted by permission of Prab Robots, Inc., Kalamazoo, Mich.)

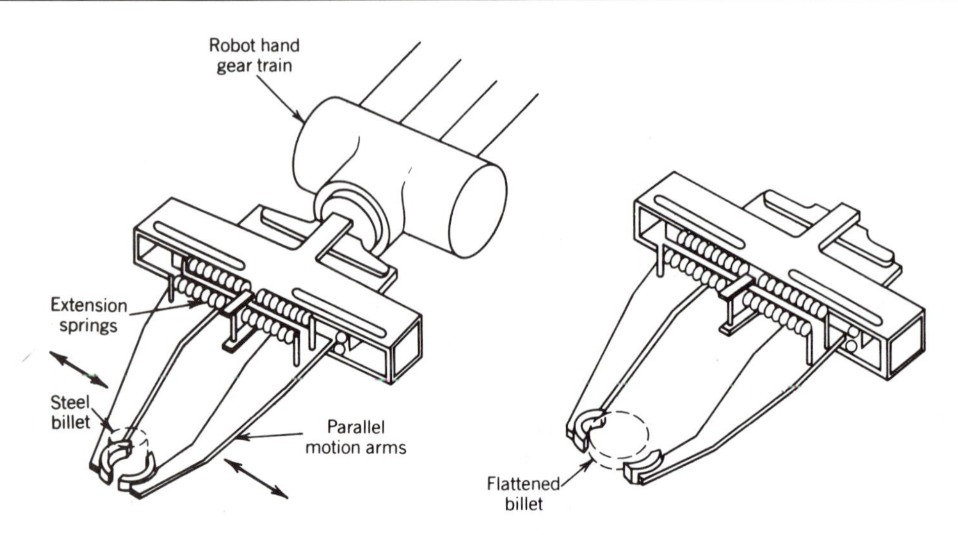

FIGURE 10.9 Specialized robot gripper design for forging applications. (From *Robotics in Practice,* Joseph F. Engelberger, copyright 1980 by AMACON Division of the American Management Associations. Reprinted by permission.)

configuration of the robot fingers that is appropriate for providing necessary grip strength without conducting excessive heat from the hot forging blank back to the working parts of the gripper and the robot arm.

SPRAY PAINTING

The challenge of building a spray-painting booth in conformance with all safety and health codes makes spray painting a natural application for industrial robots. In addition, robots are able to achieve a level of consistency that is difficult to expect human spray painters to duplicate. Although a skilled human spray painter is required the first time to teach the robot a painting task, once taught, the robot will do the spray-painting operation repeatedly, with a consistency unattainable by the same expert who taught the robot the first time through. A comprehensive spray-painting line for automobile bodies uses robots not only for the spray-painting operation but also for opening the doors of the automobile bodies to facilitate the operation. Such a spray-painting line is a dramatic sight to watch, as are spot-welding lines; an exciting film entitled *Ballet Robotique** sets such scenes to ballet music. Spray-painting robots are being utilized on a more modest scale in industries other than the automobile industry.

Chapter 4 warned of the impact upon system productivity caused by the malfunction or low reliability of a single-unit function in continuous automatic production lines. The impact is especially severe upon spray-painting lines. A unit malfunction on a multiple-unit spray-painting line can result in a very costly stoppage of the line, possibly resulting in a whole series of product quality defects due to the interruption of the continuous spraying process. Correcting the malfunction and resetting the equipment can also be a time-consuming undertaking, much more so than the simple unjamming of assembly lines, as studied in Chapter 4.

Ballet Robotique is available in both 16-mm movie and videotape from Pyramid Film and Video, P.O. Box 1048, Santa Monica, CA 90406.

The most practical strategy in the event of unit failure in a spray-painting line is usually to continue operation if possible and to correct any product deficiencies in a manual touch-up area subsequent to the automatic line. Unfortunately, however, some robotic paint sprayer malfunctions inside the automatic spray area can cripple the line. Two examples would be (1) machine/product interference failures, or (2) failure to shut off sprayer at a critical point in the operation. The impact of such malfunctions now will be examined among other factors in case studies of the economics of robot spray-painting lines.

EXAMPLE 10.7 (CASE STUDY) Automated Spray-Painting Line Economics

A firm anticipating major sales increases is contemplating a $1.7 million capital investment in a system of 11 robotic paint sprayers (including one spare) with associated automatic spray line conveyor equipment to replace a manual spray line employing 14 skilled human spray painters.

The existing manual system produces 90 units per hour and achieves an actual daily operation time of six hours per day, employing the spray-painting personnel a full eight-hour shift including breaks, line servicing, maintenance, cleanup, and protective equipment dress up/down time. The total cost of each human spray painter, including salary, benefits, and overhead, is $16 per hour. Additional maintenance costs of the manual system amount to $400 per month.

The new $1.7 million robotic spray line simultaneously operates ten robotic paint sprayers and achieves a production rate of 150 units per hour of operation, working a full eight-hour shift per day. Amortizing the capital investment considering interest, investment credit, depreciation, and salvage values of both the new system (projected) and the retiring manual system results in an estimated equivalent annual cost of $800,000 per year. Not included in this figure are the costs of service, repair, and two skilled human operators to supervise the automatic system. These service, repair, and supervision costs are expected to amount to $75 per hour of system operation. Compare the unit production costs using the current manual system and the proposed robotic system.

Solution

Manual System

$$\begin{aligned}
\text{hourly production cost} &= \frac{14 \text{ workers} \times \$16/\text{hr} \times 8 \text{ hr/day/worker}}{6 \text{ hr/day}} \\
&\quad + \frac{\$400/\text{mo} \times 12 \text{ mo/yr}}{50 \text{ wk/yr} \times 5 \text{ days/wk} \times 6 \text{ hr/day}} \\
&= \$299/\text{hr} + \$3.20/\text{hr} = \$302/\text{hr} \\
\text{hourly production} &= 90 \text{ units/hr} \\
\text{cost/unit} &= \frac{\text{hourly production cost}}{\text{hourly production}} \\
&= \frac{\$302/\text{hr}}{90 \text{ units/hr}} = \underline{\$3.36 \text{ per unit}}
\end{aligned}$$

Robotic System

$$\begin{aligned}
\text{hourly production cost} &= \frac{\$800{,}000 \text{ yr}}{2000 \text{ hrs/yr}} + \$75/\text{hr} = \$400/\text{hr} + \$75/\text{hr} \\
&= \$475/\text{hr} \\
\text{hourly production rate} &= 150 \text{ units/hr} \\
\text{cost/unit} &= \frac{\text{hourly production cost}}{\text{hourly production}} = \frac{\$475/\text{hr}}{150 \text{ units/hr}} = \underline{\$3.17 \text{ per unit}}
\end{aligned}$$

The foregoing analysis shows the new robotic system to be worthwhile, but the comparison is close—too close to warrant the risk of capital investment in some firms even though interest at an appropriate rate of return already had been considered in the $800,000 per year figure for amortization of the robotic system investment.

Adding to the misery of the eager robotics and automation engineer proposing the new robotic paint spray system in Case Study 10.7 might be a skeptical manager who perceives the hazards of failure of one of the ten robots within the paint spray system with potential line stoppage and damage to product as well as production losses. Case Study 10.8 adds this consideration to the analysis.

EXAMPLE 10.8 (CASE STUDY) Paint Spray-Line Analysis Considering Potential Robot Failure

In the robotic paint spray system proposed in Case Study 10.7 suppose each of the ten robots were subject to the possibility of a failure of the type that would stop the line. The mean time between failure (MTBF) for each of these robots for failures of this magnitude is reported by the robot manufacturer to be approximately 2000 hours, or one full year of operation on a standard eight-hour shift basis. In the event that such a failure should occur, and they obviously will occasionally, the estimated loss of production time is four hours, and an additional material and damage expense is expected to amount to $5000 per occurrence. Recompute the production cost per unit, allowing for the occasional inevitable occurrence of such damaging failures of the system.

Solution

Since all ten robots are quoted as having the same MTBF of 2000 hours, the system MTBF would be one-tenth that amount or 2000 hr/10 = 200 hrs. If the ten individual robots had had various MTBFs, the system MTBF computation would have been more complex [ref. 58]. There are two costs to be considered: (1) the loss of system production time, and (2) the additional material and damage cost. The loss of the production time of four hours per occurrence results in the system's production availability being computed as:

$$\frac{200}{200 + 4} = \frac{200}{204} = 0.98 \text{ or } 98\%$$

Thus, in a standard 2000-hour year, only

$$0.98 \times 2000 = 1960 \text{ hr}$$

can be considered productive time for the robotic spray system. This will increase the production cost per hour, because the $800,000 robotic system equivalent annual capital cost must be spread over 1960 productive hours instead of 2000 hours.

The other element of cost is the $5000 material and damage cost per system failure. Since the system MTBF is 200 hours of *productive* time, the system will fail at the rate of not quite ten times per year:

$$\text{system failures/year} = \frac{1960 \text{ productive hr}}{200 \text{ hr (MTBF)}} = 9.8 \text{ failures/year}$$

An easier way to consider this material and damage cost is to prorate it per *productive* hour as follows:

$$\text{cost/hr (material and damage)} = \frac{\$5000\text{/system failure}}{200 \text{ hr between failures}} = \$25\text{/hr}$$

Summarizing the calculation:

Robotic System

$$\text{hourly production cost} = \frac{\$800{,}000\text{/yr}}{1960 \text{ hr/yr}} + \$75\text{/hr} + \$25\text{/hr}$$
$$= \$408\text{/hr} + \$75\text{/hr} + \$25\text{/hr}$$
$$= \$508\text{/hr}$$

$$\text{hourly production rate (per productive hour)} = 150 \text{ units/hr}$$

$$\text{cost/unit} = \frac{\text{hourly production cost}}{\text{hourly production}} = \frac{\$508\text{/hr}}{150 \text{ units/hr}} = \underline{\$3.39\text{/unit}}$$

Thus, Case Study 10.8 shows that the recognition of the possibility of robot failure in the ten-robot paint spray system is sufficient to turn the economic analysis in favor of the manual system, at least from the production-cost-per-unit criterion. The evidence against the new robotic system is far from conclusive, however.

To begin with, there is the capacity criterion. At the beginning of Case Study 10.7, it was stated that the firm anticipated major demand increases. The increased productivity of the proposed robotic paint spray system might be just what is needed to meet capacity requirements without introducing major plant expansion to accommodate multiple lines. Comparing line capacities of the two systems:

Manual System

$$\text{annual capacity} = 90 \text{ units/hr} \times 6 \text{ hr/day} \times 5 \text{ days/wk} \times 50 \text{ wk/yr}$$
$$= 135{,}000 \text{ units/year}$$

Robotic System

$$\text{annual capacity} = 150 \text{ units/hr} \times 1960 \text{ hr/yr}$$
$$= 294{,}000 \text{ units/year}$$

Thus, the robotic system achieves more than double the annual production of the manual system.

Another consideration in the comparison of the proposed robotic paint spray system with the manual system is the possibility of a two- or three-shift operation. The robotic system option benefits from increased hours of production per day because it is capital-intensive, whereas the manual system is labor-intensive. We shall see the effect of these influences in Example 10.9.

EXAMPLE 10.9
Paint Spray Line with Two-Shift Operation

Refer to Case Studies 10.7 and 10.8 and recompute the production cost per unit for both the proposed robotic paint spray system and the current manual system on the basis of a two-shift operation for each. Assume a 10 percent shift differential for all operator personnel costs associated with the second shift.

Solution

Manual System

$$\text{avg. hourly production cost} = \frac{\text{1st shift costs} + \text{2nd shift costs}}{2} + \text{maintenance costs}$$

$$= \frac{\$299/\text{hr} + \$299/\text{hr} \times 1.10}{2} + \$3.20/\text{hr}$$

$$= \$314 + \$3.20$$

$$= \$317/\text{hr}$$

$$\text{hourly production} = 90 \text{ units/hr}$$

$$\text{cost/unit} = \frac{\text{avg. hourly production cost}}{\text{hourly production}} = \frac{\$317/\text{hr}}{90 \text{ units/hr}} = \underline{\$3.52/\text{unit}}$$

Robotic System

$$\text{hourly production cost} = \frac{\$800{,}000/\text{yr}}{1960 \text{ hr/yr/shift} \times 2 \text{ shifts}} + \frac{\$75/\text{hr} + \$75/\text{hr} \times 1.10}{2} + \$25/\text{hr}$$

$$= \$204 + \$78.75 + \$25$$

$$= \$308/\text{hr}$$

$$\text{hourly production} = 150 \text{ units/hr}$$

$$\text{cost/unit} = \frac{\text{hourly production cost}}{\text{hourly production}} = \frac{\$308/\text{hr}}{\$150 \text{ units/hr}} = \underline{\$2.05/\text{unit}}$$

Thus, in a two-shift operation the robotic system emerges as a substantially better alternative. This advantage is available only if the firm actually needs the additional capacity made available by the second-shift operation.

A lesson can be learned from the foregoing economic analysis of the spray-painting operation that can be applied to automation projects in general. A decision to automate a manual operation is a decision to make a capital investment in equipment, an investment that will in general pay greater returns if the equipment is used more hours per day. Therefore, if a robotics or other automation application is not economically attractive on a one-shift basis, it will likely be more attractive on a two-shift basis. Three shifts will likely be better than two, and three shifts seven days per week (approximately equivalent to four shifts) will likely be the best alternative, provided the demand exists for the type of production volume such an operation would represent.

FABRICATION

As was seen in the sections on welding and spray painting, the robot end-effector can act as a tool itself so that the robot can accomplish fabrication directly instead of loading another machine, which in turn accomplishes the work. In this section we explore various ways in which a robot can accomplish fabrication in general.

Drilling

Drilling is a simple machining process in that the feedrate is unidimensional, in this case tool-z. Compare drilling with milling, a process that can involve feedrates in three dimensions. Drilling is a widely used fabrication process and should be considered a candidate application for robotics whenever the drilling is done by a handheld drill. Most drilling operations are done by fixed drill presses, but in the fabrication of large products such as aircraft, space vehicles, ships, cranes, railroad locomotives, and other transportation equipment, handheld drilling operations are standard procedure. The reason is that the workpieces are so large and unwieldy that it is more feasible to bring the tool to the workpiece than to fixture the workpiece in a drill press.

An industrial robot equipped with a drill as its end-effector has many of the capabilities of a human operator with a handheld drill. It can move about and it can position its work platform in an infinite number of planes. This positioning capability is essential in such tasks as drilling rivet holes in aircraft skins, the surface of which may be curved and therefore is no consistent plane. The human operator with a handheld drill or an industrial robot can adjust the plane of the tool platform such that the direction of the feed is orthogonal to the surface of the aircraft skin. The human operator typically manually positions the handheld drill in an orthogonal orientation by visual determination. In such an operation, the robot can be positioned more precisely than a human-held drill, and thus quality is enhanced even when using a template. Not only quality, but economics and safety are benefits to be gained from robotic drilling operations, as the following case study attests.

EXAMPLE 10.10 (CASE STUDY)

At the Forth Worth plant of General Dynamics one robot drills 550 holes in the vertical tail fins of F-16 fighters. This application was begun with a hydraulic robot and was successful for five years and 2600 aircraft before the process was updated by replacing the hydraulic model robot with an electric model, both from Cincinnati Milacron (see Figure 10.10). This application not only saved worker hours but also eliminated a hazard by preventing worker exposure to composite dusts created by the drilling operation [ref. 117].

ASSEMBLY

The most challenging arena for robot application is robotic assembly. Assembly requires precision, repeatability, variety of motion, sophisticated gripper devices, and sometimes compound gripper mechanisms in which the gripper plays an active and primary role in one of the assembly steps, besides simply holding the piece-parts.

The basis of a decision to employ industrial robots in automatic assembly is usually to save labor costs. This places a particular obligation upon robotic assembly operations to be fast and efficient. Even a slow and inefficient robotic application in die casting or arc welding might be acceptable if it eliminates hazards or improves weld quality, but robotic assembly does not enjoy such luxury. For the most part, assembly operations are simply not dangerous, dirty, hot, or unhealthful. Therefore, the assembly robot must be fast and efficient to compete with human assembly operators without consideration for the robot's special capabilities for working in inhospitable environments.

FIGURE 10.10 Cincinnati Milacron $T^3$776 Robot drills rivet holes for vertical fin skins in F-16 aircraft. (Photo courtesy of General Dynamics.)

In cases where speed and efficiency are not the criteria for employing an assembly robot, accuracy may be the motivation. Some assembly operations, such as electronic circuit board assembly, are susceptible to human errors of omission or substitution. The industrial robot is not likely to commit these types of errors. Also, some defective or marginally defective assemblies are the result of *intentional* omissions on the part of the human assembly operator to speed up production and achieve incentive awards. Some assembly defects such as missing or loose screws may not be obvious to assembly inspectors or final inspectors. In these cases, unethical human assembly operators may occasionally get by with contributing to product inferiority, whereas the robot is programmed to perform the operation exactly the same way every time.

Engine Assembly

The automobile industry is the primary user of spot-welding and spray-painting robots. To a lesser extent, assembly robots are also playing a role in the automotive industry. Figure 10.11 illustrates a pilot robot station along an assembly line for automobile engines that was developed at Fraunhofer-Institute for Production Automation in the Federal Republic of Germany. Note that the main engine block assembly travels on a pallet transfer mechanism. The robots can easily access subassemblies delivered by small auxiliary conveyors. Vibratory bowls feed and orient small parts such as screws and bolts.

In most complicated assembly stations such as the one illustrated in Figure 10.11, it is necessary for the robot to equip itself automatically with various grippers, each designed

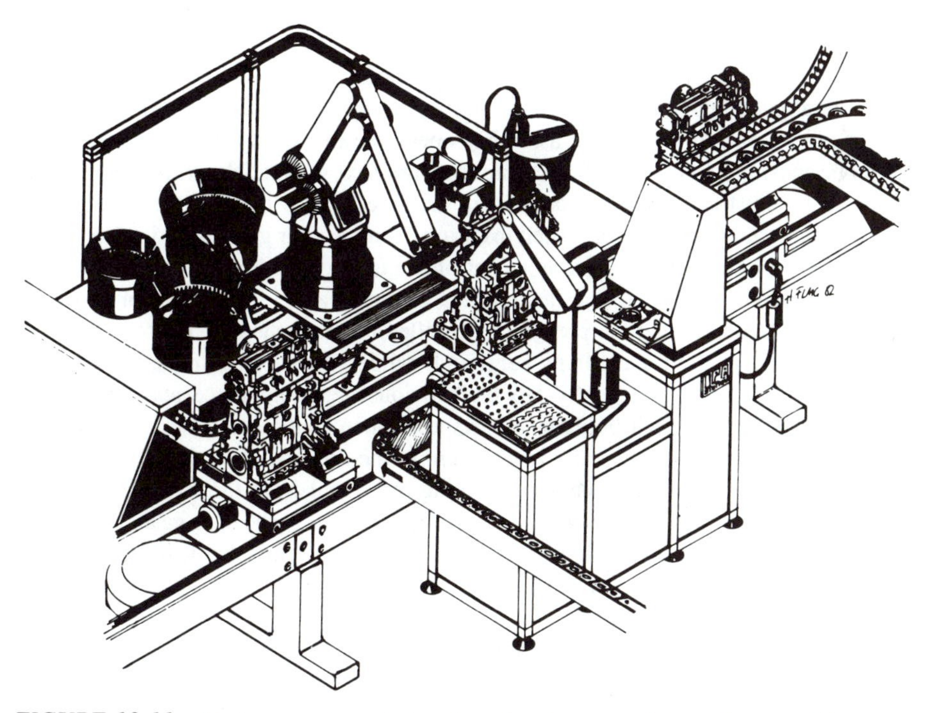

FIGURE 10.11 Pilot automotive robot work station for automobile engine assembly. (Reprinted with permission of the Society of Manufacturing Engineers, Proceedings of the 13th International Symposium on Industrial Robots and Robots 7, April 1983, pp. 5–39.)

for a specific task. This feature can be programmed into the robot's flexible memory system with the different grippers arranged in a nearby rack or magazine. One such tool is an automatic screwing device of which there may be several types at a single station. The robot thus provides automatic position and orientation flexibility to the automatic screw mechanisms described in Chapter 4.

Electrical/Electronic Machine Assembly

The indexed, palletized carrier type of transfer mechanism in robotic assembly is not limited to heavy assemblies such as automobile engine blocks. A small cassette tape deck mechanism assembly is handled by similar carriers in a highly automatic robotic assembly line at the Kofu, Japan, plant of Sankyo Seiki Manufacturing Company, Ltd. A series of 12 robot stations is interrupted in only two places by manual operations between robots 9 and 10 and again between robots 11 and 12. Most of the robot stations exhibit compound motions and multiple grippers. The automatic line is fascinating to watch and is made even more remarkable by the wide variety of intricate assembly operations requiring compound robot motions.

A successful robot assembly operation usually incorporates a variety of tactile and/or visual feedback loops to permit the robot to monitor results and take action accordingly. To see why this is necessary, compare the manual operation of drilling a hole versus the manual operation of assembling two mating parts. The assembly operation requires more "feel" for the objects being handled, some orientation of parts, perhaps some restarting or repositioning, and even perhaps some groping around for lost or missing parts. For these reasons, it is often necessary to install video camera vision systems over the assembly work station for providing input to the robot controller, and even more important may be the provision of tactile force sensors upon the robot hand itself. The controller inputs from these force sensors can be monitored to provide the "feel" to tell the robot when a part is misaligned and when to back up and try again. The robot may be programmed even to index back to a previous assembly operation and gently tap a misaligned part into proper position so that it will accommodate the next sequential part in the assembly. Figure 10.12 illustrates such a capability in the assembly of meshing gears at an IBM assembly plant in Austin, Texas. When inserting a mating spur gear, a tactile force sensor on the robot gripper signals the robot controller that too much resistance is being encountered in the assembly. If resistance is encountered early in the insertion stroke, the correct interpretation is that the previously installed mating spur gear is not in its proper position. The robot then can lift its hand and, while still holding the spur gear to be assembled, move back and gently tap the previous gear into place and then resume assembly of the spur gear it holds in its gripper. When trying a second time, the robot may again encounter resistance at a lower point in the downward stroke. By monitoring the position of the robot hand while monitoring the signal from the tactile force sensor on the gripper, the robot controller can interpret correctly that the spur gears to be mated are positioned correctly but a slight rotation is needed to permit the gear teeth to align. Accordingly, the robot hand is rotated a small amount and inserted until the hand position indicates that mating has occurred.

A natural consequence of the incorporation of visual and tactile sensory capabilities in the assembly robot is to use this capability also for quality control. The robot can sense a defective assembly and make note of it for signaling corrective action in a future process. A convenient trick for doing this has been used by the IBM Corporation (Figure 10.13) when typewriter ribbon cartridge assemblies are processed on a palletized indexing conveyor. When the robot senses a defective assembly, it merely withdraws its hand,

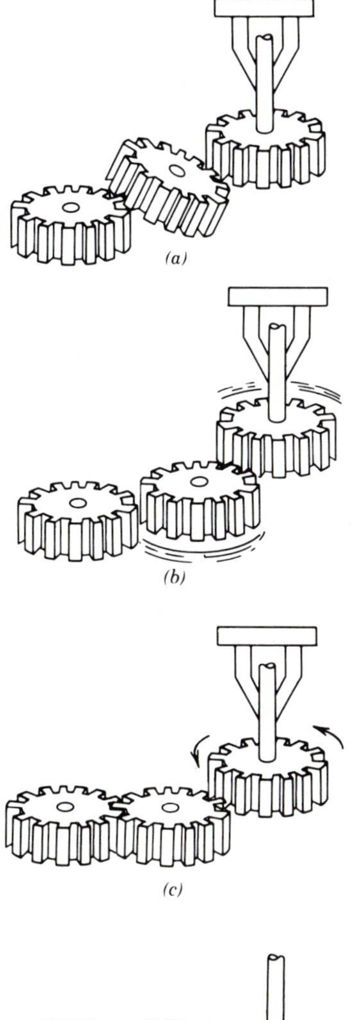

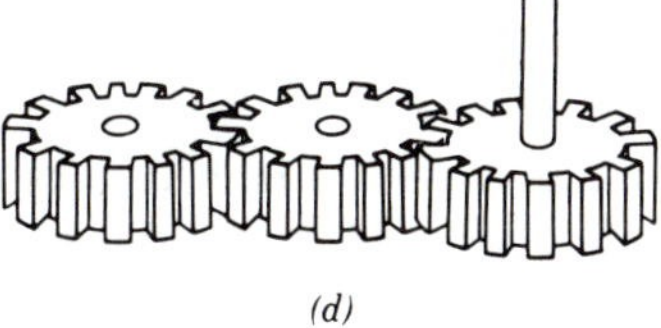

FIGURE 10.12 Tactile force sensors on the gripper of an assembly robot monitor the position of mating spur gears. The robot controller is programmed to permit the robot to take a variety of corrective actions. (*a*) Force sensor on robot gripper indicates resistance in the downward stroke; the previous spur gear is not in position. (*b*) The robot backs away, shifts back over the previous gear, and gently taps, nudges, and perhaps slightly rotates the previous gear until it falls into position. All of this is done while the third spur gear is still held in the gripper. (*c*) The robot now proceeds to the installation of the third gear. When the force sensor again encounters resistance, but lower in the stroke, the robot merely rotates the gear slightly so the teeth will mesh. (*d*) Assembly of the mating spur gears is complete.

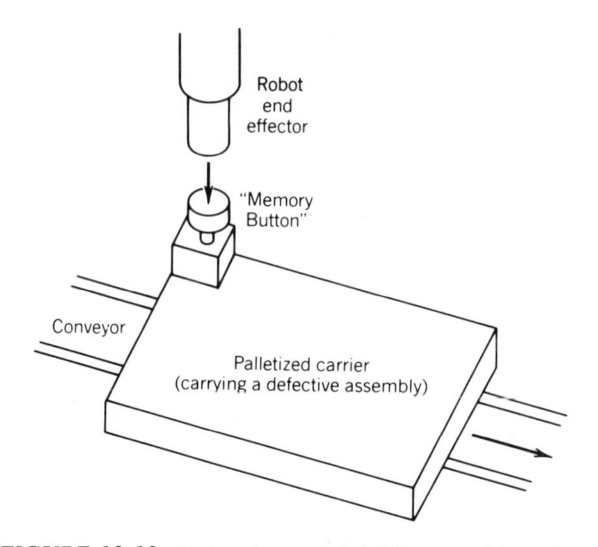

FIGURE 10.13 Robot detects a defective assembly and proceeds to the corner of the palletized conveyor and presses a memory button for future reference and corrective action.

advances to a programmed position near the corner of the palletized carrier, and mechanically pushes down a plunger or "memory button" that is used to signify a defective assembly on this particular carrier. The information provided by the memory button can be accessed and brought to bear upon subsequent automatic assembly stations down the line or it can be used to switch the carrier to a rework spur on the conveying system.

Typewriters are themselves an intricate assembly task, and the degree of variation among keyboard styles and character sets from model to model places a premium upon flexible automation in their assembly. Especially for the assembly of the keys on terminal keyboards is the flexible approach needed. Vanderbrug, Wilt, and Davis [ref. 116] report an automatic keycap assembly system using a video camera to provide vision capability for verifying the correctness of the keycaps and for detecting major defects. The vision system assembles the typewriter keycaps into magazines. Then a Cartesian style robot equipped with a multiple-fingered end-effector is used to assemble several keycaps at a time from a part presentation array onto a keyboard base that is transported on a rail-mounted shuttle. The shuttle removes the necessity for operator entry into the hazardous work envelope of the robot.

A key feature in the design of this system by Automatix Incorporated was to separate the system into two stations: the magazine assembly station and the robot station. The magazine assembly station contains the bowl feeders and vision system for orienting and checking keys when making up the magazines. The assembled magazines are transported manually to the parts presentation array by a cart that can handle all magazines for a single job at one time. The carts thus become a convenient buffer storage between the two stations, the value of which should be obvious from the concepts studied in Chapter 4. Another advantage of separating the station is that at some future date when automation is more prevalent, a vendor may supply keycaps preassembled into magazines, eliminating the need for the magazine assembly station. The following example describes the analysis needed to balance two stations in an automatic assembly system with buffer storage between the stations as in the robotic keycap assembly system.

EXAMPLE 10.11
Balancing Two Stations in Robotic Keycap Assembly

A magazine assembly station for a robotic assembly operation consists of bowl feeder and vision systems for orienting and inspecting piece-parts before loading them into 250-piece magazines—one magazine for each different part type. The magazines then are transported to a parts presentation array, where they are loaded for access by the robot. A vibratory bowl/vision system can orient, inspect, and assemble a magazine in ten minutes. The robot can assemble an 80-piece product at a production rate of about 50 assemblies per hour. How many bowl feeder/vision systems are needed to be operated in parallel to keep pace with the robot?

Solution

To orient, inspect, and assemble 80 magazines would require $80 \times 10 = 800$ min. The robot could deplete these magazines in $250/50 = 5$ hr $= 300$ min. Two parallel stations could assemble the magazines in 400 minutes, but the robot would still wait a substantial amount of time. Three stations would reduce the magazine assembly time to $800/3 = 267$ min, enough to keep pace with the robot under ideal conditions.

To allow time for processing defective, misoriented, or incorrect parts in the magazine assembly operation of Example 10.11, it might be prudent to supply four parallel vibratory bowl/vision systems to keep pace with the robot. Relative cost of the magazine assembly equipment and the robot station could also influence the decision between three and four parallel stations.

General Assembly

The ultimate in assembly flexibility would be a system that could assemble virtually any product of appropriate size given that the assembly of the product is basically feasible—that is, that the product is "assemblable." This is an ambitious goal considering that assembly jigs and fixtures might render the system too product-specific. Undaunted by such problems, however, researchers at Stanford Research Institute International, with the support of the National Science Foundation, have proceeded to develop such a generalized assembly station consisting of a station controller, a binary vision module with three video cameras, and two manipulator modules called arm 1 and arm 2, respectively. The end-effector of arm 1 consists of a plastic remote-center compliance device, a video camera, and a two-fingered hand. The end-effector of arm 2 consists of a six-axis wrist force/torque sensor and a two-fingered hand. Smith and Nitzan [ref. 99] report that the generalized station has exhibited the following capabilities:

- No special-purpose fixtures—some parts are presented to the manipulator on a pallet, others are on a slide. The assembly is assisted by a simple fixture with "V" notches for centering cylindrical workpieces.
- Utilization of medium-level module commands, such as MOVETO, FIND, and GRASP.
- Binary vision to locate and identify parts on a pallet for acquisition. Three cameras are used during the assembly—two fixed in the ceiling and one attached to the end-effector of one of the manipulators.
- Force feedback to actively control compliance and insertion forces in snap-together, part-mating operations.

- A stationwide calibration scheme for two manipulators and multiple cameras, enabling part locations to be given with respect to a common reference frame.
- Concurrent operation of two manipulators in a central assembly area, while the binary-vision system recognizes and locates parts elsewhere.
- Synchronization of manipulator motions to avoid collisions in the central assembly area.
- Simple sensors to verify operation, including the use of finger separation to ascertain the presence of a part in the hand, and a click detector (employing a microphone mechanically coupled to the assembly area) to verify that semirigid parts have snapped together successfully, or to detect a drop of a part.

UNUSUAL APPLICATIONS

At least 90 percent of all robot applications fall into the four categories already covered in this chapter: welding, machine loading, spray painting, and assembly. It is worthwhile, however, to consider a small number of unusual applications that are opening the field to tasks previously considered impossible for a robot. Chapter 8 showed the potential for machine vision systems to be applied to inspection tasks. Most remarkable among these applications are the sorting operations performed on such small and unlikely items as coffee beans, peanuts, and rice. Agricultural products have such unquantifiable variation that they pose the greatest challenge to mechanization and automation. The motivation to overcome this challenge is the large and stable demand for agricultural products. We now focus upon one agricultural application of robots, perhaps the most unlikely application ever attempted by the field of robotics.

Sheep-Shearing Robots

The shearing of sheep is a large-volume operation in many countries of the world, especially in Australia and New Zealand, where sheep populations far outnumber the human population. The local demand for this operation, which is not considered a desirable job for humans, created the impetus for a research effort undertaken by the University of Western Australia to design and build a robot that is capable of shearing sheep. This application first astonished the world in the early 1980s (see Figure 10.14) but continues to be perfected. Trevelyan [ref. 109] describes the operation as requiring "a delicate touch, yet fast reactions, a large workspace, substantial load carrying capabilities, and adaptive control." The principal task is to keep the cutter close to the skin, and this is accomplished by tactile and proximity sensors providing real-time information with respect to a continuously changing workpiece. However, path and trajectory adaptation may not be sufficient, according to Trevelyan. The researchers are now experimenting with machine vision systems to generate geometric models of the sheep's surface to permit online strategy planning, much as a human operator would do. The ability to monitor a wide variety of variables permits inputs to a real-time expert system that is capable of replanning the overall strategy in real-time.

Robots in Construction

Adverse working conditions, heavy lifting loads, and workplace hazards are commonplace on construction sites and motivate the applications of robotics if possible. Despite these

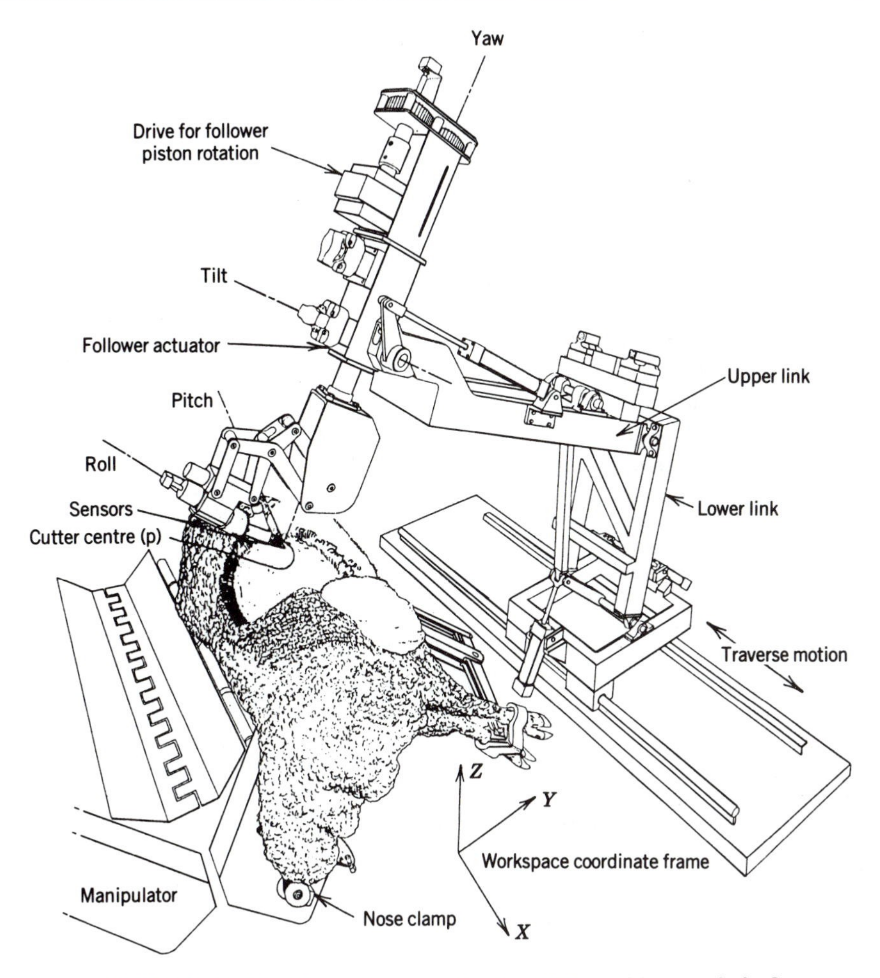

FIGURE 10.14 Sheep-shearing robot [ref 109; used by permission].

motivations, progress in applying robots to construction has been slow, at least in the United States. Japanese construction companies have made much more progress [ref. 79], having already developed robotic machines to spray fireproofing insulation, assist in the erection of steel girders, and automatically skim concrete floors.

Related to construction applications are operations to survey and clean up hazardous material sites and to decommission nuclear power plants. The Three Mile Island nuclear plant crisis made visible the need for robots to perform hazardous cleanup and repair operations. William L. Whittaker at Carnegie Mellon University applied remote work systems to the exploration and remediation of the Three Mile Island reactor containment basement [ref. 79]. Carnegie Mellon's research in construction robotics includes the development of a Portable Pipe Mapper (PPM) for inferring utility line location for excavation and other construction operations. Also developed at Carnegie Mellon is the Site Investigation Robot (SIR) that includes both an all-terrain mobile robot for collecting data and a software system for analyzing the sensed data to produce a numerical model of the site, which may be a hazardous site here on earth or the terrain of another planet.

SUMMARY

The primary application of industrial robots is in welding: first spot welding and then arc welding. The automotive industry put robots in the spot-welding business, and this industry is the primary user of spray-painting robots, too. Recognizing the importance of the automotive industry in the application of industrial robots already in place, the current growth rates are in industries other than automotive. The recent rise in arc-welding robots has importance to the shipbuilding, structural steel, and agricultural equipment industries as it does to the myriad industries that use arc-welding in the manufacture of smaller metal products. Due to the nuisance of dealing with personal protective equipment in hazardous arc-welding operations, considerable productivity increases are possible with the use of arc-welding robots that can average a much higher arc-on time.

Machine loading is probably the most important application of industrial robots outside of the automotive industry. Since more readers of this text are expected to be outside the automotive industry than inside, machine-loading applications have been emphasized in this chapter. One of the driving forces in the automation of machine-loading operations has been worker safety, especially for punch presses. The key to many successful and efficient machine-loading applications for industrial robots is to equip the robot with a double-handed gripper. When one robot must attend several machines in a sequence, some planning is often necessary to prevent costly machine-idle time while waiting for the robot. A special motivation to use industrial robots to load machines exists when the parts to be loaded are hot, as in forging and die casting.

Perhaps the most elegant application of industrial robots is in spray painting. Spray painting certainly requires some of the most sophisticated robot types, the continuous-path models. The synchronized, simultaneous movement of an entire bank of spray-painting robots has been likened to a ballet performance. But when a crippling malfunction occurs, costly damage and restart procedures can be the result. In economic case studies, multiple-shift operation favors the robotic solution to spray painting and other automation applications.

The ultimate challenge in industrial robot application is in automatic assembly. In this application, the robot usually cannot rely upon its capabilities to work in harsh environments and must compete with human operators in dexterity, efficiency, and speed. The human usually wins this competition, but some robots are beginning to gain acceptance in automobile engine assembly and especially in electrical and electronic equipment assembly. What manufacturing engineers are learning from robotic assembly research is probably helping manual assembly as well. The ultimate in robotic assembly, a completely flexible and general robotic assembly station without special-purpose jigs and fixtures, is still not on the factory floor. However, it is significant to note that such general and flexible assembly stations are currently under development in research laboratories.

EXERCISES AND STUDY QUESTIONS

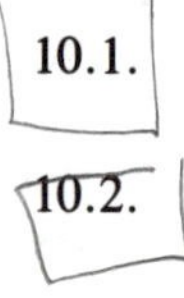

10.1. What type of application employs the largest number of industrial robots today? In what industry is the largest number of robots found?

10.2. Suppose a new arc-welding robot is able to achieve a 90 percent arc-on percentage while at the same time increasing weld speed 25 percent during arc-on time. If three of these robotic welding units can be simultaneously operated at peak efficiency with only a single, experienced human welder to tend them, what is the potential increase in productivity per human welder over average conventional arc-welding methods?

10.3. What problem is associated with using an axis-limit robot for feeding a machine when the material to be fed is located in a vertical stack?

10.4. A machine with an operation cycle time of 3 sec is loaded and unloaded by an industrial robot having the following elemental times:

Load part into machine	.7 sec
Unload part from machine	.6 sec
Move finished part to discharge conveyor	1.3 sec
Drop part onto discharge conveyor	.2 sec
Move from discharge conveyor to infeed conveyor	2.4 sec
Pick up part at infeed conveyor	.8 sec
Move part to work station	1.1 sec

Calculate the total ideal cycle time per piece and the ideal production rate, assuming that the robot is equipped with a single gripper.

10.5. In Exercise 10.4 suppose the robot is equipped with a double-handed gripper. The rotation of the hand between unloading and loading increases the load time from .7 sec to .9 sec. Calculate the total ideal cycle time per piece and the ideal production rate.

10.6. In a process design improvement suppose the machine in Exercise 10.4 is equipped with a pneumatic parts ejection system for removing finished pieces from the die area. Suppose that the pneumatic ejection adds $\frac{1}{2}$ sec to the 3-sec operation cycle. The sole task of the robot in this new configuration is to fetch new parts and load them into the machine. Calculate the total ideal cycle time per piece and ideal production rate in this improved configuration. Would the presence or absence of a double-handed gripper make any difference in this configuration? Why or why not?

10.7. Suppose that it is possible to feed the machine of Exercise 10.4 using a coil feeder so that the feeding portion of the total cycle is automatic. Suppose that the automatic feed requires one second in addition to the three seconds required by the machine operating cycle. The robot's task in this new configuration is simply to unload the finished parts and take them to the discharge conveyor. Calculate the total ideal cycle time per piece and ideal production rate in this improved configuration. Would the presence or absence of a double-handed gripper make any difference in this configuration? Why or why not?

10.8. What is the principal motivation for using industrial robots in machine loading and unloading of punch presses?

10.9. What particular motivation exists to use robots for machine loading for forging and die-casting operations?

10.10. An industrial robot has a mean time between failure (MTBF) of 1000 hours. Each failure results in an average downtime of 16 working hours. What is the percentage availability of this robot? If the standard work-year for this robot is 4000 hours, how many hours per year will it actually be available for operation?

10.11. Five robots load and unload five machines, respectively, served by a single continuous conveyor that runs alongside the five machines. If the operation is a five-station, sequential production line, and each of the machine cycle times is 12 seconds including machine loading and unloading, what is the daily *ideal* production capability per eight-hour shift?

10.12. In the five-robot/five-machine production line described in Exercise 10.11, suppose each robot has an MTBF of 1000 hours and that a robot failure results in an average downtime of 16 working hours. With no storage between machine stations, assume that a failure of any of the robots stops all five stations until the failure is corrected. Using the same station cycle time of 12 seconds, what is the daily eight-hour-shift production rate considering potential robot failure?

10.13. Five robots load and unload five machines, respectively, served by a single conveyor. Although all five machines are performing the same operation, they are parallel and independent of each other's operation. If the machine cycle time including loading and unloading is 12 seconds, what is the daily *ideal* production capability per eight-hour shift?

10.14. In exercise 10.13 assume that each robot has an MTBF of 1000 hours and that a robot failure results in an average downtime of 16 working hours. Using the same cycle of 12 seconds, what is the daily eight-hour-shift production rate considering potential robot failure?

10.15. A single robot serves the machines that process workpieces sequentially as in the layout of Figure 10.7. The robot time for each station is 0.1 min for pick-up and 0.1 min for release, plus any necessary move time. Robot move times are in accordance with Table 10.1. Assume machine times as follows:

Machine 1 (station B) time = 6 min
Machine 2 (station C) time = 9 min
Machine 3 (station D) time = 5 min

Assuming an ideal robot availability of 100 percent, determine the cycle time for this system. What is the production rate?

10.16. If the robot in Exercise 10.15 has an MTBF of 600 hours and time to correct a failure is 10 hours, what is the revised production rate?

CHAPTER 11

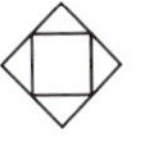

INDUSTRIAL LOGIC CONTROL SYSTEMS

The entry of industrial robots as the "steel-collar" workers of today's factories has been dramatic, but a closer look at the field of automation reveals the quieter but even more powerful role being played by automatic industrial logic control systems. At the heart of this development lie the programmable logic controllers (PLCs), a subject dealt with specifically in Chapter 13.

The word *robot* is familiar to virtually everyone, but words seem to fail us when we talk about "programmable logic controllers," "digital electronics," or "logic control systems." However, all three of these terms refer to the basic elements that control automatic systems throughout the factory, including many robots. The purpose of this chapter is to explain what industrial logic control is and why it is so important to manufacturing automation and robots.

An industrial plant consists of machines, processes, people, raw material, and products all functioning together according to plans, objectives, specifications, and schedules. But opposing these plans and objectives are defective materials, machine breakdowns, hazards to people and processes, defective product, delays, and other upsets common to the real world. The manufacturing system must deal with these upsets by making decisions and applying corrective action, whether these decisions be made by people or machines. Besides dealing with the upsets, decision systems must control the ordinary and natural sequence of when to do what next.

Industrial decisions are sometimes divided into two broad categories:

1. Decisions by attributes
2. Decisions by variables

The field of inspection and quality control identifies attributes decisions as "Accept/Reject" or "GO/NOGO," whereas variables decisions consider such matters as:

1. How high is the temperature?
2. What is the weight?
3. How long is the piece-part?

Quality control is not the only field that deals with attributes and variables decisions. Consider the operations implied by the following questions for which decisions must be made by people or machines:

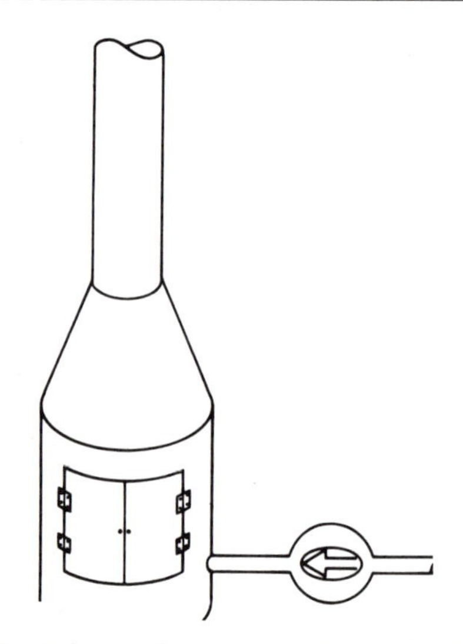

FIGURE 11.1 Small furnace for preheating forging blanks for processing.

1. Is the in-feed conveyor on the number 2 fill line operating?
2. What is the stock count for containers used on the packaging line?

If the question can be answered yes or no, it is the "attributes" type and is a candidate for automation by an industrial logic control system. Even if the question is the "variables" type, which is answered by a number or quantity instead of by yes or no, it still may be broken down into a series of yes/no decisions that establish the desired value. Thus, the industrial logic control system can be used directly for attributes-type decision systems and indirectly for variables-type decisions, too, as will be seen in examples later in this chapter.

A simple example will help to remove industrial logic control systems from the realm of the abstract. Figure 11.1 is a sketch of a small furnace used to preheat forging blanks for processing. The furnace is fueled by natural gas, and in event of flame failure the system must be reignited, or, if the reignition is unsuccessful, the fuel must be turned off to prevent buildup of a serious hazard. In a crude manual system, a worker might monitor the flame and operate a manual valve to control the fuel. But accepted standards of efficiency and safety demand an automated system to sense system status and take immediate action as required. Of course, manual overrides would be necessary, and the control system would have to be governed by an ON/OFF switch. Timer circuits might be used on an automatic ignition cycle, and the complexity grows. Before long, this complexity demands some sort of shorthand or convention for explicitly stating all of the logical relationships between variables. Several tools for this purpose will be explained in this chapter, beginning with "truth tables."

TRUTH TABLES

A truth table is a matrix that relates all feasible combinations of logical input conditions to the desired output conditions. To illustrate the truth table we will use another example that is even simpler than the furnace control system discussed earlier.

Console Lock Example

Suppose a production process is operated by a console and we desire to limit console access by means of a lock and key for authorized persons only. For simplicity in the extreme, suppose the console has only one control, an ON/OFF switch, in addition to the lock (see Figure 11.2). The ON/OFF switch is capable of being physically manipulated regardless of the state of the lock, but the function of the ON/OFF switch is disabled if the lock is locked. The console lock system is an industrial logic control system with discrete attribute-type variables, each of which can take on two and only two values or states. These variables and their states are defined as follows:

S represents the ON/OFF switch:

$$S = 0, \text{ if the switch is in the OFF position}$$
$$= 1, \text{ if the switch is in the ON position}$$

L represents the console lock:

$$L = 0, \text{ if the key is absent or turned left (console locked)}$$
$$= 1, \text{ if the key is present and turned right (console unlocked)}$$

P represents the process:

$$P = 0, \text{ if the process is shut down}$$
$$= 1, \text{ if the process is running}$$

The digits 0 and 1 customarily are used to represent the two states of logic variables. Other common usages are

True/False

T/F

ON/OFF

After the variables are defined, the truth table is used to completely define their interrelationships. Input variables are tabulated in the columns on the left side of the matrix, and output variables are tabulated on the right.* The truth table for the console lock example is displayed in Table 11.1

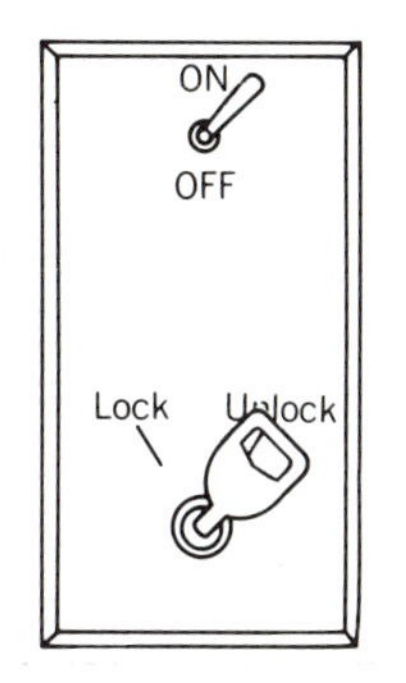

FIGURE 11.2 Console lock. The switch controls the process only if the lock is unlocked.

*A variation in truth table format uses rows instead of columns to represent variables. In such tables, the top portion of the matrix represents input variables and the bottom portion represents output variables.

TABLE 11.1 Truth Table for Console Lock Example

Inputs			*Output*
S	*L*	*P*	*P*
0	0	0	0
0	0	1	1
0	1	0	0
0	1	1	0
1	0	0	0
1	0	1	1
1	1	0	1
1	1	1	1

The console lock example has three variables: three are input and one is output. This appears to be a contradiction because of course three plus one is *four,* not three. The explanation is that the variable *P,* representing the process, is both an input variable and an output variable. This makes sense because if the key is not in the lock, we want the process to remain shut down if it was already shut down or to continue running if it was already running, regardless of the position of the ON/OFF switch. In other words, the current status of the process *P* (input) may affect its future status (output).

The truth table is interpreted row by row. For instance, the third row in Table 11.1 is interpreted as follows:

IF the ON/OFF switch is in the OFF position

AND IF the key is in the lock

AND IF the process is shut down

THEN the process should remain shut down

In general, truth tables may have any number of inputs and outputs, and the general interpretation of any row is

IF (input 1) AND IF (input 2) . . . AND IF (input *n*)

THEN (output 1) AND (output 2) . . . AND (output *m*)

where the logic control system has *n* inputs and *m* outputs.

Row 4 of Table 11.1 seems to present a paradox: How can the variable *P* take on both values 1 and 0 *on the same row?* The resolution of this seeming contradiction is that one tabulation of the variable *P* is as an input and the other is as an output. A verbal interpretation of the fourth row of Table 11.1 is as follows:

If the ON/OFF switch is in the OFF position

AND IF the console is unlocked

AND IF the process is running

THEN the process should shut down (immediately)

It is best to think of the values of the input variables as existing an instant in time *before* the corresponding output values on the *same* row of the truth table. In Chapter 13, we shall see that a finite but very short time is required for conversion of inputs to outputs in a real-world application of industrial logic control systems. Sometimes this tiny difference between input and output values of the same variable is denoted by calling the input P and the output P', or vice versa.

One thing to remember about truth tables is that the sequence of the rows (or columns) is completely arbitrary. No meaning is necessarily attached to the choice of row sequence. Sometimes it is helpful to use a natural sequence according to the normal operation of the system from state to state through time. But such a strategy has the disadvantage of sometimes overlooking unusual states that might be taken on by the system. Another strategy is to use an orderly sequence that exhaust every possible configuration of system inputs. This was the strategy used in Table 11.1. Since each variable can take values of either 0 or 1, it is convenient to view each combination of inputs as a number in the binary number system. Thus, an orderly sequence that will eventually exhaust all input combinations is simply to count in binary. Such a strategy always begins with a row of zeroes in the input columns and ends with a row of ones. Note that this strategy was used in Table 11.1. Also note that in Table 11.1 the initial row of zeroes extends to include the output column; the same can be said for the final row of ones. However, this was merely a coincidental characteristic of logic underlying the console lock application. Other applications will not necessarily follow this rule.

In a real industrial application, a truth table will often present a long list of input variables but have only a small set of feasible combinations of values to be taken on by the input variables, recognizing the constraints of the industrial application it is being used to analyze. In these cases, it would be senseless tedium to attempt to exhaust all input combinations by counting in binary. Such applications are best developed in the truth table using the natural sequence strategy described earlier.

Push-Button Switch

Most large industrial machines are turned off by means of separate spring push buttons for ON and OFF (see Figure 11.3). This has safety implications in that the OFF push button(s) can be given priority to shut down the machine in an emergency regardless of the status of the ON push button. The logic of an ON/OFF push button scheme is shown here as another example of the use of truth tables. Inputs and outputs are defined as:

G represents the ON push button (G for GO):

= 1 when the ON push button is depressed

= 0 when the ON push button is released

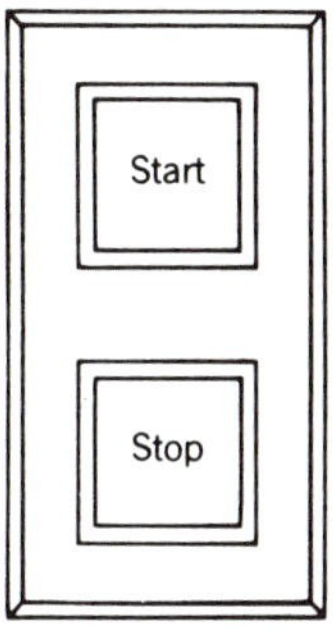

FIGURE 11.3 Push-button switch. Push "Start" to turn system on. System stays on until "Stop" button is pushed. Stop takes precedence over Start.

TABLE 11.2 Truth Table for ON/OFF Push-button System

Inputs			*Output*
G	*S*	*R*	*R'*
0	0	0	0
0	0	1	1
0	1	0	0
0	1	1	0
1	0	0	1
1	0	1	1
1	1	0	0
1	1	1	0

S represents the OFF push button (*S* for STOP):

= 1 when the OFF push button is depressed

= 0 when the OFF push button is released

R represents the state of the machine (*R* for RUN):

= 0 when the machine is not running

= 1 when the machine is running

The truth table for the push-button switch example is displayed in Table 11.2.

Comparison of Table 11.2 with Table 11.1 reveals some interesting contrasts between the push-button switch example and the console lock example discussed earlier. First note that Table 11.2 uses R' to distinguish the output variable from input variable R. Note that both examples used the same orderly strategy (binary counting) for exhausting all input combinations in the truth table. But study the differences in the outputs. The differences are the result of the logical differences between the operation of the console lock system and that of the push-button switch. The variations possible with the truth table are limited only by the imagination of the automation engineer in dreaming up practical applications of logic control. A final observation is that the last row of Table 11.2 is not all 1s, as was true of Table 11.1. Readers should think through the logic of the push-button switch to satisfy themselves that the logical output for a 1-1-1 input should be 0, not 1, for the push-button switch.

BOOLEAN ALGEBRA

In addition to truth tables, algebraic expressions can be used to define logical outputs as algebraic expressions of input variables. The algebra of logic is called Boolean algebra* and is similar to but different in important ways from ordinary algebra. The basics of Boolean algebra are presented here with a minimum of theory. The purpose is twofold:

1. To permit the automation engineer to express logical relationships between systems in a convenient shorthand
2. To take advantage of the theorems and laws of Boolean algebra to simplify industrial logic control systems

*After nineteenth-century English mathematician and logician George Boole.

Basic Operators

Only three basic operators are needed to construct Boolean lo
are defined in Table 11.3. It may be confusing that the plus sig
used for the logical OR, instead of AND. Further, the dot, which
in ordinary algebra, is used for AND in Boolean algebra. It is conv
of logic to use the words *factors* and *terms* in their conventional mean
user is aware of the differences between the algebras.

In interpreting a Boolean logic expression, one way to rationalize the lo
is to think of each factor as a *requirement* for obtaining the output and eac
acceptable *alternative* for obtaining an output.

Some examples will help to illustrate the meaning of the logical algebraic o
Equation 11.1 is an elementary logical algebraic equation in which an output (dep
variable Y is written as an expression of two input (independent) variables A and

$$Y = A \cdot B \quad (11.$$

The equation states that *both* inputs A and B must be on—that is, 1 or true—in order to obtain an ON condition—or logic 1 or true—for the output. Since there is only one term in the expression, there are no alternatives: Only the specified combination of both A and B can produce an output Y.

Equation 11.2 adds an alternative for obtaining an output Y:

$$Y = (A \cdot B) + C \quad (11.2)$$

When Eq. 11.2 holds, output Y will be on when input C is on, even if A and B are not both on. Equation 11.2, which reads (A and B or C), describes a "more permissive" arrangement for output Y than did Equation 11.1.

The NOT operator merely reverses the value of an expression—that is, if the expression has a value of 0, the NOT operator makes the value become 1, and vice versa. Parentheses are used in Boolean expressions with the same interpretation as in ordinary algebraic expressions.

Relation to Truth Table

A Boolean logic expression can be completely described by a truth table. All variables named in the expression are listed as input variables in the table, and the value of the entire expression is an output variable in the table. Thus, in Boolean equations of the type exhibited in Eqs. 11.1 and 11.2, the variable named on the left-hand side of the equal sign would be the output variable of the truth table. Tables 11.4 and 11.5 are truth tables for Eqs. 11.1 and 11.2, respectively.

TABLE 11.3 Boolean Operators

Name	*Symbol*	*Example*	*Meaning*
AND	$\cdot$	$A \cdot B$	Both A and B must be ON or true for the entire expression to be true.
OR	$+$	$A + B$	If either A or B or both are ON or true, then the entire expression is true.
NOT	$-$	$\overline{A}$	If A is true, the expression is false; if A is false, the expression is true.

xpressions for any truth table. Returning
Boolean logical expression representing
on between input variable P and output
P'. Note from Table 11.1 that output
ne possible alternative for achieving
ed with four terms in the Boolean
in the configuration specified in
nd row of Table 11.1 represents
P'. To obtain that possibility,

	0
	0
	0
1	1

TABLE 11.5 Truth Table for Equation 11.2

Inputs			*Output*
A	*B*	*C*	*Y*
0	0	0	0
0	0	1	1
0	1	0	0
0	1	1	1
1	0	0	0
1	0	1	1
1	1	0	1
1	1	1	1

TABLE 11.6 Truth Table with Various Example Boolean Outputs

Inputs			*Outputs*					
A	*B*	*C*	ABC	$A + B + C$	$\overline{ABC}$	$\overline{A + B + C}$	$A\overline{B}$	$A(B + \overline{C})$
0	0	0	0	0	1	1	0	0
0	0	1	0	1	1	0	0	0
0	1	0	0	1	1	0	0	0
0	1	1	0	1	1	0	0	0
1	0	0	0	1	1	0	1	1
1	0	1	0	1	1	0	1	0
1	1	0	0	1	1	0	0	1
1	1	1	1	1	0	0	0	1

inputs S and L must both be off or 0, and input P must be on or 1. The entire Boolean expression is shown in Eq. 11.3*a*:

$$P' = (\overline{S} \cdot \overline{L} \cdot P) + (S \cdot \overline{L} \cdot P) + (S \cdot L \cdot \overline{P}) + (S \cdot L \cdot P) \qquad (11.3a)$$

As in ordinary algebra, the · symbols and parentheses are often dropped, resulting in the following:

$$P' = \overline{S}\overline{L}P + S\overline{L}P + SL\overline{P} + SLP \qquad (11.3b)$$

Equation 11.3*b* can be reduced to a simpler form using the theorems of Boolean algebra as will be shown in the next section.

For review, Table 11.6 exhibits a three-input truth table with six different output columns representing various example Boolean expressions.

Algebraic Simplification

Boolean algebra has theorems and laws similar to the laws of ordinary algebra but different in some significant ways. The only reason for introducing these laws and theorems in this text is to provide a very practical way for automation engineers to simplify the logic, and in turn the hardware, necessary for their industrial applications.

To see how Boolean theorems can be used, let us begin with a simple example by reducing the Boolean expression on the right side of Eq. 11.4.

$$X = A \cdot (B + \overline{B}) \qquad (11.4)$$

In words, the equation says:

Output X will be ON IF

1. Input A is ON

AND

2. Input B is either ON or OFF

But Eq. 11.4 can quite obviously be made simpler if one thinks about it for a moment. Remember that there are only two states that can be taken on by a logical variable: on or off (i.e., 1 or 0). Thus, input B must be either on or off. But Eq. 11.4 has the expression:

$$(B + \overline{B})$$

The OR operator (+) gives us the alternative of either having B or not having B. This is no restriction at all because we will always either have B or not have B. So this portion of the expression is eliminated as a restriction upon output X, and Eq. 11.4 has been reduced to

$$X = A \qquad (11.5)$$

Thus, if input A is on, output X will be on; if input A is off, output X will be off. Reduction of the Boolean expression in Eq. 11.4 was a simple implementation of two elementary Boolean theorems:

1. The inclusion theorem: $B + \overline{B} = 1$
2. The characteristic theorem: $A \cdot 1 = A$

Most real-world logic expressions are not as simple as Eq. 11.4, but with the help of Boolean laws and theorems many of them can be reduced considerably.

Table 11.7 lists most of the commonly used laws and theorems of Boolean algebra for reference in simplifying industrial logic control applications described in this text. As with any algebra or science, fundamental axioms are used to prove theorems that in turn are used to prove other theorems and laws. It is not the subject of this text to delve into mathematical proofs, but to demonstrate the method one proof will be shown:

TO PROVE

The absorptive law
$X + XY = X$

PROOF

$X + XY = X \cdot 1 + X \cdot Y$	Characteristic theorem
$= X(1 + Y)$	Distributive law
$= X \cdot 1$	Characteristic theorem
$= X$	Characteristic theorem

Proof complete.

To sharpen one's familiarity with the Boolean laws and theorems, it is excellent practice to attempt to prove all the theorems and laws in Table 11.7. The more fundamental laws and theorems are either obviously intuitive or should be accepted as axiomatic. In this category are the commutative, associative, and distributive laws, and the characteristic, inclusion, idempotent, and negation theorems. The reader has probably already noticed that many of the basic laws of ordinary algebra are duplicated in Boolean algebra (e.g., the associative law, commutative law, and distributive law). But one should be very careful about leaning too heavily upon the old familiar proficiency with ordinary algebra. For example, examine closely the four versions of the characteristic theorem set forth in

TABLE 11.7 Theorems and Laws of Boolean Algebra for Industrial Logic Controls

CHARACTERISTIC THEOREMS

1. $X \cdot 0 = 0$
2. $X \cdot 1 = X$
3. $X + 0 = X$
4. $X + 1 = 1$

COMMUTATIVE LAW

1. $X + Y = Y + X$
2. $X \cdot Y = Y \cdot X$

ASSOCIATIVE LAW

1. $X + Y + Z = X + (Y + Z) = (X + Y) + Z$
2. $X \cdot Y \cdot Z = X \cdot (Y \cdot Z) = (X \cdot Y) \cdot Z$

DISTRIBUTIVE LAW

1. $X \cdot Y + X \cdot Z = X(Y + Z)$
2. $(X + Y)(W + Z) = XW + XZ + YW + YZ$

IDEMPOTENT THEOREMS

1. $X \cdot X = X$
2. $X + X = X$

NEGATION THEOREM

$\overline{(\overline{X})} = X$

INCLUSION THEOREMS

1. $X \cdot \overline{X} = 0$
2. $X + \overline{X} = 1$

ABSORPTIVE LAWS

1. $X + XY = X$
2. $X(X + Y) = X$

REFLECTIVE THEOREMS

1. $X + \overline{X}Y = X + Y$
2. $X(\overline{X} + Y) = XY$
3. $XY + \overline{X}YZ = XY + YZ$

CONSISTENCY THEOREMS

1. $XY + X\overline{Y} = X$
2. $(X + Y)(X + \overline{Y}) = X$

DEMORGAN'S LAWS

1. $\overline{XY} = \overline{X} + \overline{Y}$
2. $\overline{X + Y} = \overline{X}\,\overline{Y}$

Table 11.7. The first three appear to be ordinary algebra, but the fourth, $X + 1 = 1$, does *not* follow the rules of ordinary algebra. The automation engineer must be careful not to fall into the trap of using ordinary rules of algebra on logic expressions without first checking to see whether the given rule is valid for Boolean algebra.

Setting aside the proofs of the Boolean theorems, let us put the theorems to work in simplifying the logic for the console lock problem discussed earlier in this chapter. Equation 11.3*b* specified a Boolean expression for the console lock system output as a function of the inputs and was derived directly from the truth table for the console lock. But the expression was unnecessarily long and complicated; with the help of Boolean laws and theorems the expression can be reduced significantly as follows:

$$
\begin{aligned}
P' &= \overline{S}\overline{L}P + S\overline{L}P + SL\overline{P} + SLP && \text{from Eq. 11.3}a \\
&= \overline{L}P(\overline{S} + S) + SL(\overline{P} + P) && \text{Distributive law} \\
&= \overline{L}P(1) + SL(1) && \text{Inclusion theorem} \\
&= \overline{L}P + SL && \text{Characteristic theorem}
\end{aligned}
$$

KARNAUGH MAPS

Boolean algebra is an effective method for reducing a logic expression to its simplest form, but it can become quite tedious in real applications having several variables involving numerous logic expressions. A convenient graphical analysis technique utilizes "Karnaugh maps," which enumerate the logical terms and group them for elimination from the overall logical expression. The authority for using the Karnaugh maps technique is the Law of Complements, which states that

$$A + \overline{A} = 1$$

and the Characteristic Theorem, which states that

$$X \cdot 1 = X$$

Therefore, if terms of a logical expression can be grouped such that all variables of a given pair of terms are identical except one, and that one variable exists as a complement in the pair, then the complementary variable can be eliminated from the expression and the pair of terms reduced to a single term consisting of the identical variables.

Reduction of Terms

EXAMPLE 11.1

Reduce the following Boolean logic expression for output Y:

$$Y = A\overline{B}CD + \overline{A}\overline{B}CD + \overline{A}\overline{B}C\overline{D}$$

Solution

Note that the second and third terms are identical except for variable D, which is complementary in that the variable appears as D in the second term and $\overline{D}$ in the third term. Therefore the second and third terms become a pair in which D is eliminated, and the pair reduces to the expression $\overline{A}\overline{B}C$. Therefore,

$$Y = A\overline{B}CD + \overline{A}\overline{B}C$$

In the above exercise Karnaugh maps can be used to easily spot the pair of terms to be reduced, and the reduction can be performed without invoking the theorems of Boolean algebra. The procedure is to array the terms of the expression in a matrix such that adjacent elements in the matrix represent a complementary pair. Adjacent terms are then circled and the pair of terms is then reduced to a simplified single term. An example consisting of two variables best illustrates the method.

EXAMPLE 11.2

Reduce the following Boolean logic expression for output X:

$$X = A\bar{B} + AB$$

Solution

The expression is arrayed as a matrix (Karnaugh map) accommodating all possible combinations of logic variables A and B in a single term. The Karnaugh map is as follows:

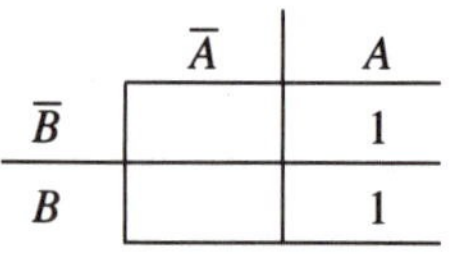

The matrix is 2 × 2, so it can accommodate four terms, the total possible in a two-variable Boolean logic expression. In our example only two terms exist in the logic expression for output X, and these two terms are each represented by "1" in the appropriate element box in the matrix. The other two boxes are blank, indicating that terms $\bar{A}\bar{B}$ and $\bar{A}B$ are both missing from the given expression for output X. Since the two boxes containing "1" are adjacent, they can be grouped (circled) as follows, resulting in the elimination of the complementary variable B:

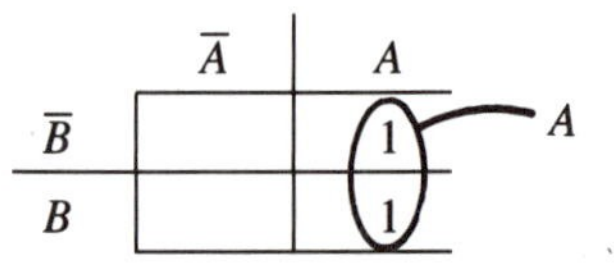

The reduced expression A is equivalent to the pair represented by $A\bar{B} + AB$. Note that A is the variable common to (and identical in) the two terms in the given pair.

Suppose the Karnaugh map had appeared as follows: What would be the reduced expression? Answer: $\bar{B}$

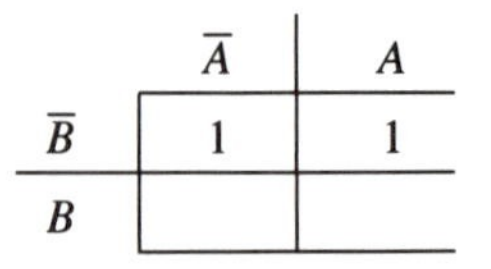

And for the following Karnaugh map the answer is $\bar{A}\bar{B} + AB$.

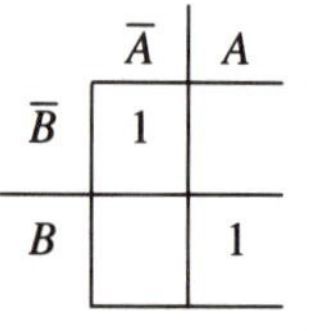

In this last case there would have been no reduction possible because no adjacent boxes in the matrix contained "1", and thus no complementary variable existed.

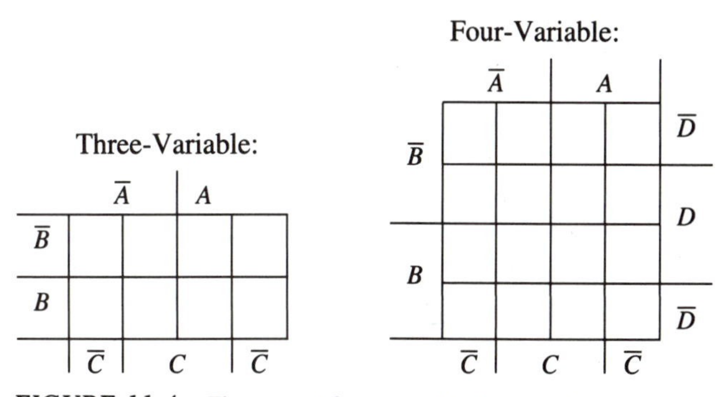

FIGURE 11.4 Three- and four-variable Karnaugh map formats.

Three- and Four-Variable Matrices

Karnaugh maps are of trivial benefit when only two logic variables are to be considered, but the technique is also applicable to three-variable and four-variable expressions. The three- and four-variable matrices can be represented by two-dimensional matrices by imbedding the third variable in the columns and the fourth variable in the rows as shown in Figure 11.4.

By paying attention to the row and column labels, the analyst can determine which Boolean logic variable is the appropriate complementary variable to eliminate from the pair of adjacent terms. Some examples will illustrate:

EXAMPLE 11.3

Reduce the following Boolean logic expression for output Z:

$$Z = \bar{A}B\bar{C} + \bar{A}BC$$

Solution

Karnaugh map:

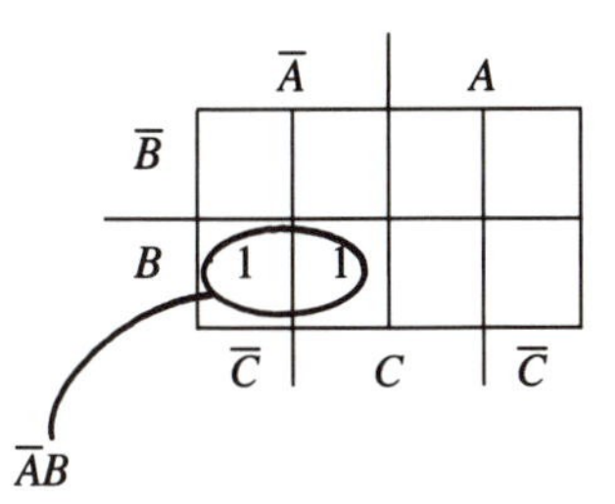

Note from the column labels that Boolean logic variable C is the complementary variable eliminated from the expression. Therefore:

$$Z = \bar{A}B$$

EXAMPLE 11.4

Reduce the following logic expression for output X.

$$X = \overline{A}\,\overline{B}C\overline{D} + A\overline{B}C\overline{D} + \overline{A}B\overline{C}D + \overline{A}B\overline{C}\,\overline{D} + ABC\overline{D} + AB\overline{C}\,\overline{D}$$

Solution

Karnaugh map:

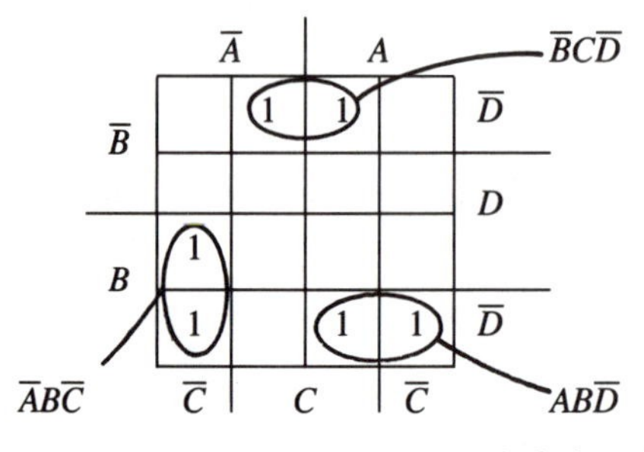

Eliminating complementary variables from pairs of terms as circled on the Karnaugh map:

$$X = \overline{B}C\overline{D} + \overline{A}B\overline{C} + AB\overline{D}$$

Pattern Elimination

The benefit to be derived from Karnaugh maps should become apparent from Example 11.4, but more can be done with the map if certain patterns appear. Consider the following example.

EXAMPLE 11.5

Reduce the following logic expression for output Y.

$$Y = ABCD + AB\overline{C}D + ABC\overline{D} + AB\overline{C}\,\overline{D}$$

Solution

Karnaugh map:

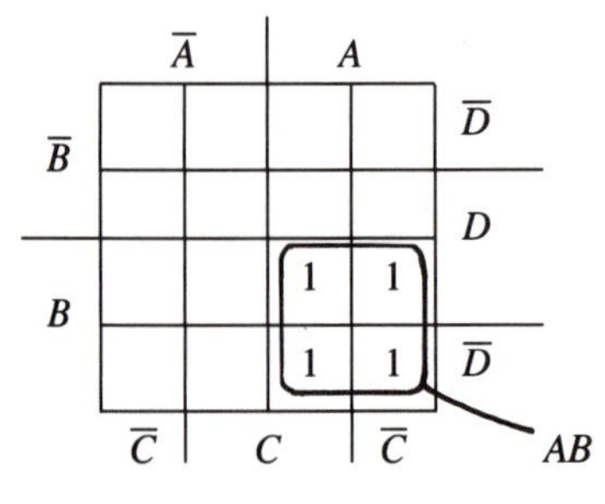

In this case all four terms are adjacent in the Karnaugh map, making possible the elimination of both variable C and variable D, leaving only AB. The idea can be extended to eliminate three and even all four variables as follows:

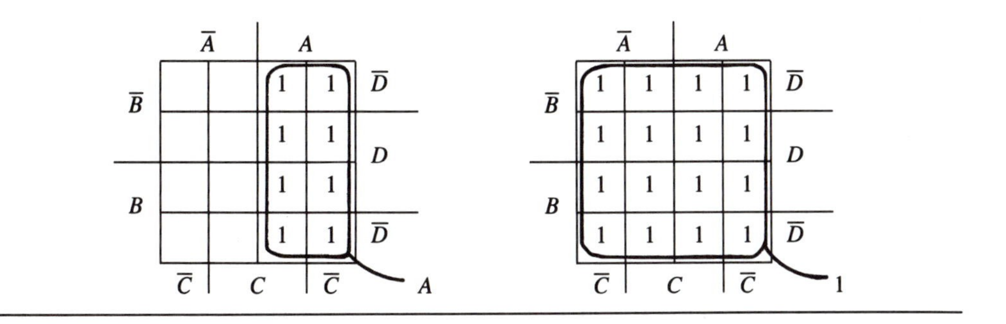

Overlapping Patterns

In each of the examples thus far each element in the Karnaugh map has been circled only once. However, it is entirely permissible to circle a single element more than once if a simpler logic expression can be achieved in this way. The authority for circling an element twice in the Karnaugh map is the Idempotent Theorem of Boolean algebra:

$$A + A = A$$

Consider the following example.

EXAMPLE 11.6

Reduce the following logic expression for output W.

$$W = \bar{A}\bar{B} + \bar{A}B + AB$$

Solution

Karnaugh map:

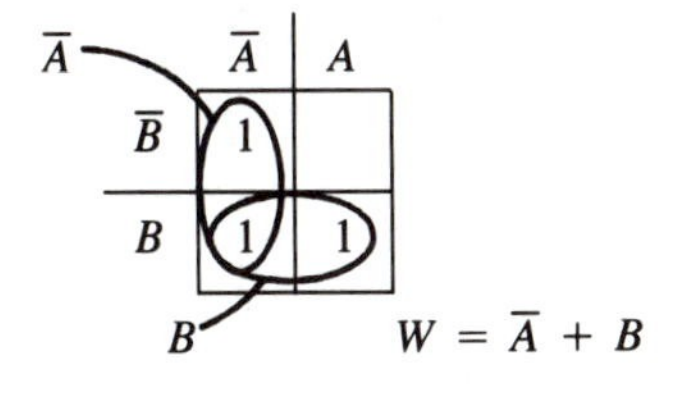

In this example the matrix element $\bar{A}B$ was circled twice, because it was convenient to do so to achieve maximum reduction in the expression. The reader can verify that the result is correct using Boolean algebra, invoking the Reflective Theorem.

Several examples will now be used to illustrate the versatility of the Karnaugh maps.

EXAMPLE 11.7

Reduce the following logic expression for output V.

$$V = \bar{A}\bar{B}\bar{C}D + \bar{A}\bar{B}CD + \bar{A}BCD + A\bar{B}CD + ABCD + A\bar{B}\bar{C}D + AB\bar{C}D + AB\bar{C}\bar{D}$$

Solution

Karnaugh map:

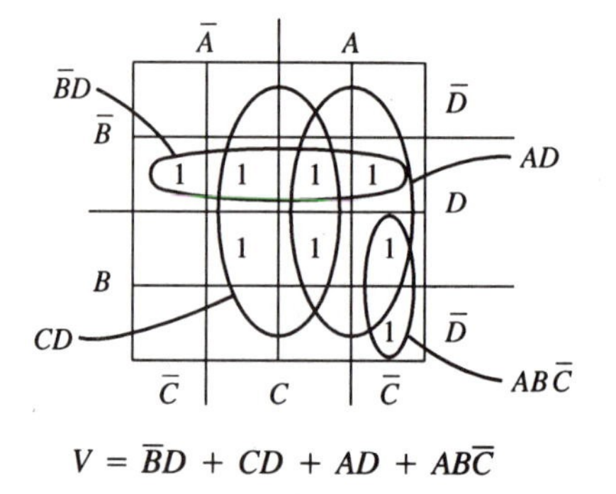

$$V = \bar{B}D + CD + AD + AB\bar{C}$$

EXAMPLE 11.8

Reduce the following logic expression for output X.

$$X = \bar{A}\bar{B}\bar{C}\bar{D} + \bar{A}\bar{B}C\bar{D} + A\bar{B}C\bar{D} + A\bar{B}CD + ABCD + \bar{A}BCD + AB\bar{C}D + AB\bar{C}\bar{D}$$

Solution

Karnaugh map:

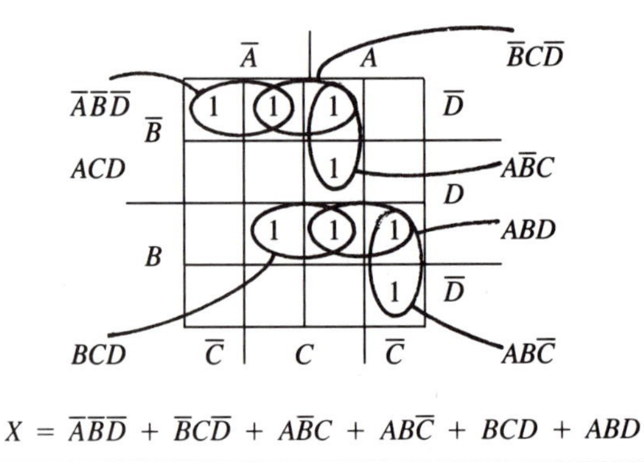

$$X = \bar{A}\bar{B}\bar{D} + \bar{B}C\bar{D} + A\bar{B}C + AB\bar{C} + BCD + ABD$$

EXAMPLE 11.9

Reduce the following logic expression for output Y.

$$Y = \bar{A}\bar{B}C\bar{D} + A\bar{B}C\bar{D} + A\bar{B}\bar{C}\bar{D} + \bar{A}\bar{B}\bar{C}D + \bar{A}\bar{B}CD + A\bar{B}CD + A\bar{B}\bar{C}D + \bar{A}B\bar{C}D + \bar{A}BCD + ABCD + AB\bar{C}D + ABC\bar{D} + AB\bar{C}\bar{D}$$

Solution

Karnaugh map:

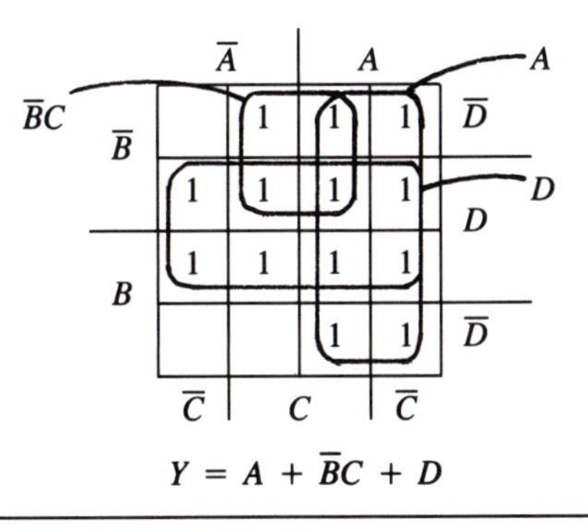

$$Y = A + \overline{B}C + D$$

In Example 11.9 can be seen the real power in using Karnaugh maps. The simplified expression for output Y would have been difficult to achieve using Boolean algebra alone.

Adjacent Exterior Columns and Rows

One characteristic of Karnaugh maps that has been deliberately avoided in the examples of this chapter so far is that the exterior columns of the matrix should be considered adjacent and the exterior rows should also be considered adjacent. Thus the matrix should be considered as a cylinder, either column-wise or row-wise. Therefore it is possible to eliminate variables by circling the outside elements that would be adjacent if the matrix were seen as a cylinder. This capability is illustrated by Example 11.10.

EXAMPLE 11.10

Reduce the following logic expression for output Z.

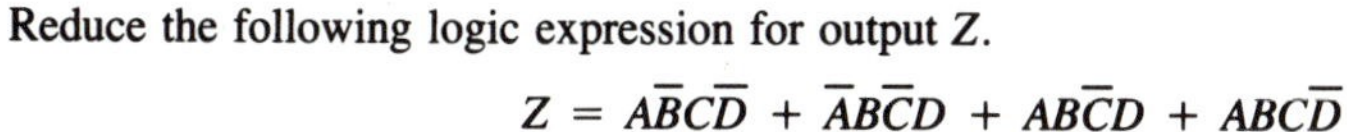

$$Z = A\overline{B}C\overline{D} + \overline{A}B\overline{C}D + AB\overline{C}D + ABC\overline{D}$$

Solution

Karnaugh map:

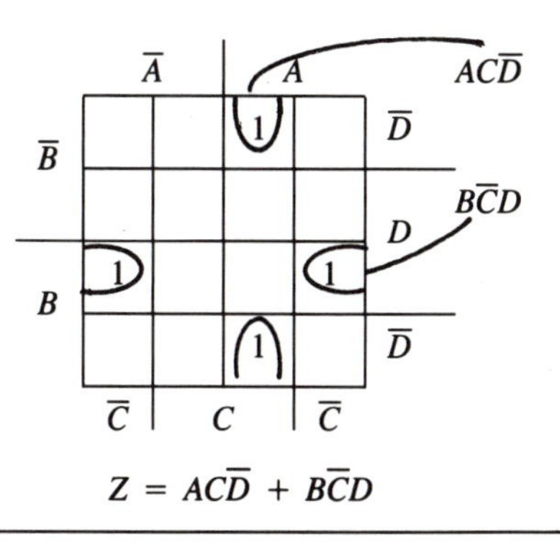

$$Z = AC\overline{D} + B\overline{C}D$$

The solution for Example 11.10 was achieved by two circles on the Karnaugh map. One circle recognized the adjacency of columns; the other circle recognized the adjacency of rows. It takes a little practice to recognize this feature of Karnaugh maps, but some logic expressions cannot be reduced in any other way.

A corollary to the principle of adjacent exterior columns or rows is the adjacency of all four corners, making possible the elimination of two variables at once by circling all four corners. This feature is illustrated in Example 11.11.

EXAMPLE 11.11

Reduce the following logic expression for output W.

$$W = \bar{A}\bar{B}\bar{C}\bar{D} + A\bar{B}\bar{C}\bar{D} + \bar{A}B\bar{C}\bar{D} + AB\bar{C}\bar{D}$$

Solution

Karnaugh map:

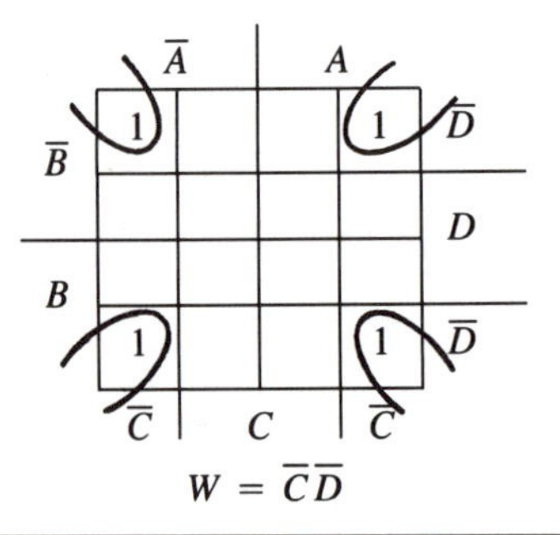

$$W = \bar{C}\bar{D}$$

"Don't Care" States

In most real industrial applications of logic expressions, a complete enumeration of all possible combinations of the input variables includes some input combinations that are either infeasible, inappropriate, or irrelevant to the extent that the designer does not care whether the output variable is ON or OFF for these input combinations. Such combinations of the input variables are called "don't care" states, and these states can be utilized to simplify the output logic expression using the Karnaugh map. The procedure is to place an "X" in each matrix position in the Karnaugh map for which the output can be either 0 or 1. Then when circling patterns in the matrix the analyst is free to consider the "X" positions as equivalent to "1" positions. Thus some patterns can be achieved that would have been incomplete if only the "1" positions were considered. An example will illustrate.

EXAMPLE 11.12

Reduce the following logic expression for output W.

$$W = \bar{A}\bar{B}CD + A\bar{B}CD + A\bar{B}\bar{C}D + \bar{A}B\bar{C}D + \bar{A}BCD + ABCD$$

The following input combinations describe states for which we "don't care" whether output W is ON or OFF:

$$\bar{A}\bar{B}\bar{C}D,\ AB\bar{C}D$$

Solution

Karnaugh map: (without recognizing "don't care" states)

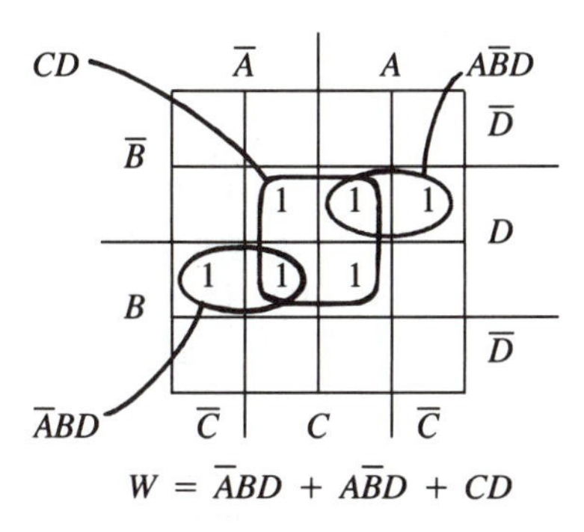

$$W = \bar{A}BD + A\bar{B}D + CD$$

Alternate Solution

Karnaugh map: (recognizing and taking advantage of "don't care" states)

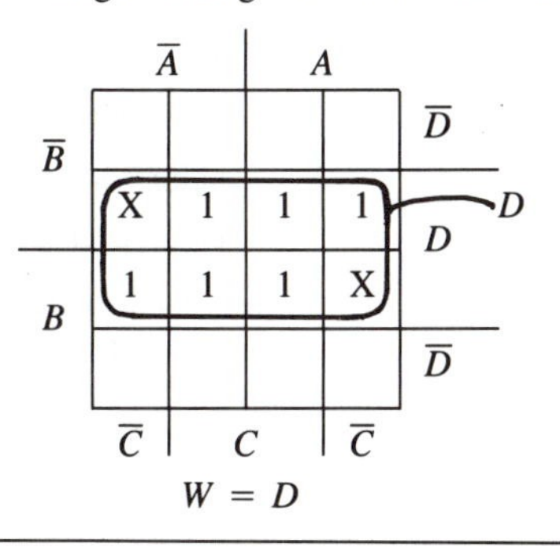

$$W = D$$

From Example 11.12 can be seen the powerful advantage afforded by recognizing "don't care" states when analyzing and simplifying logic expressions. It is important to keep in mind that both solutions shown for Example 11.11 are correct, but obviously the "alternate solution" is much simpler to implement than the solution that fails to take advantage of "don't care" states.

It should be recognized that Boolean algebra is not equipped to handle "don't care" states. Boolean algebra deals only with terms in the expression for which the output is ON or for which the logic is true. For any conditions unsatisfied by the terms in the Boolean logic equation the output is forced OFF. There is no "maybe" about it. But many real-world industrial applications do have states in which "maybe" is appropriate. For these "don't care" states Karnaugh maps can be very useful in obtaining the maximum reduction.

Truth tables often simply omit "don't care" states. In a large problem involving several logic input variables it would be tedious to completely enumerate all possible combinations, including combinations for which the analyst does not care whether the output is ON or OFF. The Boolean algebra approach simply assumes that omitted combinations in the truth table are to be considered as generating an OFF condition for the output. Such an assumption limits the analyst in attempts to simplify the logic expression. When a truth table is used with a Karnaugh map, all missing rows in the truth table are entered as Xs in the Karnaugh map matrix. The output 0s are left blank in the matrix, and the output 1s are entered as 1s, of course. A case study will now be used to illustrate the utility of Karnaugh maps when dealing with a real problem in industrial automation.

EXAMPLE 11.13 (CASE STUDY)

An automatic work station cycles whenever the "cycle ready" sensor is ON, provided that the power is ON, and the emergency STOP button is not actuated. The input and output definitions are summarized as follows:

Inputs

$G = 0$	when POWER ON spring push button is released
$= 1$	when POWER ON spring push button is depressed
$S = 0$	when the emergency STOP spring push button is released
$= 1$	when the emergency STOP spring push button is depressed
$P = 0$	when power is OFF
$= 1$	when power is ON
$C = 0$	when cycle-ready sensor indicates "not ready"
$= 1$	when cycle-ready sensor signals "ready"

Outputs

$P = 0$	power is OFF
$= 1$	power is ON
$Y = 0$	station idle
$= 1$	station cycles

The logic of the system is defined by the truth table in Table 11.8. Since this case study has four input variables the truth table would have $2^4 = 16$ rows if all combinations of the input variables were exhausted. However, note that Table 11.8 shows only 14 rows in the truth table. The following input combinations have been omitted from the table:

G	S	P	C
0	0	0	1
1	0	0	1

TABLE 11.8 Truth Table for Case Study 11.13

G	S	P	C	P'	Y
0	0	0	0	0	0
1	0	0	0	1	0
0	0	1	0	1	0
0	0	1	1	1	1
0	1	1	1	0	0
0	1	1	0	0	0
1	0	1	0	1	0
1	0	1	1	1	1
1	1	0	0	0	0
1	1	1	0	0	0
1	1	0	1	0	0
1	1	1	1	0	0
0	1	0	0	0	0
0	1	0	1	0	0

These combinations were omitted because it does not make sense for the cycle-ready sensor to indicate "ready" if the power is OFF. Thus, the designer has determined that these two input combinations would never occur, and so they have been ignored in the truth table. In the Karnaugh map to be derived from the truth table, these two missing states are to be entered as "X," because the designer does not care whether the outputs are to be turned ON or OFF by the logic system for these two irrelevant states. It will be seen that these "don't care" states will be useful in reducing the logic expressions for the outputs. A Karnaugh map is now constructed for each output (P' and Y).

Karnaugh map for P'

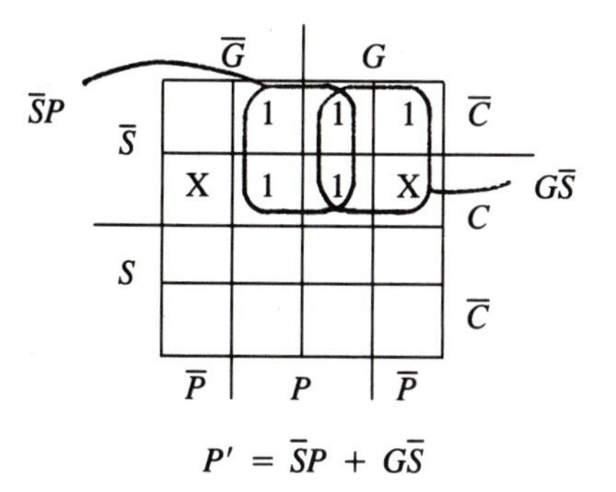

$$P' = \bar{S}P + G\bar{S}$$

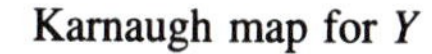
Karnaugh map for Y

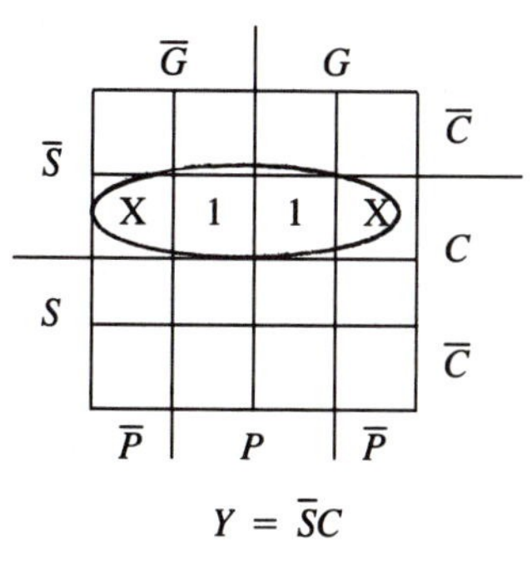

$$Y = \bar{S}C$$

Upon closer observation of the truth table in Table 11.8 the reader might wonder whether additional rows might have been omitted as "don't care" states. For instance, it would seem that a state as inappropriate as those already omitted from the table would be the following state:

G	S	P	C
0	1	0	1

Although it is true that this state would hardly be feasible, it was still included in the truth table as a matter of safety. Note that the emergency STOP spring push button S is depressed in this input combination. To be absolutely safe it seems prudent to specify that both outputs be set to zero for this input state because both outputs should be at logic 0 if the STOP push button is depressed. If the row had been omitted in the truth table, it would have been possible for the Karnaugh map analysis to have circled that element in the matrix, causing one or more outputs to be set ON even if the emergency STOP button is pressed! Thus the analyst should take care to be sure that rows omitted from the truth table really do represent "don't care" states for the outputs, in order to avoid an unlikely but dangerous situation.

Limitations of Karnaugh Maps

At this point it may seem that as an analysis technique Karnaugh maps are superior to Boolean algebra in every way. But there are two major limitations to the use of Karnaugh maps. First, Karnaugh maps are suitable for reducing logic expressions of at most four input variables. Beyond four variables a way has not been found to write out the table on two-dimensional paper, much less analyze it.

The other major limitation of Karnaugh maps is that they must begin with the enumeration of all basic combinations of inputs, the same as the form taken directly from the truth table. This is not a problem if the analyst is actually working with a truth table. But if the logic expression contains parentheses or shows reductions from the basic forms, it is difficult to apply Karnaugh maps to achieve the desired output states. Consider the following logic expression for output V:

$$V = A(B + C) + \overline{CD}$$

Until the above logic expression can be rewritten in terms of the form $ABCD$, the Karnaugh map cannot be used to analyze the expression.

In summary, Karnaugh maps are a useful tool for analyzing logic expressions to reduce their complexity. Karnaugh maps can be drawn up directly from the truth table and have an advantage over Boolean algebra in the consideration of "don't care" states. However, Karnaugh maps are limited to applications involving four input variables or less. Also Karnaugh maps must deal with logic expressions in the basic form of terms consisting of products of all input variables associated with the problem.

To further illustrate the concepts presented in this chapter, let us examine a simplified problem in machine control: the logic control of an automatic press.

EXAMPLE 11.14 (CASE STUDY)
Automatic Press Control

The press diagrammed in Figure 11.5 is a light-duty bench model that is powered either by compressed air or electric solenoid. The machine is turned on by an ON push button and for safety is turned off by any of three OFF push buttons located conveniently on various parts of the machine. Once turned on, the press automatically actuates the ram whenever limit switch 1 is deflected upward. The press sustains the power to the ram during the downward stroke until limit switch 2 is deflected downward. At this point, power to the ram is removed, and the ram returns to its full-

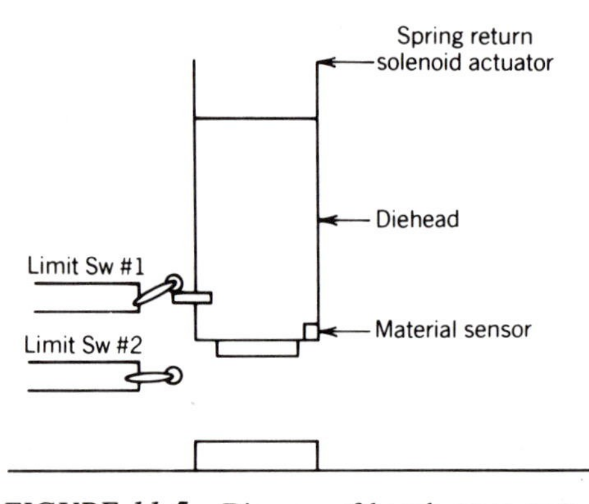

FIGURE 11.5 Diagram of bench press system for case study in automatic press control.

up position by spring action until it deflects limit switch 1 upward again. The ram is then actuated again and continues to oscillate up and down automatically until one of the stop push buttons brings the machine to a halt.

The automatic press just described really has only one output, which we will call "ram actuate." There are several inputs, however, that are tabulated and defined in Table 11.9. All of the variables listed in Table 11.9 are input variables, but variable A (ram actuate) is both an input variable and an output variable. Using the same convention that we used in the console lock example, we will use A to refer to the input variable and A' to refer to the output variable.

TABLE 11.9 Input/Output Variables for Automatic Press System

Variable Description	*Name*	*Definition*
Start push button	G	= 1 when pressed = 0 when released
Stop push button i (i = 1, 2, or 3)	S_i	= 1 when pressed = 0 when released
Limit switch (i = 1 or 2)	L_i	= 1 when deflected = 0 when not deflected
Ram actuate	A	= 1 ram actuate = 0 ram de-actuate

Constructing a truth table to accommodate the entire set of variables listed in Table 11.9 would be a formidable task. There are at least seven inputs, and a tabulation of all of them in a truth table would require $2^7 = 128$ rows in the truth table! We can simplify the problem by dividing it into three smaller problems.

For the first smaller problem, let us address all those stop push buttons. Table 11.10 presents a truth table that introduces an intermediate variable S, which summaries the condition of the three variables S_1, S_2, and S_3.

TABLE 11.10 Truth Table for Stop Push Buttons

Inputs			*Output*
S_1	S_2	S_3	S
0	0	0	0
0	0	1	1
0	1	0	1
0	1	1	1
1	0	0	1
1	0	1	1
1	1	0	1
1	1	1	1

The logic expression for Table 11.10 is developed as follows:

$$\bar{S} = \bar{S}_1 \bar{S}_2 \bar{S}_3 \qquad \text{from truth table}$$

$$\bar{S} = \overline{S_1 + S_2 + S_3} \qquad \text{De Morgan's law} \tag{11.6}$$

$$S = S_1 + S_2 + S_3$$

Note that in this example it was easier to start with $\bar{S}$ because the initial expression for $\bar{S}$ has only one term; only one row of the truth table results in an output $\bar{S}$.

The next step is to create a truth table for describing whether or not the machine is running. For this truth table, we will introduce another variable, R, as follows:

$$R = 1 \text{ if the machine is running}$$
$$= 0 \text{ if the machine is not running}$$

R is both an output and an input variable, so we will use R' to present the output. Table 11.10 is a truth table for machine running status. Note that one of the *input* variables (S) in Table 11.11 was the *output* variable in Table 11.10.

TABLE 11.11 Truth Table to Determine Machine Running Status

Inputs			*Output*
G	S	R	R'
0	0	0	0
0	0	1	1
0	1	0	0
0	1	1	0
1	0	0	1
1	0	1	1
1	1	0	0
1	1	1	0

The logic expression for table 11.11 is developed as follows:

$$\begin{aligned}
R' &= \overline{G}\,\overline{S}R + G\overline{S}\,\overline{R} + G\overline{S}R && \text{from truth table} \\
&= \overline{G}\,\overline{S}R + G\overline{S}(\overline{R} + R) && \text{Distributive law} \\
&= \overline{G}\,\overline{S}R + G\overline{S}(1) && \text{Inclusion theorem} \\
&= \overline{G}\,\overline{S}R + G\overline{S} && \text{Characteristic theorem} \\
&= \overline{S}\,(\overline{G}R + G) && \text{Commutative and distributive laws} \\
&= \overline{S}(R + G) && \text{Commutative law and reflective theorem}
\end{aligned} \tag{11.7}$$

Finally, we will construct a larger truth table (Table 11.12) that incorporates the Table 11.11 output variable R' as an *input* variable, along with the remaining three input variables of the problem. Since there are four input variables in Table 11.12, there are $2^4 = 16$ rows in the table. The logic expression for Table 11.12 is developed as follows:

TABLE 11.12 Truth Table for Ram Actuate Logic

Inputs				*Output*
R'	L_1	L_2	A	A'
0	0	0	0	0
0	0	0	1	0
0	0	1	0	0
0	0	1	1	0
0	1	0	0	0
0	1	0	1	0
0	1	1	0	0
0	1	1	1	0
1	0	0	0	0
1	0	0	1	1
1	0	1	0	0
1	0	1	1	0
1	1	0	0	1
1	1	0	1	1
1	1	1	0	0
1	1	1	1	0

$A' = R'\overline{L}_1\overline{L}_2A + R'L_1\overline{L}_2\overline{A} + R'L_1\overline{L}_2A$	from truth table
$= R'\overline{L}_1\overline{L}_2A + R'L_1\overline{L}_2(\overline{A} + A)$	Distributive law
$= R'\overline{L}_1\overline{L}_2A + R'L_1\overline{L}_2(1)$	Inclusion theorem
$= R'\overline{L}_1\overline{L}_2A + R'L_1\overline{L}_2$	Characteristic theorem
$= R'\overline{L}_2(\overline{L}_1A + L_1)$	Commutative and distributive laws
$= R'\overline{L}_2(A + L_1)$	Commutative law and reflective theorem (11.8)

The logic from all three tables now can be combined as follows:

Substitute Eq. 11.7 into Eq. 11.8:

$$A' = \overline{S}(R + G)\overline{L}_2(A + L_1)$$

Substitute Eq. 11.6:

$$A' = \overline{S_1 + S_2 + S_3}(R + G)\overline{L}_2(A + L_1) \tag{11.9}$$

Equation 11.9 summarizes all the logic used in the system for controlling the automatic press. In Chapters 12 and 13 we will see how this logic equation can be used as a model to construct a physical system to control the machine.

SUMMARY

In this chapter, we have examined the basic building blocks of logical expressions and have seen how these can be used to model industrial automatic logic control systems.

The most explicit method for detailing all relationships between input and output variables is the truth table. A systematic method is needed to exhaust all combinations of input variables in the truth table. In both the console lock and the push-button switch examples, we saw that the state of an output variable can be used as an input variable in the determination of the next logical state of the output variable.

The introduction of Boolean algebra provided both a shorthand for logical expressions and a shortcut for reducing their physical complexity. Such simplification provides direct benefit in both economics and system reliability in applications of industrial automation.

Karnaugh maps were introduced as a convenience for analyzing industrial logic control systems involving four or fewer input variables without using the theorems of Boolean algebra. Although Karnaugh maps have limitations, they have the advantage of recognizing that the automation engineer may want the option of choosing whether or not to turn an output on or off for those combinations of logic inputs in which the outcome does not matter.

In Chapter 12 we learn how to construct schematic diagrams of industrial logic systems, getting us closer to actual physical representations of industrial logic control systems.

EXERCISES AND STUDY QUESTIONS

11.1. A machine is turned on by an ON push button and turned off by either of two OFF push buttons. Construct a truth table and develop a logic expression for the output "machine run." Use Boolean algebra to simplify the logic expression as much as possible.

11.2. Construct a logic expression for the truth table for the push button switch described in Table 11.2. Reduce the expression to lowest terms using Boolean algebra.

11.3. Which of the following Boolean equations are true and which are false?

(a) $A + B = A + \overline{A}B$
(b) $\overline{A}\,\overline{B} = \overline{A}\,\overline{B}$
(c) $A = A + AB + ABX$
(d) $X = X(Y + \overline{Y}) + X(Z + \overline{Z})$
(e) $Y = Y(X + \overline{X}) + XZ\overline{Z}$

11.4. Construct a truth table for the following output expressions of inputs A, B, and C.

(a) $AB\overline{C}$
(b) $(A + B)(B + C)$
(c) $(A + B) + (B + C)$
(d) $A + AB + ABC$
(e) $(A + \overline{A})(B + \overline{B})(C + \overline{C})$

11.5. Which of the following expressions is most restrictive? Which is most permissive? Explain your choices.

$A + \overline{A}$
A
$A + B + C$

11.6. Design an industrial logic control system for an exhaust air system that filters the contaminated air and returns it to the room. However, if the filter is too dirty, as determined by the differential pressure across the filter, the control system will turn on an alarm light. Even while the filter is still dirty and the alarm light on, the blower will continue to operate and a filter bypass valve will be opened, provided outside air is within prescribed limits of contamination. However, if the outside air is found to be contaminated at the same time that the filter is dirty, the blower will shut down and the filter bypass valve will close. When the power is turned off, the blower is off, the alarm is turned off, and the bypass valve is closed, regardless of conditions of the filter or outside air. For a diagram of the system see Figure 11.6.

11.7. An industrial logic control system is being used to control a walking beam conveyor in a refrigerator manufacturing line. The operation has the following inputs:

1. Power-on push button (momentary)
2. Emergency stop push button (momentary)
3. Sensor to detect "robots home"

and a single output: conveyor in operation. The conveyor will start if the power is on, the robots are home, and the emergency stop push button is not depressed, and it will stay in operation, once started, unless the emergency stop push button is pressed.

(a) Generate the truth table that describes the operation of the system.
(b) Write the logic equation from the truth table, and then reduce this equation using Boolean algebra.

11.8. Figure 11.7 displays a popular digital display consisting of seven LED segments. The display can be used to output any decimal digit corresponding to its BCD

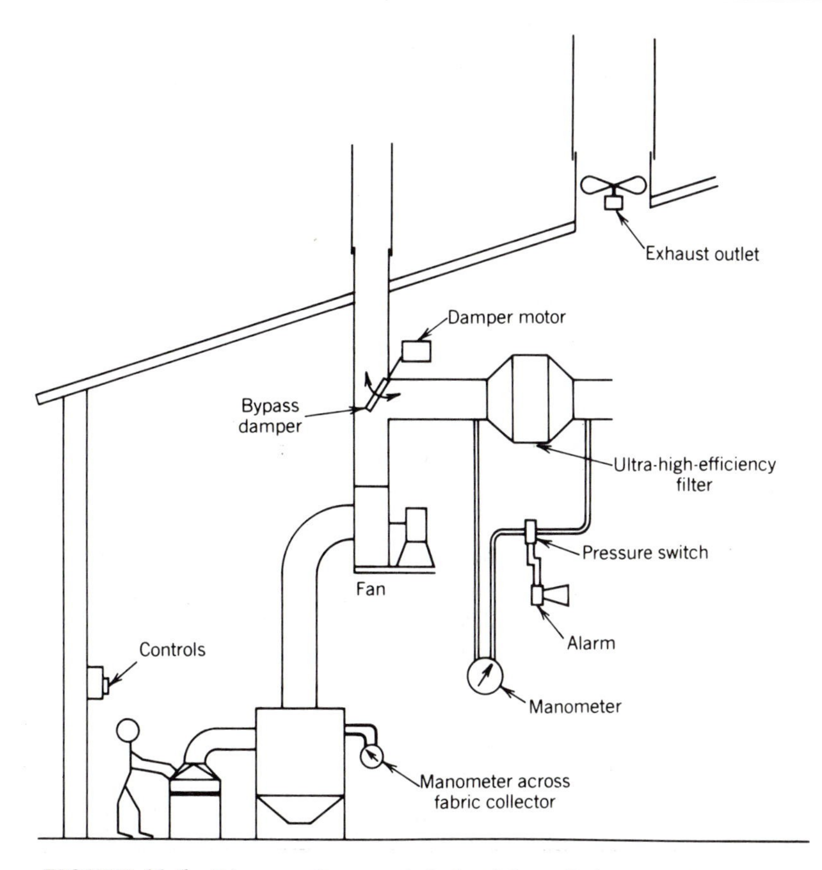

FIGURE 11.6 Diagram of automatic industrial ventilation control system for Exercise 11.6. (Reprinted by permission of American Conference of Governmental Industrial Hygienists, Committee on Industrial Ventilation, PO Box 16153, Lansing, Mich.)

equivalent as specified in the truth table (Table 11.13). The seven Y_i outputs in the truth table correspond to the seven LED segments in the diagram. Draw a Karnaugh map for each output, marking "don't care" states with an X, and then write a simplified Boolean expression for each, based upon the reduction marked on the Karnaugh map. Explain your rationale for the "don't care" states.

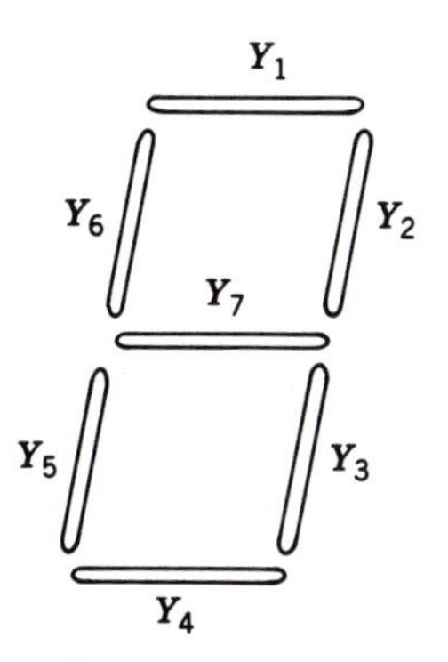

FIGURE 11.7 "Seven-segment" LED display.

TABLE 11.13 Truth Table for BCD-to-Seven-Segment Decoder

BCD Inputs				*Outputs*							*Decimal Equivalents (for reference)*
X_4	X_3	X_2	X_1	Y_1	Y_2	Y_3	Y_4	Y_5	Y_6	Y_7	
0	0	0	0	1	1	1	1	1	1	0	0
0	0	0	1	0	1	1	0	0	0	0	1
0	0	1	0	1	1	0	1	1	0	1	2
0	0	1	1	1	1	1	1	0	0	1	3
0	1	0	0	0	1	1	0	0	1	1	4
0	1	0	1	1	0	1	1	0	1	1	5
0	1	1	0	0	0	1	1	1	1	1	6
0	1	1	1	1	1	1	0	0	0	0	7
1	0	0	0	1	1	1	1	1	1	1	8
1	0	0	1	1	1	1	0	0	1	1	9

11.9. ***(Design Case Study)*** Suppose it is desired to invent an automatic multiplying machine that will instantly multiply two two-digit numbers in binary $A \times B$ to result in a product P. An industrial logic control system to handle this problem would have four inputs and four outputs arranged as follows:

$$\begin{array}{r} A_1A_2 \\ \times\ B_1B_2 \\ \hline P_1P_2P_3P_4 \end{array}$$

Note that the product can have four digits. Explain why. Set up a truth table. Construct logic expressions for each of the outputs. Use Boolean algebra to reduce each output expression to lowest terms.

CHAPTER 12

LOGIC DIAGRAMMING

Industrial logic control systems, introduced in the previous chapter, can be envisioned, designed, and explained more easily with any of a variety of diagramming tools. Diagramming becomes especially important when the logic system becomes complex—the usual case for real industrial applications. Indeed, most real industrial logic systems are impossible to describe solely by means of Boolean logic expressions. The reason for this difficulty is that real logic systems usually employ timers, memories, counters, or delays; these useful devices will be explained later. Upon completion of this chapter, the reader should have insight into the tremendous potential of logic systems when applied to manufacturing automation problems. This insight should enable students to start creating their own industrial logic solutions to applications of automation that they can envision themselves. They also should see how to design and build simple, practical industrial robots from standard building blocks of automation.

LOGIC NETWORKS

The first diagramming tool we shall consider is a direct extension of the Boolean logic expressions studied in Chapter 11: the logic network. The logic network is constructed of diagramming elements that directly compare to the logic operators AND, OR, and NOT, as studied earlier. For example, Figure 12.1 displays the logic network diagram for the simple console lock example that we studied in Chapter 11. For the reader's convenience, the meanings of the logic variables are repeated in Figure 12.1 so that the logic network can be interpreted. Some students will be able to decipher the network symbol code by comparing the diagram with the companion logic expression in Figure 12.1. Note that the logic expression is the same as was developed using the theorems of Boolean algebra in the previous chapter.

Figure 12.2 provides a key to the symbols used in the logic network diagram. In addition to the standard symbols, special elements can be defined simply by using a box and naming the element inside the box "timer," "counter," or whatever title is appropriate for the element. In addition to the preferred symbols that are used throughout this text, other symbols sometimes seen in practice are also listed in Figure 12.2.

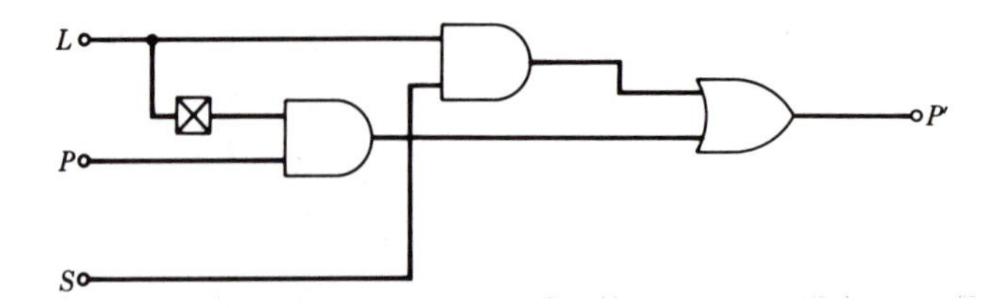

FIGURE 12.1 Logic network diagram for simple console lock example.

$$P' = \bar{L}P + SL$$

where L represents the console lock
P represents the process
and S represents the ON/OFF switch

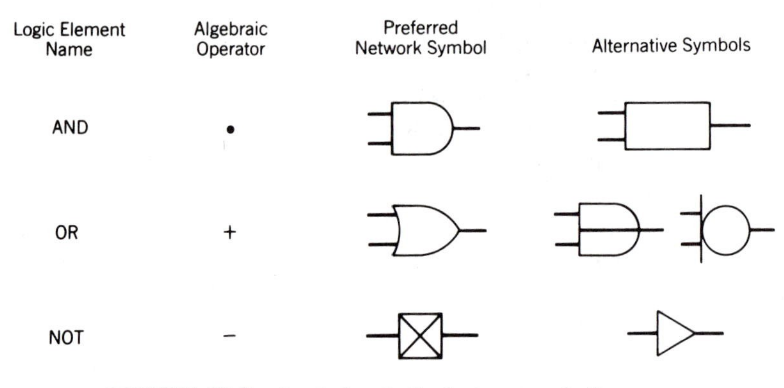

FIGURE 12.2 Symbol code for logic network diagrams.

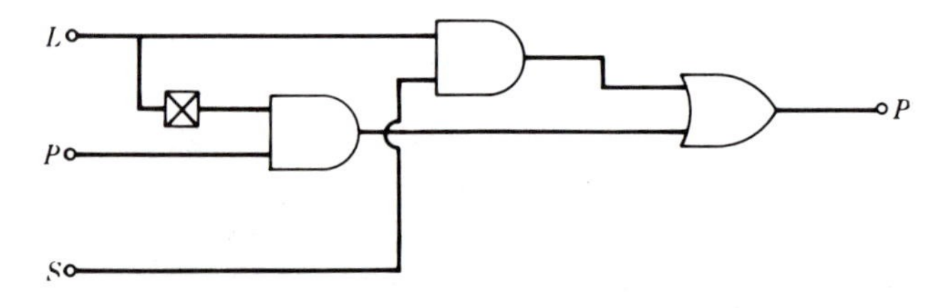

FIGURE 12.3 Alternate diagramming format for representing the logic network for the console lock.

It is convenient to think of a logic network as an electrical diagram in which the inputs supply voltages to the outputs. When making the analogy to electrical diagrams, however, remember that only two levels of voltage or states should be considered to retain the binary nature of logic systems.

The diagram of Figure 12.1 uses the convention in which a black dot indicates a line junction, and absence of the black dot indicates that crossing lines are *not connected*. (The black dot suggests the use of solder to make a connection.) An alternate convention is to interpret all crossing lines as direct connections unless a small semicircular arc is used to indicate that one line crosses over the other without contacting it (see Figure 12.3). The black-dot format used in Figure 12.1 seems to be preferable.

A series of two-way ANDs can be linked together to form a series of logic variables connected by AND operators. However, it is possible to have n-way ANDs to achieve the same purpose with one device. The same can be said for n-way ORs. Figure 12.4 illustrates this equivalence.

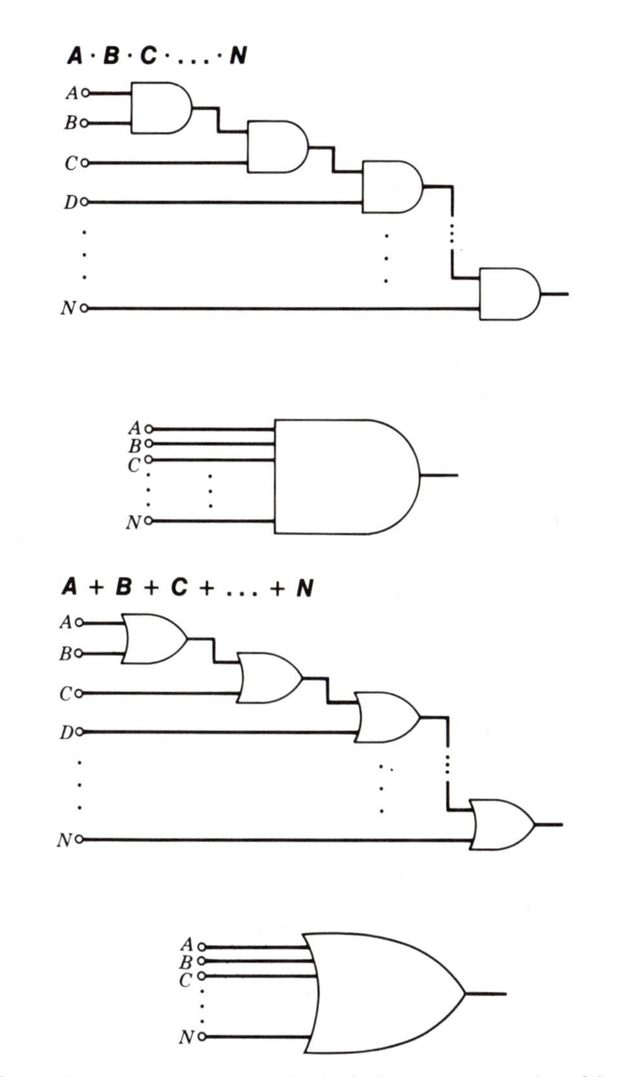

FIGURE 12.4 Equivalent usage of n-way logic devices versus a series of 2-way logic devices.

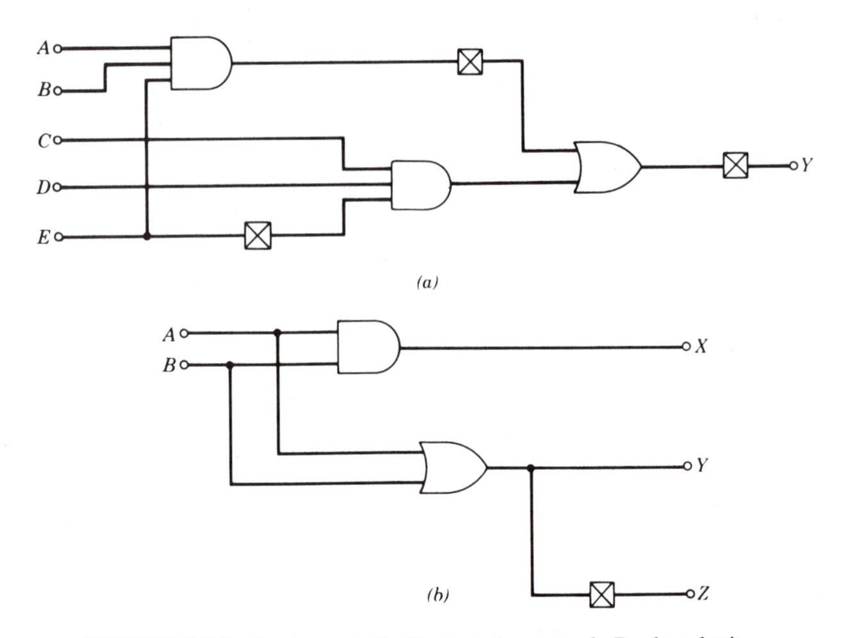

FIGURE 12.5 Logic network diagrams for example Boolean logic equations. (*a*) Example logic equation: $Y = \overline{\overline{ABE} + CD\overline{E}}$. (*b*) Example logic equations: $X = AB$; $Y = A + B$; $Z = \overline{A + B}$.

To further illustrate the use of logic networks to represent Boolean logic expressions, Figure 12.5 provides example equations with their corresponding logic networks. Note in Figure 12.5*b* that more than one equation is represented in a single logic network diagram.

LADDER LOGIC DIAGRAMS

Logic network diagrams may be the simplest and most direct extension of Boolean logic expression, but they are not the most familiar format to equipment maintenance technicians. For such an audience, the *ladder logic diagram* excels; furthermore, ladder logic is the language of *programmable logic controllers,* the subject of Chapter 13.

Ladder logic diagrams take advantage of the analogy of *series* circuits to the logical AND operator and parallel circuits to the logical OR operator. An example of a ladder logic diagram is shown in Figure 12.6. In this simple example, the ladder has only one (or perhaps two) rungs, but a vague resemblance to a ladder can be seen. In case the reader has not already guessed, the application is the familiar console lock example. Figure 12.6 should be compared with Figure 12.1 to understand the ladder logic diagramming format as contrasted with the logic network diagram format.

The basic elements of the ladder logic diagram are contact points (inputs) and loads (outputs). Figure 12.7 summarizes the element symbols and format for ladder logic diagrams. The left siderail of the ladder in a ladder logic diagram represents the hot line of an electrical circuit (usually energized at 120 volts ac to ground, but at much lower voltages for most logic circuits such as 5 volts dc for logic one and .7 volts dc for logic zero.) The right siderail represents the neutral line of an electrical circuit. Accordingly,

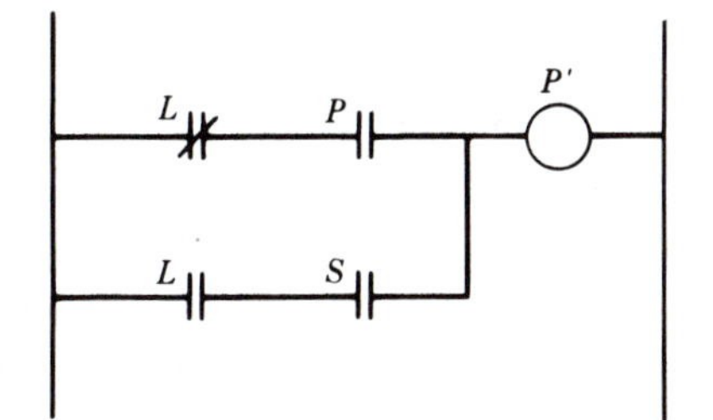

FIGURE 12.6 Example ladder logic diagram.

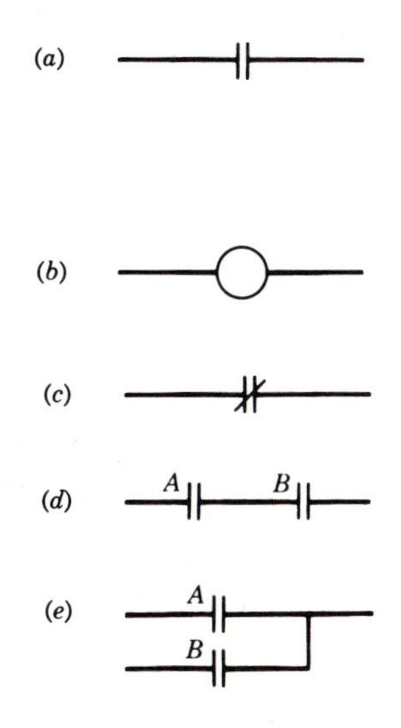

FIGURE 12.7 Symbol format for ladder logic diagrams.
(*a*) Input contacts, such as switches, relays, photoelectric sensors, limit switches, and any other off/on signal from the industrial logic system. (Do not confuse this symbol with the familiar electrical symbol for capacitors!).
(*b*) Output loads, such as motors, valves, alarms, bells, lights, actuators, or any other electrical load to be driven by the industrial logic system.
(*c*) The logic inverse of an input; this symbol permits the use of the logic *NOT* operator.
(*d*) $A \cdot B$ input contacts connected in *series* implies the logic *AND* operator.
(*e*) $A + B$ input contacts connected in *parallel* implies the logic OR operator.

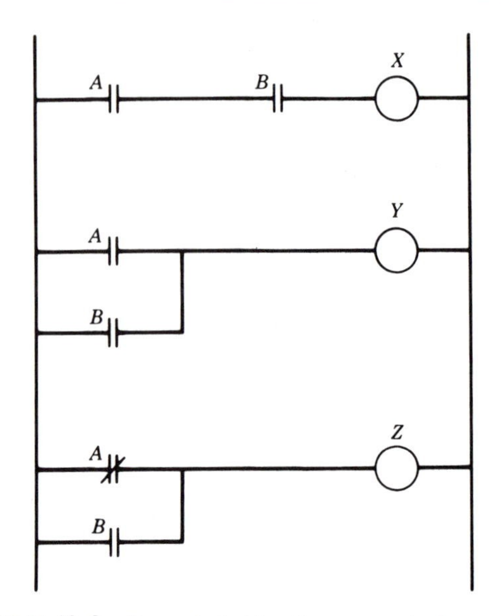

FIGURE 12.8 Example ladder diagram; note the repetition of input A in all three rungs of the ladder.

it is appropriate to place the switches or inputs toward the left side of the diagram and the output loads on the right, just before reaching the neutral. Obviously, every ladder rung or compound rung must have some sort of output load before connecting to the neutral or there will be a short circuit. In effect, a no-load ladder rung would act as a bus from the hot to the neutral, which should blow a fuse or throw a breaker.

One of the most difficult things to learn about ladder diagrams is that input contacts can be repeated as often as desired throughout the diagram. Figure 12.8 illustrates this point by referring to input A in every rung of a ladder diagram. The interpretation is as follows:

$$X = A \cdot B$$
$$Y = A + B$$
$$Z = \overline{A} + B$$

Another thing to remember about ladder diagrams is that the ON/OFF condition of a given *output* load can be a useful bit of information input to the logic system and can thus be used as an *input* contact on another rung of the ladder diagram. Figure 12.9, illustrating this point, should be interpreted as

$$Z = (Y + X) \cdot \overline{A}$$
$$\text{where} \quad Y = X \cdot D$$
$$\text{and} \quad X = (A + \overline{C}) \cdot B$$

Substituting:

$$Z = [(A + \overline{C}) \cdot B \cdot D + (A + \overline{C}) \cdot B] \cdot \overline{A}$$

Using the Boolean theorems studied in Chapter 11, the student should be able to simplify the expression readily for output Z, which in turn would lead to a simplification of the ladder diagram. This will be left as an exercise.

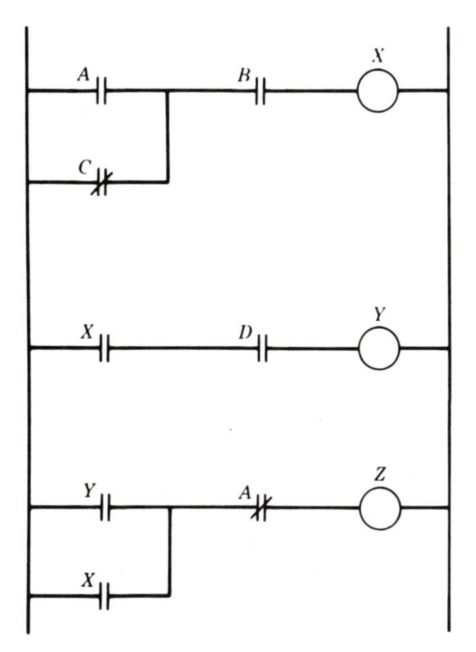

FIGURE 12.9 Example ladder diagram in which the condition of an output load is used as an input contact on another rung of the ladder.

It is essential to understand that the logic of the ladder diagram, for all practical purposes, is executed *simultaneously*. The sequence of the rungs in the ladder is inconsequential, except in some programmer controller applications, as will be seen in Chapter 13. At this point, however, it is best to ignore the few exceptions and think of the entire ladder diagram as being executed simultaneously—that is, all rungs of the ladder are parallel branches of the same electrical circuit.

Returning to our original ladder diagram example in Figure 12.6, the reader should now be equipped to interpret the ladder diagram as representing the logic of the now familiar console lock example. Note the use of P' to designate output P as contrasted with P to designate input P. This notation was used to correspond to the notation for the logic network diagram in Figure 12.1 and to conform to the notation format used in Chapter 11. However, in ladder diagrams, from the symbology it is obvious which components are inputs and which are outputs, so the use of "prime" or other modification of the variable name is unnecessary to distinguish between outputs and inputs.

The push-button switch is another good application to illustrate the construction of ladder diagrams. Let us begin this illustration with the simplest form of switch and then add elements to our ladder diagram one step at a time. Figure 12.10*a* illustrates a simple switch in which input *A* is a switch that turns on output *Y*. As soon as input *A* returns to off or 0, the output turns off. This of course would not be a satisfactory push-button switch because as soon as the push button is released, the output would return to its OFF state. Figure 12.10*b* adds an OR condition that keeps the output on if it is already on. In other words, the first input, *A*, turns the output on; the second input, *Y*, keeps the output *Y* on after it is turned on by *A*. This is an improvement, but now there is a new problem: once turned on, output *Y* will never be turned off by the logic system! As laid out in Figure 12.10*b*, there is no way to turn off output *Y* except to deenergize the entire

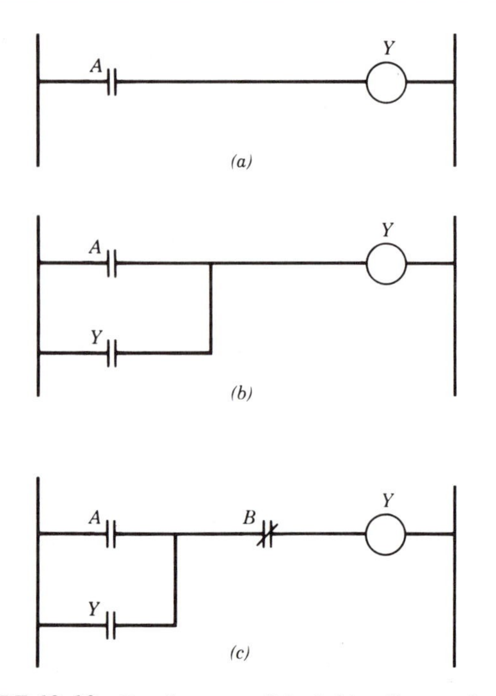

FIGURE 12.10 Development of the ladder diagram for a push-button switch. (*a*) Input switch *A* turns on output *Y*. (*b*) Input switch *A* turns on output *Y*, and input *Y* keeps output *Y* on. (*c*) Input switch *A* turns on output *Y;* input *Y* keeps output *Y* on until input *B* turns it off.

system (the entire ladder). To complete our push-button switch system, we add another input switch (*B*) in Figure 12.10*c*. Note that the input contact is "inverted"—that is, the output will receive voltage only if input contact *B* is *not* on. If input *A* is a spring push button designated as the ON button and input *B* is a spring push button designated as the OFF button, the ladder diagram displayed in Figure 12.10*c* represents a push-button switch, with the same features as held by the push-button switch developed in Chapter 11 (Figure 11.3).

The push-button switch ladder diagram can also be called a "memory" device. In the logic network format a separate element called a "memory" is usually represented by a square, but to be sure of identification it is wise to further identify the device in the diagram by writing "MEMORY" inside the square. There are many varieties of memories, having various requirements for turning them on or off. Some memories are "off-return"—that is, they return to an off state whenever power to the system is removed. Other memories are "retentive"—that is, they remember and return to the same state as before whenever power is turned on again.

Figure 12.11 illustrates the use of a memory device in a logic network diagram. The function of the network is to act as a push-button switch, the same as was true of the ladder diagram in Figure 12.10*c*. The memory device of Figure 12.11 is an OFF-dominant type—that is, if both inputs are satisfied, the output will be off. This feature is desirable from a safety standpoint on most push-button switch systems. Some memories may be designed to allow the current state (static) to dominate in case both inputs are competing to switch the state at the same time. This serves to illustrate the variety possible with

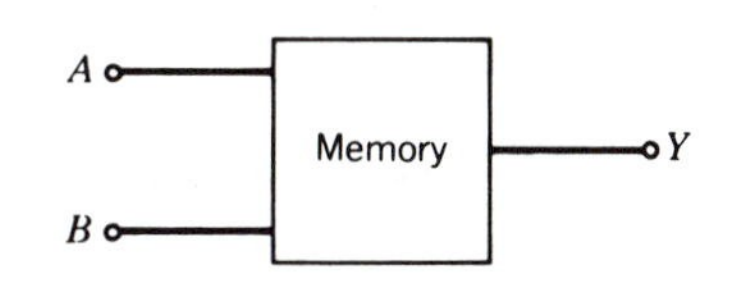

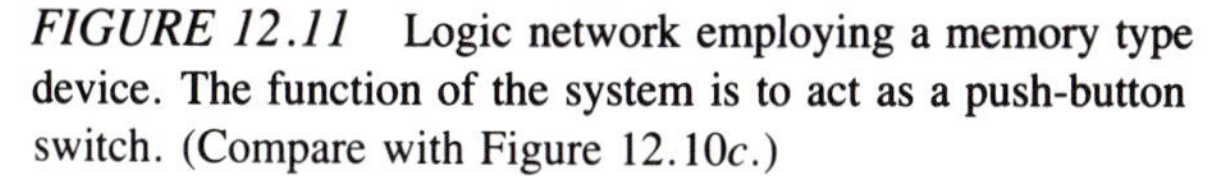

FIGURE 12.11 Logic network employing a memory type device. The function of the system is to act as a push-button switch. (Compare with Figure 12.10*c*.)

memory devices. Finally, the memory illustrated in Figure 12.11 should be the off-return type to assure that the system will remain off in the event of return of power after a power outage. With most systems, it is safer to require a restart procedure after power returns to avoid an unsafe sudden, automatic start or start attempt.

TIMERS

Up to this point in our discussion of logic systems, the logic outputs have been dependent upon the simultaneous condition of the inputs. Thus, any change in any of the inputs calls for an immediate reevaluation of the logic for immediate change of output variables if appropriate. But for many industrial applications it is appropriate for the logic system to cause the output variables to wait a predetermined time before reacting to a change in the inputs. Such a situation calls for timers or delay circuits. Industrial applications that require timers or delays are extremely useful in the field of robots and automated manufacturing; it is ironic that systems with timers are more difficult to analyze and design than are ordinary logic systems.

Ordinary Boolean algebra is not equipped to handle timers and delays. Even the truth table presents difficulties when one attempts to analyze a logic system with delays and timers. The best strategy is to use truth tables and Boolean algebra to become very familiar with time-static logic systems. Then when time-dynamic systems are encountered, the analyst can proceed directly to a logic network diagram or a ladder logic diagram.

The symbol format for a timer is usually a square with the word "TIMER" lettered inside to distinguish the device from counters and other special-purpose devices. The square (or rectangle) is used whether the timer is diagrammed in a logic network diagram or a ladder logic diagram. Also in the square somewhere should be a designation of the regular period or duration of the timer. Most industrial timers have adjustable times depending upon the application; the user sets the timer with a screwdriver, thumbwheels, or other mechanical means. Computer or programmable timers are set by programmers during the initial setup of the logic system.

There are many varieties of industrial timers, and one must read the product description for each to use it properly. Figures 12.12*a* and *b* compare equivalent usages of a popular type of industrial timer in a logic network and a ladder diagram, respectively. The figure serves to illustrate the use of the timer and gives the reader another comparison for familiarization with the formats of logic networks and ladder diagrams. But Figure 12.12 is more than an academic comparison; both diagrams are useful in a variety of real applications in manufacturing automation. Consider the conveyor system of Figure 12.13 in which an automatic weigh station activates a trap door or diverter in event an overweight item passes over the weigh station. The trap door opens immediately and remains open for four seconds to allow sufficient time for the item to drop through to the overweight

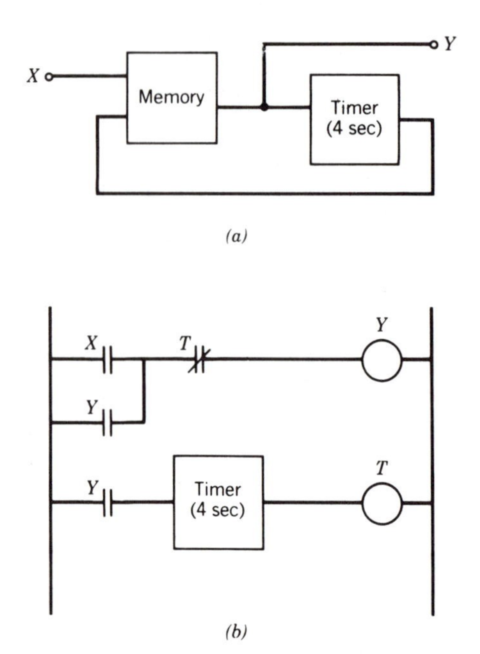

FIGURE 12.12 Equivalent usages of a timer in a logic system using a logic network and a ladder diagram. The logic for the output at Y is identical in both diagrams. (a) Logic network employing a four-second timer. (b) Ladder diagram employing a four-second timer.

track. For the system to work properly, it is necessary for successive items on the conveyor to be separated by distances of at least five seconds or so. The reader can no doubt think of other manufacturing automation applications for the logic system described by Figure 12.12 besides the automatic weigh station.

RESPONSE DIAGRAMS

To compensate for the inability of truth tables and Boolean algebra to deal effectively with logic systems employing timers, another diagramming technique is available that is helpful for all types of logic systems: the *response diagram*. With the response diagram the automation engineer can vary any of the inputs and show transitions from an input state to another with subsequent effects upon outputs. The response diagram is also an effective tool for describing exactly the behavior of various types of memory and timer devices.

The response diagram is simply a series of graphs, each representing a logic variable, in which the horizontal axis is time and the vertical axis is logic state—that is, 0 or 1. The graphs are placed so that their time axes are synchronized; in this way, a vertical line at any point on the graph describes a point in time, and all input and output variables can be evaluated at that point.

Figure 12.14 is a response diagram for the automatic weigh-station conveyor system illustrated in Figure 12.13 and for which the logic system was provided in Figure 12.12.

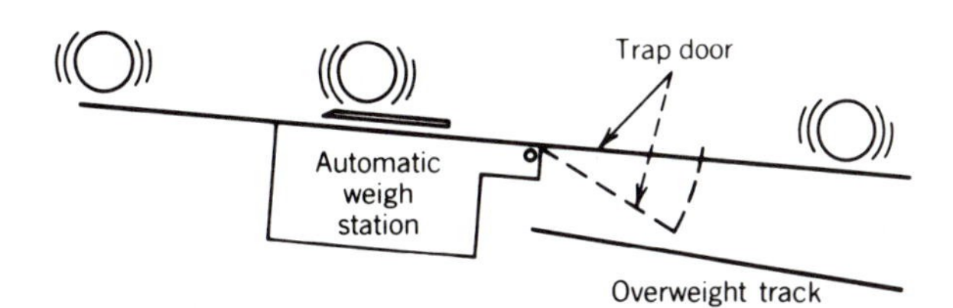

FIGURE 12.13 Automatic weigh station on a conveyor system; an overweight item passing over the weigh station causes the diverter to open for four seconds.

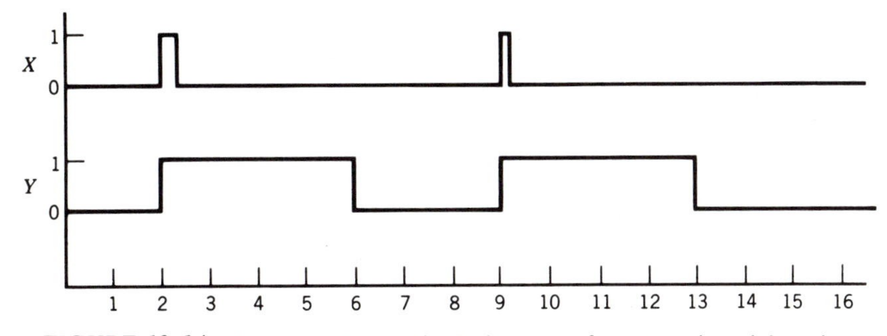

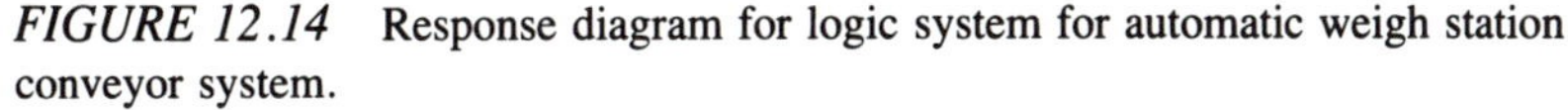

FIGURE 12.14 Response diagram for logic system for automatic weigh station conveyor system.

Note that each of the two logic variables plotted has only two states: 0 and 1. In the real automation system represented, these two states are likely to be represented by voltages, but pneumatic pressure, height, or other physical media could be used to represent logic state. Also note that the selection of various logic states for the *input* variable *X* was the choice of the person drawing the diagram. A two-second quiet period was selected at the beginning, and then an overweight item passes over the weigh station at time points 2 and 9. By contrast, the graph of the *output* variable *Y* is determined by the structure of the logic system and of course the pattern of the input.

The response diagram can also be used to illustrate the operating characteristics of ordinary logic systems that do not have timers. For example, an AND system is diagrammed in Figure 12.15. As in all response diagrams, the selection of the pattern of the input states was the choice of the person drawing the diagram. In Figure 12.15, a pattern was chosen that illustrates every meaningfully different combination of inputs for presentation to a simple AND. The behavior of the AND then is completely described in the response diagram for output *C*.

Although the response diagram is usually more tedious to prepare than are the other tools for describing logic systems, its power in describing all states of a logic system is unsurpassed. This power is very useful in the case of timer systems, and extremely so in comparing the features of various types of memories that are almost identical. Figure 12.16 illustrates three different types of memories by means of their respective response diagrams.

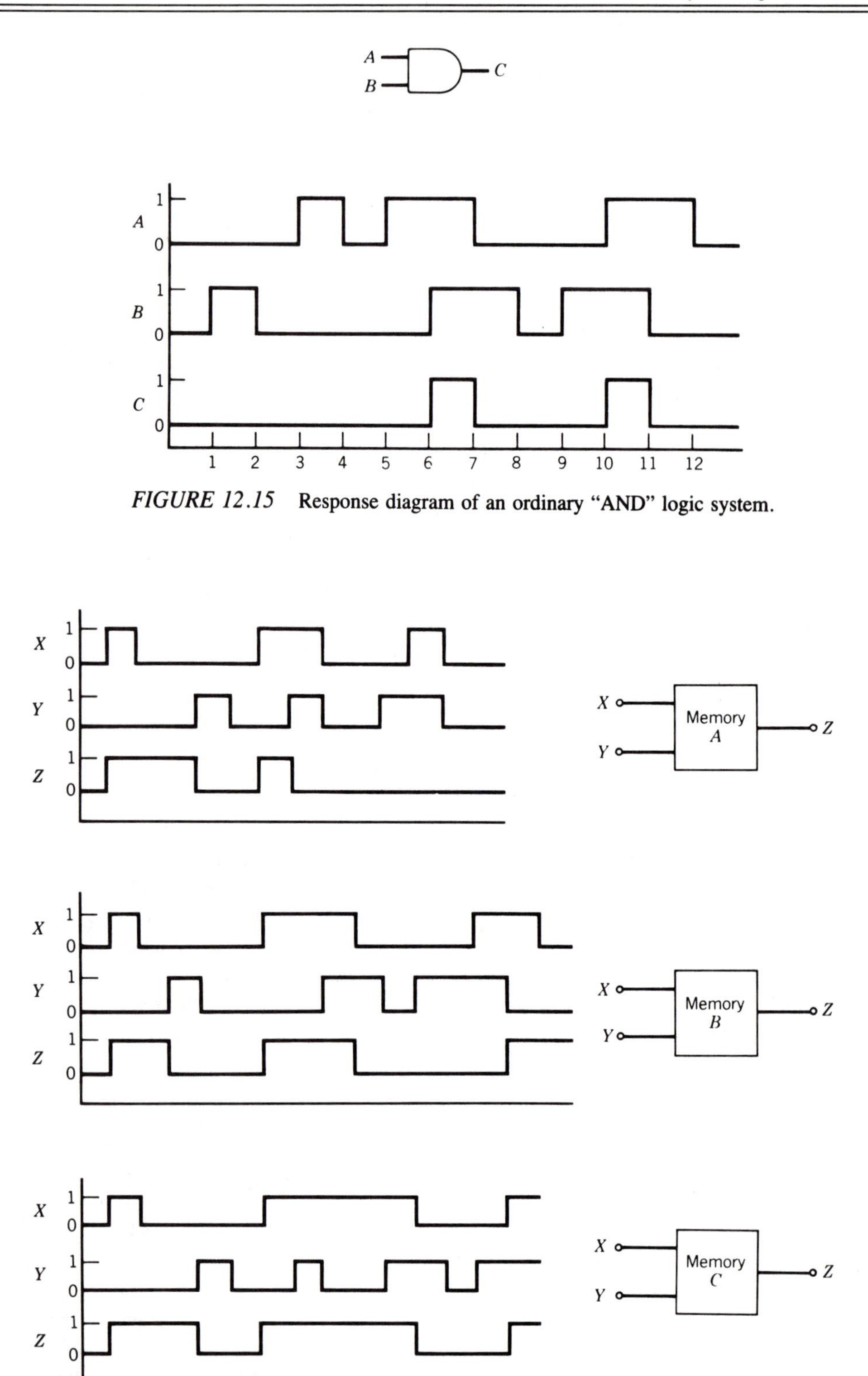

FIGURE 12.15 Response diagram of an ordinary "AND" logic system.

FIGURE 12.16 Three types of memory flip-flops with differences described by response diagrams. (*a*) "OFF-dominant" memory. (*b*) "Static-dominant" memory. (*c*) "ON-dominant" memory.

EXAMPLE 12.1 (CASE STUDY) **Automatic Ventilation System**

The opportunity to use timers greatly expands the applications of logic systems to the field of robots and manufacturing automation. A simplified manufacturing automation system will now be studied to illustrate the use of response diagrams. Many machines or processes require ventilation while the machine is running and also for a brief time period (for example, three minutes) after the machine has been shut down to purge the system completely. This is an ideal simple example for illustrating the tools for constructing logic systems developed in this chapter.

The need for a timed ventilation cycle after the machine has been shut down demands the use of a timer in the logic system, and that fact makes the use of truth tables and Boolean algebra virtually infeasible. The approach then will be to use the diagramming tools described in this chapter: response diagrams, logic network diagrams, and ladder logic diagrams. The sensible first step is to use a response diagram to describe the relationship between the output and input variables. The automatic ventilation system described in this case study really has only one input—the system ON/OFF switch. This switch turns on the machine and the ventilation fan, both of which are outputs. The ventilation fan remains on for three minutes, however, after the system switch has been turned off. Summarizing the logic variable definitions:

INPUT

A = 1 if the system switch is turned on
= 0 if the system switch is turned off

OUTPUTS

B = 1 if the machine is on
= 0 if the machine is off

D = 1 if the ventilation fan is on
= 0 if the ventilation fan is off

Choosing an arbitrary but variable pattern for the input A, we construct a response diagram in Figure 12.17 to document the desired response of the two output variables to various sequences of positions taken by the input variable. In the input sequence chosen for Figure 12.17, note that the system is turned on at time 2. The output variable B (machine) is identical to A throughout the diagram, but output variable D (the ventilation fan) remains on an additional three minutes (to time 10) after the system has been turned off at time 7. At time 14, a short ON cycle is used to demonstrate that the run length of the ventilation fan after the system has been shut down remains

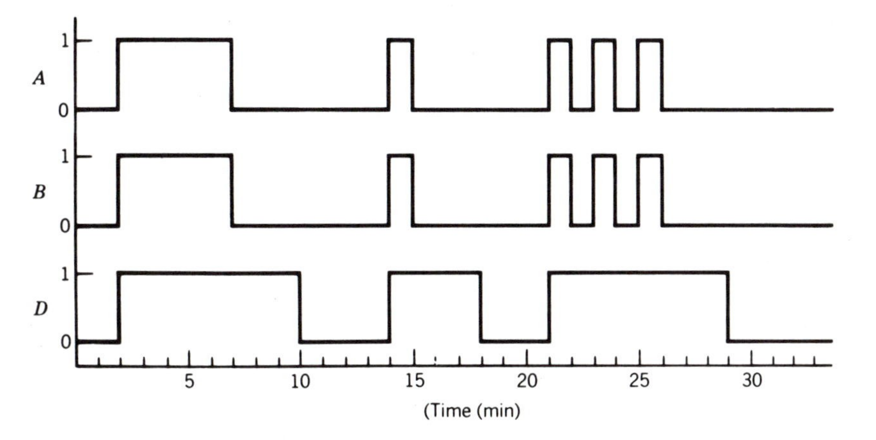

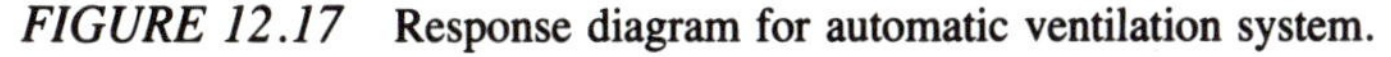

FIGURE 12.17 Response diagram for automatic ventilation system.

constant at three minutes regardless of time ON for the input variable A. The series of short ON cycles at time 21 demonstrates that the ventilation fan remains on throughout this period and continues on to time 29 after the system is finally shut down at time 26. The response diagram has thus been used to *specify* in the design phase what we want our logic system to do. Later, in the development and *checkout* phase, the response diagram can be used again to see if the logic system performs as specified.

After documenting the desired system performance characteristics with the time diagram, either a logic network diagram or a ladder logic diagram can be used to specify the logic system required to achieve the objectives specified. We will begin with the logic network diagram. In the absence of a truth table or Boolean algebra, we must analyze the input and output variables and attempt to create the appropriate logical relationships between them. A good way to start is with the output variables and then work backwards. Since output variable B must follow input variable A, it can be connected directly. Output variable D (the ventilation fan), however, has *two alternatives* for its operation: Either input A is on *or* the timer has not yet expired after input A has been turned off. The word "alternatives" should be a key to the analyst that a logic OR device is needed. The partially completed logic network diagram is thus constructed as shown in Figure 12.18. Note that the lone input is shown to the left of the diagram and the two outputs are shown to the right.

Working further backwards through the diagram, we attempt to connect the input conditions to the OR device. The first connection is easy; one of the alternatives for having the ventilation fan is simply to turn the system (input variable A) on. The other connection must come from the timer, but note that it is when the timer has *not* yet expired that we want the ventilation fan to be turned on. This suggests the use of a logic NOT on the output of the timer before connecting it to the logic OR. Figure 12.19 updates the logic network to the configuration explained so far.

The only question that remains is what logical input should be used to trigger the timer. The timer selected is a simple one that behaves as illustrated in the response diagram shown in Figure 12.20. Note that the timer is not cumulative; each time the input X returns to logic 0, the time count is reset to zero, and the timer must begin the count toward three minutes at zero again. Thus, if input X is interrupted by OFF periods at intervals of less than three minutes, output Y will never turn on. What we want is really the reverse. If the system is interrupted by OFF periods of less than three minutes, we do not want the ventilation system to ever turn *off*. That is why we selected a logic *NOT* on the output of the timer before we connected it to the logic OR. On the input side, we need a logic NOT also because it is only when the system is turned OFF that we want the three-minute time count to begin. The logic network diagram is thus completed in Figure 12.21.

A ladder logic diagram can also be constructed for the automatic ventilation system. This can be done by analyzing the logic for each output as was done earlier for the logic network diagram or by converting element-for-element from the logic network diagram. We shall do the latter, using the former as a check on our method. The ladder diagram will need at least two rungs, corresponding to the two outputs B (machine) and D (ventilation fan). The input switch A will be repeated wherever it is needed to represent the correct logic for a given rung. We will also include a third rung to set up logic for the timer. The completed ladder diagram is shown in Figure 12.22, and the reader should study it carefully to assure that the logic is identical to that of the corresponding logic network diagram of Figure 12.21.

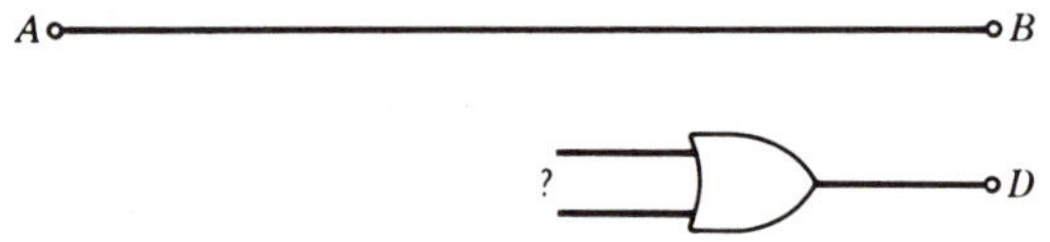

FIGURE 12.18 Partially completed logic network diagram for ventilation system.

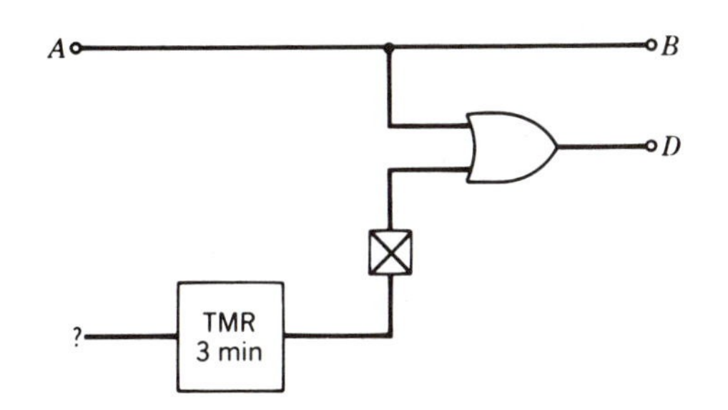

FIGURE 12.19 Second phase of development of the logic network diagram for the automatic ventilation system.

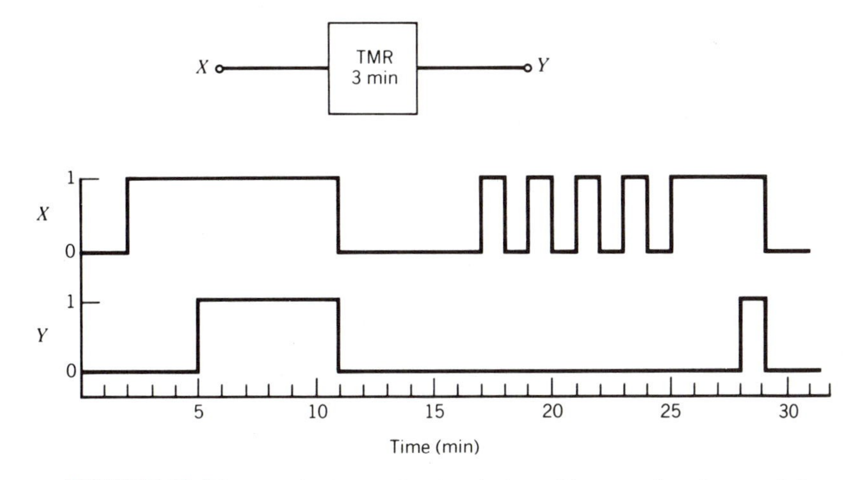

FIGURE 12.20 Simple timer with description of its operating characteristics.

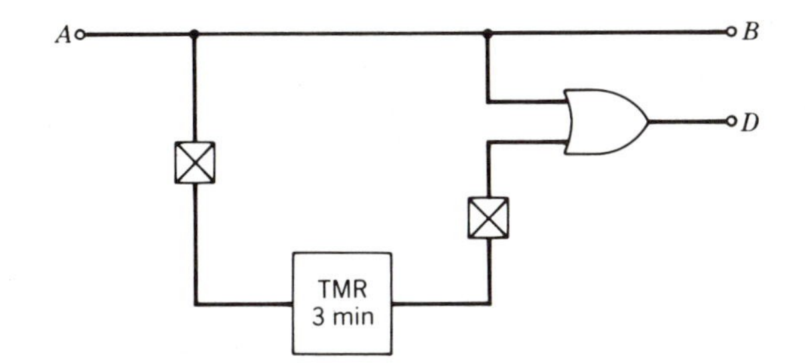

FIGURE 12.21 Completed logic network diagram for the automatic ventilation system case study.

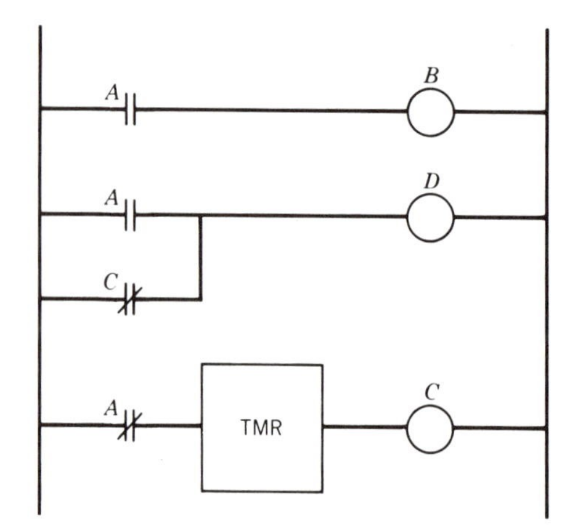

FIGURE 12.22 Completed ladder logic diagram for automated ventilation system case study.

Before leaving the ventilation case study, some observations should be made. Considerable latitude existed in the choice of arrangement of both the logic network diagram and the ladder logic diagram. It should be remembered that there are many correct solutions. Also note the sequence of rungs in the ladder diagram in Figure 12.22. Some analysts might have interchanged the sequence of the second and third rungs, believing this would be a more natural sequence. Such a switch would have been entirely satisfactory because, from a practical standpoint, the sequence of the rungs makes no difference. Finally, the choice of names of the input and output variables was completely arbitrary. We will see in Chapter 13 that manufacturers of logic hardware may use a naming convention for the input and output variables, but adherence to such conventions is the choice of the automation engineer.

To gain further confidence with logic system diagrams, we shall examine another case study of slightly more complexity.

EXAMPLE 12.2 (CASE STUDY)
Furnace Control

Figure 11.1 illustrated a small furnace to preheat forging blanks before the blanks are placed in the dies of a drop-forge hammer. The furnace is fueled by natural gas, and if the flame goes out, it must of course be reignited. In fact, there is a potential hazard if the flame goes out and the gas valve remains open. Prior to any automation of such systems, workers had to watch the flame and either reignite it or turn off the gas valve if the flame went out. Modern systems monitor the flame automatically and take action accordingly. Manual overrides are needed to overcome the automated system when necessary. It should be apparent that the various conditions to be sensed and actions to be taken in this system are of a binary nature; thus, an industrial logic control system would be appropriate. The reader can see perhaps the need also for at least one timer in this system to give an ignition system an opportunity to do its work and to shut down the system for safety if the flame does not ignite or reignite. Describe the logic control system for this furnace using a response diagram, a logic network diagram, and a ladder logic diagram.

Let us define the following pertinent input and output variables for Case Study 12.2.

INPUT

Power: $P = 1$ if power is on
$= 0$ if power is off

Flame: $F = 1$ if flame is lit
$= 0$ if flame is not lit

OUTPUT

Fuel: $G = 1$ if gas valve is open
$= 0$ if gas valve is closed

Ignition: $S = 1$ if sparking ignition system is on
$= 0$ if sparking ignition system is off

Alarm: $A = 1$ if alarm is flashing
$= 0$ if alarm is not indicated

Many other input and output variables could be conceived, such as "manual override" and "fuel pressure," as the complexity and capability of the automation system grow. However, with the five essential logic variables named and defined above, an interesting though simple logic system can be studied.

Examining the five input and output variables more closely, one might question why "ignition" was considered an output variable and "flame" an input variable, instead of vice versa. After all, does not ignition cause flame? But remember that we are designing a *logic control system* and we must interpret the words "input" and "output" as they relate to the *logic control system,* not the overall furnace system. Ignition is an automatic *action* (output) taken by the logic control system to deal with an *input* condition (no flame). The system might employ a photoelectric or temperature sensor to detect the presence of the flame. Fuel is an output variable because the industrial logic system decides whether to open or close the fuel gas valve. Power is an input variable that should be requisite to all of the output variables.

The timer is needed to keep the fuel on *even without a flame* for a sufficient time period to allow the ignition system to ignite the fuel but to shut off the fuel and activate the alarm in event the flame fails to appear after a reasonable time interval, perhaps five seconds.

As in Case Study 12.1, the response diagram is probably the best tool to begin with because it enables us to describe the results we expect from the furnace logic control system. Figure 12.23 is one such response diagram. Referring to the figure, let us step through each of the points in time at which a significant change in system states occurs. At time 1, the power is turned on, at which time it is appropriate for the fuel and ignition systems to turn on also. Three seconds later, at time 4, we achieve a flame, and this is cause for turning off the ignition system but not the fuel. The system operates normally until time 9, at which time we introduce a momentary loss in flame to initiate an appropriate reaction on the part of the ignition system. The ignition comes on at time 9, and we achieve a flame again at time 10, so the ignition turns back off. At time 13, a period of extended loss of flame begins. The ignition system comes on immediately, but the flame does not reappear. Finally, at time 18 the emergency procedure is initiated that consists of turning off the fuel, stopping the ignition system, and activating the alarm. The system will remain in this state indefinitely with the power on. At time 22, we introduce an operator action of turning off the system power. This manual action causes the alarm to turn off. The system is thus completely deactivated and will remain so until the power is turned back on. At time 24, we introduce a spurious one-second signal of the flame sensor. Theoretically, this should not be possible because the fuel is turned off, but we want to be sure that our logic control system can deal with the unexpected (such as a malfunctioning sensor or an ambient light source). Appropriately, the system does nothing. At time 27, we turn the power back on, starting a repeat of the procedure begun at time 1. This time, however, the flame does not ignite, demonstrating at time 32 that the emergency shutdown procedure functions at system startup the same as it does when the flame is accidentally

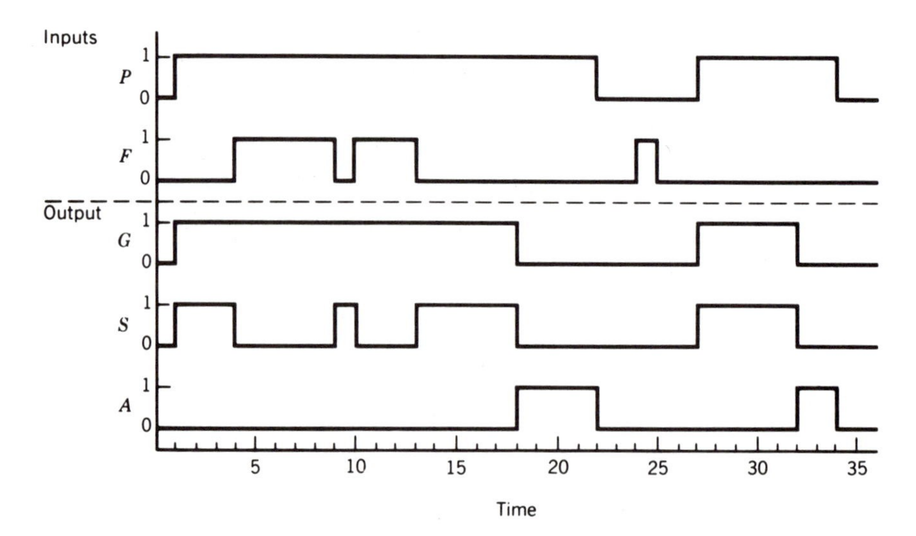

FIGURE 12.23 Response diagram for furnace ignition logic control system.

lost during normal operation. The action at time 34 is the same as at time 22. Thus, all of the feasible possibilities and sequences have been demonstrated by the response diagram. The diagram serves as a specification of the precise way we want our logic system to behave under various input circumstances. Later, the response diagram can be used as a test procedure to see if the furnace control system as constructed actually performs the way the response diagram says it should.

The next step is to construct either a logic network diagram or a ladder diagram to achieve the logic specified by the response diagram. This can be a difficult step, especially without the benefit of truth tables. Remember that the presence of the timers in the industrial furnace control system makes the use of the truth table awkward or even unusable in this case study. As in Case Study 12.1, let us tackle the logic network diagram first and start with the outputs and work backwards.

The furnace ignition control system has three outputs: fuel (G), ignition (S), and alarm (A). There are two possible situations for which we would want the fuel to be on:

1. Both the power and flame are on and the system is operating normally.
2. The power is on, and the ignition is attempting to ignite the fuel, but the safe time limit (five seconds) has not yet been exceeded.

These two possibilities for the fuel output suggest the use of a logic OR with the fuel as an output from the OR. This part of the logic network is summarized in Figure 12.24.

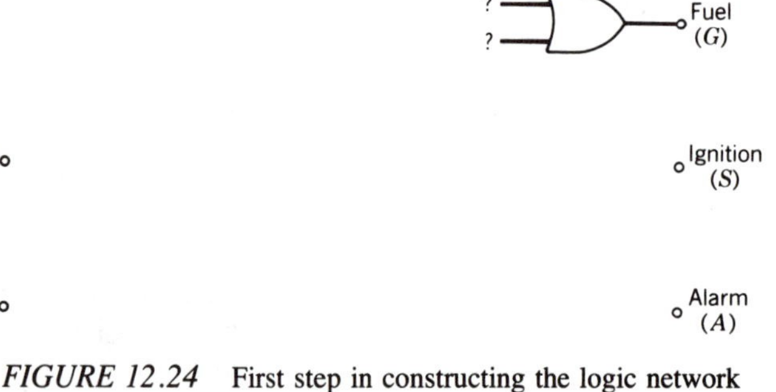

FIGURE 12.24 First step in constructing the logic network diagram for the furnace control system; there are two possibilities for obtaining "fuel on."

Working further backwards into the logic network, we see that both the power and the flame must be present to achieve the first condition for fuel. The words "both" and "must" in the foregoing statement suggest the use of a logic AND connected to inputs P and F. The other alternative for fuel ON contains three requirements:

1. Power
2. Ignition attempting to light flame (no flame yet)
3. Time limit not exceeded

The word "requirements" suggests the use of another AND, this time a three-input AND connected to input P, an inverted (NOT) flame input, and the output from a five-second timer. At this point, the logic network diagram can be drawn to the degree of completion shown in Figure 12.25.

Working further backwards in the diagram, we need an input for the timer. The timer selected is of the same simple type as was used in the ventilation system case study, but this time it is set for five seconds instead of three minutes. Reviewing the operating characteristics of the simple timer, note that if the timer input is interrupted by OFF periods at intervals of less than five seconds, the timer output will never turn on. This is exactly what we want in the furnace control application. If the flame always appears before five seconds, we do not want the alarm to be activated. The appropriate period to be timed, then, is that in which the power is on but the flame is not present. This suggests a two-input AND connected to inputs P and NOT F.

To complete the logic network diagram, we need to work backwards from outputs S and A. This process will be easier than it was for output G because considerable logic has already been

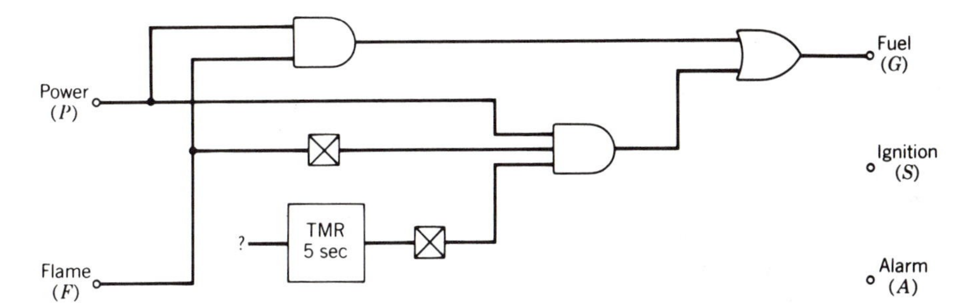

FIGURE 12.25 Second step in constructing the logic network diagram for the furnace control system.

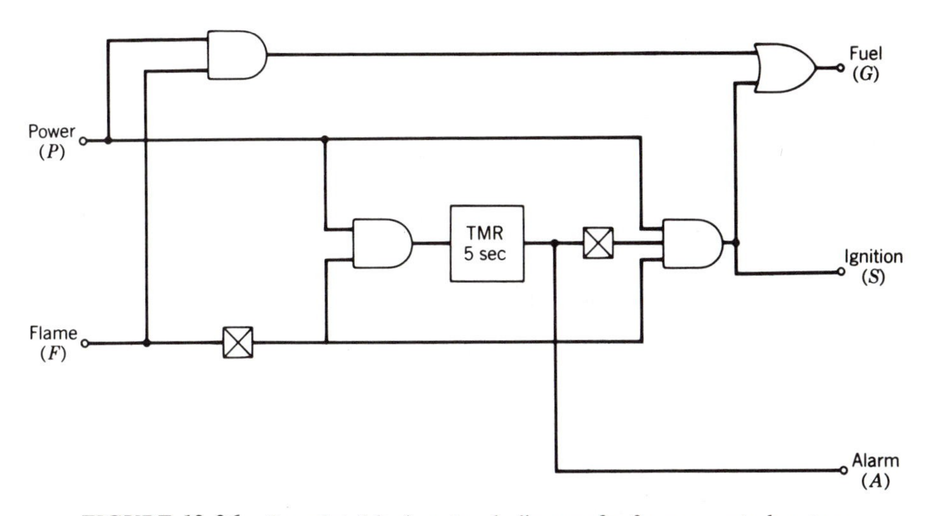

FIGURE 12.26 Completed logic network diagram for furnace control system.

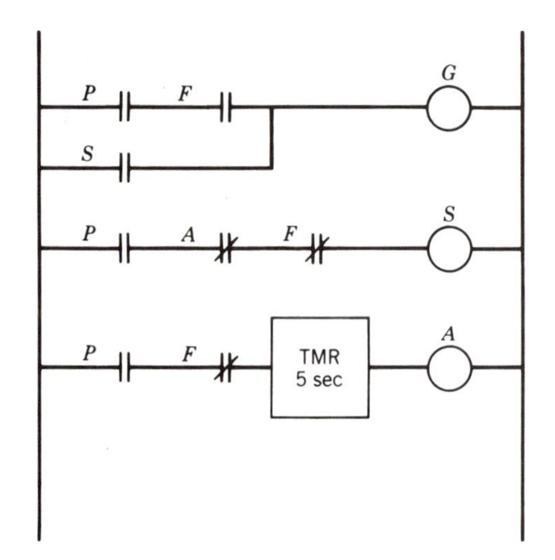

FIGURE 12.27 Ladder logic diagram for furnace control system.

developed in the diagram and will not have to be repeated. For instance, output *S* (ignition) should be identical to the second input for the OR leading to output *G* (fuel). Output *A* (alarm) is obviously the timer output. The completed logic network diagram for the furnace control system is shown in Figure 12.26. This circuit could be built up with standard electrical components representing the various logic devices and then tested to see whether the system behaves according to the response diagram of Figure 12.23.

To complete our case study of the furnace control system, let us construct a ladder diagram. The ladder diagram will need at least three rungs, corresponding to the three outputs: *G* (fuel), *S* (ignition), and *A* (alarm). The inputs *P* (power) and *F* (flame) would represent contact points that would be repeated wherever they are needed throughout the ladder diagram. The completed ladder diagram is shown in Figure 12.27.

This completes the case study of the ignition control system for the furnace for preheating forging blanks. In retrospect, it would be easy to discount the study as superficial and not considering such real-world problems as purge periods before restart, intermittent noise in the input signals, and failures in the output valves or actuators. But we must walk before we run, and a simplified case serves to illustrate the principles without bogging down in the detail required in dealing with the real-world system.

In Chapter 13, we shall see that standard hardware is readily available to convert from a practical idea in manufacturing automation directly to a logic control system to make that idea a reality. No longer must the analyst be a wizard in electrical or electronic circuitry to make a real-world logic control system work. It is simply a matter of visualizing what is desired for the automation system to do, documenting the logic by constructing the logic network or ladder logic diagram, and then programming a piece of standard hardware to execute the logic. Chapters 11 and 12 have provided the analytical tools, and Chapter 13 will cover the implementation of these tools using actual hardware.

SUMMARY

This chapter has added pictures to the words, tables, and equations used to describe industrial logic control systems. Some industrial logic control systems, especially those employing timers or delays, defy expression using words and equations; thus, some form of diagram becomes essential.

The response diagram is an excellent tool for describing *what* an industrial logic system should do but does not specify *how* the system will accomplish its objective. The response diagram is thus useful in planning and specifying an automated system and later in putting the logic system through a test to see whether it performs as specified.

The logic network and ladder logic diagrams are two methods of expressing *how* the logic of an automated system will function. Both are suggestive of electrical circuits, but it must be remembered that electrical circuits are only one of several media for executing logic circuits.

Two examples of simple automated systems were used to illustrate the diagramming tools of industrial logic control systems. Although the applications were idealized and would be much more complicated in real-world systems, their simplicity permitted concentration upon the techniques of expressing and analyzing industrial logic control system.

EXERCISES AND STUDY QUESTIONS

12.1. What limitation of truth tables and Boolean algebra makes the use of logic networks and ladder diagrams very important?

12.2. What is the difference between the words *timer* and *delay?* Are they interchangeable?

12.3. Figure 12.28 illustrates a two-input cumulative timer in which one input acts as an enable switch and the other acts as the timer input. Thus, the timer input can be interrupted, and the timer will remember and keep a running total of the time elapsed while the timer input is on. Every time the enable input is switched off, the timer is rezeroed and is ready for a restart whenever the enable input is turned back on. Construct a response diagram to describe the operating characteristics of this timer.

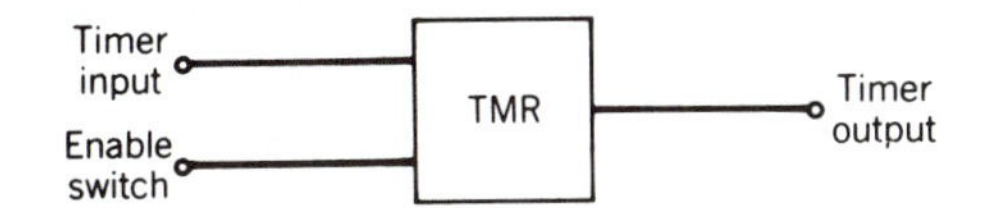

FIGURE 12.28 Two-input cumulative timer.

12.4. Explain how the two-input timer of Figure 12.28 could be connected to represent the simple timer described in Figure 12.20.

12.5. Redraw the logic network diagram for the automated furnace control system to utilize a two-input cumulative timer instead of the simple timer shown in Figure 12.26.

12.6. Redraw the ladder logic diagram for the automatic ventilation system to incorporate the two-input timer instead of the simple timer shown in Figure 12.22.

12.7. Construct a logic network diagram for the automatic recirculating industrial ventilation control system described in Exercise 11.6.

12.8. Draw a ladder logic diagram for the automatic recirculating industrial ventilation control system described in Exercise 11.6.

12.9. Draw the logic network diagram for the automatic binary multiplying machine described in Exercise 11.7.

12.10. Construct a ladder logic diagram for a logic system that will accomplish the purpose of the automatic binary multiplying machine in Exercise 11.7.

12.11. One process for making blueprints employs ammonia gas, which, although necessary for the process, must be purged from the system after the machine is stopped to prevent the remaining gas from further contaminating the surrounding atmosphere. In one installation, it is the operator's responsibility to turn on the purge fan and allow two minutes for the ammonia gas to be purged from the system. The purge fan must *not* be on while the system is operating because the ammonia gas is needed by the process. How would you automate this system? How does this system differ from the automatic ventilation system case study described earlier in this chapter? Demonstrate with a response diagram.

12.12. Construct a logic network diagram for the automated system designed for the blueprint machine of Exercise 12.11.

12.13. Construct a ladder logic diagram for the automated system designed for the blueprint machine of Exercise 12.11.

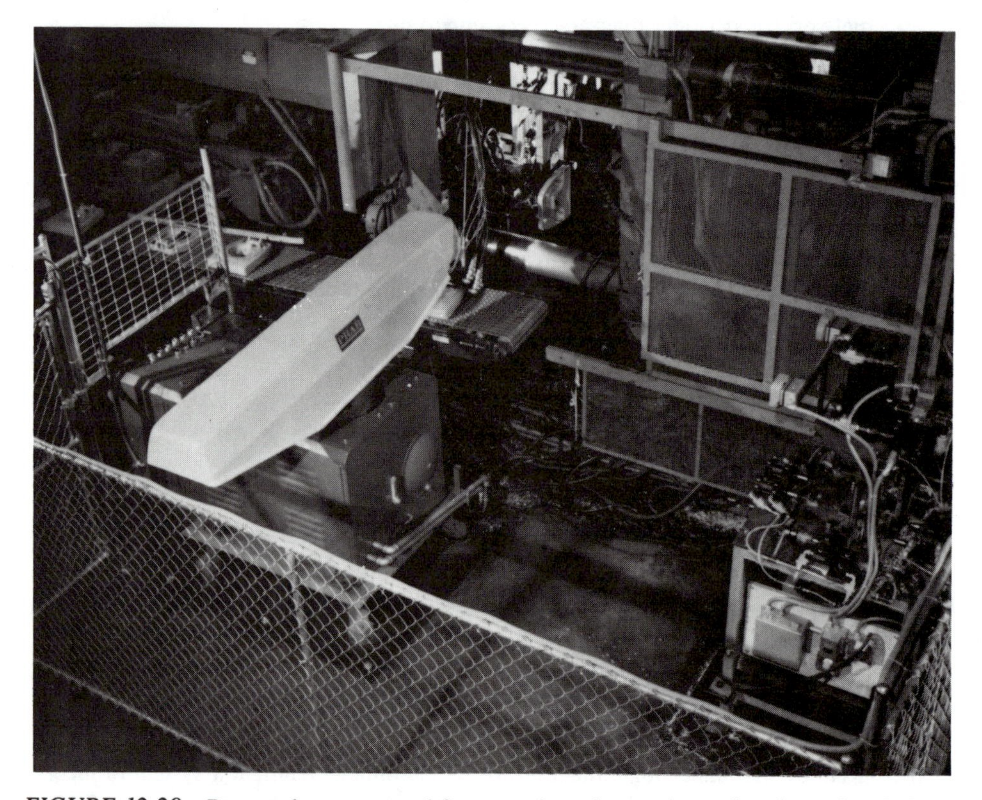

FIGURE 12.29 Personnel are prevented from entering robot work envelope by an interlocked barrier. (Reprinted by permission of Prab Robots, Inc., Kalamazoo, Mich.)

12.14. Compare the logic network diagram of Figure 12.5*b* with the ladder logic diagram of Figure 12.8. Note that the logic is not quite identical. Redraw Figure 11.5*b* so that its logic matches Figure 12.8. Redraw Figure 12.8 so that its logic matches Figure 12.5.

12.15. The robot pictured in Figure 12.29 has an interlocked barrier that will interrupt power to the robot if an access gate is opened. Design a logic system that will not only interrupt power to the robot but also will sound an alarm if any of five access gates are opened. The alarm should sound for 30 seconds, but the power to the robot should remain off until the access door(s) is (are) again closed and a system restart button is pressed.

CHAPTER 13

PROGRAMMABLE LOGIC CONTROLLERS

This chapter climaxes a three-chapter sequence on the subject of industrial logic control systems. Chapter 11 introduced the analysis of industrial logic control systems, Chapter 12 added the diagramming tools, and this chapter shows how these systems can be implemented using one of the most ingenious devices ever devised to advance the field of manufacturing automation. So versatile are these devices that they are employed in the automation of almost every type of industry. The device, of course, is the programmable logic controller, or PLC, and thousands of the these devices go unrecognized in manufacturing plants—quietly monitoring security, managing energy consumption, and controlling machines and automatic production lines.

The relationship between programmable logic controllers and industrial robots is an interesting one. It is easy to visualize an industrial logic control system that signals to a robot the cue to start its operating cycle based upon logic inputs sensed from the process. It also is possible to visualize a robot output signal to a logic controller, which in turn signals a machine process to commence. But besides these applications, conventional programmable logic controllers are widely used as robot controllers themselves. By this we mean that the detailed operational movements of the robot such as REACH–GRIP–WITHDRAW–ROTATE–EXTEND–RELEASE–WITHDRAW and so on are controlled directly by general-purpose, commercially available programmable logic controllers. Thus, it is possible to build an industrial robot (especially the axis-limit type) with conventional pneumatic cylinders, motors, actuators, and limit switches and use a programmable logic controller as the system's brain.

WHAT IS A PC?

PCs and PLCs

When people use the term "PC," they usually mean "personal computer," but it was not always so. In the early 1980s, the IBM Corporation popularized the term *personal computer* to refer to its entry into the small, low-cost computer field; most users and other manufacturers had called these small desktop models "home computers." In those days the term PC in industry meant "programmable controller." So successful were IBM's small computers that the term PC became synonymous with "personal computer," and since the general public was more familiar with personal computers than with programmable controllers, the latter began to lose their identity with the term PC. In the same time frame, a popular manufacturer of programmable controllers, the Allen-Bradley Company, began to name its controllers "PLCs" to represent "programmable logic controller." The name PLC has since been picked up by most industrial users to refer to any make of programmable controller, and that is the terminology that we use in this book, although the reader should be aware that the term PC is still sometimes used in the industry to refer to programmable controllers instead of personal computers. By whatever name it is called, the programmable logic controller is the key element in most manufacturing automation applications.

Figure 13.1 diagrams the essential elements of a typical programmable logic controller. Note that it operates on ordinary house current, although it may control circuits of large amperage and voltages of 440 V and higher. Also note that the unit has a *detachable* programmer module. One programmer can be taken easily from one PLC to another, attached to construct a new program or modify an existing one, and then taken to other PLCs or perhaps stored in an office. The interface units and input/output contact terminals

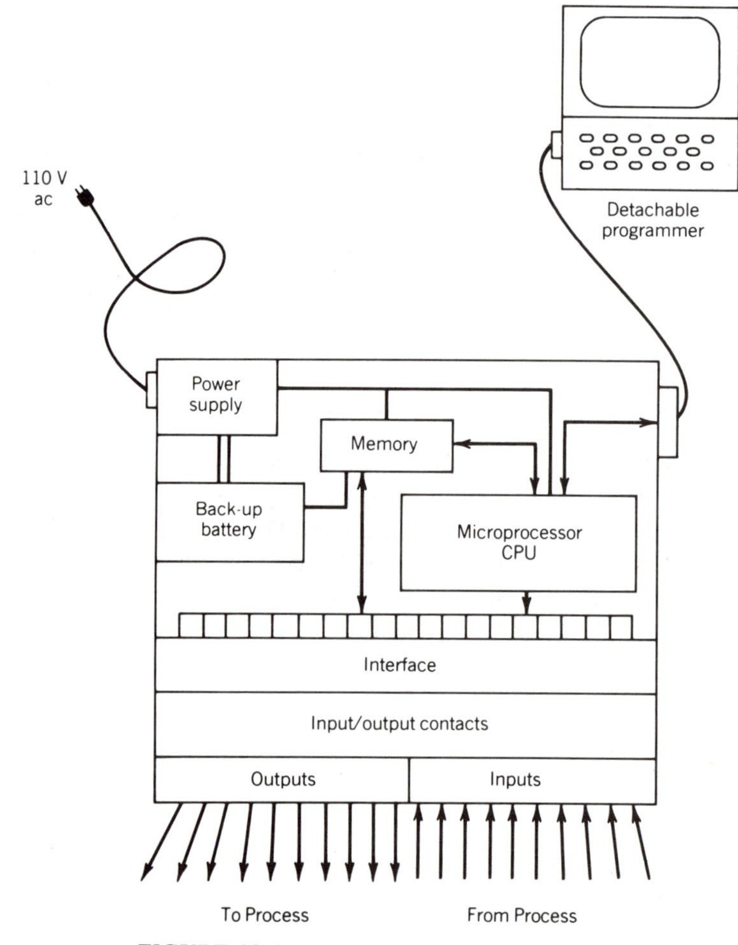

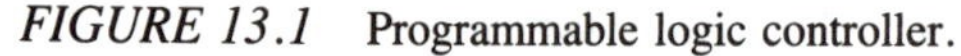

FIGURE 13.1 Programmable logic controller.

shown as part of the controller unit in Figure 13.1 can be expandable modules separate from the controller in most models.

Comparison of PLCs with Computers

The components of the main unit, such as memory and CPU (central processing unit), suggest that the PLC is just another computer. It is true that, like the personal computer, the programmable logic controller has a microprocessor type computer as its basic component. But it is best not to think of a programmable logic controller as a computer. Why? Almost everyone knows that a *computer* performs a complicated series of operations very rapidly. The individual steps in this series are executed in microseconds or even fractions of microseconds, but the complete program of steps in the series can require seconds, minutes, or even hours to complete, depending upon the complexity of the computer program. But such an operation would be out of the question for a typical *programmable logic controller*. The programmable logic controller's *complete program* is executed in a tiny fraction of a second. The individual steps are processed no faster than in conventional computers, so the program must be shorter and simpler. This is no problem because the basic purpose of the programmable logic controller is to make yes/no logic decisions, not to manipulate long series of complicated arithmetic computations upon mountains of data. The sequence of steps in the programmable logic controller is executed so quickly that it is typically appropriate to consider them as *simultaneous*. Thus, a programmable logic controller continuously and virtually simultaneously analyzes process input conditions, makes logic decisions, and dispatches outputs back to the process.

Another basic difference between conventional, data-processing computers and programmable logic controllers is in the management and presentation of programs for execution. A conventional computer loads a program upon command, executes it once or twice, loads another, executes it, and so on. A good part of the computer's time is occupied in compiling new programs presented to it by various users. Each day the computer is usually commanded to execute a different array of programs depending upon the needs of its users. In short, the conventional computer is a general-purpose device. By contrast, the programmable logic controller is a special-purpose device. It will typically execute its tiny program continuously *hundreds of millions* of times before being interrupted to introduce a new program. The programmable logic controller may be reprogrammed hourly, daily, upon setup of a new production lot or model year, or maybe only once in its entire lifetime! This difference from computers in mode of operation was perhaps obvious from Figure 13.1. Note that there are no printers, typewriters, tape readers, floppy disks, CRTs, or keyboards, and even the programmer is detachable. The programmer may have a keyboard or a CRT, but these features are not needed during the execution phase. Later in this chapter, we shall see that such peripheral equipment can be added to the programmable logic controller for data collection and reporting, but to add these features makes the unit a hybrid between a computer and a programmable logic controller.

PLC CYCLE

It was stated earlier that the sequence of program steps in the PLC is executed very quickly over and over. New executions may be triggered by the line cycle of the alternating current power supplied to the programmable logic controller. In these models a new cycle

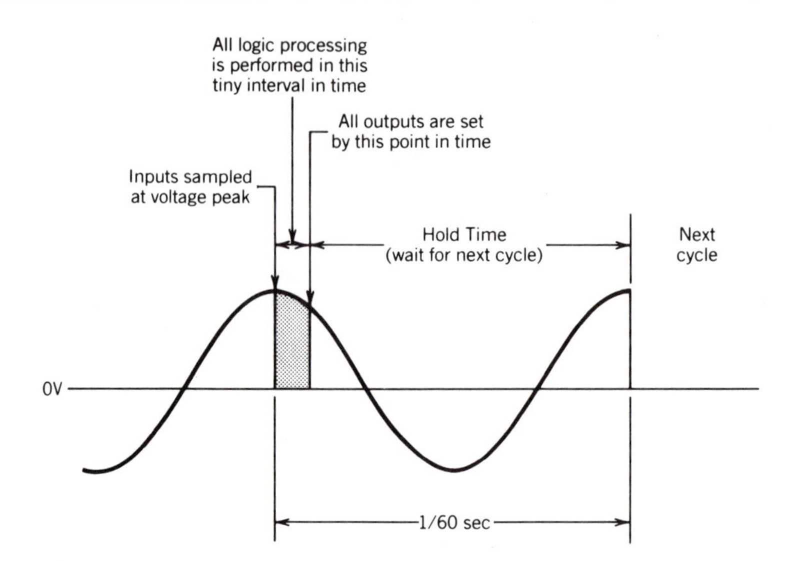

FIGURE 13.2 Process cycle for an example programmable logic controller. Note that the logic process time is a small fraction of the overall ac cycle.

begins each time the voltage peaks as shown in Figure 13.2. Note how short the logic program execution period is when compared to the overall line cycle. The program execution time period (the shaded area in Figure 13.2) was not dimensioned in Figure 13.2 because the actual time would vary dependent upon the number of ladder diagram rungs, the number of outputs, and the number of inputs. In other words, the length of the logic-execute cycle (the shaded area in Figure 13.2) is dependent upon the complexity of the automation application. The length of the overall cycle (the ac line cycle), however, remains constant at a closely controlled frequency, which in the United States is 60 cycles per second or 60 Hz. In other countries, the ac cycle may be different. Other PLCs may cycle at frequencies other than line frequency, but the point is that PLC cycle frequencies usually remain fixed.*

At this point, the reader should be wondering what might happen if a process input to the programmable logic controller should happen to change in that tiny slice of time represented by the PLC cycle. In this event, it is indeed possible to obtain an erroneous input, which is inconsistent with the programmed logic. In reality, however, this usually is not a problem. Remember that an opportunity exists to correct the logic a fraction of a second later in the next cycle through the programmed logic. Most real-world applications will never exhibit any discernible anomalies during execution of the program over millions of cycles. It is possible, however, to construct a ladder diagram in which the ladder rungs are highly interdependent and create an industrial logic control system that will result in obvious logic errors, even catastrophic errors. These situations can be avoided or their adverse effects minimized by being aware of their causes and by careful planning of the PLC program. PLC programming concepts will make this point clearer later in this chapter.

*Most modern PLCs cycle at frequencies much higher than 60 Hz, and frequencies may be program-size dependent.

Logic Control versus Sequencing

There are basically two approaches to setting up a programmable logic controller to control a series of steps in an automatic process. In the *logic control* approach, some process condition is sensed that signals the completion of a given step in the automatic process and also is used to trigger the execution of the succeeding step in the process. The other approach is the *sequencing* approach, which employs internal timers in the PLC to trigger the completion of one step and the start of its follower in the sequence.

The principal difference between the two approaches is closed-loop control (the logic control approach) versus open-loop control (the sequencing approach). The closed loop method has the advantage of assuring that the completion of previous steps actually is achieved before starting the next step in the sequence. The closed-loop approach also can be faster because the sequencing approach demands that a sufficient fixed time be allocated for each step, the actual execution of which may be variable. In other words, sufficient time must be allocated to accommodate the longest time expected to be required to complete that step. Thus, when the actual execution is faster than expected, some system idle time is inevitable with the sequencing approach.

With the obvious advantages of closed loop, it would seem that the sequencing approach would not be used. But the sequencing approach has the advantage of convenience and the even more important advantage of not requiring expensive sensors to detect the completion of each step. Some programmable logic controllers have special internal features such as simulated multiple cam timers that make them ideal for sequencing control. We will now examine some of these special features.

PLC INTERNAL FEATURES

The basic functions of a programmable logic controller are to scan a set of sensor inputs rapidly and repeatedly, evaluate their logical relationships to defined outputs according to a logic program based upon a ladder diagram, and set the outputs according to the programmed logic. The programmable logic controller can perform these functions without timers, counters, analog-to-digital converters, and other special devices. Even if these devices were needed for a given application, they could be external to the PLC and connected to the PLC's outputs and inputs. However, in actual practice many of these special features have been built into the PLC. The manufacturer of the PLC programs the software to accommodate these special features into the read-only memory of the PLC so that they are available to the user for whatever applications are desired.

One of these special features, available in virtually every programmable logic controller, is the timer. Timers can time up or down and can be cumulative or single usage. Since timers can be external to the PLC, they can be used in conventional relay-logic control systems too. As such, timers were discussed in Chapter 12, and all of the features of timers discussed in Chapter 12 are also available internally in programmable logic controllers. For brevity, timers will not be discussed in detail in this chapter. Closely related to the timer is the counter; in fact, a timer is a counter of clock pulses. Counters were introduced in Chapter 2 as independent units, but they are built-in capabilities in virtually all programmable logic controllers and will be explained in more detail here.

Counters

Production control, inspection, and material handling all involve the counting of discrete components and assemblies. When these operations can be done automatically, there are many benefits to be derived besides reduction in direct labor costs. Automatic counting is more accurate because operators usually have other important tasks to do at the same time and forget to count. Often the count is the basis for incentive pay, and when done manually, opportunities exist for incentive cheating or collusion in the cheating of other workers. It is naive to rely totally upon the honesty of all workers. Even relatively honest workers may rationalize that they have forgotten to count occasionally and justify adding to the count to avoid erring on the short side.

A distinction can be made between counting static stock as in warehouse inventory and counting active production stock as it moves through a process. A human inventory analyst can make use of volume computations, samples, and estimates in assessing an inventory count, with some advantage over automatic counting machines. Even then the human analyst usually will use partially automated computation aids. But counting of active production stock in motion happens to be one of those tasks that is easily automated and costly to relegate to a human job.

So far, we have discussed automatic counters that replace the human function of counting wherever it is appropriate and advantageous to do so. But the counters housed within programmable logic controllers as well as external-device solid-state counters are capable of counting voltage pulses at speeds much faster than humanly possible, opening up opportunities for automation that otherwise would have been impossible. Even the logic timers introduced in Chapter 12 are merely counters of clock pulses—that is, voltage pulses emitted by a pulse generator at precisely constant intervals in time. The extremely high-speed clocking and counting functions of solid-state electronics have made feasible the bar code reader, the encoder, and many other subsystems that have made industrial robots and manufacturing automation possible.

The counters in bar code readers and encoders are usually dedicated to that single function, and programmable logic controllers usually are not used for such specialized service. The typical PLC internal counter is used to count voltage pulses emitted from some external process sensor, and when a programmed limit count has been reached, an output signal is triggered. This output signal from the counter can be programmed then to switch directly on or off one of the PLC outputs that is transmitted back out to the process, or it can be used internally as a "control relay" that acts as an input contact on another rung of the ladder diagram. Some PLC counters can be set up to count up or down, with one input triggering a count up and the other a count down.

Figure 13.3 illustrates an up/down counter in the ladder diagram of a programmable logic controller; the counter happens to be set for a limit count of 4. In the figure, contact

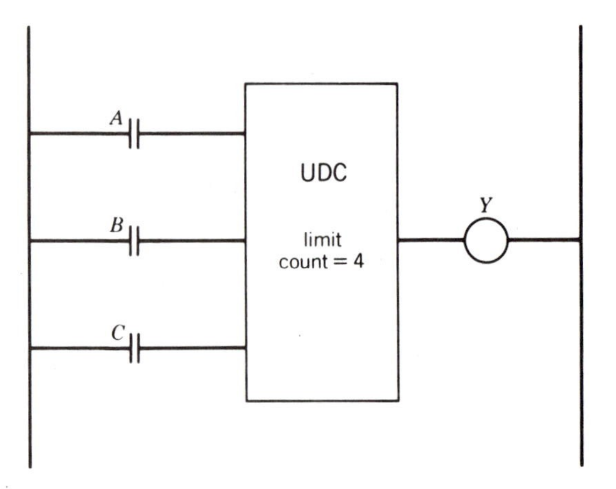

FIGURE 13.3 Up/down counter with limit count set at 4.

A represents the count-up input to the counter. Contact *B* represents the count-down input. Contact *C* resets the count to zero whenever it is *opened;* therefore it acts as an *enable* device. The counter will not function unless contact *C* is closed. Note that the up/down counter in Figure 13.3 has only one output, namely *Y*. This output might indicate when the up count reaches or exceeds a specific value and might also indicate when the total count is zero or negative, as a result of down counting or a reset operation in which the enable contact is opened. Other counters might have two outputs so that a count-down limit and a count-up limit can be distinguished.

Drum Timers

A familiar timer/sequencer, open-loop controller is the cam timer pictured in Figure 13.4. A rotating shaft acts as a clock that advances multiple cams arranged along the shaft, each tripping its respective limit switch according to its programmed setting. The entire assembly is often called a *drum timer*. The programming of these drum timers is simply a matter of obvious mechanical adjustments and the use of setscrews as a diagrammed in Figure 13.5. But even more quickly programmed are the built-in simulators for these mechanical drum timers within commercially available programmable logic controllers. No real mechanical drum or cam exists inside the PLC: the operation is totally solid state, with the mechanical timer being simulated by the software resident in the controller. The user merely accesses the software simulation by identifying:

1. Output switches to be tripped by each "cam."
2. The speed of rotation of the simulated shaft or drum (this can be made variable).
3. The output on/off configuration (mask) for each discrete step of the simulated drum rotation.

FIGURE 13.4 Multiple cam timer or "drum" timer. (Reprinted by permission of Eagle Signal Controls, Davenport, Iowa.)

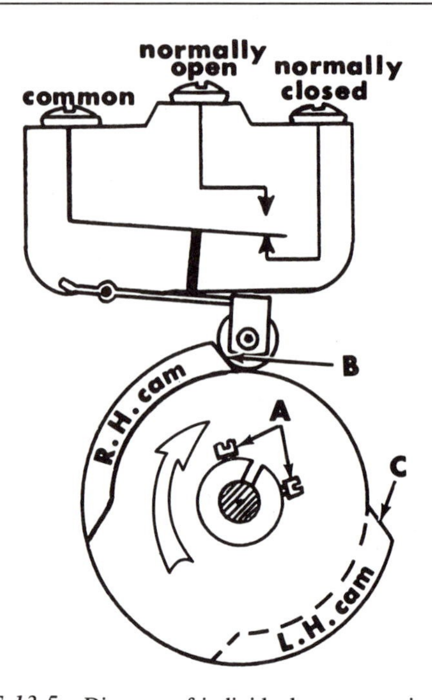

FIGURE 13.5 Diagram of individual cam operation on a drum timer. Setscrews *A* can be used to permit adjustment of cam point *B* ("turn-on" position for normally open contact) and cam point *C* ("turn-on" position for normally closed contact). (Reprinted by permission of Eagle Signal Controls, Davenport, Iowa.)

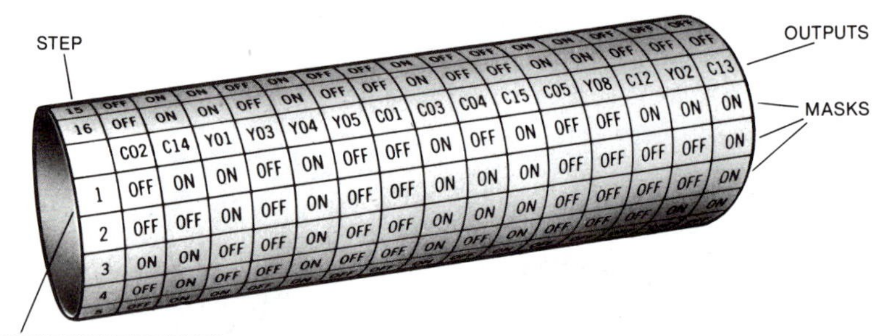

FIGURE 13.6 Physical organization of simulated drum timer in a programmable controller. (© 1983 Texas Instruments. Reprinted by permission.)

Even the discrete steps do not have to be of uniform length (in time), as the number of counts per step can be specified for each step. The "counts" have a fixed relationship to the internal PLC "scan rate" or PLC cycle. For instance, if the PLC cycle is set at the line frequency of 60 Hz, the fastest drum timer step possible would be 1/60 sec or 16.67 ms. To achieve this short an interval of time on the drum timer would require a "counts/step" specification of one. A specification of 60 counts/step would create a drum timer step of one-second duration, and a specification of 3600 counts/step would create a one-minute step. Thus, by varying the counts/step, it is possible to simulate variable-speed drum rotation.

The physical relationship between steps, counts/step, and output masks can be visualized more clearly by studying Figure 13.6. If the drum pictured were really a mechanical one instead of a simulation, the drum would rotate *away* from the viewer (counterclockwise when viewed from the left end of the drum). The identification of the outputs that the programmer desires to associate with each cam ring appears in the row on the drum that has a blank box on the left for "mask no." Each of the 16 steps around the drum represents a different mask for the 15 cam rings on the drum. The first step turns on outputs C14, Y1, Y4, C3, C15, C12, Y2, and C13. The program continues around the drum to the sixteenth step, during which outputs C14, Y1, Y4, C3, C5, and Y8 are turned on.

When planning for a drum timer application for a programmable logic controller, it is wise to first lay out a multiple activity chart describing the desired start and end points of each activity. Since the sequencing method of programming is the mode used with drum timers, the start times of each activity are known and are preplanned without the necessity of waiting for a closed-loop signal from the previous step of the process. After the start and end times of each operation are planned on the multiple activity chart, a matrix can be drawn up that will specify the masks for the drum timer. Every point in time at which one or more of the desired outputs change states (from on to off or off to on), a new step must be established at that point on the drum's revolution. Of course, several outputs can be specified to change either way at the same point in time. All of these points are illustrated in Case Study 13.1.

EXAMPLE 13.1 (CASE STUDY)
Automatic Machine Cycle under Drum Timer Control

The objective of this case study is to utilize a conventional machine tool in an automatic cycle under the control of a simulated drum timer in a programmable logic controller. The machine tool is loaded and unloaded by an industrial robot that operates on a preprogrammed two-second load cycle and a similar two-second unload cycle. Various activities transpire at scheduled sequence times, and several of these activities overlap. The entire operation is summarized by the multiple activity chart in Figure 13.7. The analyst selects a PLC output to energize each of the activities, and for identification in this case study these outputs are designated A through H. These outputs later become identified with the cam rings on the drum timer. To specify the drum timer steps and the output masks for each step, a matrix is used as illustrated in Table 13.1. It is entirely permissible to use fewer outputs than the cam ring capacity of the drum timer. It is also permissible to use fewer than the maximum allowable number of drum steps. Both of these points are illustrated in the matrix of Table 13.1, which specifies only eight outputs and seven (of a possible 10) steps. The unused steps 8, 9, and 10 are relegated to dummy status by the insertion of zero in the counts/step column. In effect, zero counts per step cause the step to be processed instantly and hence to be ignored. The seven necessary steps are determined by examining the start and end point of every activity listed in the multiple activity chart in Figure 13.7. Every time something happens—that is, one or more activities either begin or end—a new step must be identified.

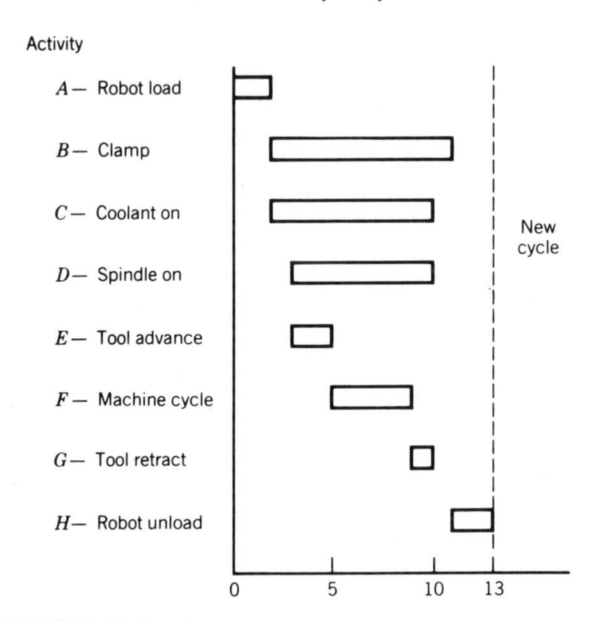

FIGURE 13.7 Multiple activity chart for Case Study 13.1.

TABLE 13.1 Drum Timer Matrix for Case Study 13.1

Drum Number 1 (Automatic Machine Cycle) SCANS/COUNT = 60									
		Outputs							
Step	*Counts/Step*	*A*	*B*	*C*	*D*	*E*	*F*	*G*	*H*
1	2	1	0	0	0	0	0	0	0
2	1	0	1	1	0	0	0	0	0
3	2	0	1	1	1	1	0	0	0
4	4	0	1	1	1	0	1	0	0
5	1	0	1	1	1	0	0	1	0
6	1	0	1	0	0	0	0	0	0
7	2	0	0	0	0	0	0	0	1
8	0	0	0	0	0	0	0	0	0
9	0	0	0	0	0	0	0	0	0
10	0	0	0	0	0	0	0	0	0

The scans/count specification for Case Study 13.1 is 60, which suggests a 60-Hz PLC cycle frequency and a one-second time interval represented by each count. This is a very convenient selection for Case Study 13.1 because the activity end points are specified in whole seconds. Thus, the counts/step specifications in Table 13.1 represent whole seconds. The output masks are displayed using 0s and 1s in the conventional logic sense: 0 represents "off," and 1 represents "on." Note that the transition from step to step is continuous, permitting outputs to remain on from step to step. A good example of this is output *B*, which remains on throughout steps 2 through 6 regardless of the changes taking place in the other outputs.

The preceeding discussion of Case Study 13.1 and the study of counters, timers, and drum timers should be awakening the reader's awareness of the immense possibilities for programmable logic controllers for application to the field of robots and manufacturing automation. We now turn to the subject of PLC programming and learn how easy it is to turn these possibilities into realities.

PLC PROGRAMMING

PLC programming methods vary from manufacturer to manufacturer, but the basic ladder logic approach appears to be standard throughout the industry. For the most part, then, the iterative cycle in the programmable logic controller is ignored by the analyst, who visualizes the logic system as a simultaneous execution of the rungs of the ladder.

Most PLC manufacturers provide a graphic display on a CRT screen that appears as a ladder diagram. This ladder diagram takes form while the programmer builds it up using the keyboard. Figure 13.8 displays a popular model PLC programmer keyboard and CRT with a real ladder diagram displayed on the screen. Note that the keys themselves have symbols such as ⊣⊢ and ⊣/⊢, which are interpreted exactly as explained in Chapter 12.

The popularity of personal computers has caused a movement among PLC manufacturers to furnish software to permit off-line programming of ladder logic on a personal computer for subsequent down-loading onto the PLC. Figure 13.9 provides a sample printout using the Texas Instruments software package TISOFT. Many readers erroneously assume that the principal advantage of programming ladder logic off-line on a personal

FIGURE 13.8 PC programmer keyboard and CRT display. (Reprinted by permission of Gould Inc., Programmable Control Division, Andover, Mass.)

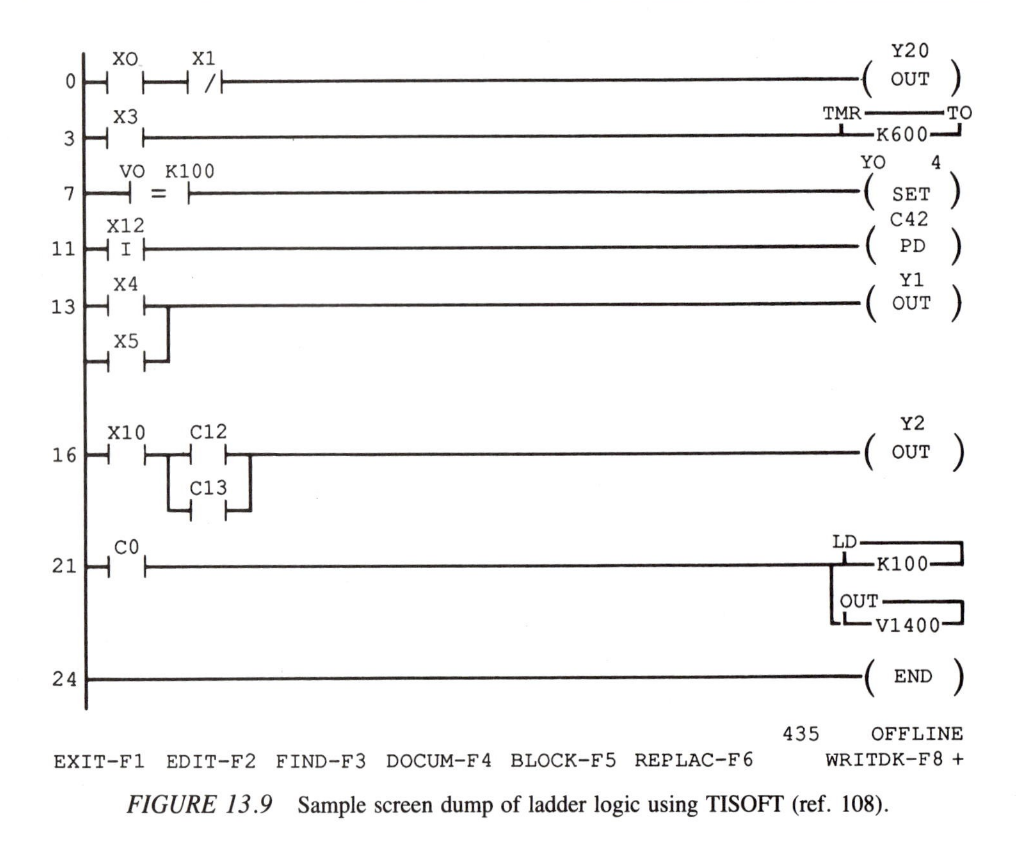

FIGURE 13.9 Sample screen dump of ladder logic using TISOFT (ref. 108).

computer is speed and convenience. However, the real advantage is documentation and, perhaps, more complete access to comprehensive diagnostics. For sheer speed of entering a simple program into a PLC, the simple, detachable programmers to be described in the next section may be faster than off-line programming. But for long and complicated programs, detailed documentation is essential.

At the other end of the PLC programmer spectrum is a small, detachable programmer, as illustrated in Figure 13.10. This programmer uses keys with two- or three-letter abbreviations to write programs that bear some resemblance to computer coding. A display at the top of the programmer exhibits the PLC instruction located in working memory. Error messages are also displayed here.

On the face of the programmer in Figure 13.10 can be seen keys that correspond to the familiar logical operators discussed in Chapter 11. These keys are used to build up the logic of a given rung of the ladder diagram. If the user is familiar with the logic operations of AND, OR, and NOT, programming with the detachable programmer is very quick, especially for small programs. The beginning of the ladder rung is signaled by the STR key, which can be interpreted as either START of STORE. This key is followed by a SHF key (representing "SHIFT"), so that a numeric entry can be made, and then the numeric name of the first input contact on the ladder rung. The controller has designated input and output variables that are hardwired to terminals for connection to external devices. Besides these designated contacts, the user can select internal "control relays" that have no electrical connection to the outside world but are useful in describing the logic relationships within the ladder diagram. The typical ladder rung is ended by an

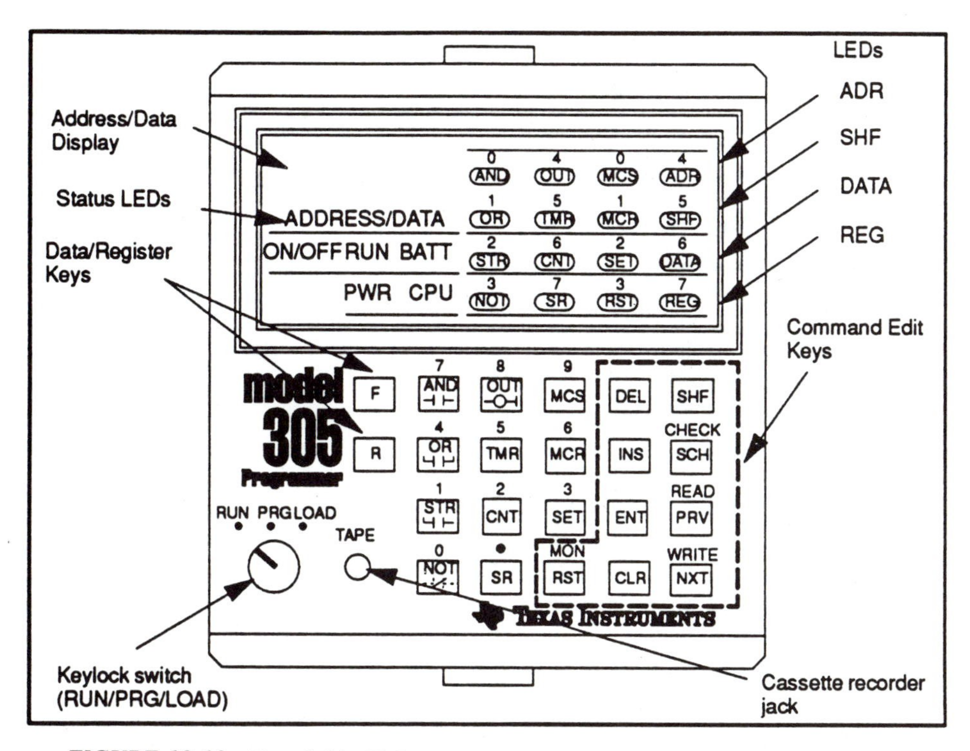

FIGURE 13.10 Detachable PLC programmer. (Copyright © 1989 Texas Instruments. Reprinted by permission.)

OUT command followed by SHF and an appropriate number representing an output variable.

The Texas Instruments TI-305 Series programmable logic controller is the system represented by the programmer in Figure 13.10. Since this controller is a popular, low-cost system suitable for college and university laboratories, its programming scheme will be exhibited in this text to provide an example of PLC programming in general. Figure 13.11 displays the controller with programmer detached. The terminal strip on the top of the PLC is for connection of inputs.

A similar terminal strip on the bottom edge of the front of the PLC is for PLC outputs and also for the ac power connection. The basic TI-315 accommodates 15 inputs and nine outputs labeled as can be seen on the face of the controller in Figure 13.11. Expansion slots are available for additional I/O. Characteristics of representative small, medium, and large programmable logic controllers are shown in the table in Appendix A.

Now to see how easily and conveniently the PLC can be programmed, let us examine some simple example logic systems. Figures 13.12 through 13.15 display ladder diagrams with their corresponding PLC programs. Each line of these programs is entered into memory by means of the ENT key after coding that line. The first two examples (Figures 13.12 and 13.13) illustrate the most basic logic circuits. Any input contact can be inverted to a NOT condition simply by inserting the NOT key just prior to keying in the name of the input contact. This is illustrated in an assortment of input contacts in both Figures 13.14 and 13.15. Note in particular the use of OR STR and AND STR to connect *assemblies* or *networks* of logic contacts logically. Every time a new assembly of input contacts is begun, the STR key is used. It is even possible to build *nests* of networks by

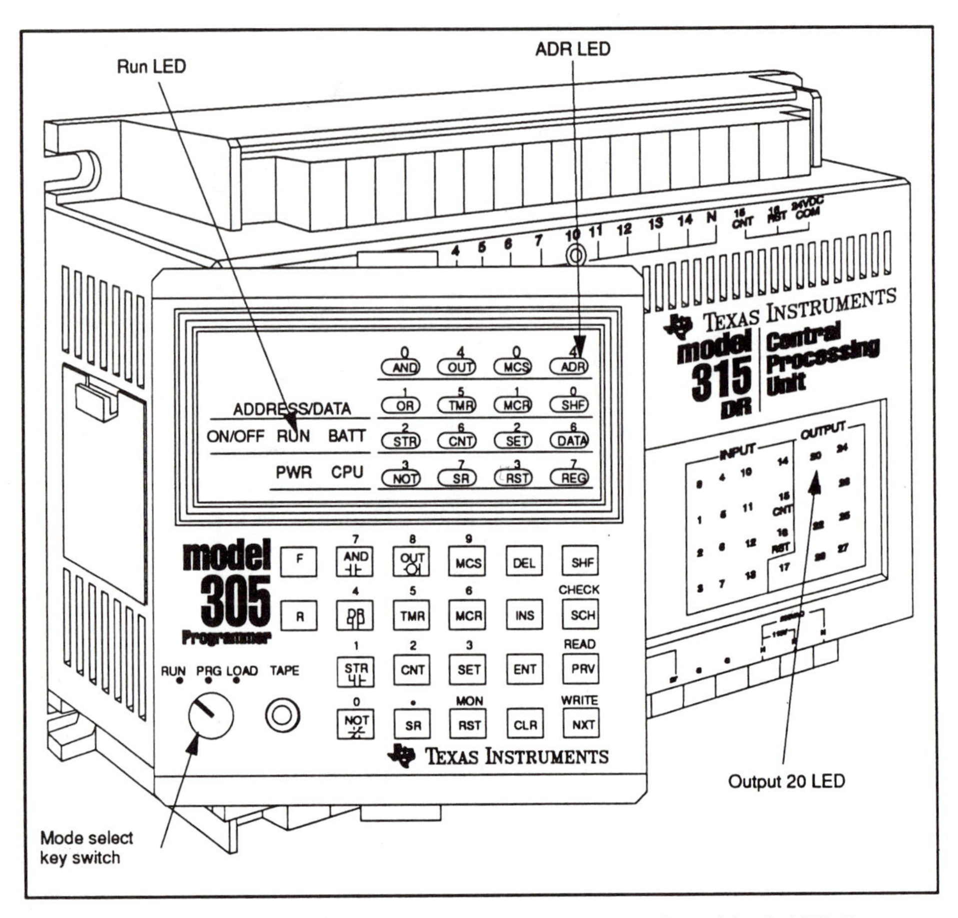

FIGURE 13.11 TI-315 PLC with detachable programmer. (Copyright © 1989 Texas Instruments. Reprinted by permission.)

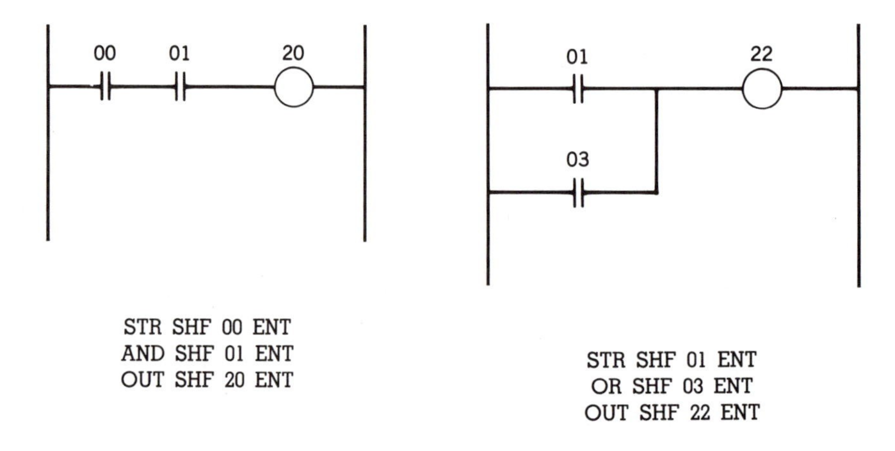

STR SHF 00 ENT
AND SHF 01 ENT
OUT SHF 20 ENT

FIGURE 13.12 PLC program for simple AND circuit.

STR SHF 01 ENT
OR SHF 03 ENT
OUT SHF 22 ENT

FIGURE 13.13 PLC program for simple OR circuit.

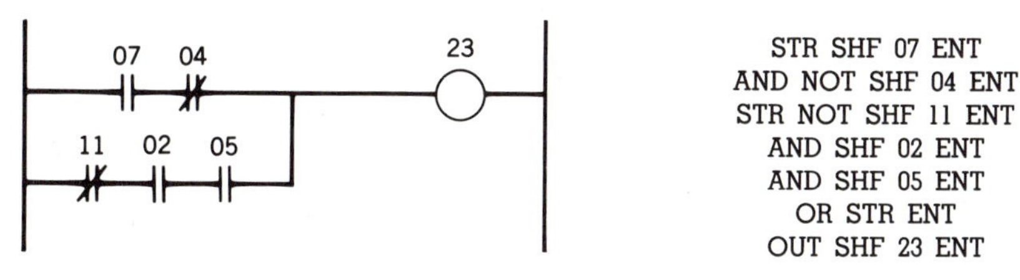

FIGURE 13.14 Linking two ladder rung series in parallel to one output using OR STR.

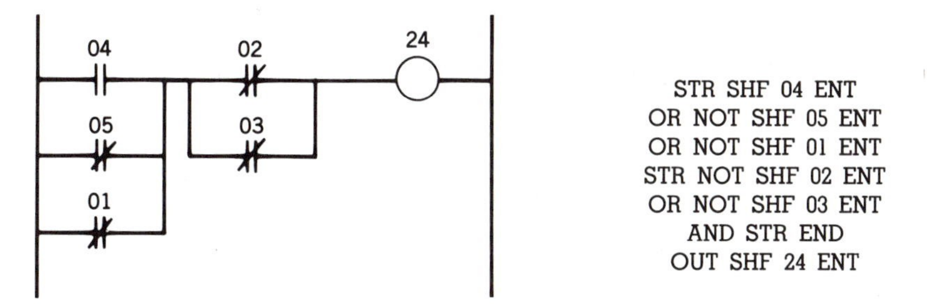

FIGURE 13.15 Linking two parallel networks in series to one output using AND STR.

using OR STR and/or AND STR successively. Figure 13.16 illustrates such a nest. In this example, the OR STR links the 02 and 03 branch to the 04 and 05 branch; then the AND STR links this entire assembly to the 00 or 01 branch.

In the examples used thus far in this chapter, only PLC inputs have been used as ladder input contacts. However, it is possible to use the outputs of other ladder rungs as inputs in a ladder rung. This is illustrated in Figure 13.17, which can be recognized as the simple push-button switch example discussed in Chapters 11 and 12. Can you identify which input contact has been selected as the ON push button and which has been selected as OFF? Since the selection of inputs and likewise the selection of outputs are arbitrary, it is possible for a technician to *prewire* the system to any available inputs and outputs as desired and for the PLC program to be set up later by another technician or engineer to accommodate whatever wiring selection the technician has made.

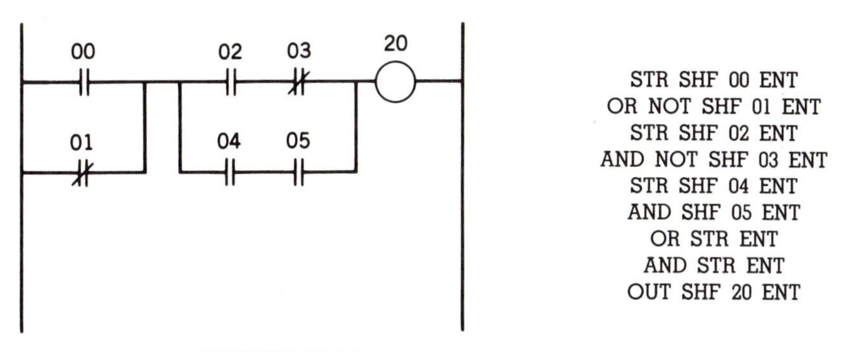

FIGURE 13.16 Building a nest using OR STR and AND STR.

The logic pattern seen in Figure 13.17 is so frequently encountered that it has a mechanical equivalent. The reader should be able to figure out that this is the logic represented by the "latching relay" discussed in Chapter 2. "Latching relays" have widespread application in manufacturing automation and thus programmable logic controller manufacturers have provided the software to simulate latching relays in construction of the ladder logic. The convention is to use the mnemonic "SET" in place of "OUT" when specifying the output of a given ladder rung. Then, even if the rung logic becomes logic 0 subsequent to the setting of this output, the output remains on until it has later been "RST" (reset) by another rung in the ladder logic. The convention is illustrated in Figure 13.18, for which the logic is identical to the arrangement depicted in Figure 13.17.

If the output to a rung on the ladder diagram has no physical significance to some external device, an output Y terminal need not be connected. Thus, any available internal control relays can be used to represent a ladder rung output. These control relays can then be used as input contacts on other or even the same ladder rung(s) of the diagram.

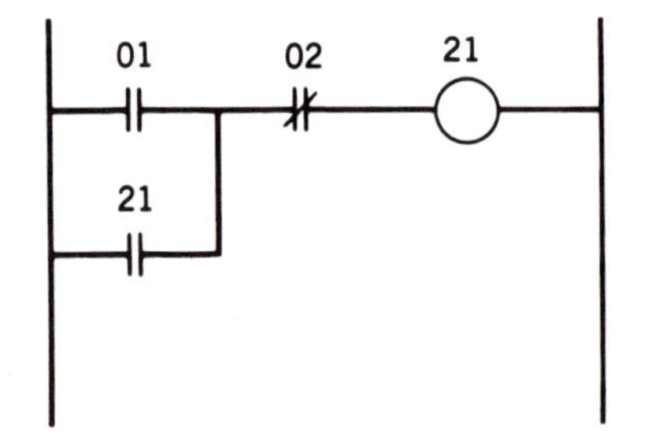

STR SHF 01 ENT
OR SHF 21 ENT
AND NOT SHF 02
OUT SHF 21 ENT

FIGURE 13.17 Use of an output as an input contact; simple push-button switch example.

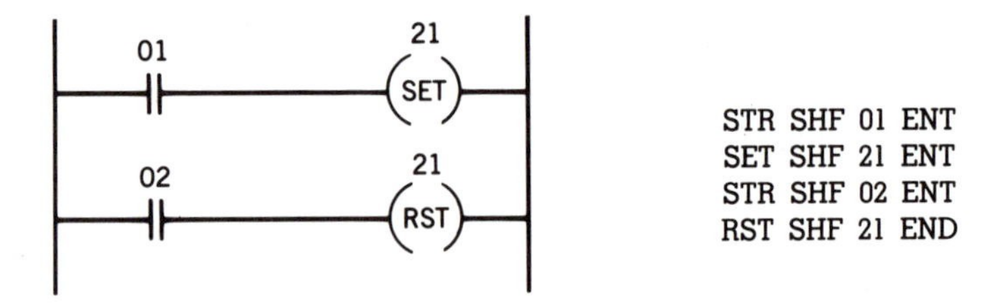

FIGURE 13.18 Programming of a "latching relay" using SET and RST. The logic is the same as in Figure 13.17.

Indeed, when control relays are used as "outputs" on a ladder diagram, their *only* purpose is to be employed as input contacts elsewhere in the ladder diagram. Control relays are useful in triggering next steps in an industrial logic sequence after verifying completion of the previous step in the sequence using the "logic control approach" discussed earlier in this chapter.

A timer is set up in one rung of the ladder diagram and then referenced in other rungs to turn on or turn off outputs. A sample diagram with associated programming steps is shown in Figure 13.19. Note that this timer has a single input and is not cumulative as was the two-input timer described in Chapter 12. Different controller manufacturers set up the software for timers in different ways, but by studying the documentation the user can usually achieve the same purposes with different controllers. The controller programmed in Figure 13.19 has a scan time of 0.1 seconds. The timer count or "PRESET" is 15, so the timer will energize 1.5 seconds after input 01 becomes logic 1. If input 01 becomes logic 0 again, the timer is reset back to 0 and is ready to begin the timing sequence again whenever input 01 becomes logic 1.

A counter is coded in a fashion similar to that used for the timer. The up-counter shown in Figure 13.20 has two inputs, the first to register counts whenever the input makes the transition from logic 0 to logic 1, and the second to act as an enable/reset. When input 02 is a logic 1, the counter is reset; when 02 is a logic 0, the counter is enabled. While the enable/reset is a logic 1, the count input 01 is ignored by the counter. When the counter reaches its PRESET limit of 4, output 22 is energized. The counter can also be monitored and used to turn on outputs at intermediate points in the count, as is indicated in the subsequent rung of the ladder diagram. The result is that output 23 is turned on when counter 601's count reaches 3. The logic of Figure 13.20 is illustrated

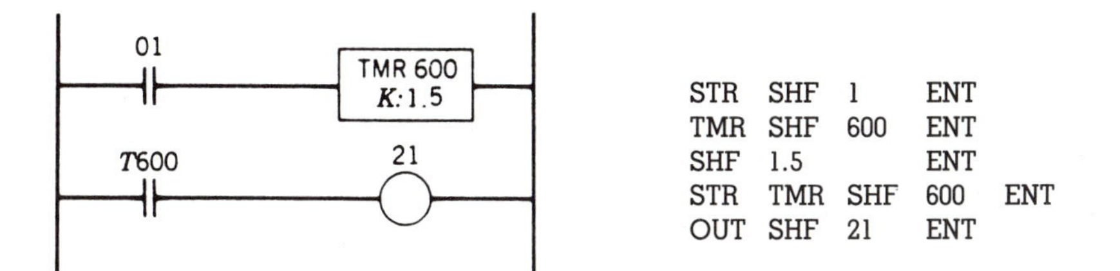

FIGURE 13.19 Programming the controller for a time interval of 1.5 seconds.

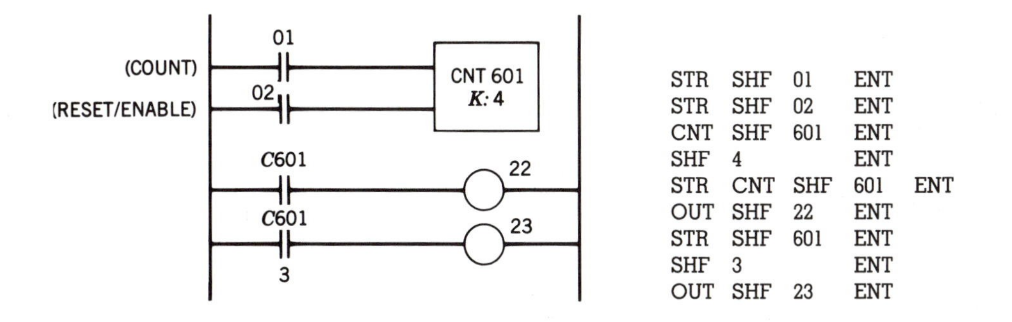

FIGURE 13.20 Programming the controller for a count of 4. Output 22 turns on when the count reaches the limit of 4. Output 23 turns on when the count reaches 3.

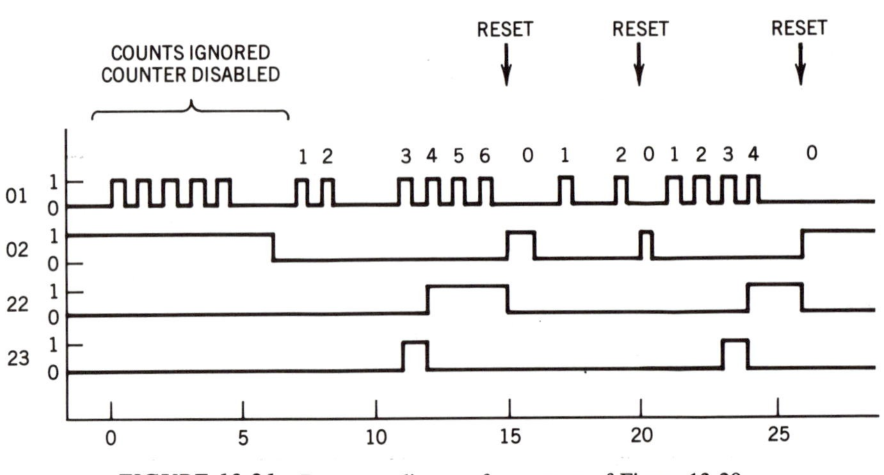

FIGURE 13.21 Response diagram for counter of Figure 13.20.

by the response diagram in Figure 13.21. The reader can perhaps now visualize a method using a combination of timers and counters to simulate the drum timer discussed earlier in this chapter.

Although coding a complete drum timer is necessarily detailed, in general the coding of a programmable logic controller is a simple task. Details of editing and testing programs have not been covered here and would generally be machine specific. Many PLCs have sophisticated features for handling feedback of analog process variables, but only the basics of PLC programming have been covered here. With an understanding of even only the basic fundamentals of PLC programming contained in this section, the reader should now be ready to tackle the construction of PLC programs to handle manufacturing automation applications or even to control an industrial robot.

APPLICATION PROGRAMS

The PLC programming examples in the previous section were instructive, but their programs had no real-world function. Let us now exercise our PLC programming tools on some real applications in industrial automation.

EXAMPLE 13.2
Automatic Assembly Line Conveyor Speed Control

Suppose we want to use a programmable logic controller to handle the control circuitry to require an assembly line to be started in low speed before permitting operation in high speed. The following inputs and outputs are defined:

11 = momentary spring push button to signal to start the line in low speed

00 = momentary spring push button to signal to switch the line to high speed

01 = momentary spring push button to stop the line

20 = low speed

21 = high speed

The ladder diagram and PLC program to accomplish this function are illustrated in Figure 13.22.

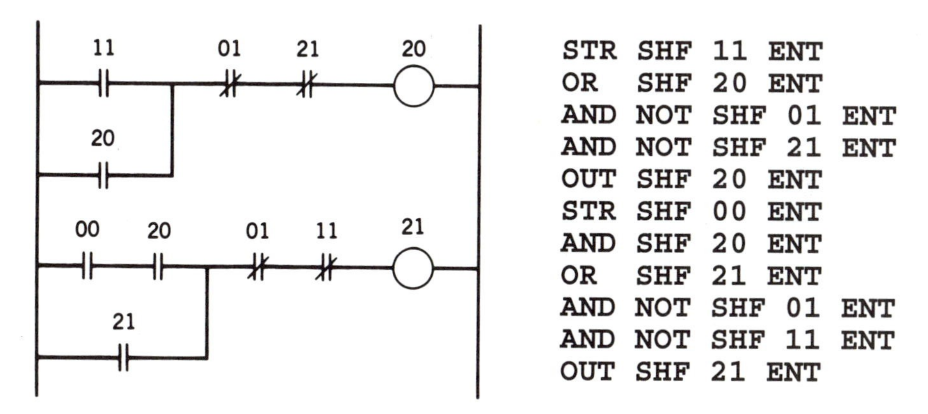

FIGURE 13.22 PLC program for controlling assembly line conveyor speed (Example 13.2).

EXAMPLE 13.3
Automatic Packing Control

A small, bottled pharmaceutical is packed in 12-bottle cartons in a 144-bottle gross. The automatic packing machine receives a signal (for 3 sec) from a production counter every time 12 bottles have come down the automatic conveyor and an additional signal (5 sec) when the 144-bottle gross has been completed.

The following inputs and outputs are identified for PLC programming:

00 = system power on
15 = production count sensor
20 = carton packaging signal (3 sec)
21 = gross packaging signal (5 sec)

The ladder diagram and PLC program for this case study appear in Figure 13.23 on page 370. At first glance, Figure 13.23 may look somewhat formidable, but do not be intimidated by it. It is typical for real-world applications to have long and complicated ladder diagrams. But usually these diagrams can be analyzed and broken down into chewable bites. Note that the *form* of the bottom half of Figure 13.23 is a carbon copy of the top half. Only the variable names and numbers have been changed. Also note that each rung is of a simple type studied earlier in this text.

Observe that in the numbering of the counters and timers that no numbers are repeated. Some PLCs use the same memory allocations for both timers and counters, and it is thus inappropriate to refer to a timer and a counter by the same number.

CONTROL OF INDUSTRIAL ROBOTS

At the beginning of this chapter, it was stated that the relationship between PLCs and industrial robots is an interesting one. There are two types of PLC control over robots: supervisory and direct control. We shall examine both.

Supervisory Robot Control

Programmable logic controllers can be used to control an overall automated manufacturing process, of which the industrial robot is one part. Thus, the robot is signaled by the PLC

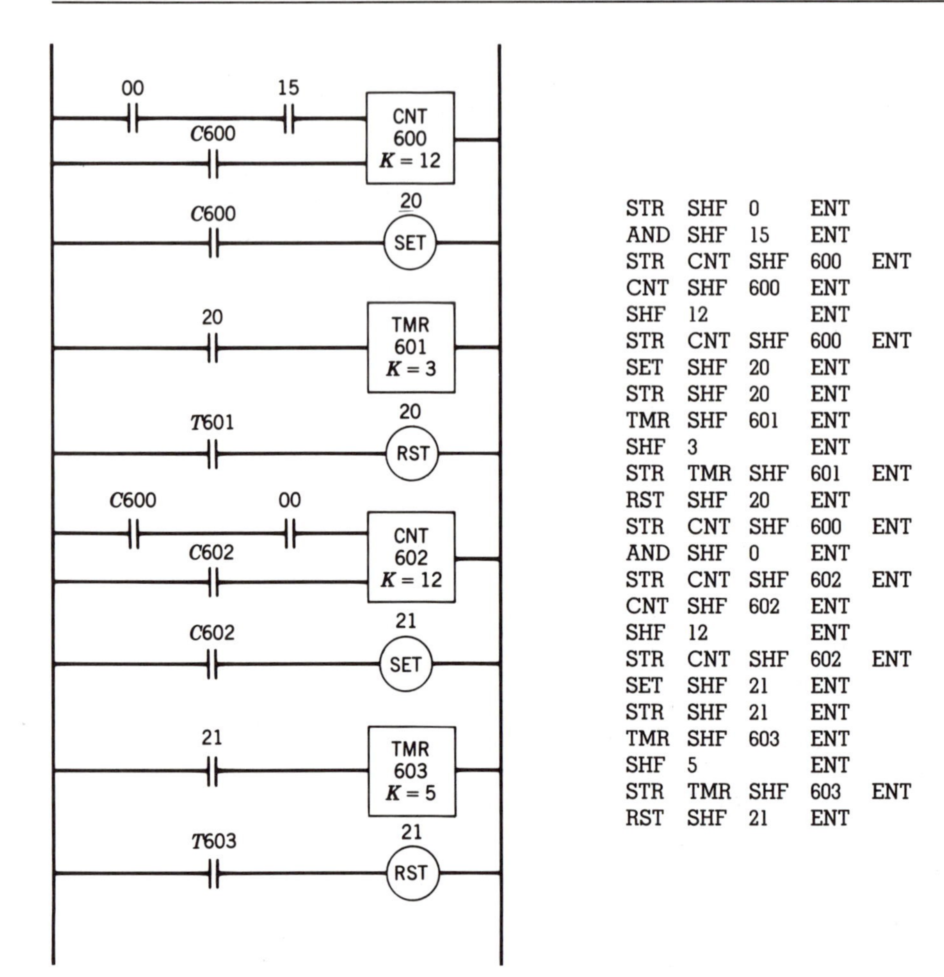

FIGURE 13.23 Ladder diagram and PLC code for Example 13.3, Automatic Packing Control.

to perform the assigned robot *sequence* of operations at the appropriate time. The PLC also may monitor various functions, interactions with process equipment, and abnormal or alarm conditions and take action accordingly. In the supervisory mode of PLC control over industrial robots, it should be remembered that in addition the robot has its own special-purpose controller, which is programmed to give the robot the desired robot motions. But the PLC has a broader role than the robot's own controller, as can be seen in the diagram of Figure 13.24.

Direct Robot Control

Not only are programmable logic controllers useful in linking and synchronizing robots with the manufacturing processes they serve (supervisory control), a PLC actually can serve as the robot controller itself. A look inside the control panel of a commercially available robot will often reveal that a standard programmable logic controller (see Figure 13.25) has been used as the robot controller. PLCs are ideal for the simple axis-limit varieties of robots and are seen most often used with pneumatic pick-and-place robots. In fact, with the knowledge you already have from Chapters 2, 7, 11, 12, and this chapter,

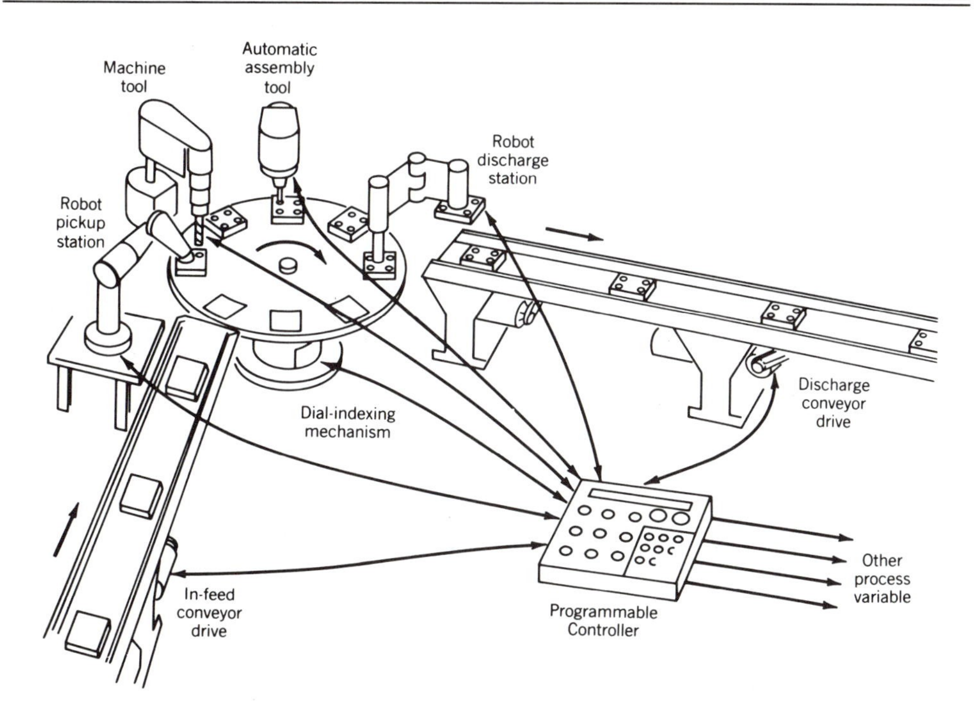

FIGURE 13.24 Supervisory PLC control of an automatic process that utilizes industrial robots.

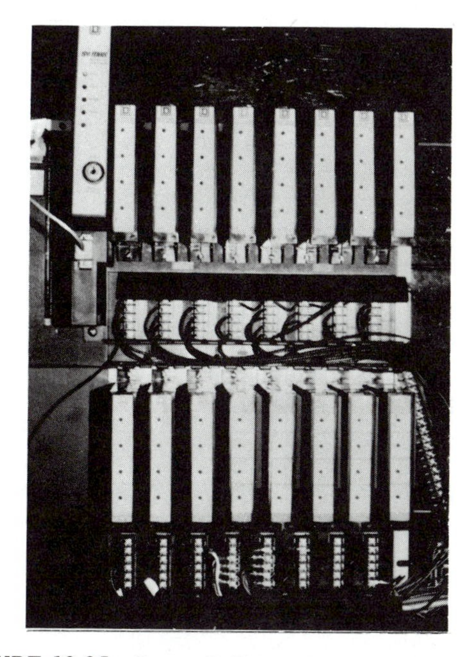

FIGURE 13.25 Square D SY-MAX programmable logic controller employed to control an industrial robot.

you should be able to *build your own* pick-and-place robot using a PLC, standard pneumatic cylinders, valves, and limit switches. Two examples of pick-and-place robots will now be illustrated, one using the *sequencing control* approach and the other using the *logic control* approach; both approaches were introduced earlier in this chapter.

EXAMPLE 13.4 (CASE STUDY) PLC Sequence Control of a Pick-and-Place Robot

Consider the simple robot application pictured in Figure 13.26. The knowledgeable professional will recognize this robot as a typical cylindrical configuration most likely having axis-limit motion control. The diagram in Figure 13.26 displays three robot axes of motion. We shall add a fourth, gripper closure, to complete the axes of motion necessary for this case study. Set up such a robot for control by a programmable logic controller.

The first step would be to identify each robot motion with a unique PLC output so that the PLC could signal the robot to move in that axis whenever desired. We will assume in this case study that the robot will hold its position in each axis unless commanded to change position. Therefore, we need two outputs for each of the four axes of motion because we need to be able to command the robot to travel *both* ways in each axis. Thus, to control this robot we need $2 \times 4 = 8$ output circuits. We shall choose the following identifications for our PLC outputs in this case study.

20	Rotate base counterclockwise	24	Extend arm
21	Rotate base clockwise	25	Retract arm
22	Lift arm	26	Grasp (close gripper)
23	Lower arm	27	Release (open gripper)

Notice that we have used our Chapter 7 convention of beginning with the base and working outward toward the gripper in identifying our robot axes. Also note that we have grouped our outputs by robot axis of motion and that the odd-numbered outputs represent one orientation of motion and the even-numbered outputs represent the opposite orientation. All of this selection has been arbitrary,

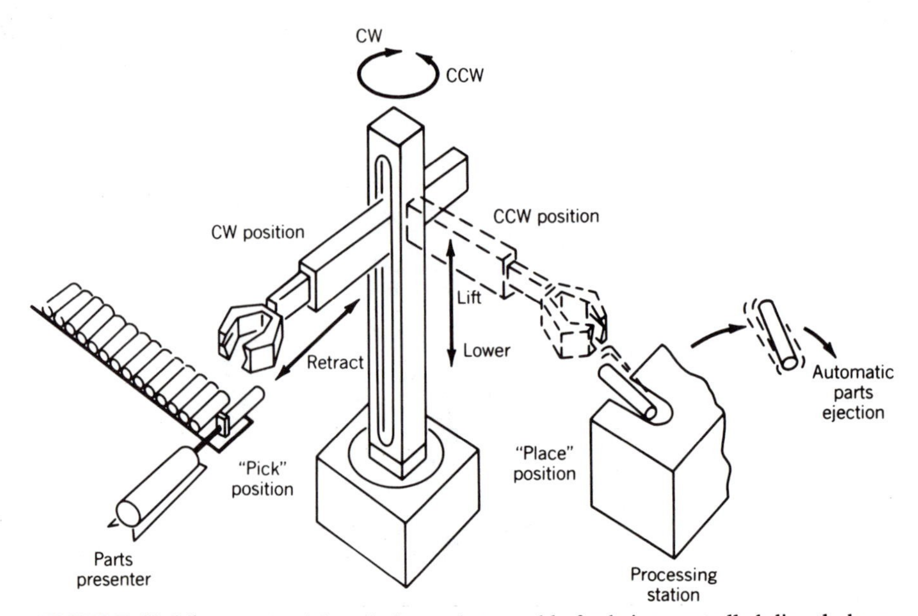

FIGURE 13.26 Simple pick-and-place robot capable for being controlled directly by a programmable logic controller.

but some logical scheme for identification has been used to help us remember which output belongs to which motion.

In this case study, we demonstrate sequence control, which was defined earlier in this chapter as requiring no feedback to determine that each desired action in the sequence was actually performed satisfactorily. All we need to do in sequencing control is turn on the selected outputs at the appropriate time and leave them on long enough to allow the robot to complete its axis motion and encounter a mechanical stop, or perhaps even a lock, at the end of its stroke in that axis. A PLC drum timer is ideal for this purpose.

As we did in our first example application for a drum timer, in Case Study 13.1, we will again construct a multiple activity chart or "cycle layout." To do this, we need to move the robot of Figure 13.26 mentally or physically through the desired sequence of motions. Since the operation is open-loop, we need to allocate an appropriate time period for each axis to make its move to its mechanical stop. It may be appropriate to vary the times since some robot axes may be faster than others, but it makes sense to keep times constant for identical motions. Usually, there is some uncertainty concerning how much time each axis will need to make its move because times can vary due to pneumatic pressure variations, resistance, payload variations, and other factors. Therefore, it is wise to allow a little margin of extra time for each axis to be sure a given operation is completed before the next operation in the sequence begins. It is entirely possible for axis motions to *overlap* in a robot sequence, but the applications engineers or designers must decide at what points in the robot sequence such overlaps will be permitted. In this example, we shall specify that no overlaps will be permitted.

To lay out the actual sequence, we also need to decide what point in the cycle we shall call our cycle start. This is really an arbitrary choice, but with whatever we decide we should be

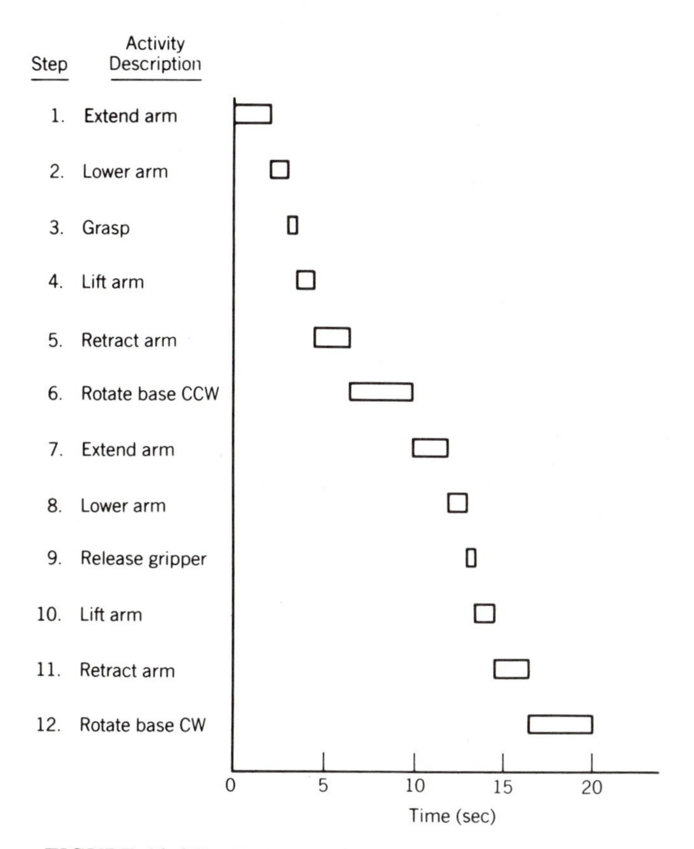

FIGURE 13.27 Multiple activity chart for Case Study 13.4.

TABLE 13.2 PLC Drum Timer Matrix for Pick-and-Place Robot of Case Study 13.4

	Counts/	*Outputs*								*Count Frequency*
Step	*Step*	*20*	*21*	*22*	*23*	*24*	*25*	*26*	*27*	*2 Counts/Sec*
1	4	0	0	0	0	1	0	0	0	
2	2	0	0	0	1	0	0	0	0	
3	1	0	0	0	0	0	0	1	0	
4	2	0	0	1	0	0	0	0	0	
5	4	0	0	0	0	0	1	0	0	
6	7	1	0	0	0	0	0	0	0	
7	4	0	0	0	0	1	0	0	0	
8	2	0	0	0	1	0	0	0	0	
9	1	0	0	0	0	0	0	0	1	
10	2	0	0	1	0	0	0	0	0	
11	4	0	0	0	0	0	1	0	0	
12	7	0	1	0	0	0	0	0	0	

consistent. A logical place to start seems to be when the robot is positioned with its base rotated clockwise, poised in a lifted, retracted position, ready to pick up a new part, somewhat as pictured in Figure 13.26.

Figure 13.27 displays the cycle layout with times allocated for each robot motion in the sequence. The base-rotate operations are moving the most mass, so the longest times are allocated for these motions due to inertia and due to distance moved. The quickest steps are for opening and closing the gripper.

To set up the PLC drum timer to accomplish the robot sequence, we need to select the count frequency. The smallest step times in Figure 13.27 were the $^1/_2$-second gripper operations; therefore, a count frequency of 2 counts per second should give the desired precision. The entire PLC drum matrix is laid out in Table 13.2. The ladder diagram and PLC program can be found in Appendix B.

EXAMPLE 13.5 (CASE STUDY)
PLC Logic Control of a Pick-and-Place Robot

This case study takes the same problem as in the previous case study but uses the logic control approach instead of the sequencing approach. Whereas the previous case study used a PLC drum timer and was open-loop, this case employs sensors to provide *feedback* at each step of the robot's motion. Mechanical limit switches or proximity switches can be used to sense the point of completion of each robot motion and provide input to the programmable logic controller. The PLC can then be programmed not to begin the next step until confirmation has been received that the predecessor step has been completed, regardless of how long this might take. Therefore, in this case study we not only need to identify *output* variables 20 through 27 as was done in the earlier study, but we also need to identify *input* variables associated with each output to represent sensed completions of each robot motion. Our arbitrary convention for this will be to name 00 as the input from the limit switch that signals completion of "base rotate counterclockwise" (output 20); likewise input 01 will be associated with output 21, 02 with 22, and so on, and finally 07 with 27. The entire scheme along with necessary prior inputs for each robot motion are tabulated in Table 13.3. Note in Table 13.3 that the same sequence of robot motions was used that was used in Case Study 13.4. However, this time there is no drum timer and no timing of the outputs. Instead, each output is triggered by key inputs completed in previous steps.

TABLE 13.3 Cycle Sequence for Pick-and-Place Robot Under Direct Logic Control of a Programmable Logic Controller (Case Study 13.5)

Robot Motion	*PLC Output*	*Limit Travel (PLC Input)*	*Necessary Prior Conditions (PLC Inputs)*
Extend arm	24	04	01, 07
Lower arm	23	03	04, 01, 07
Grasp	26	06	03, 01
Lift arm	22	02	06, 01
Retract arm	25	05	02, 06, 01
Rotate base CCW	20	00	05, 06
Extend arm	24	04	00, 06
Lower arm	23	03	04, 00, 06
Release gripper	27	07	03, 00
Lift arm	22	02	07, 00, 04
Retract arm	25	05	02, 00, 07
Rotate CW	21	01	05, 00, 07

In this case study, it might seem at first that it is necessary only to input the completion of the step just prior to a given step to trigger the execution of that step. Such a scheme would work if every robot motion were used *only once* in the robot motion sequence. But for this case study each ladder rung in the ladder logic diagram must contain additional logic inputs to distinguish that robot step from other similar steps in the sequence. For instance, the second step in Table 13.3 is "lower arm," for which completion is indicated by 03. If 03 were to be used as the sole logic input to trigger the next step in the sequence, "grasp," the robot would *always* grasp after it executes "lower arm." However, in the eighth step of the program we also "lower arm," but we do *not* want to "grasp" after we "lower arm." So it is necessary to provide other essential inputs in the logic to be sure that the correct steps and *only* the correct steps are taken at each point in the sequence. This is why additional inputs are used to describe "necessary prior input conditions" for each step of Table 13.3.

It bears repeating here that all of the rungs on the ladder diagram are executed essentially *simultaneously*. Therefore it makes no difference in what sequence we enter the ladder rungs into the programmable logic controller, at least not for this example. Figure 13.28 on page 376 displays the ladder diagram corresponding to Table 13.3. Input 10 is used to represent a power-ON switch. With the inclusion of input 10 as a necessary condition on every ladder rung, the ladder diagram follows directly from the table. Note how the ladder rungs are *not* sequential as were the lines in Table 13.3. In fact, logic ORs (parallel rungs) were used to provide alternative logic paths to outputs that are repeated in the robot sequence. Finally, to implement the ladder diagram, the PLC program follows in Figure 13.29.

STAGE PROGRAMMING

Case Study 13.5 revealed one of the most difficult aspects of the programming of real industrial logic control systems. The industrial robot described in Case Study 13.5 was required to perform a sequence of operations. It is difficult to maintain a sequence of operations in an arrangement in which all the rungs of the ladder diagram are executed simultaneously. That is why a variety of input combinations were used to alternately block out or give the go-ahead for an output to be activated. This is sometimes called "interlocking" and is one of the most complex and tedious elements of the task of programming a PLC for a real application. Likewise, in the debugging phase of the program preparation, it is difficult to trace the problem in an extremely complex and

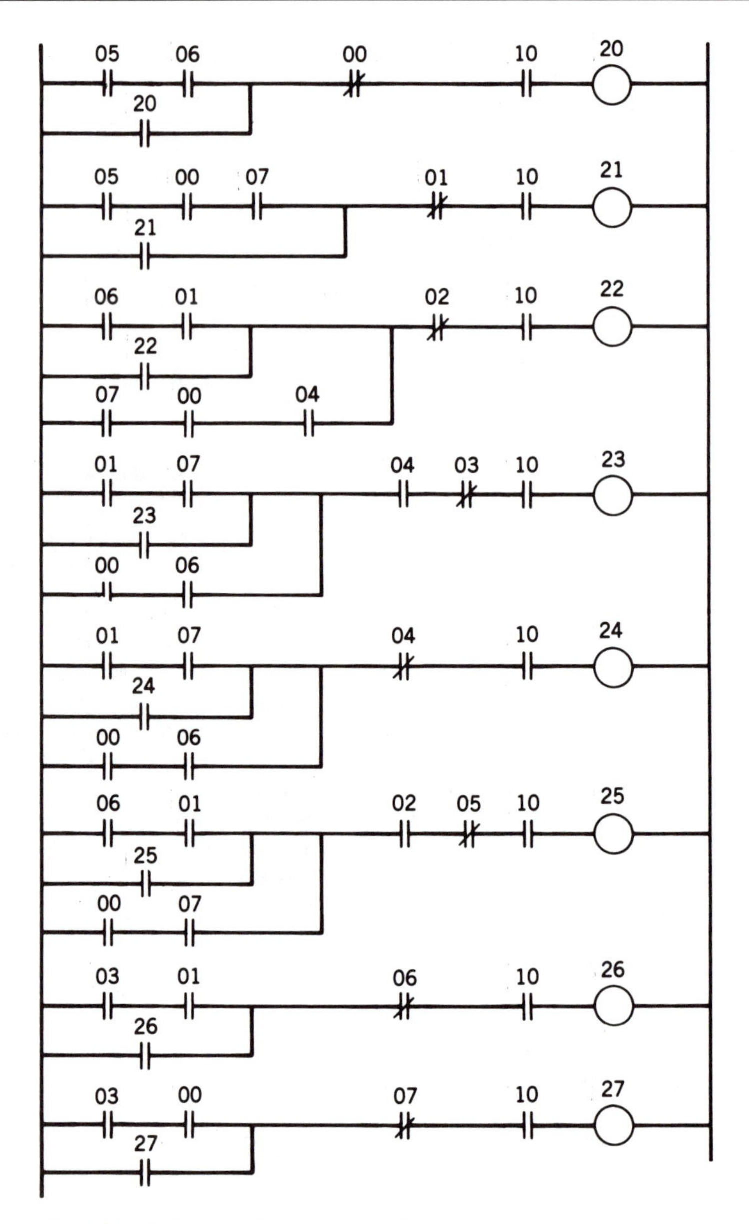

FIGURE 13.28 PLC ladder logic diagram for pick-and-place robot (Case Study 13.5).

STR	SHF	05	ENT	AND	SHF	10	ENT	STR	SHF	06	ENT
AND	SHF	06	ENT	OUT	SHF	22	ENT	AND	SHF	01	ENT
OR	SHF	20	ENT	STR	SHF	01	ENT	OR	SHF	25	ENT
AND NOT	SHF	00	ENT	AND	SHF	07	ENT	STR	SHF	00	ENT
AND	SHF	10	ENT	OR	SHF	23	ENT	AND	SHF	07	ENT
OUT	SHF	20	ENT	STR	SHF	00	ENT	OR STR	ENT		
STR	SHF	05	ENT	AND	SHF	06	ENT	AND	SHF	02	ENT
AND	SHF	00	ENT	OR STR	ENT			AND NOT	SHF	05	ENT
AND	SHF	07	ENT	AND	SHF	04	ENT	AND	SHF	10	ENT
OR	SHF	21	ENT	AND NOT	SHF	03	ENT	OUT	SHF	25	ENT
AND NOT	SHF	01	ENT	AND	SHF	10	ENT	STR	SHF	03	ENT
AND	SHF	10	ENT	OUT	SHF	23	ENT	AND	SHF	01	ENT
OUT	SHF	21	ENT	STR	SHF	01	ENT	OR	SHF	26	ENT
STR	SHF	06	ENT	AND	SHF	07	ENT	AND NOT	SHF	06	ENT
AND	SHF	01	ENT	OR	SHF	24	ENT	AND	SHF	10	ENT
OR	SHF	22	ENT	STR	SHF	00	ENT	OUT	SHF	26	ENT
STR	SHF	07	ENT	AND	SHF	06	ENT	STR	SHF	03	ENT
AND	SHF	00	ENT	OR STR	ENT			AND	SHF	00	ENT
AND	SHF	04	ENT	AND NOT	SHF	04	ENT	OR	SHF	27	ENT
OR STR	ENT			AND	SHF	10	ENT	AND NOT	SHF	07	ENT
AND NOT	SHF	02	ENT	OUT	SHF	24	ENT	AND	SHF	10	ENT
								OUT	SHF	27	ENT

FIGURE 13.29 PLC Program for Case Study 13.5.

interlocked ladder diagram when an output is found to be unexpectedly on or off when it should have been behaving the opposite. Also, a small change to a ladder diagram can have far-reaching effects in a long and complicated program of ladder logic.

Stage programming is a method of dealing with the problem of sequencing in a long and complicated PLC program. The STAGE™ scheme enables the programmer to alternatively activate or deactivate portions of the ladder diagram to follow a desired sequence of operations. The STAGE™ software works with the Texas Instruments TISOFT™ package for generating ladder logic programs for PLCs. Other schemes might employ master control relays to activate and deactivate portions of the ladder logic at the appropriate time to achieve the same purpose as stage programming.

Case Study 13.5 represents an ideal study for the application of the concept of stage programming. In Table 13.3 notice that the process calls for a motion "Retract arm" in two places. After one of the "Retract arm" operations the base is to be rotated *clockwise*. After the other "Retract arm" operation the base is to be rotated *counterclockwise*. In the conventional ladder logic approach the logic must be interlocked to prevent the controller from attempting to rotate the base both ways at once. Other steps are repeated within the sequence also, and the system must be programmed such that these steps are interlocked also to prevent unintended occurrence. With stage programming it becomes no longer necessary to keep track of the various interlocking input conditions to be sure that the outputs maintain the correct sequence of operation. The completion of each output action is signaled by the making of a limit switch, which in turn signals the start of a new phase of the control program. Only the current stage is active; thus, interlocking is not needed in the other stages to prevent unintended activation of outputs. Figure 13.30 shows the scheme of a stage programming approach to the problem of Case Study 13.5, solved previously by using conventional ladder logic programming. Notice in Figure 13.30 that,

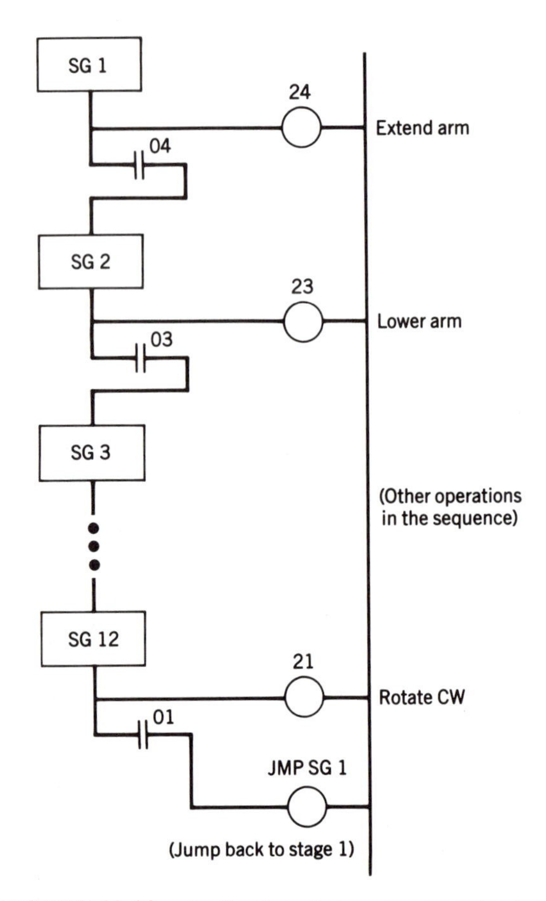

FIGURE 13.30 Application of stage programming to the logic control of a pick-and-place robot as described in Case Study 13.5.

unlike the ladder diagram of Figure 13.28, a complicated set of input conditions is not required to trigger each output. The sequence of operations is controlled by the linking of stages 1 through 12 in a series before jumping back to stage 1 to start the sequence over again. Note that the sequence of operations is exactly the same as was shown in Figure 13.27, to be adhered to in both Case Studies 13.4 and 13.5.

The stage programming example in Figure 13.30 shows only the main sequence of operations of the robot. It is permissible to have parallel stages active at the same time to permit monitoring of alarm conditions while operational stages carry on the required sequence of operations. During system start-up, an initial stage can be used to bring about the desired initial state of the system before activating the operational states.

In a long and complicated program, a PLC can be efficiently programmed in STAGE™ not only to sequence a series of steps but also to activate or deactivate whole sections of an operation or even a plant. Stage programming is merely a way of organizing and controlling the activation of various parts of a ladder logic diagram so that selected rungs of the logic with their corresponding outputs can be held *inactive* when desired. Stage programming can significantly reduce the number of steps required in a PLC program. Program debugging is also made easier by the ability to trace active and inactive stages within a program.

ADVANTAGES of PLCs

All the industrial logic control systems studied so far, in case studies in this chapter as well as those in Chapters 11 and 12, could have been implemented using individual logic components or relays described in Chapter 2. But setting up a programmable logic controller to do the same thing has some distinct advantages that should not be overlooked by the robotics and automation engineer.

Flexibility

The most important advantage of the programmable logic controller over conventional, hard-wired relays is the ease with which the logic system design can be changed—that is, the PLC's *flexibility*. Any time a model changes, a new production lot is set up, or perhaps a decision is made to change control strategies, the programmable logic controller software can be changed in a matter of minutes. No wires need to be disconnected or rerouted, and no components need to be changed. It is a classic comparison between software and hardware changes.

The flexibility of programmable logic controllers offers additional advantages, even if the major process or control strategy remains constant. Sometimes it is beneficial to fine-tune a robotic application or other automatic process by making tiny adjustments to the cycle sequence to make it more efficient. Sometimes a working process can be made slightly better by speeding up the index cycle or perhaps by shortening a process dwell. With hard-wired relay systems, such changes to optimize the process simply may not be worth the trouble. But such tinkering with the process is easy with a programmable logic controller and can even be *fun* as the automation engineer attempts to squeeze the last drop of productivity out of a process.

One way that processes change is that they *grow*. They may grow in sheer production volume, or they may grow in the sophistication of their automation. The flexibility of the programmable logic controller is helpful when such growth occurs, but some PLCs are better than others in this regard. When planning robotics or other automation processes, the engineer should consider the prospects of growth, and if appropriate, buy a modular PLC that can be expanded as the need arises.

Setup Speed

Although it is a direct consequence of a PLC's flexibility, setup speed should be mentioned as a separate advantage. Imagine being able to ask an electrician to hook up process wiring to an industrial logic control system even before the control system has been designed. As seen in Figure 13.1, the programmable logic controller has a number of inputs and outputs. The selection of these I/O contacts is arbitrary, so it does not matter how the electrician connects them to the process. Usually the programmer decides which inputs and outputs will represent what and merely hands an I/O identification list to the electrician. Of course, the voltages and power delivery of the various I/O circuits will vary, but this is done with conventional relay circuitry. PLC outputs are typically 110 V ac, fused for low-power delivery. These low-power circuits then trip ac solenoids, motor starters, or contactors to close the major power circuits. Although the power delivery of the output circuits is low, 110 V ac circuits are selected over lower voltage dc circuits to minimize the effects of induced voltages from nearby heavy power circuits typical of the industrial environment. It is not unusual at all for low-voltage control circuits to be plagued by nuisance tripping as a result of electrical noise from adjacent electrical machinery. The programmable logic controller I/O voltages are selected to minimize this problem. The PLC *inputs* also may be designated to accept 110-V signals for the same reasons, but sometimes inputs are accepted at lower voltages, such as 24 V dc. Once these standard I/O voltages have been determined, the electrician may go ahead and connect the PLC to the process without further specification.

The other part of the PLC system setup is of course the programming of the controller itself. As was seen in the previous section of this chapter, PLC coding is simple and powerful and can be accomplished very quickly by anyone well founded in the principles of industrial logic control systems covered in Chapters 11 and 12 of this book.

While stressing the advantage of ease of programming when setting up a PLC application, the task of laying out, designing, and implementing an industrial logic control system must not be minimized. The programming may be easy, but the automated system design may not be. Automation engineers must be able to decide exactly what the process should do, when, and under what conditions. Real-world applications are large and complex and may have dozens of inputs and outputs and hundreds of rungs in the ladder diagram. An example is the special-purpose robotic control and handling system illustrated in Figure 13.31. This axis-limit system is driven by a programmable logic controller and follows a sequence of robot motions consisting of 53 steps. The ladder diagram has 200 rungs. The plans for such a real-world system must be laid out in complete detail, and

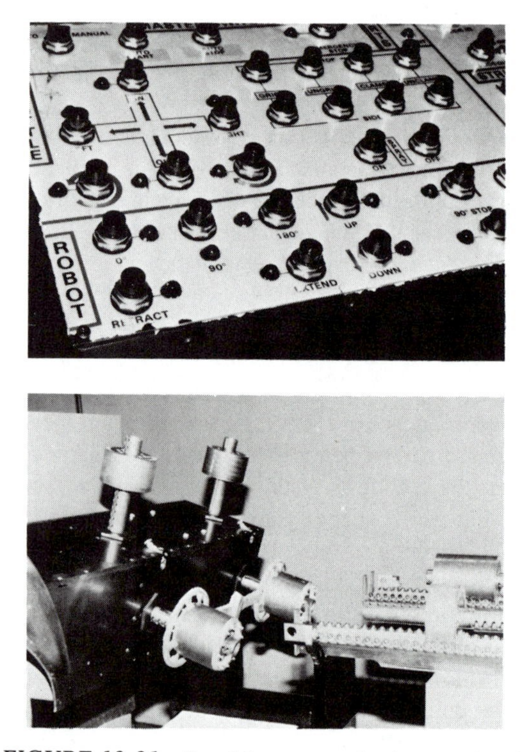

FIGURE 13.31 Special-purpose robotic control and handling system. (Reprinted by permission of Garrison Engineering, Rogers, Ark.)

there can be no inconsistencies or ambiguities in the specification of inputs and outputs. But once such a plan is correctly laid out in a ladder diagram, the actual PLC coding is a simple task that can usually be accomplished in minutes.

The last phase in a system setup is to test the program and make any corrections of errors that could not be discerned in the planning and layout. In a real-world system, the system complexity almost guarantees that some minor changes will have to be made. So the advantage of ease of programming software changes once again favors the programmable logic controller over hard-wired relays.

Reliability and Maintainability

Machine breakdowns are always a disappointment, but it should be clear from Chapter 3 that the penalties for such breakdowns are much more severe when automation and robotics are involved. Unlike mechanical relays, programmable logic controllers are of solid-state construction and thus have no moving parts. The result is greatly enhanced reliability and a lifetime that is virtually independent of the number of cycles executed by the various logic elements. This independence of failure rate over the lifetime of solid-state electronic equipment is illustrated by the flat portion of the familiar "bathtub curve" for reliability of such equipment. Industrial logic control systems often are called upon to operate around the clock, resulting in severe demands upon the system's reliability.

The frequency of system failures is of course only half the story when evaluating overall system availability or uptime; the other half is maintainability. When a conventional relay panel experiences a malfunction, the diagnostic procedure can be a long journey through the system wiring with a voltmeter. But with a CRT and programmer keyboard, the technician can easily check various portions of the PLC program for power flow and verify the logic. Even the lower cost systems that do not have CRTs usually have some method of verifying power flow to each logic element.

Data-Collection Capability

Programmable logic controllers have many of the advantages of conventional process control computers in their ability to automatically tabulate parts counts, instances of quality rejections, machine hours, and other data of interest to the overall management of the process. The programming flexibility of the PLC permits selection of periodic or continuous monitoring of process variables.

Input/Output Options

The advance of robots and manufacturing automation devices driven by stepper motors creates a demand for controllers that can deliver series of pulsed outputs. The programmable logic controller with its solid-state construction is much better equipped to deliver such outputs than are conventional relay systems. PLC output options also include CRT displays and hard-copy printers, although this degree of input/output sophistication is usually associated with process control computers, which will be discussed in Chapter 14. Indeed, as the sophistication of a programmable logic controller grows, it becomes very difficult to distinguish it from a general-purpose process control computer. Like process control computers, PLCs can be linked to central computers to achieve overall computer control of a factory.

Cost

For the very simple logic systems involving only a few input contacts, conventional relays still may be more economical than programmable logic controllers, at least from an initial-cost standpoint. But when the complexity of the system increases to approximately ten relays or more, the economy of the PLC begins to dominate. With very complex logic systems, there is no comparison between the costs of the two approaches.

In addition to initial cost, one must consider the compactness of programmable logic controllers, which can greatly decrease the floorspace required, and also the weight, which is important for portable automated manufacturing systems and robots. Energy efficiency can also be a consideration since the solid-state medium of the programmable logic controller requires much less energy than do conventional relays.

DISADVANTAGES OF PLCs

For all the power, elegance, flexibility, and convenience associated with programmable logic controllers, there are some drawbacks to be considered. The ease of programming and flexibility of PLCs make catastrophic programming errors more likely than in the deliberate, painstaking construction of conventional, hard-wired relay circuits. When a system is so easy to change, more changes will be tried, and along with the beneficial changes it will also be found that some of the changes will be bad ones.

A frequent warning issued for the designers of control systems using solid-state programmable logic controllers is to add a *master control relay* that can deenergize all PLC outputs at the flip of a switch. It is strongly recommended that such master relays be of conventional, mechanical construction—*not* solid-state. Solid-state devices are extremely reliable, but the addition of a master control relay that is of different construction provides a safety device that is not subject to the same types of physical perils that may affect the PLC. Of course, mechanical relays fail, too, for different reasons, and this fact must also be considered in the design of conventional relay systems. The physical hazards to solid-state systems will be discussed further in Chapter 15.

Sometimes the sudden loss of power to a given output can be even more catastrophic than the opposite problem. The PLC programmer should exercise care in programming outputs to remain on in the face of certain rare circumstances. Most PLCs have the capability of *forcing* an output on or off regardless of the status of programmed logic. The forcing of outputs is also useful for troubleshooting. The use of forced outputs permits the user of the PLC to avoid undesirable consequences of malfunctions.

SUMMARY

The programmable logic controller is an ingenious device that has greatly simplified the large banks of complicated electrical control panels that formerly consisted of electrical logic relays. Along with that simplification came much greater flexibility and reduced hardware, in terms of both volume and cost.

Programmable logic controllers and personal computers are related in that they both have microcomputer central processing units and computer memories. But the programmable logic controller executes the same short program rapidly and repeatedly thousands of times an hour, whereas a personal computer or a data processing computer executes a variety of comparatively longer programs. Also, the inputs and outputs of the programmable logic controller are industrial logic circuits, not the input data and output reports typical of conventional computers. So fast is the processing of these industrial logic circuits that for most industrial applications the programmable logic controller can be dealt with as a continuous logic control system, the same as an industrial logic ladder circuit.

Understandably, the basis for programmable logic controller programming is the ladder diagram. The programming device itself is usually detachable because once the ladder diagram is programmed, the programmable logic controller will operate continuously until a change is desired. Counters, timers, and even simulated mechanical cam timers or drum timers are all available through the preprogrammed software resident in the programmable logic controller.

Programmable logic controllers are intimately associated with industrial robots. PLCs can control robots one at a time or in groups. They are ideal for controlling a manufacturing workcell of which the robot is a part, assisting in coordinating the motions of the robot with the machines with which it works. So versatile are programmable logic controllers that they can act as the robot's brain and enable the automation engineer to build an industrial robot directly from the basic components of automation studied in Chapter 2.

In this chapter, case studies were used to demonstrate the control of industrial robots using both strategies of control by programmable logic controllers: sequencing and logic control. Sequencing is the open-loop type typified by the use of drum timers. Logic control is the closed-loop type and requires sensors to generate inputs to signify completion of process steps.

The advantages of flexibility, ease of setup, system checkout, reliability, maintainability, and cost become quickly apparent to anyone who learns the basics of a programmable logic controller's operation. Not so apparent are a few of the pitfalls. As reliable as programmable logic controllers and their solid-state logic systems are, it is often unwise to rely totally on solid-state logic. Master control relays of mechanical construction are suggested to provide a means of breaking power to all outputs in event of a rare failure of the solid-state logic system.

EXERCISES AND STUDY QUESTIONS

13.1. Describe the type of control application for which the concept of stage programming is appropriate.

13.2. Is a programmable logic controller a computer? Explain why or why not.

13.3. What is meant by the term *PLC cycle*? How long is a typical PLC cycle?

13.4. Compare the advantages of the two principal control strategies for PLCs: sequencing versus logic control. Which is closed-loop?

13.5. Construct a ladder diagram and write a PLC program to flash a warning signal off and on automatically at a frequency of one flash per second.

13.6. Construct a ladder diagram and write a PLC program to turn on a plant heating system automatically to operate from 7:00 A.M. to 6:00 P.M. daily.

13.7. Construct a ladder diagram and write a PLC program to turn on a plant heating system automatically to operate from 6:00 A.M. to 9:00 P.M. *only on weekdays*; on weekends, the heating system should remain off.

13.8. An axis-limit industrial robot under PLC control feeds a two-step sequential automatic process. Following is the sequence of steps executed with PLC outputs and associated motion completion inputs. Using feedback loops to verify completion of each step in the sequence before proceeding to the next, construct a ladder diagram and PLC program to execute this robot sequence. Assume that each robot axis will hold its position until an output from the controller commands that axis to move to its alternative position.

	PC Output	*Motion Completion Sensor*
GRIP	27	07
LIFT	23	03
EXTEND	25	05
PAUSE	10 seconds	
RETRACT	24	04
BASE CCW	21	01
EXTEND	25	05
PAUSE	6.5 seconds	
RETRACT	24	04
LOWER	22	02
RELEASE GRIP	26	06
BASE CW	20	00

13.9. Use open-loop control methods employing a drum timer to solve Exercise 13.8, allowing two seconds for completion of each robot motion. Set up the drum timer matrix, but do not write out the PLC program.

13.10. A single industrial robot serves to load and unload four different machines on demand. The industrial robot has its own direct controller, but overall supervisory control is by programmable logic controller. While all four machines are running, the robot waits in a ready position that is sensed by PLC input 00. Whenever a machine completes its cycle, it automatically signals to the robot that it needs unloading and reloading. Machine 1's signal is PLC input 01, machine 2's is 02, machine 3's is 03, and machine 4's is 04. If the robot is not busy with another machine, it will service the requesting machine immediately, restart that machine's cycle, and return to the ready position, all under the command of the PLC. But if the robot is busy with another machine, the machine requesting service will have to wait. If more than one machine happens to be waiting when the robot becomes available, it makes no difference which waiting machine is serviced first. To command the robot to service a machine, the PLC merely energizes a unique output to the robot for each machine to be serviced: 21 for machine 1, 22 for machine 2, etc. Each output $2i$ remains on until that machine's service request $0i$ is turned off. Construct a ladder diagram and write a PLC supervisory program to control this robot.

13.11. In the four-machine servicing system of Exercise 13.10, suppose it is desirable to permit *manual* loading and unloading of a machine. Once a machine has been manually serviced and restarted, there is no need for the robot to duplicate the manual service on that particular cycle. Construct a ladder diagram and write a PLC program to accommodate manual machine servicing while the robotic system is operating automatically.

13.12. Identify safety hazards that could arise from permitting manual servicing of machines that are served by an automatic robotic system. What provisions could be made in the construction of the PLC ladder diagram to prevent these hazards?

CHAPTER 14

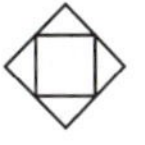

ON-LINE COMPUTER CONTROL

The first digital computers had one-track minds. Programmers submitted programs that were stacked in batch mode and processed one at a time. Within each program processed were individual instructions, also processed one at a time by the computer's central processing unit. Program execution times were short, typically a few seconds or perhaps a few minutes, but during that short interval of time the user had the computers' undivided, or almost undivided, attention.

Today's computers still accept the same type of batch-mode computer programs written in the same languages, such as PASCAL, FORTRAN, and COBOL, but the computer hardware and task organization have changed. There are still a few "one-track mind" computers around, but these have become very small and inexpensive. Large, general-purpose (mainframe) computers today are never dedicated to a single, uninterruptable program sequence.

Before going further, it will be assumed in this chapter that the reader is somewhat familiar with batch-type computer programming and that most readers of this text will already have computer programming proficiency in PASCAL or FORTRAN or some other user-oriented, batch-type language. But those readers who are not already programmers should not feel inadequate, because in truth this chapter and this book are not really about computer programming, at least not the usual PASCAL and FORTRAN kind of programming.

PROCESS CONTROL COMPUTERS

The subject of this chapter is the on-line computer, which almost continuously monitors and controls a large number of robots or automated manufacturing processes concurrently. Some background on the evolution of today's process control computer will help to clarify matters for conventional batch-type computer users.

A key word used in the foregoing is "concurrently." Another word, "simultaneously," has a slightly different meaning and therefore was not used. Modern digital computers will execute only one coded instruction in the same instant in time, but their speed enables them to jump quickly from program to program using *time-sharing,* to permit them to continuously monitor several processes concurrently.

Even the old one-track mind computers of the 1950s and 1960s had some capability for monitoring the status of their peripheral equipment. It was essential to interrupt program control and execute alarm routines when a mechanical failure occurred in a tape drive, disk drive, or, worse still, the central processor itself. Someone noticed that if the one-track mind computer could be interrupted for essential hardware alarm conditions, it could also be interrupted for other high-priority tasks, and the concept of time-sharing was born (in the early 1960s). The emphasis still was upon data-processing types of applications, but with time-sharing a long, involved program could be interrupted for a few milliseconds to execute completely a short program for someone sitting at an on-line terminal screen and keyboard. With time-sharing, the long waits for batch-job turn-arounds virtually were eliminated for the short-run programmer. The central processor could handle a large number of these short-run jobs and still give the impression to the user that he or she had exclusive control of the computer. This especially was true when the principal task of the computer was to handle short-run jobs from remote terminals. Among the first systems typifying this mode were the on-line reservations systems augmented by the airlines in the 1960s.

Time-sharing was implemented to handle efficiently batch-type data processing jobs of varying length and priority. The next discovery was a marriage of the concept of timesharing with the concept of monitoring the computer system's own hardware. It was reasoned that if the computer could be interrupted for some hardware malfunction of the computer system itself, why could it not also be automatically interrupted to service some urgent condition external to the computer system? The correct response to the condition might be to print or sound alarms or, to take automation a step further, to send to the process some control signals to equipment that would take corrective action *automatically.* Taking advantage of the speed of the computer, it is possible to improve the speed of response to process upsets by putting the computer into a direct, closed loop with the process. An added advantage is that the computer can perform complicated calculations upon the process data monitored by the computer in order to take *more effective* corrective action in real time than would have been possible for a human operator.

Thus, the digital computer has become a significant tool for factory automation in addition to its contribution to the fields of science and data processing. It is even possible to think of this type of computer as a robot, since it possesses certain anthropomorphic characteristics and is doing a job that was performed earlier by humans.

LEVELS OF IMPLEMENTATION

To appreciate the various levels or degrees to which computers can be used to assist in the automation and control of factory processes, let us begin with the manual process, the case in which no computer at all is used. Figure 14.1 portrays the conventional manual control process in which a human operator observes the process variables by means of chart recorders, meters, and annunciators and takes action by console switches and controls. Not to be overlooked is the information flow via the process instruction manuals, because without a process control computer, the human operator must have greater knowledge of the process operation and control strategies. Also note that the operator must maintain the operating log manually, although this can be supplemented by circular or linear graphs from the chart recorders.

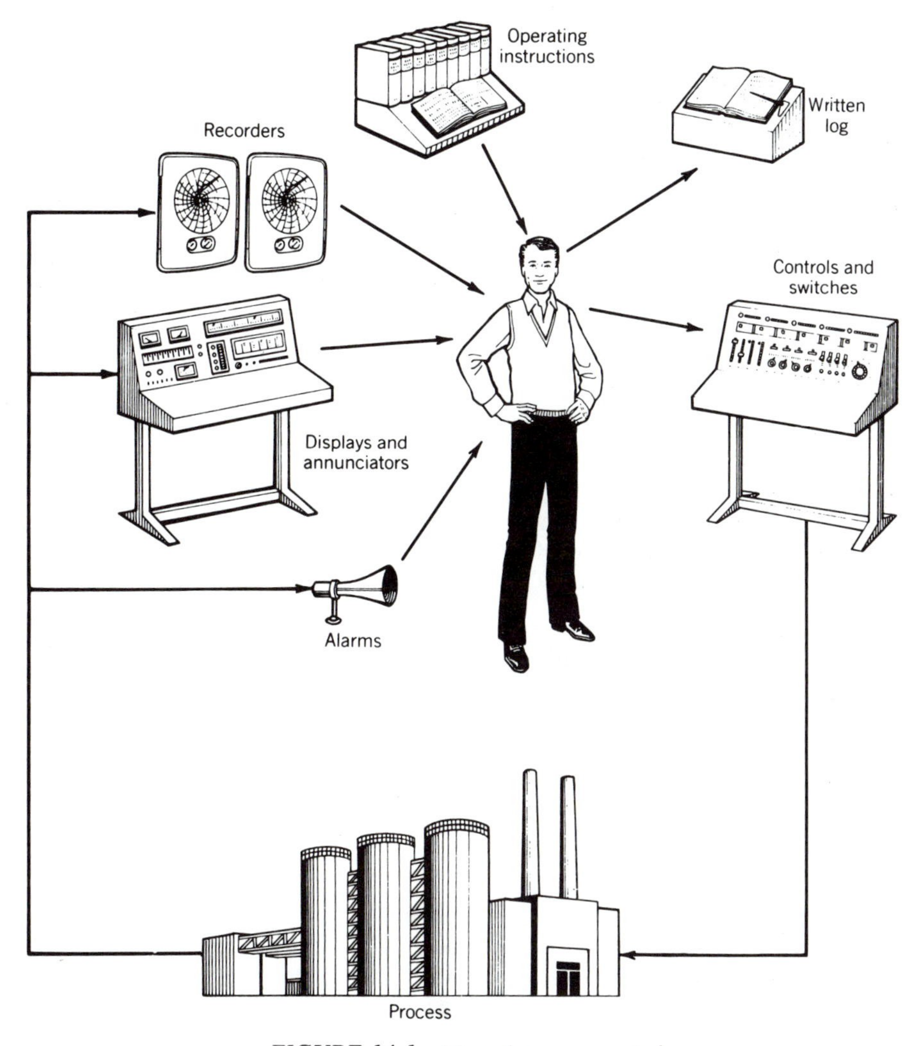

FIGURE 14.1 Manual process control.

Off-Line Processing

As illustrated in Figure 14.2, a data-processing type of computer can be of some benefit to the human operator in controlling a process by facilitating analysis of the observed process variables. Statistical quality control and process optimization studies are possible with off-line data processing, but there is a serious drawback to the off-line approach: time delay. By the time log data are converted into a machine-readable form and are processed by the computer into batch-mode, it may be too late to do the process operator any good. This mode of computer-assisted control of factory processes was popular in the time when large batch-mode computers were still the principal type in use, but the mode is nearly obsolete today. Its redeeming quality—which keeps it alive in some factories—is the fact that it does not require expensive sensors and interfacing hardware.

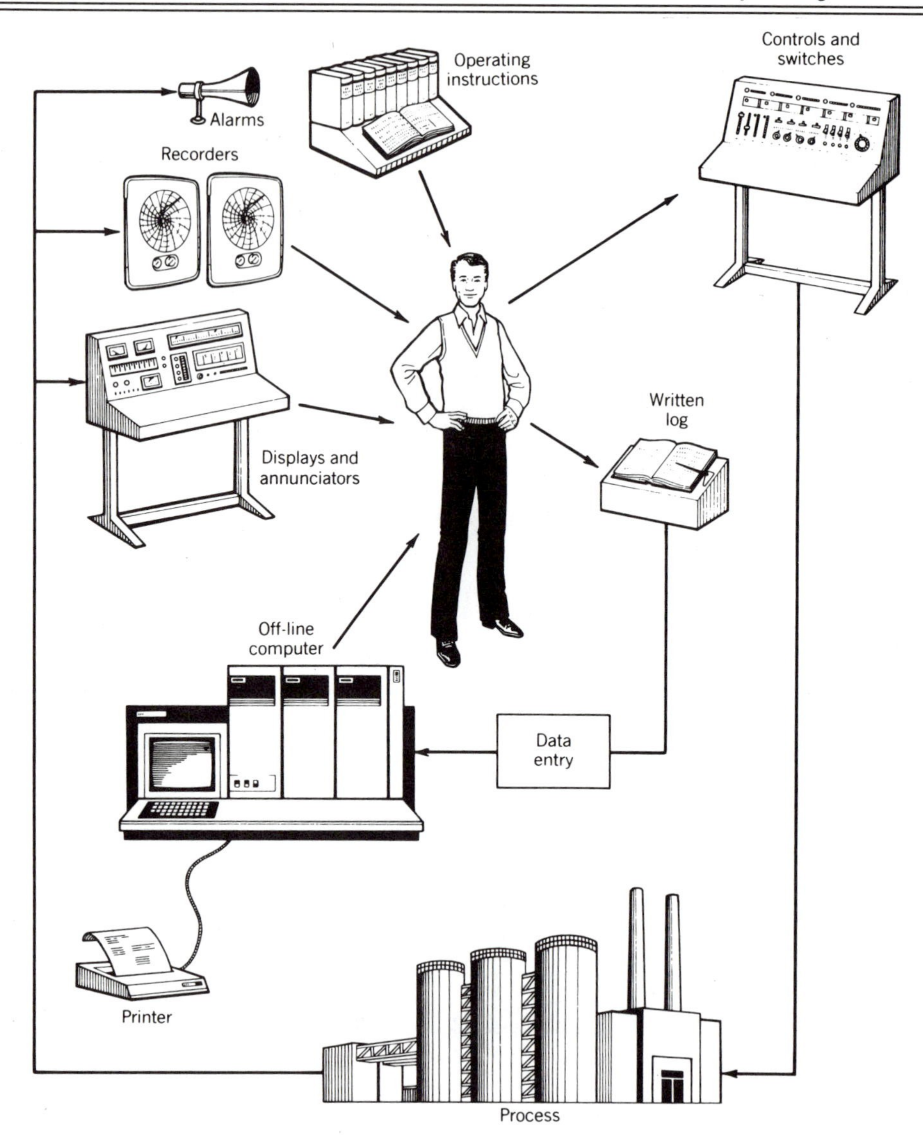

FIGURE 14.2 Process control assisted by off-line data processing type computer.

Process Monitoring

The next step up from off-line analysis is process monitoring (Figure 14.3), in which the computer is connected on-line to certain process sensors or counters to determine what is going on. In process monitoring, the computer has no outputs to the process and is powerless to correct any problem that may arise. The objective of this level of automation is to keep management informed, usually on a periodic basis, of the production progress of a particular process. Note that the operator has been relieved of the task of manually recording entries in the log. Besides the periodic reporting, process monitoring computer systems may also immediately notify operators or other personnel of various conditions, such as the current temperature of a vat. Note, however, that the operator still must refer to process instructions and have sufficient knowledge of the process to know what to do

FIGURE 14.3 Processing monitoring with an on-line process control computer.

about the "temperature of the vat" or any other process information provided by the computer in process-monitoring mode.

On-Line Assist

The next step up in on-line computer automation is on-line assist, in which the computer not only monitors what is going on but also makes specific suggestions to the operator for actions to be taken to control the process properly. For example, whereas the process-monitoring computer might record the temperature of a vat and print out values periodically for the operator's reference, the on-line assist computer might display a message stating:

> Vat temperature at 187° is too high; turn valve K-9 down 1/4 turn to reduce fuel flow to burners.

Such a message obviously requires considerably more sophistication than did the process-monitoring mode, but it is not closed-loop control. The human operator is still in complete control and has the final decision-making power over what will be done to correct a given condition.

Figure 14.4 illustrates the on-line assist mode of process control computer. Note that the process operating instruction manuals were omitted in Figure 14.4. Probably the most significant advantage of the on-line assist mode over the process monitoring mode is that the *operating instructions* are coded into the computer logic instead of being in written manuals available to the operator. Written manuals are good for documentation, but as an operating tool they usually are not quick enough. Human operators usually must memorize the essential elements of the control of the process or they will not be able to respond to process upsets quickly enough to resolve the problem. For instance, one would never order an inexperienced operator to start operating a boom crane without training and hand him or her an operating manual to use if problems arise. Thus although the manuals in an on-line assist mode would obviously be retained for occasional reference, they no longer would be needed in the day-to-day operation, and operating personnel would not require as much training or process knowledge. This can be a significant advantage of the use of on-line assist.

FIGURE 14.4 "On-line assist" process control computer.

Closed-Loop Control

The highest level of computer automation by on-line process control is closed-loop control, as illustrated in Figure 14.5. Note in this figure that the operator is sitting down because the operator is no longer in the feedback loop for correction of process problems. The human operator can remain for various unusual or unforeseen developments as a process is rarely if ever completely automated to accommodate every disturbance that might confront its operation.

Human analogies to the computer function can be constructed for each level of computer process control. The process monitoring level can be compared to a clerk or analyst who gathers data. The on-line assist level is similar to the function of a staff assistant in a human organization. The staff assistant has authority to advise and recommend action by line organization prersonnel, but has no authority to take direct action. The closed-loop control level is comparable to a person in the line organization chain of command. Line organization personnel have authority to take action to correct problems, and so does the closed-loop computer. Now let us examine in more detail the hardware characteristics of computers that perform process control functions.

FIGURE 14.5 Closed-loop process control computer.

CONTROL STRATEGIES

The on-line computer's capability for closed-loop process control makes possible sophisticated control strategies for continuous variables without the investment in conventional analog control hardware for each control loop. The objective of such control systems for continuous process variables is to apply a measured corrective action to compensate for variations from the desired level. There are many external disturbances that might cause a process variable to deviate from the desired level. In addition, the operator or other controller suddenly might desire to change the desired level of a process variable by a control switch or other process input. The process control system would then be required to react to this sudden change to bring the process variable to a new level to correspond to the new desired "set-point." In control theory terminology, any deviation from set-point is called "error," and error can be considered to be positive if the process variable is higher than the set-point and negative if the process variable is lower than the set-point.*

In Figure 14.6, a continuous process variable, robot shoulder axis location, is shown to match its set-point nicely until at time t_1 a sudden change in the desired position of the robot causes a large negative error to exist. If the control system were ideal, it would change the process variable to follow the set point exactly with no error. But real-world variables cannot be forced to change instantly especially not large, hydraulic, industrial robot arms. So the process variable corrects a short time later at time t_2, and error is reduced to 0. However, at this point it is impossible to immediately stop the correction, and overshoot occurs that peaks at time t_3. During the overshoot, the control system is sensing that the error has changed to positive and accordingly reverses its corrective action. Eventually, overshoot oscillations diminish to zero, and the process variable is following its new set-point quite nicely again.

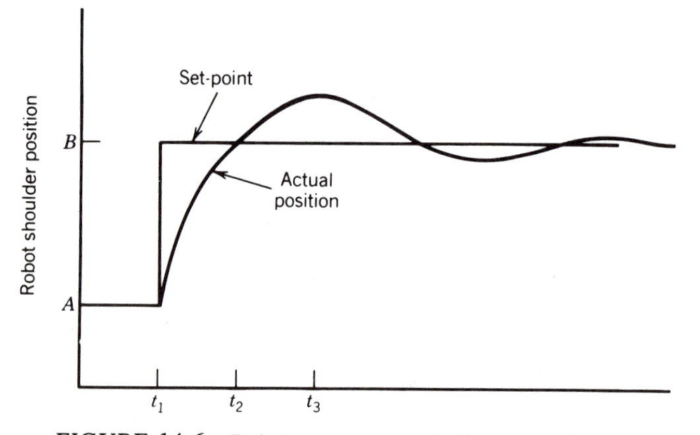

FIGURE 14.6 Robot computer controller responds to a sudden desired change in the robot shoulder position from position *A* to position *B*.

*It is acknowledged that this sign convention is the opposite to the convention commonly adopted by the control industry. However, the convention adopted here will make subsequent diagrams easier to understand by the reader unfamiliar with feedback control theory analysis.

Naturally we want our control systems to react quickly to any error that develops because the sooner we reach the new set-point the sooner we reach our new objective. However, there is a trade-off, because quick reaction to error also means that the process will overshoot its mark. Overshoot can be so large as to cause the system to oscillate wildly between overshoots, and successive overshoots can even increase rather than diminish through time. Such a system is said to be "unstable," and the remedy is to apply "damping." On the other hand, some processes are so slow as to be considered over-damped. Figure 14.7 compares control systems of various response speeds.

Proportional Control

The most obvious control strategy is to apply correction to a process variable as soon as any error is detected and to make the correction proportional to the magnitude of the error. This is called *proportional control* and is described mathematically in Eq. 14.1.

$$C_p(t) = K_p e(t) \tag{14.1}$$

where K_p = proportion constant

$e(t)$ = process error (function of time)

$C_p(t)$ = proportional corrective action (function of time)

If the proportional ratio K_p is high, the response will be fast and the system is said to have high "gain." Of course, if the gain is too high, the process can become unstable.

Figure 14.8 illustrates a process under proportional control. Note that the process variable does not quite match the set-point after oscillations have died out. This is a permanent offset due to the fact that the proportional control system cannot eliminate the error completely. The reason for this is that some error must be detected for corrective action to be applied, and the error must be of sufficient magnitude to be recognized by the process sensor.

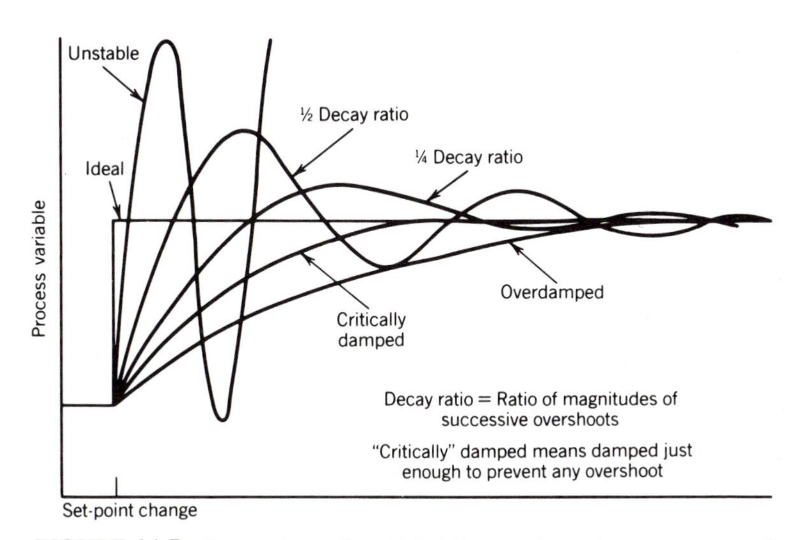

FIGURE 14.7 Comparison of control systems with varying response speeds.

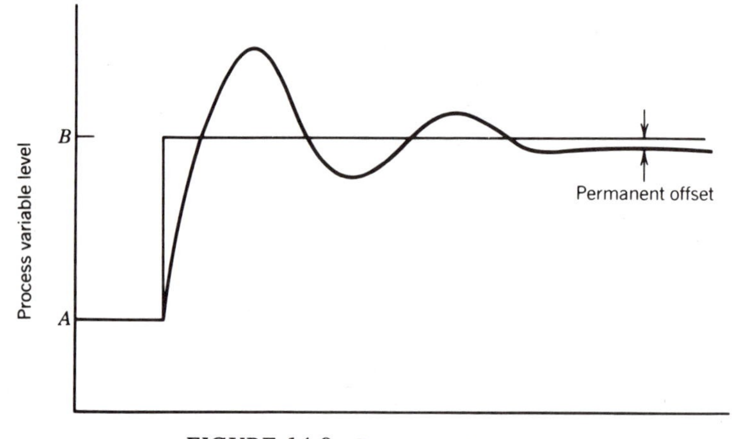

FIGURE 14.8 Proportional control.

Integral Control

Sometimes called "lag" compensation or "reset" control, the *integral control* strategy is to gather cumulative data on the error and apply corrective action proportional to the total error accumulated over time. Integral control is described by Eq. 14.2:

$$C_i(t) = K_i \int_0^t e(T)dT \tag{14.2}$$

where K_i = integral control constant

$e(T)$ = process error (function of time)

$C_i(t)$ = integral corrective action (function of time)

It is easy to see how the offset we observed in Figure 14.8 can be eliminated by integral control. Although the offset is small, as it is accumulated over time it can become significant and eventually will trigger a response from the integral controller. Thus, it is attractive to use proportional and integral control together—proportional to get an immediate reaction to the error and integral to remove the offset.

Derivative Control

The most sophisticated of control strategies also includes *derivative control,* sometimes called "lead" control or "rate" control, although this mode of control is less popular and is rarely used alone. Derivative control is described by Eq. 14.3:

$$C_d(t) = K_d \frac{de(t)}{dT} \tag{14.3}$$

where K_d = derivative control constant

$e(t)$ = process error (function of time)

$C_d(t)$ = derivative corrective action (function of time)

Derivative control ignores the level of the error and the amount of error that has accumulated in the past and concentrates upon the speed at which the error is developing. Thus, it is able to anticipate or "lead" the process somewhat. Derivative control works *against* proportional control when process correction begins, because the proportional controller "sees" the error that exists now, but derivative control "sees" where the error is headed (toward overshoot). Derivative control can be used to damp overshoot. Properly balanced, derivative control can effectively improve the level of control achievable by proportional and integral control modes.

The most sophisticated control systems apply all three modes of continuous process control: proportional, integral and derivative control. This type of control is summarized in Eq. 14.4:

$$C(t) = C_p(t) + C_i(t) + C_d(t) \tag{14.4}$$

$$C(t) = K_p e(t) + K_i \int_0^t e(T)\, dT + K_d \frac{de(t)}{dT}$$

Control engineers usually work with a variation of Eq. 14.4 that recognizes that practical methods of applying proportional control require that the process gain (G) be applied to all three effects: proportional, integral, and derivative. The physically adjustable constant for integral control is the *integral time interval,* T_i, related to the gain as follows:

$$K_i = G \frac{1}{T_i}$$

The physically adjustable constant for derivative control is the *derivative time interval,* T_d, related to the gain as follows:

$$K_d = GT_d$$

The control engineer's expression of the entire output of the controller to the controlled variable is:

$$P = P_0 + Ge(t) + G \frac{1}{T_i} \int_0^t e(t) dT + GT_d \frac{de(t)}{dT} \tag{14.5}$$

where P_0 = set-point

P = output to the controlled variable

Finally, remember that control engineers define error as "set-point minus process," not vice versa, as we have defined it here.

Equation 14.5 is more compatible with actual control system hardware, but Eq. 14.4 better explains the general concept of the three types of control modes: proportional, integral, and derivative. Figure 14.9 summarizes Eq. 14.4 by tabulating the sense (positive or negative) of each type of process control strategy for various positions of the process variable's reaction to a set-point change.

Common industrial terminology designates any control system that uses some combination of proportional, integral, and/or derivative control strategies as "PID control."

Control Optimization

The selection of the ideal mix (K_p, K_i, and K_d) of control strategies for a PID system is the subject of control system optimization. Optimization requires detailed analysis, and the convenience of Laplace transformations becomes almost essential to manage the complexity of the process variations through time. Such analysis techniques existed long before the rise of robots and manufacturing automation. Electrical engineers have long dealt with control theory to apply control to electrical devices and systems. Mechanical

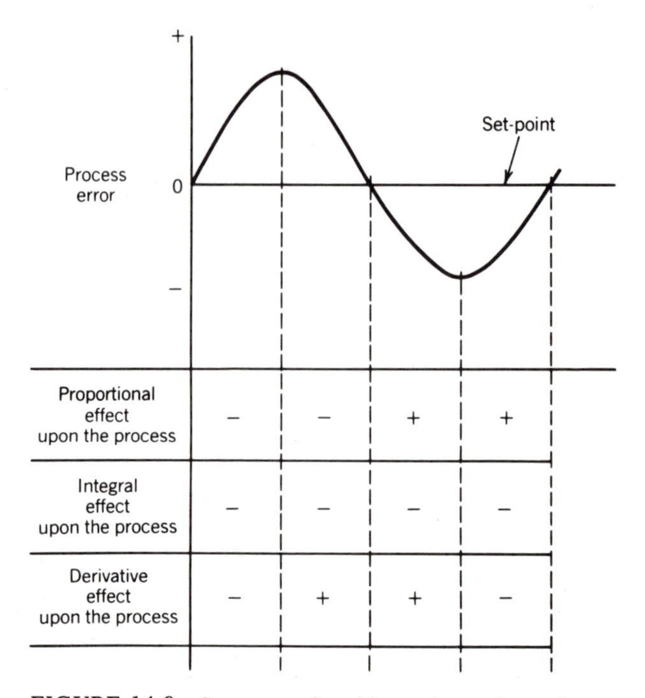

FIGURE 14.9 Summary of positive and negative actions to correct the process variable taken by each of the three major control strategies.

engineers also use control theory to design mechanisms to control mechanical system variations.

The entry of the digital process control computer gave rise to digital *sampling* of continuous process variables for analysis and control. Digital sampling added a new dimension to continuous control system optimization called *sampled data systems*. The Z transformation is more appropriate for analysis of sampled data systems than is the Laplace transformation. A detailed analysis of control system optimization using either Laplace transformations or Z transformations is beyond the scope of this book. It is not necessary to understand control system optimization to apply robots and manufacturing automation successfully. It is essential for the automation engineer to know the principal concepts of control strategies and that the on-line computer is capable of applying them. Now let us examine the characteristics of on-line computers that make them capable of being applied to robots and manufacturing automation.

PROCESS INTERFACE

Besides the ordinary input/output peripherals such as printers, CRT screens, and keyboards with which computers must have connection, a process control computer must have a hard-wire connection to process variables. For process monitoring and on-line assist modes (open-loop), only input contacts are required for sensors, switches, valves, and meters. For closed-loop control, output lines are required for motor starters, relays, actuators, and valves. Process inputs and outputs are of two basic types: digital and analog.

Digital

Digital or discrete inputs and outputs are the easier of the
computer to handle because they are already in a form with wh
operates. In fact, they are already usually in binary since most of the
Examples of digital input and output variables are as follows:

Input

1. Object present or not present
2. Voltage positive or negative
3. Temperature above limit or below limit
4. Liquid level above or below limit

Output

1. Start or stop motor
2. Open or close valve
3. Accept or reject piece-part

It can be seen that several of the characteristics in the above examples are continuous variables (such as temperature or level), but the data input or output to the computer is discrete because limits have been established.

Digital data can be packed very efficiently for on-line computer analysis. For instance, the status of eight different logic variables can be stored in a single computer register of length one byte (eight bits). However, compact, digital data are inconvenient to manipulate by programs written in high-level computer languages such as BASIC, PASCAL and FORTRAN. The tedious and detailed manipulation of digital process data is sometimes more efficiently handled using a detailed computer language such as process assembler or assembly language.

Analog

If actual levels of continuous process variables, such as temperature, flow, and pressure, must be input into the computer directly, some type of conversion is necessary because the digital computer can accept only digital data. For input, the computer always settles for a discrete approximation to the continuous variable of interest, since the computer is a digital device. For output, the process always accepts some type of step function or discrete approximation, also because the computer is unable to update the output variable on a continuous basis. This necessary approximation on the part of the digital computer is a key to the *method* of analog-to-digital conversion and vice versa. The term *analog* is used in process control because physical continuous variables must be converted into some electrical analog (usually voltage or current) before they can be interpreted and converted into digital approximations for input to a process control computer. In Chapter 8 we examined a primary example of analog-to-digital conversion in the transformation of pixel light intensity to a digitized gray-scale. In this chapter we shall generalize the concept.

Since analog-to-digital conversion is merely an approximation, one must decide how accurate that conversion must be. There is a trade-off because the greater the accuracy, the more binary digits that are required to represent the analog value. In the extreme, a continuous analog input variable represented by only one binary digit becomes identical to a digital input variable—for example, "temperature above limit or below limit" as seen two-binary-digit approximation to a of four levels of discrete values to

(14.6)

evels taken on by the n to the continuous

gits used to represent ation

the analog signal from the sensor in gnal with a reference voltage for each ts representing the outcomes of each ght (highest order to lowest order). analog-to-digital conversion process. igits are allocated to the digital result, meaning that $2^3 = 8$ digital levels are available to approximate the analog variable in this case. The analog value 6.6 then becomes digital (binary) 101. The representative value (midpoint) of the range of analog values converted by 101 is:

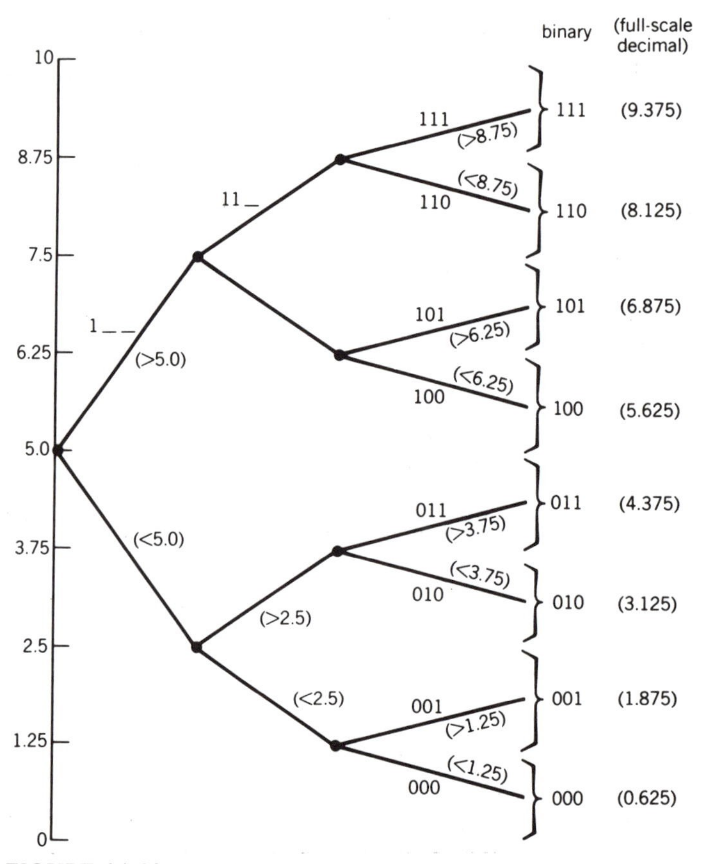

FIGURE 14.10 Decision tree for analog-to-digital conversion process.

$$\frac{7.50 + 6.25}{2} = \frac{13.75}{2} = 6.875$$

Therefore, the round-off error in the conversion is:

$$6.875 - 6.600 = 0.275$$

The following example demonstrates the calculation for a different sample analog value.

EXAMPLE 14.1
Analog-to-Digital (A/D) Conversion

A process analog variable that has a physical range of 0 V to 10 V is at level 2.9 V. What digital approximation will the A/D converter transmit to the process control computer input register?

Solution

$$\text{process variable } X = 2.9 \text{ V}$$
$$\text{digital approximation } Y = ?$$

Using the decision tree of Figure 14.10:

$$X < 5.0 \rightarrow \ Y = 0\_\ \_$$
$$X > 2.5 \rightarrow \ Y = 01\_$$
$$X < 3.75 \rightarrow Y = 010$$

The digital (binary) approximation 010 is used to represent the analog value 2.9 in this case. The representative value (midpoint) of the range of analog values covered by digital 010 is:

$$\frac{3.75 + 2.50}{2} = \frac{6.25}{2} = 3.125$$

Therefore, the round-off error in the conversion is:

$$3.125 - 2.900 = 0.225 \text{ V}$$

Besides the number of binary digits to be assigned to the analog variable, there is the question of range, which also affects accuracy. In general, a real process variable will convert to an electrical analog, which will need amplification or suppression to a prescribed range of voltages that matches the range of the A/D converter. A popular industry standard range for A/D converters is 0 V to 10 V. The amplification or suppression of the signal voltage to the standard 10-V range is a hardware consideration and must take into consideration the expected range of the process variable. Later in the internal calculations the process control computer program can multiply the digital approximation times a scaling factor to give it a numerical representation corresponding to the full-scale range of the original continuous variable.

The resolution of an A/D converter can be calculated as a function of the number of binary digits assigned to the digital approximation and the full range of the continuous variable as follows:

$$\text{resolution} = \frac{R}{2^n} \tag{14.7}$$

where R = full-scale range of the continuous process variable

n = the number of binary digits assigned to the digital approximation

The maximum error that could arise due to the A/D conversion would be one-half the resolution or:

$$\text{maximum error} = \pm\frac{1}{2}\left[\frac{R}{2^n}\right] = \pm\frac{R}{2^{n+1}} \tag{14.8}$$

EXAMPLE 14.2
A/D Resolution and Maximum Error

An automatic process control computer monitors the approach of a robot gripper to an object by means of a radio frequency proximity device that is capable of sensing distance. The range of robot distance to the object is 48 in. The A/D converter uses a 6-bit register for the digital approximation to the analog value, which can be assumed to be accurate and linear throughout the range from 0 to 48 in. What is the resolution of the A/D converter for this process variable, and what is the maximum error?

Solution

$$\text{resolution} = \frac{R}{2^n} = \frac{48\text{ in.}}{2^6} = \frac{48\text{ in.}}{64} = \frac{3}{4}\text{ in.}$$

$$\text{maximum error} = \pm\frac{R}{2^{n+1}} = \pm\frac{48\text{ in.}}{2^7} = \frac{48\text{ in.}}{128} = \pm\frac{3}{8}\text{ in.}$$

$$\left(= \frac{1}{2}\text{ resolution}\right)$$

The process of A/D conversion is called *encoding,* and to be useful again the digitized variable must be subjected to "decoding," according to the following decoding formula:

$$\begin{aligned} Y &= R\sum_{i=1}^{n}\frac{B_i}{2^i} + \frac{1}{2}\left[\frac{R}{2^n}\right] + Y_{\min} \\ &= R\left[\frac{B_1}{2} + \frac{B_2}{4} + \cdots + \frac{B_n}{2^n}\right] + \frac{1}{2}\left[\frac{R}{2^n}\right] + Y_{\min} \end{aligned} \tag{14.9}$$

where Y = the approximation to the continuous process variable

R = the full-scale range of the continuous variable

B_i = $(n + 1 - i)$th order binary bit value

n = number of binary digits allocated to the A/D conversion

$Y_{\min}$ = the minimum value the process variable is expected to assume

The formula is derived from the standard binary-to-decimal conversion relation plus an offset to place each digital increment in the midpoint of its range with the entire quantity scaled to the full range of the continuous variable.

Equation 14.8 can be simplified further as follows:

$$Y = R\left[\sum_{i=1}^{n}\frac{B_i}{2^i} + \frac{1}{2}\left[\frac{1}{2^n}\right]\right] + Y_{\min}$$

$$Y = R \sum_{i=1}^{n+1} \frac{B_i}{2^i} + Y_{\min}, \qquad \text{where } B_{n+1} = 1 \tag{14.10}$$

$$= R\left[\frac{B_1}{2} + \frac{B_2}{4} + \cdots + \frac{1}{2^{n+1}}\right] + Y_{\min}$$

Equation 14.10 is more computationally compact but loses some of the intuitive value of Eq. 14.9. The student may validate Eqs. 14.9 and 14.10 by using Figure 14.10. The following case study illustrates decoding.

EXAMPLE 14.3 Decoding A/D Coded Data

A binary computer register contains a digitized approximation to a continuous process variable that has a range of 1200 to 1800 psi. The register currently reads 1011. What process pressure does this represent?

Solution

Using Eq. 14.10,

$$Y = R \sum_{i=1}^{n+1} \frac{B_i}{2^i} + Y_{\min}, \qquad \text{where } B_{n+1} = 1$$

$$= (1800 - 1200 \text{ psi}) \left[\frac{1}{2} + \frac{0}{4} + \frac{1}{8} + \frac{1}{16} + \frac{1}{32}\right] + 1200 \text{ psi}$$

$$= 600 \text{ psi } [0.71875] + 1200 \text{ psi}$$

$$= 1631.25 \text{ psi}$$

An application of decoding is when the process control computer must provide an output in the form of a continuous variable. This type of decoding is called D/A output conversion and is the reverse of A/D input conversion. It should be emphasized, however, that D/A output is not nearly as important as A/D input. The process itself is often equipped to respond to digital controller outputs. An example can be seen in the robot stepper motor. The robot controller computer outputs a discrete number of voltage pulses, and the stepper motor motion itself approximates the desired move. Add to this the fact that process sensing and monitoring constitutes a larger part of the process computer's function than does output, making A/D conversion more important than D/A conversion.

An analog variable to be monitored often has a statistical distribution approximated by the normal distribution as described in Chapter 1 (see Figure 1.11). A difficulty with such variables is that it is possible to get a spurious reading or, at least, a very high or very low reading that cannot be accommodated by the selected range of the digitized variable. If a very large range is selected for the variable to accommodate all possible data points, the sacrifice is in resolution, according to Eq. 14.7. A practical solution to this dilemma is to select a reasonably small range to preserve resolution and to minimize error for the majority of readings and then to interpret the highest and lowest digitized values as representing all outlying values of the analog variable. This is to say that in the process analysis the low and high values are not to be trusted to lie within the subinterval ranges assigned to their digitized approximations. A design case study will

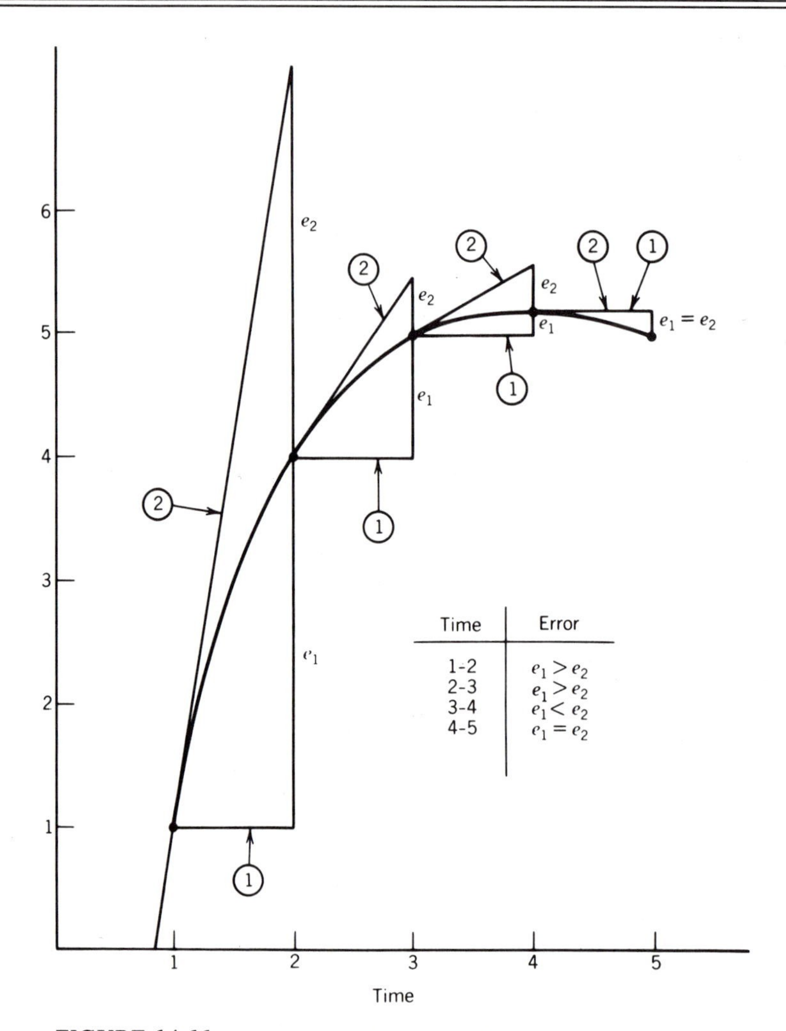

FIGURE 14.11 Comparison of strategies for approximating a sampled continuous variable:
1. Level assumed constant between samples.
2. Rate of change assumed constant between samples.

illustrate the concept of this design trade-off while reviewing the entire process of A/D and D/A conversion.

EXAMPLE 14.4 (DESIGN CASE STUDY) Forging Temperature Digital Monitor

This case study is to design an appropriate digital monitoring system for a hot steel forging operation. The design should take into consideration the approximate temperature range of interest to the desired level of resolution, but should also deal appropriately with temperatures out of range on either the high or low side. Digital registers should be large enough to accommodate the monitoring system requirements, but the design should not represent costly overkill. Compare the design with a corresponding full-range monitoring system with respect to resolution and accuracy.

Design Solution

Background information on the process of hot forging can be found in a good reference text on manufacturing processes, such as DeGarmo [ref. 29].

Hot forging is done when the forging blank is at a temperature sufficiently high to cause the material to flow elastically without damage to the die. For all metals, the hot forging temperature is somewhat below the melting point, which for steel is approximately 3500°F. A good range of interest for steel would be approximately 2200°F ± 300°F, with a desired resolution of approximately 100°F.

A 3-digit binary register would be sufficient to represent temperatures from 1850°F to 2650°F with the desired resolution, shown as follows:

$$\text{resolution} = \frac{R}{2^n} = \frac{2650°\text{F} - 1850°\text{F}}{2^3} = \frac{800°\text{F}}{8} = 100°\text{F}$$

In such a system all temperatures below 1850°F would be interpreted as being in the cell from 1850° to 1950°F, and all temperatures above 2550° would be interpreted as being in the cell 2550°–2650°F. Each cell is appropriately represented by its midpoint. In such a representation the maximum error, except for reading off-scale on the high temperature or low temperature side, would be as follows:

$$\begin{aligned}\text{maximum error} &= \pm\tfrac{1}{2}\ \text{resolution}\\ &= \pm\tfrac{1}{2}\,[100°\text{F}] = \pm 50°\text{F}\end{aligned}$$

The appropriate digital readouts for each of the 8 cells would be the cell midpoints, except for the highest and lowest cells, which may contain out-of-range temperatures. Designing the system to range from 1850°F to 2650°F conveniently places cell midpoints on the even hundreds of degrees Fahrenheit. The cell midpoints can be calculated by the standard decoding formula (Eq. 14.9):

$$\begin{aligned}Y &= R\sum_{i=1}^{n}\frac{B_i}{2^i} + \frac{1}{2}\left[\frac{R}{2^n}\right] + Y_{\min}\\ &= [2650 - 1850]\sum_{i=1}^{3}\frac{B_i}{2^i} + \frac{1}{2}\left[\frac{2650 - 1850}{2^3}\right] + 1850\end{aligned}$$

If we use this formula and keeping in mind that the highest and lowest cells may contain off-scale readings, the following digital read-outs would be appropriate for our forging temperature digital monitoring system:

Cell	*Binary*	*Digital Readout*
1	000	Lo temp—out of range
2	001	2000°F
3	010	2100°F
4	011	2200°F
5	100	2300°F
6	101	2400°F
7	110	2500°F
8	111	Hi temp—out of range

This system as designed covers the approximate range desired while maintaining the desired resolution of 100°F and, at the same time, the use of a 3-digit binary register does not represent overkill.

To evaluate our design, we compare it with a corresponding full-range monitoring system with respect to resolution and accuracy. Practical limits for a full-range system would be minimum ambient temperature, which we can conveniently place at approximately 0°F and, at the upper limit, the melting point of steel, which is approximately 3600°F. The resolution for such a full-range system could be calculated as follows:

$$\text{resolution} - \frac{R}{2^n} = \frac{3600°\text{F} - 0°\text{F}}{2^3} = \frac{3600°\text{F}}{8} = 450°\text{F}$$

$$\text{maximum error} = \pm\tfrac{1}{2}\ \text{resolution}$$

$$= \pm\tfrac{1}{2}\,[450°\text{F}] = \pm 225°\text{F}$$

With the full-range monitoring system it would no longer be necessary to deal with out-of-range temperatures, but this advantage would be achieved at the price of a loss in resolution and accuracy (maximum error) of 450 percent:

$$\text{full range resolution loss} = \frac{450°\text{F}}{100°\text{F}} = \frac{225°\text{F}}{50°\text{F}} = 450\%$$

Therefore, by designing our monitoring system to deal with temperatures in the range of interest and declaring other temperatures to be out-of-range, we preserve the desired resolution of 100°F using only a 3-digit binary register.

In either A/D or D/A the conversion must be done on an intermittent basis. In the first place, a finite amount of time is required to make the conversion, and besides that, the process computer has other functions to accomplish between conversions. This raises the question of what value the variable should be given between conversions. An obvious strategy is to hold the variable at its previously calculated level until the next conversion takes place. However, another logical strategy is to take into consideration the rate of change of the variable, especially if the rate of change is great. Figure 14.11 compares these two methods. The error thus incurred between conversion instants may be greater than the error due to poor resolution. This has given rise to the theory and analysis of sampled data systems, an interesting but detailed and complicated subject beyond the scope of this text.

To put matters further into perspective, analog input/output is not as important to robots and manufacturing automation as is digital input/output. It is the combination of digital *logic* conditions in a process that most often calls for automated corrective action, not the degree of variation of continuous variables. And in the output phase, process control actions usually are discrete logic functions, not continuously variable responses. It is true that continuous process plants, such as chemical plants, refineries, power plants, and cement plants, have many critical *continuous* process variables that need close monitoring. But for applications of robots and manufacturing automation, the products, the processes, and the control variables are for the most part discrete.

INTERRUPTS

Picture an automation process with dozens of variables under control of an on-line computer. An orderly procedure must be set up to give attention to appropriate variables at the right times. Some of the variables or conditions are of such urgency that when they assume certain values, they interrupt the computer's other activities and are processed before any other activity takes place. Less urgent conditions can wait until the computer checks upon their status on a periodic basis. This is called polling. Both types are usually serviced by interrupts, however—the urgent conditions by direct interrupt and the polled conditions by timer-generated interrupt.

Real-Time Clock

The timer-generated interrupts are governed by the real-time clock within the process control computer. The clock enables the computer to keep pace with the outside world, especially with the process it is monitoring and controlling. The clock usually oscillates at a high frequency, but counters can be used in the fashion described in Chapters 12 and 13 to produce slower moving subclocks that may be of interest in the program logic. Since the clock is real-time, it counts continuously, independent of program control. To be usable in the control program the clock must be accessible by some addressable register. The convention is to set the clock register to some positive integer value and wait for the counter to automatically count down to zero. Upon reaching zero, the clock will automatically trigger an interrupt, which the programmer can choose to recognize by programming a routine intended to respond to the interrupt. One such routine would be to poll external process variables to determine whether action is required for any abnormal conditions.

Priorities

An on-line computer and its program are designed to be responsive to the process they are controlling. Therefore they may be busy or idle, dependent upon conditions that exist in the process. At a time in which one or more abnormal conditions exist in the process, these conditions may cause other things to go wrong while the computer is attempting to deal with the first problems encountered. Despite the speed of the computer, it is possible for it to become overwhelmed by a large number of conditions, each triggering interrupts and attempting to get the computer's attention. Picture a typical application in which the computer is subject to potential interrupt at any time by any or all of 50 process interrupts. Add to this a regular clock interrupt for polling another 200 process variables at intervals of perhaps five seconds. Although the time required to service any of the interrupts might be measured in milliseconds, a large number of abnormal conditions may cause the computer to be unable to complete the service of all interrupts. This characteristic of on-line computer operation has led to complicated software schemes for assigning priorities for various process interrupts.

The detailed analysis of interrupt priorities gives the automation engineer such intimate knowledge of the process and what could happen to it that sometimes design improvements to the process itself can be made. This can be considered a fringe benefit of automation. For instance, suppose a process computer monitors the levels of material in the in-feed positions for a robotic assembly station. If the variety of in-feed positions is large and the buffer size for each is small, the computer may become very busy monitoring levels and servicing the in-feed positions. A design improvement might be to increase the buffer sizes.

Once priority levels have been assigned to computer interrupts, the computer can be instructed to ignore lower priority interrupts while servicing higher priority interrupts. The term *interrupt masking* is used to describe the inhibition of lower priority interrupts on a selective basis. An interrupt is completely ignored while it is selectively masked. A system that recognizes various levels of interrupt priorities is called a *multilevel interrupt system*. Comparison of such a system with a single-level interrupt system is seen in Figure 14.12.

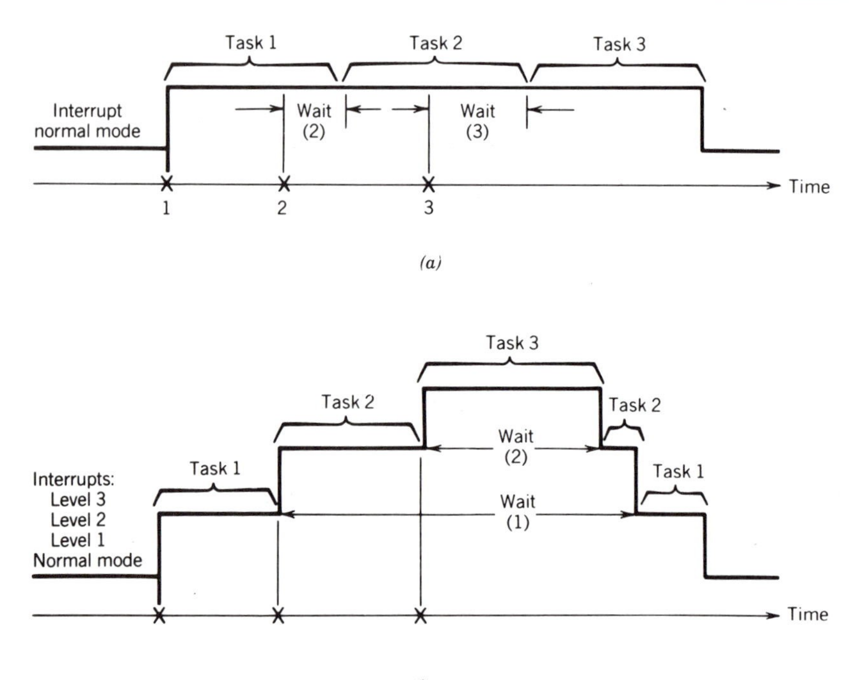

FIGURE 14.12 Comparison of computer responses to single-level interrupt system versus multilevel interrupt system. (Mikell P. Groover, *Automation, Production Systems, and Computer-Aided Manufacturing,* Copyright © 1980, p. 321. Reprinted by permission of Prentice-Hall, Englewood Cliffs, N.J.)

The assignment of appropriate priority levels is trickier than some people imagine. For instance, consider the following list of interrupt conditions:

- Robot hydraulic fluid level low
- Empty in-feed hopper
- Parts jam at discharge chute
- Message to be output to the CRT
- Operator pushes console switch

Which of the above conditions seems to be of highest priority? Which is lowest? Most would consider "robot hydraulic fluid level low" to be an urgent condition of high priority, and "operator pushes console switch" to be a low-priority operation input. In fact, a closer look at the crisis potential of each would likely place "operator pushes console switch" at highest priority and "robot hydraulic fluid level low" at the lowest priority position. In a crisis, it may be imperative for the operator to regain manual control of the process. Picture an industrial robot that refuses to acknowledge an operator input because it is too busy dealing with a multitude of process problems that it may not be equipped to handle. Unfortunately, some humans seem to exhibit this same distribution of priorities.

Referring back to the list of potential interrupts, another condition that may seem to be of low priority is "message to be output to the CRT." When things go wrong it may seem that the computer is too busy to bother with printing messages, but what good does it do to discover problems if there is no time to describe what is going wrong? It is entirely possible to program a process control computer to be so busy checking on serious problems in a crisis that it has no time to do anything to alleviate the crisis. Can you see some similarity here with the way some people react to a crisis? Maybe more will be learned about human behavior as we learn to assign priorities in the programming of industrial robots and their control computers.

We have examined the concerns of priorities in event of a crisis; now let us look at the opposite problem: Suppose the computer has serviced all interrupts and has nothing to do! This should be the normal condition for an on-line computer so that sufficient time will be available to service the interrupt routines if trouble does occur. There are two solutions to the problem of keeping the computer busy:

1. Background data processing in batch mode
2. Do-nothing routines

Background data processing is done on a time-shared basis and is sometimes called "free-time" processing. This is not to imply that background processing is of less importance, but from an urgency standpoint, such processing can usually wait a few seconds, minutes, or hours, depending upon the seriousness of the on-line interrupts encountered and what the computer must do to remedy the situation.

Of perhaps greater curiosity is the programming of do-nothing routines. This concept is the most difficult to accept by conventional batch-type data processing programmers. Consider the following FORTRAN program segment:

```
C  TWIDDLE THUMBS
   12 GO TO 13
   13 GO TO 12
```

or in BASIC:

```
12 GOTO 13     :REM TWIDDLE
13 GOTO 12     :REM THUMBS
```

The above program segment is a very tight endless loop, the bane of all freshman programmers. But endless loops are entirely acceptable in on-line computer programming and are even necessary at times to keep the computer active in RUN mode while it awaits the occurrence of an interrupt. Remember that the interrupt will cause the computer to drop what it is doing and branch, jump, or "GOTO" a specified memory address for the next instruction to be executed. The branch is permanent, and the computer will never return to the do-nothing routine unless specifically instructed to do so by the program.

PROCESS COMPUTER PROGRAMMING

At this point, the reader should recognize that on-line process computer programming is somewhat different from conventional batch-type programming. Still, it is possible to program in the same or similar compiler languages. Instead of FORTRAN, the language may be Process FORTRAN; instead of BASIC, the language may be Power BASIC. Instead of assembly language, the language may be PAL (Process Assembly Language). Whatever language is used, some facility must be available in the language for recognizing (and inhibiting) process and/or real-time clock interrupts. The language also must be capable of accepting process inputs, manipulating logic bits, and delivering process outputs. The needs of high efficiency in programs and the manipulation of binary logic tend to favor assembly languages in many applications, although both compiler and assembly languages are used.

The computer program flowchart of Figure 14.13 describes a portion of a system of routines for on-line monitoring of process variables. The purpose of the routine shown is to read an analog variable on a real-time sampling basis, average results, and alarm for any overload conditions indicated by analog readings that are out of limits.

Actual computer programs for executing the flowchart of Figure 14.13 are omitted in this text for brevity. The actual coding will be dependent upon available hardware and compiler/assembler software available for the application. The concepts studied in this chapter are intended to provide the automation engineer with an understanding of the overall approach to on-line computer programming and application.

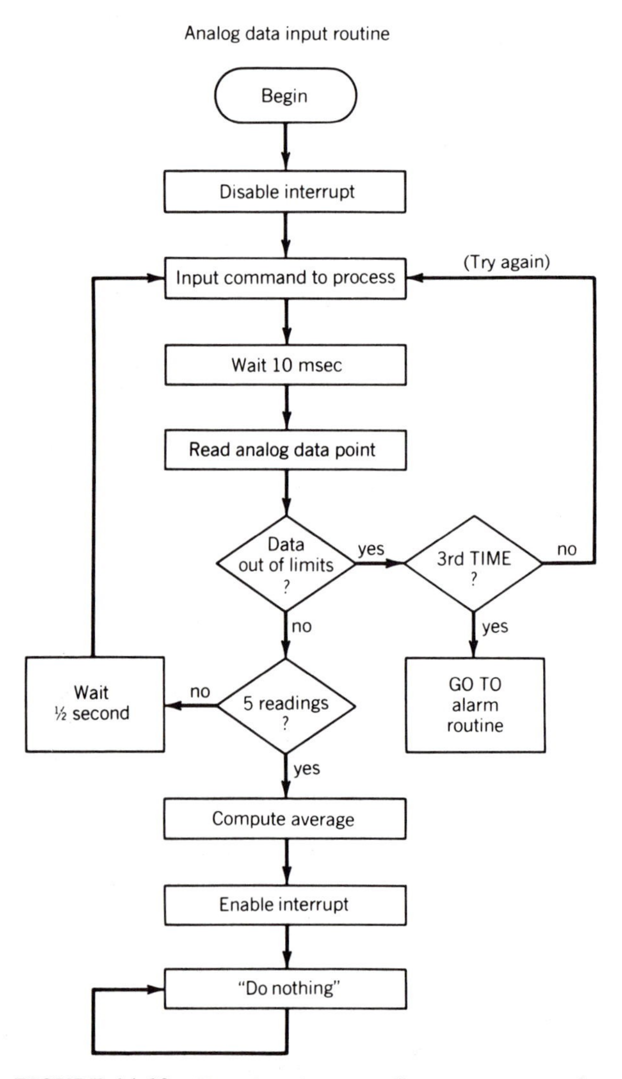

FIGURE 14.13 Flowchart for an on-line, process control computer program.

SUMMARY

On-line process control computers are similar to conventional batch-mode computers with an extended set of capabilities. Principal among these capabilities are the ability to read hard-wired inputs and write or send hard-wired outputs and the ability to respond to external interrupts. Several levels of computer automation of processes can be identified from off-line processing to closed-loop control of the process.

Three principal control strategies are used to minimize error between a continuous process variable's actual value and its desired level. These are proportional control, which acts upon existing error, integral control, which acts upon the accumulation of error through time, and derivative control, which anticipates error by acting upon its rate of change, either positive or negative. Combinations of these three strategies are usually used in practical control systems for continuous processes. Such systems are called PID control systems.

Two kinds of process data are accepted by the on-line process control computer: digital and analog. The same is true of computer outputs back to the process. Since the computer is digital, all analog data must be converted (encoded) by an A/D converter to digital format. Any digital representation of analog data is an approximation, the precision of which is determined by the range of the analog variable and the number of binary digits allocated to its digital representation. The translation of digital data back to its full-scale analog counterpart is called decoding. Despite the importance of analog data, especially for the continuous process industries, digital input and output remain the more important type of data in applications of robots and manufacturing automation.

The recognition and response to process and timer-generated interrupts are essential functions of the on-line computer software. In a crisis situation, the computer easily can become extremely busy servicing interrupts, too busy to recognize the more essential tasks. It becomes necessary to assign priorities to various interrupts and to inhibit or mask lower priority interrupts during the servicing of higher priority interrupts. The analysis and assignment of interrupt priorities are important and sometimes tricky maneuvers. The process calls upon the skill, judgment, and experience of the automation engineer.

The actual programming of on-line computers is similar to batch-type data processing programming with added instructions for dealing with interrupts and process input/output. The demand for speed and high efficiency from the on-line computer program tends to favor assembly languages in some instances although both assembly and compiler languages are used with on-line computers. With understanding of the basic concepts and application of on-line process control computers, the automation engineer should be ready to learn machine-specific process languages wherever they are needed to control systems for robots and manufacturing automation.

EXERCISES AND STUDY QUESTIONS

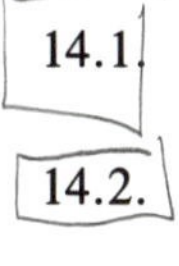

14.1. What are the principal differences between a process control computer and a conventional data processing computer?

14.2. Describe at least three levels or degrees of on-line computer implementation for an automatic process.

14.3. What is the principal advantage of the on-line assist mode over the process monitoring mode of on-line process control computer application?

14.4. Which is more important to robots and manufacturing automation, A/D converters or D/A converters, and why?

14.5. What is the most serious drawback to off-line data processing assistance to control a process?

14.6. Which of the major feedback control modes tends to correct accumulated residual error in the system? Which tends to minimize overshoot?

14.7. Explain the concept of "decay ratio."

14.8. When a sudden change in a process set-point occurs, which of the three major feedback control modes has the greatest immediate impact on the process variable? Which mode immediately works *against* the corrective action of this mode?

14.9. Why is there a limit to how fast a control system should force a correction in a process variable? In the extreme, what will happen if process gain is too large?

14.10. A robotic inspection station is monitored by an on-line computer, and the variable of interest is product weight. The expected range of weights is 100–130 lb, but weights outside this range, either high or low, are sometimes encountered. For a binary register length of five digits, set up the cell widths to accommodate an extra cell at each end of the range to account for out-of-limits points. Then calculate the resolution and maximum error.

14.11. An industrial robot gauges insertion resistance for an assembly operation by using a transducer that converts resistance in pounds to an analog voltage for input to the robot's on-line control computer. The range of assembly resistances is 20–50 lb. Five binary digits are allocated to the conversion.

(a) What is the resolution of the A/D converter for this process, measured in pounds?

(b) What is the maximum conversion error?

(c) For an assembly resistance of 34 lb, determine the encoded digital equivalent.

(d) Decode a digital value of 10111 for this system.

14.12. Normal limits for welding temperatures for a certain arc welding robot are 6000 to 8000°F. A transducer is capable of producing an analog to the temperature and is sensed by the robot's on-line computer controller via an A/D converter. Three binary digits are allocated to the digitized variable with upper and lower out-of-limits indicators provided at the two extreme values 111 and 000, respectively.

(a) Calculate the resolution of the A/D converter in degrees Fahrenheit.

(b) Determine the maximum conversion error.

(c) Determine the digital equivalent of 7500°F for this variable.

14.13. The following table displays time of occurrence of each of four interrupts and the computer time required to service each of them:

Interrupt	*Time of Initiation*	*Time Required to Service*
1	20	6
2	24	4
3	29	9
4	35	2

Determine the time of completion of service of interrupt 3 if

(a) Single-level interrupt mode is in effect.

(b) Multilevel interrupt mode is in effect and each interrupt is of successively higher priority.

(c) Multilevel interrupt mode is in effect and each interrupt is of successively lower priority.

(d) Single-level interrupt mode is in effect and interrupt 2 is selectively masked.

14.14. ***(Design Case Study)*** This study represents a diversion from applications pertaining to manufacturing, but many of the elements are present in this design problem for a review of the concepts of Chapters 11 through 14 of this book. A sailboat can travel at speeds of up to approximately 15 knots relative to the current (if any) in which it is traveling. It is possible, though unlikely, for a sailboat to move backwards in the water under certain extreme conditions such as remaining for a long period "in irons" (headed directly into the wind). It is desired to develop a digital readout for velocity of the sailboat as sensed by an analog flowmeter pickup from the water flowing past the hull of the boat. One can picture such a system as part of an on-line computer system monitoring various conditions on the boat.

The design problem is to select design parameters for an A/D convertor to provide a digital readout for the boat's velocity to the nearest knot. Design parameters should include:

(a) Number of binary digits required.

(b) Specification of what velocities shall represent "out-of-limits high" and "out-of-limits low."

(c) Decision points for each binary digit in a decision tree for the A/D conversion.

(d) Calculation of system resolution and maximum error.

In addition, design the logic system for decoding the digitized value to a two-digit decimal LCD display. For the display you may use a seven-segment LCD decimal display for each of the two decimal digits, or any other display scheme you describe in your system design. Include in your design a statement of what symbols you will use to represent high or low out-of-range velocities.

CHAPTER 15

MICROPROCESSORS

The microprocessor is the *fundamental element* in the construction of manufacturing automation equipment and is the most basic development we will study in this book. Then why is this Chapter 15 instead of Chapter 1? It is enough for many to know that virtually all of the robots, programmable logic controllers, and on-line computers use these tiny circuits and that microprocessors are capable of being intricately programmed to perform rapidly a wide variety of computations, monitoring operations, and other tasks. But those engineers who probe a little deeper and become proficient in applying microprocessors, themselves, to applications of robots and manufacturing automation will reap the greatest rewards this field has to offer.

What are the prerequisites to a proficiency in microprocessors? First, one needs to learn what a microprocessor is and how it differs from ordinary computers and even microcomputers. Then there is the need to learn something about integrated circuit technology and how a tiny substrate of silicon can be used to build up a circuit as comprehensive as a digital computer. Also, terminology can be a problem, and one needs to know the differences between ROM, RAM, PROM, EPROM, EEPROM, and other terms associated with the structure of microprocessors. And then there is programming—programming at its most fundamental level—in assembly language. To program effectively in assembly language, the need to deal with hexadecimal code becomes quickly apparent.

The purpose of this chapter is to introduce microprocessors so that the automation engineer can determine under what circumstances it becomes worthwhile to employ them directly. A second purpose is to show how much the principles studied in other chapters of this book are applied to microprocessors, both in their application and in their construction. We shall also view flexible automation from a new perspective and realize that not all recent development is toward more flexible automation systems.

MICROPROCESSORS OR MICROCOMPUTERS

It has become acceptable to interchange the terms *microprocessor* and *microcomputer,* but it is essential to note their difference in this text on the subject of robots and manufacturing automation. A microcomputer is a data processing system that employs a microprocessor as its central processing unit (CPU). The system typically has a keyboard and CRT for essential input and output and may have peripheral, floppy disk and printer equipment as well. But a microprocessor, in the narrow sense in which we use the term in this chapter, has none of these peripherals. The microprocessor is the basic silicon chip containing the electronic circuit that performs the CPU functions and houses some memory registers. If additional chips are used for memory or input/output, the system can be considered to be a microprocessor system, but we shall reserve the term micro-*computer* to refer to systems that have at least a keyboard and CRT for input and output.

Computer and data processing professionals will ask what use a microprocessor can have if it has no peripheral equipment to program it, feed data to it, and receive computed results from it in the form of output. But they ask this question because they see the microprocessor as a computer; it is *more than that*. The extremely low cost and versatility of the tiny microprocessor chips have given them a role in robots and manufacturing automation without the need for the peripheral input and output devices computer professionals like to attach to them.

Admittedly, at the beginning it is necessary to attach *temporarily* to the microprocessor some means of setting up its application program. But the microprocessor's cost is so low that once the application program is loaded into the microprocessor's memory, the hardware can be *dedicated* to that application for its entire life. The microprocessor has the flexibility to accept modifications or even a complete reprogramming if desired at a later date. But the point here is that the automation engineer has the *opportunity* to freeze the function of a microprocessor into a single low-cost application.

Compare the microprocessor as described in this chapter with the computers described in Chapter 14. General-purpose computers, whether mainframe, mini, or micro, have their place in the scheme of robots and manufacturing automation, but their role is decidedly different from the microprocessor's role. The pure microprocessor applications can in truth be likened to hard automation because of the extensive setup required and the relative permanence of the application setup. But the resultant system has more flexibility than hard automation and requires only a tiny fraction of the hardware investment.

The setup programming of these tiny silicon chips called microprocessors is a detailed and tedious task that we shall address, but first let us examine the construction of a microminiaturized, integrated electronic circuit of which the microprocessor is an example.

INTEGRATED CIRCUITS (ICs)

It is now routine for complicated circuits to be plated or etched upon a single chip of silicon. There is no mystery to the selective plating of conductors on a large circuit board because this is done on a scale that we can actually see. It is easy to extend this concept to the microscopic level. What is difficult to grasp, however, is that the *components* of the circuit, such as resistors, capacitors, and transistors, also can be plated microscopically upon the silicon chip. This is possible because silicon, a nonconductor, can be transformed into a semiconductor by adding impurities. Even more important is the fact that an electric field can be applied to the semiconductor to change its conductivity. The ability to throttle or control the flow of electrical current by means of a separate signal circuit is the key to the creation of semiconductor circuits that take the place of relays, amplifiers, and computer memory circuits. Figure 15.1 illustrates the principle of the operation of a transistor that can be seen to emulate the same functions as were studied in Chapter 2 for the electrical relay. The figure shows two types of semiconductor silicon: N-type and P-type, which are made by adding different impurities to each type. The flow of current

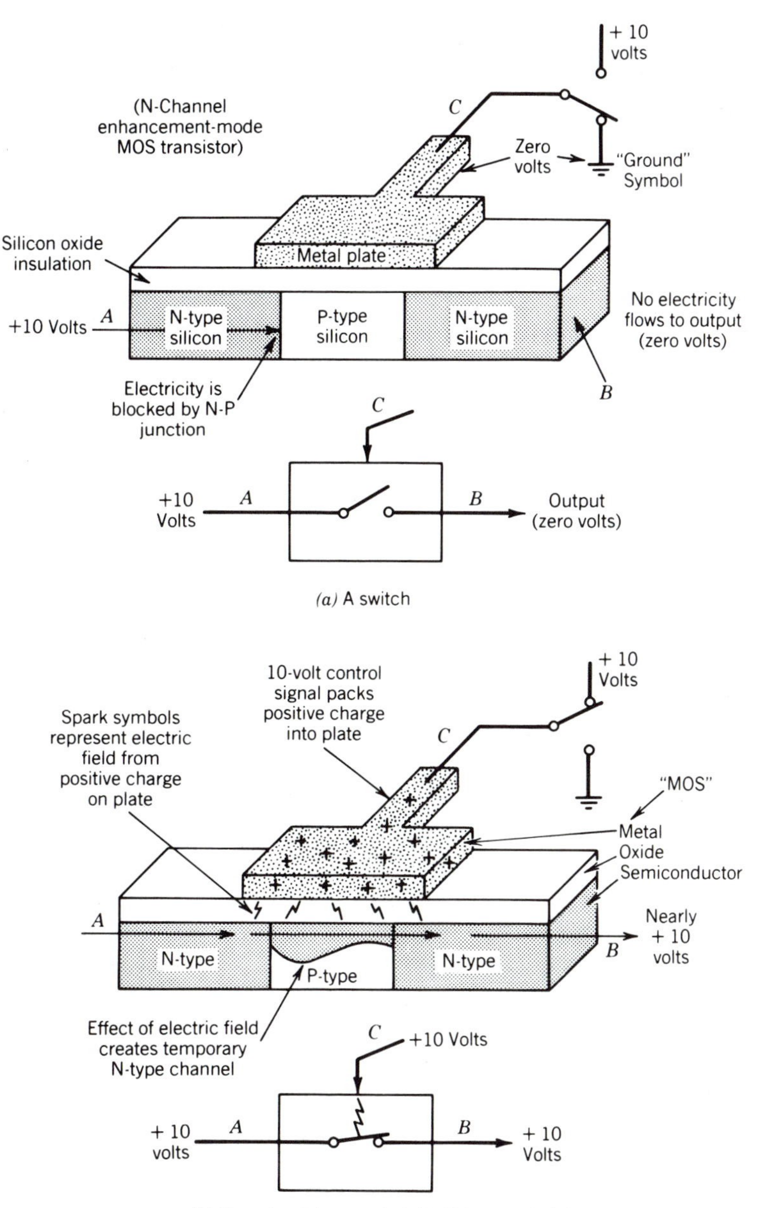

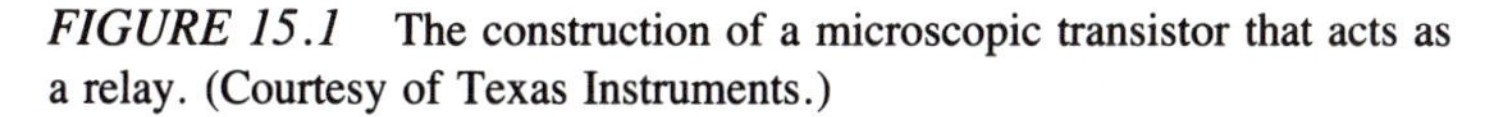

FIGURE 15.1 The construction of a microscopic transistor that acts as a relay. (Courtesy of Texas Instruments.)

through the N-type silicon is blocked by the tiny section of P-type silicon. But by applying an electric field to the small block of P-type silicon, it can be temporarily induced to admit current flow as if it were N-type silicon. The metal plate that produces the electrical field is separated from the current-carrying N- and P-type silicon layer by a silicon-oxide insulating layer. The three layers then make up a *metal-oxide semiconductor* (MOS) sandwich. Note that there are no discrete pieces to be assembled in this device, only layers to be selectively plated or etched upon the silicon base or substrate. This characteristic enables MOS sandwiches to be manufactured on a microscopic scale using photographic plating processes.

Circuit densities on semiconductor chips are phenomenal. Approximate circuit densities as of the early 1990s were as many as 1,000,000 transistors plus resistors and capacitors on a single $\frac{1}{4}$-in. (6.35-mm) square chip of silicon. Some predict 50–100 million transistors per chip in the 1990s [ref. 14]. The manufacturing method for these integrated circuit (IC) chips is to perfect the circuit design and then manufacture hundreds of identical chips on a single wafer of silicon around 3 in. in diameter. The chips are then tested because it is nearly inevitable that many of the individual chips on the wafer will be defective. Figure 15.2 illustrates the circuit testing of one tiny circuit while the chip is still intact with hundreds of other identical ICs on the silicon wafer. After testing, the wafer is scored and the individual chips are cut from the wafer. The production of the wafers, cutting apart of the IC chips, and packaging of the chips into a module suitable for wiring in a circuit board is summarized in Figure 15.3.

Circuit integration has progressively become more severely compact, and the progression is expected to continue. As of the mid-1980s the typical stage of progression was referred to as *large-scale integration* (LSI), and microprocessors were considered to be LSI chips. Figure 15.4 illustrates the degree to which a microprocessor circuit is condensed onto a single chip of silicon. Although LSI is more compact than its prede-

FIGURE 15.2 Circuit testing of an integrated circuit (IC) while it is still intact with hundreds of other identical ICs on a silicon wafer. (From *The Engineer,* copyright 1983, AT&T Technologies, Inc., New York, N.Y. Reprinted by permission.)

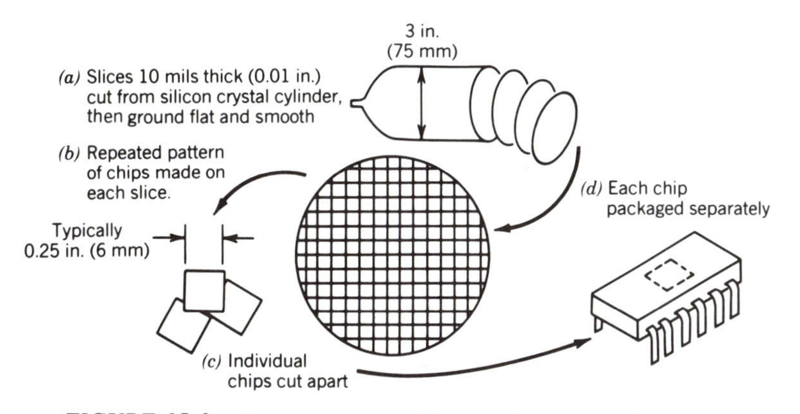

FIGURE 15.3 The manufacture of integrated circuits. (Courtesy of Texas Instruments.)

cessors, *small-scale integration* (SSI) and *medium-scale integration* (MSI), LSI chips certainly are not the ultimate in integrated circuit technology. Even more compact are *very large-scale integration* (VLSI) chips and *extra large-scale integration* (XLSI) chips. An obvious objective in miniaturizing circuits is to make them more compact so that they will fit into small devices such as watches, calculators, and some robots. But compactness per se is not really the primary objective. As circuits get smaller, they become faster and consume less energy. They also usually get cheaper.

Quality control is of paramount concern in the manufacture of integrated circuits. The first wafers produced 20 defective chips for every good chip. This dismal level of quality was tolerated as every chip was tested and the bad ones were thrown away. Improvements to the process gradually increased the chip percentages, which gradually increased the

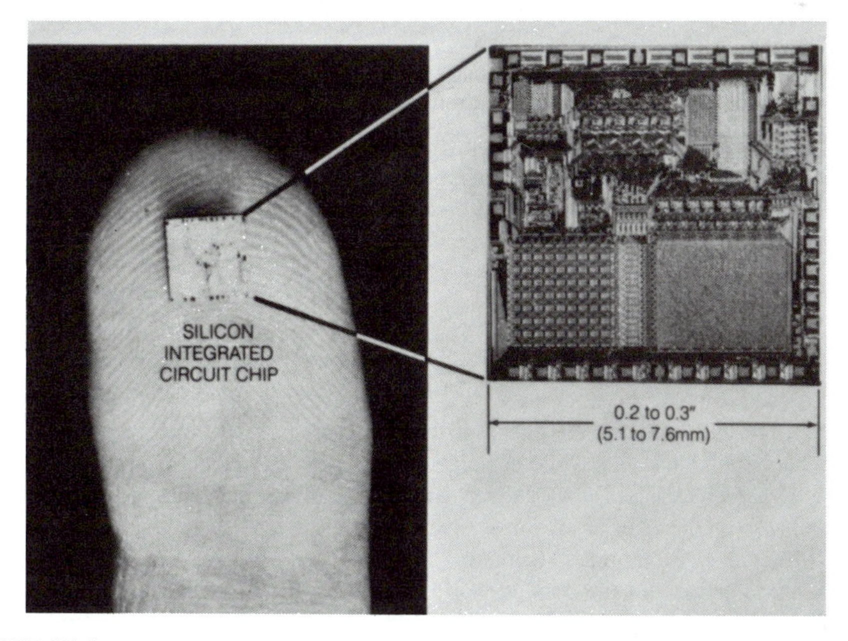

FIGURE 15.4 Integrated circuit (IC): microprocessor chip. (Courtesy of Texas Instruments.)

wafer yield in terms of good chip count, and this development began to reduce dramatically the cost of integrated circuits and the products from which they were made. A gain in wafer yield from 5 percent to 50 percent is a 90 percent reduction in IC chip cost—a change by a factor of 10. It is no wonder then that electronic calculators, watches, and microcomputers have experienced great reductions in cost near the end of the twentieth century.

Process quality problems persist, and postprocess screening and culling of defective chips remain in the method of manufacture today despite processes improvements that have been made. It is extremely important that microscopic dust does not contaminate the silicon and perhaps ruin an entire circuit. This has led to clean-room processes in which the air must contain fewer than 100 contaminant particles per cubic foot—that is, it must be 100 times cleaner than a modern hospital!

The chip-quality obstacle has blocked the development of much larger IC chips that would make large, powerful computers possible on a single chip. To grasp the significance of such a chip, consider that a "blueprint of a superchip, showing the smallest features blown up to a tenth-of-an-inch scale, would be the size of two football fields side by side" [ref. 62]. Despite clean-room procedures and advanced manufacturing methods, tiny chip imperfections are still inevitable in wafer-size chips. To compensate for these tiny flaws, wafer-size chips contain redundant circuits so that the entire circuit does not have to be discarded.

The continuing development of faster, smaller, less-energy-consuming integrated circuit chips is opening up new possibilities for microprocessor designs for robots and manufacturing automation applications. With the basic understanding of integrated circuits that has been provided by this section, let us now examine the structure of the microprocessor integrated circuit in particular because of its great importance to robots and manufacturing automation.

MICROPROCESSOR STRUCTURE

Although we have distinguished between a computer and a microprocessor, the microprocessor must have the functional elements of a computer CPU, some limited memory, and a means of passing digital information among its components and to and from the outside world (but not necessarily via peripherals, as in a microcomputer). Let us examine the structure of these elements inside the microprocessor, and while we are doing so we shall point out those features that make the microprocessor so important to the robotics and automation engineer.

CPU

The nerve center of the microprocessor or any computer is the *central processing unit* (CPU). The CPU must have digital registers for holding binary logic data. These registers hold the data temporarily while calculations or logic manipulations are performed upon them. One such register is called the *accumulator,* a name that is derived from its purpose in adding the contents of one memory location to the contents of another. The accumulator also often acts as a temporary station for binary data being input to the CPU from memory, vice versa, or to/from the outside world. Another important register is the *instruction register* or decoder, which holds and interprets a binary-coded machine instruction while it is being executed. Other registers are *address registers* for reference in locating program instructions or data in memory.

The CPU must also have an *arithmetic/logic unit* (ALU), which handles the Boolean logic to perform standard arithmetic manipulations such as addition and subtraction. The detailed construction of these logic circuits is quite complex but follows the basic methods of logic circuit construction discussed in Chapter 11.

Finally, the CPU must have some means of timing operations to provide coordination and control of all operations. The CPU timer, usually called the clock, may also be available to the programmer for counting clock pulses in a register so that microprocessor functions can be synchronized with external activities in the real world and in real time. However, the CPU clock is usually unavailable to the user, and applications that require timing employ separate support chips called "PITs" (programmable interval timers). The actual cycle timer in the microprocessor CPU operates at extremely high frequencies—for example, 33–50 MHz (megahertz).*

Memory

Some bit registers are necessary for storing binary information of two types:

1. Binary instruction code
2. Binary application data

Registers for either of these purposes are called *memory*. Memory registers in most microprocessors are 32 binary digits (bits) long. The length of the memory register is called *word length* and is a basic specification of the microprocessor.

Access to memory registers in a microprocessor is not as frequent as access to the CPU registers such as the CPU accumulator. However, access to individual memory registers wherever they are must be quite direct if they are to be useful at all for storing program instructions or application data. It would be useless to have a high-speed CPU only to have the processor wait every instruction cycle for a tedious search through a sequential file of 64,000 memory registers for the specific piece of data needed or for the program instruction to be executed. The versatility and computing power of a microprocessor demands that all memory registers be accessible immediately by means of unique memory addresses for each memory register versus the alternative of searching through a sequential file. The direct accessibility of memory registers via unique addresses is such a key characteristic of microprocessor memories that the memories themselves are designated as RAM, which means *random-access memory*. In micro*computers,* so much RAM is often needed as to require additional banks of RAM chips linked to the main microprocessor chip.

General-purpose memory registers generally are subject to change as new programs or data are introduced into the machine. This is especially true of microcomputers and to a lesser extent of microprocessors. But certain program instruction data is always needed to handle standard essential procedures common to all uses of the microprocessor. It is wise to reserve certain areas of memory for this purpose so that a mistaken programmer cannot inadvertently write over these memory registers with new instructions or data, rendering the microprocessor useless. Computer memory registers can be constructed to be permanent so that no programmer can destroy their essential contents once they have been preprogrammed. Such memory registers are designated as being part of the ROM or *read-only memory* because no programmer can write over their permanent contents.

*33 megahertz = 33 million cycles per second; the period of each cycle then is about 30 billionths of a second or 30 nanoseconds.

If the reader understood the distinction made earlier between microcomputers and microprocessors, it should be apparent at this point that ROM is more important to the microprocessor than to the microcomputer, but both have need of at least some ROM. The microcomputer is reprogrammed over and over and must necessarily have RAM for this purpose. But a microprocessor may be slaved to a single automation application and be programmed only once in its life, making the use of ROM the only appropriate memory medium in such cases.

The demands of microprocessor applications in industry have generated a need for a slight withdrawal from the absolute finality of ROM. For instance, it may be desirable to allow the user of a microprocessor to program the ROM instead of the manufacturer, acknowledging the contradiction such a statement makes (ROMs are by literal definition impossible to program). To accommodate this need on the part of the microprocessor users, a variation of the ROM is PROM, or *programmable read-only memory*. The contradiction applies only once because after the PROM has been programmed by the user once, it becomes permanent.

The industry, however, has carried the contradiction even further because sometimes the purpose of a microprocessor changes or improvements to the automatic process make the program stored in the PROM obsolete. So, compounding the contradiction is a type of memory called EPROM for *erasable programmable, read-only memory*. EPROMs cost a little more than PROMs, but they are extremely important to the automation engineer because they permit a highly efficient, assembly-language level automation or robotics program to be set up with a very low hardware cost. The procedure for reprogramming an EPROM chip is to expose it to high-intensity ultraviolet (UV) light for a few minutes. There is considerable cost to reprogramming the EPROM when the automation application changes, but such costs are considered setup, and the length of the automated production run is often long enough to justify this setup. What the EPROM provides then is a technology that serves the long production run applications for robotics and manufacturing automation without requiring the type of total hardware commitment requisite to fixed, hard automation.

Another popular type of microprocessor memory device is the EEPROM, for *electrically erasable programmable read-only memory*. The EEPROM, although more expensive than EPROMs, has the advantage of reprogrammability without the necessity of removal of the chip from the circuit board. The reprogrammability of EEPROMs is almost as efficient as for RAM. In fact, EEPROMs are so alterable as to raise the question: Why not simply use RAM? The answer is that EEPROMs can be *nonvolatile*—that is, they are not automatically erased as soon as electrical power to the chip is lost. Unfortunately, RAM is generally volatile and is usually for storage of only intermediate process calculations and data that the process can afford to forfeit if a power loss occurs. Nonvolatility is a highly attractive feature for microprocessor memory, especially for the storage of program instruction steps or for permanent routines for robots and manufacturing automation applications.

Data Transfer

We have discussed the features of the microprocessor CPU and its various memory types, but we have not yet discussed how data can move from memory to the CPU registers and vice versa, and how data can be received from or transmitted to external devices. It is obviously impractical to have a separate connection from each memory bit register to the bit registers of the CPU, so common conductors are connected between all memory locations and some of the CPU registers. Such common conductors are called *buses*. At first, this seems impossible because elemental physics tells us that only one electrical impulse (one bit) can be transferred in a single electrical conductor at one time. The loophole in the foregoing physical law is the word *time*. The extremely high speed of the CPU clock can synchronize a clock pulse at precisely the right time to coincide with the presence of data at a logic AND gate to satisfy the AND logic and effect the transfer of data. This follows directly from the principles covered in Chapter 11. Similar AND logic can be used to permit an address register to control the flow of data through a data bus. A simplified example will illustrate using a 1-bit data bus external to the microprocessor.

EXAMPLE 15.1 Control of Data Flow on a Bus

A 2-bit address register is used to select which of four bits of data will flow from an industrial robot to a microprocessor via a 1-bit data bus. Designate the two bits of the address register as logic variables A_0 (low order) and A_1 (high order) and the robot status variables as R_1 through R_4, and designate the logic status of the data bus as B. Use Boolean expressions and a logic diagram to describe the relationship between the robot status variables, the address register, and the data bus. Permit only one bit of data on the bus at one time. Specify both the binary and decimal address for each robot status variable.

Solution

Boolean Expression

$$B = \overline{A}_1\overline{A}_0R_1 + \overline{A}_1A_0R_2 + A_1\overline{A}_0R_3 + A_1A_0R_4$$

Addresses

	Binary	*Decimal*
R_1	00	0
R_2	01	1
R_3	10	2
R_4	11	3

The logic diagram is displayed in Figure 15.5.

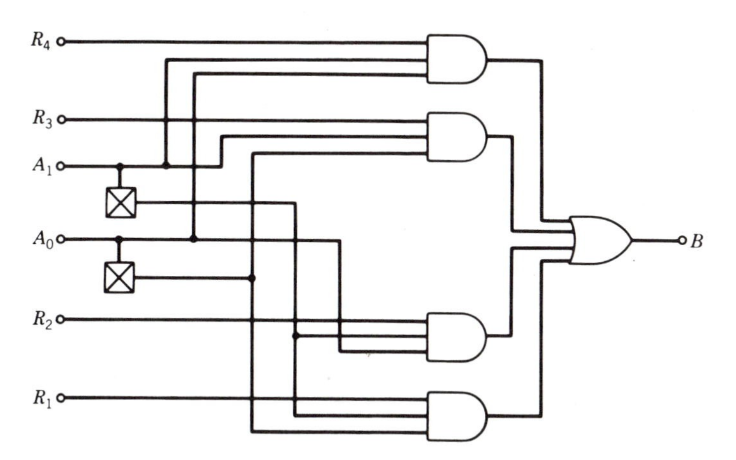

FIGURE 15.5 AND-logic used to effect data transfer on a common bus (Example 15.1).

Transfer of data to and from the outside world (input/output) is performed a word at a time and presents a different problem than internal data transfer because the outside world is not constrained to heed the clock pulse of the microprocessor. There are two schemes for word input/output: parallel and serial. Parallel I/O is the simplest because in this method a separate connector is used for every bit in the word. For an 8-bit word machine, therefore, eight separate pins are necessary for transferring each of the binary bits of information. By contrast, serial I/O uses only one connector for a word of data. In serial I/O, the individual binary bits are transferred in a series at a prescribed frequency to make up the full word. The difference between serial and parallel I/O is diagrammed in Figure 15.6 for a standard 8-bit (one-byte) word.

To summarize, it is indeed possible to house all the computer functions just described—CPU, memory, and input/output—all on a single IC chip. Figure 15.7 shows the general organization of such an integrated circuit magnified many times. For spatial reference, this circuit is the same one as was illustrated earlier on a person's fingertip in Figure 15.4. Such a circuit has tremendous potential for robots and manufacturing automation due to its high capability and small size, and especially because of its low cost. The

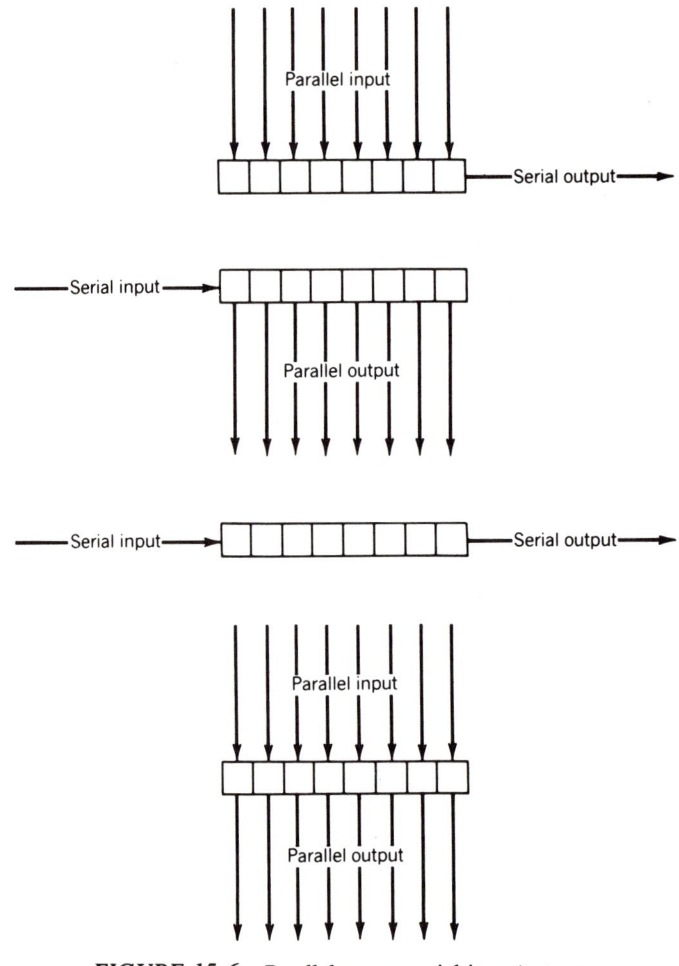

FIGURE 15.6 Parallel versus serial input/output.

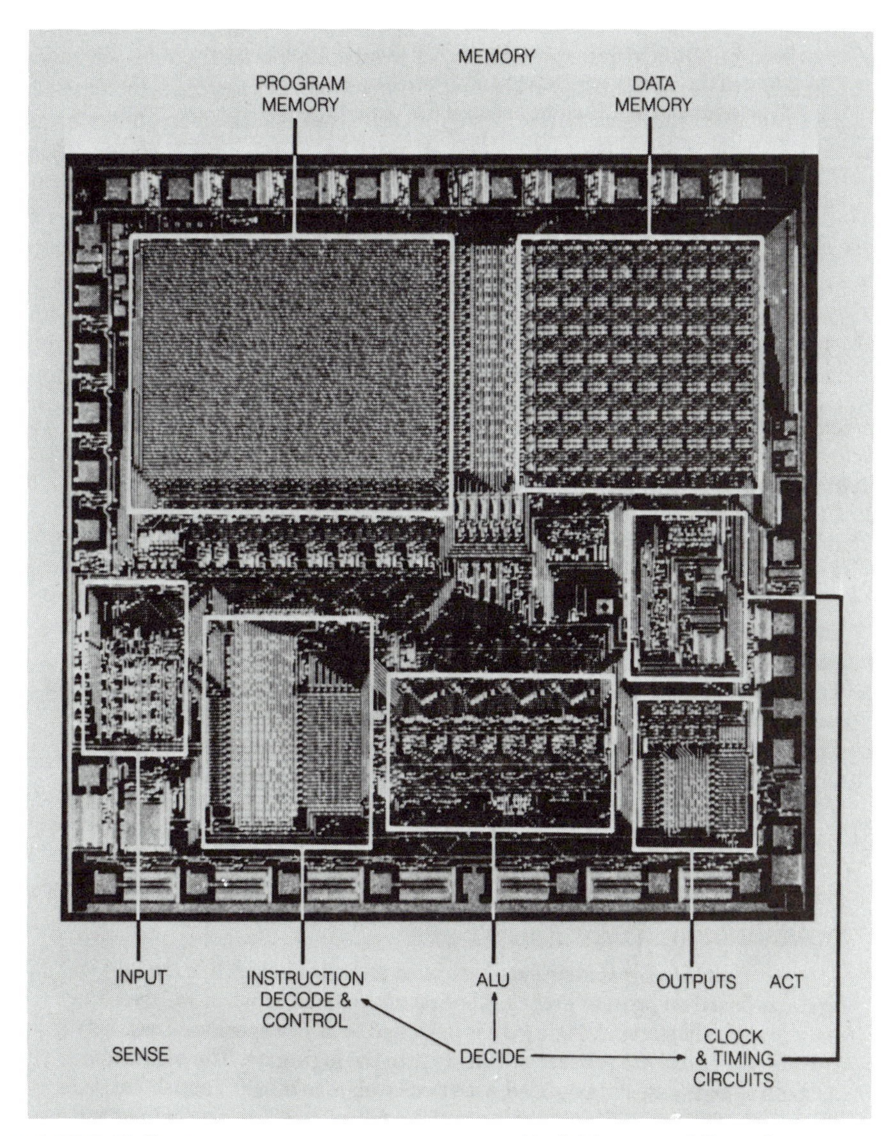

FIGURE 15.7 A microprocessor "computer-on-a-chip." (Courtesy of Texas Instruments.)

automation engineer who understands the operation and programming of these tiny chips has an expanded view of the potential of automation, a view that is much broader than that of the engineer who merely follows the operating manuals of commercially available robots and automation equipment. The next section will address the programming of these microprocessor chips.

PROGRAMMING

Programming microprocessors to control industrial robots or any other manufacturing automation device is extremely detailed. The microprocessor itself works strictly with binary forms of data and instructions, a level that is certain to befuddle the most meticulous of programmers. Higher based number systems such as octal, decimal, or hexadecimal greatly simplify the process, but when we say "simplify," we are not speaking of the degree of simplification afforded by the higher level computer languages such as BASIC, PASCAL, FORTRAN, and C. The base 10 (decimal) number system is the most familiar to the human race, and there is no question that it would be the truly universal number system if the number 10 were an even power of the number 2. In Chapter 5, we studied binary-coded decimal (BCD), which dealt with the problem created by the fact that 10 is not an even power of 2. The base 8 (octal) system is convenient because $2^3 = 8$, and octal has been a popular number system for dealing with large mainframe computers. The language of microprocessor programming, however, is expressed in hexadecimal or simply "hex," which is a base-16 number system. The importance of hex to microprocessor usage and to the programming of some commercially available robots warrants some understanding of hex on the part of the serious roboticist or manufacturing automation engineer.

Hexadecimal

Even the name *hex* for the hexadecimal number system conveys some sort of mysticism to bewitch the microprocessor programmer, but hex is really not as complicated as it may seem. Its convenience derives from the fact that $2^4 = 16$, which means that any four-digit binary number can be represented by a single hex digit. The extremely popular 8-bit (one-byte) word length of microprocessors can then be represented by two hex digits. The increasing popularity of 16-bit (two-byte) words in some models of microprocessors is also served well by hex by assigning four hex digits to a single 16-bit word.

The construction of the hexadecimal character set uses the familiar decimal characters (0 through 9) to represent the first ten characters. Six more characters are needed to complete the set, so the first six letters of the alphabet are used (A through F). Table 15.1 summarizes the hex character set by relating it to decimal and to binary.

Using Table 15.1, it is a simple matter to convert hex numbers to the microprocessor's binary system, as can be seen in the following examples:

$$2F_{(\text{hex})} = 00101111_{(\text{binary})}$$

$$CB_{(\text{hex})} = 11001011_{(\text{binary})}$$

TABLE 15.1 Hexadecimal Character Set

Decimal	*Hexadecimal*	*Binary*
0	0	0
1	1	1
2	2	10
3	3	11
4	4	100
5	5	101
6	6	110
7	7	111
8	8	1000
9	9	1001
10	A	1010
11	B	1011
12	C	1100
13	D	1101
14	E	1110
15	F	1111

Assembly Language

Most readers of this book will be familiar with one or more user-oriented procedural computer languages, such as BASIC, or FORTRAN or C. A characteristic of the computer programs expressed in these and in all other computer languages is that the instructions are executed by the computer one step at a time in a sequence. The order of the sequence may vary, and the execution of the program itself can redirect program control to cause the sequence to change, but the execution is still sequential: one step at a time. Regardless of the computer language used, all computer programs are eventually translated into the binary machine language of the computer itself prior to execution. Machine language also is sequential and is processed one step at a time. However, one sequential step in BASIC translates into a whole sequence of steps in machine language. The sequential nature is retained, but a BASIC program is a sequence of sequences of machine-language instructions. BASIC statements are therefore called "macro instructions" because one BASIC statement represents several machine-language instructions.

Most microprocessors, however, are programmed in *assembly language*. The most fundamental difference between assembly language and the higher level languages such as BASIC is that assembly language instructions relate to machine-language instructions on a one-to-one basis or nearly so, as compared to the macro instruction characteristic of the higher level languages.

The essential content of any assembly language or machine language instruction is as follows:

1. *Instruction address.* A series of binary digits (in machine language) or a smaller series of hexadecimal digits (in assembly language) that specifies the location(s) in memory (ROM or RAM) for this particular instruction. The address can be a symbolic alphanumeric label in assembly language.
2. *Operation (Op) code.* A small series of binary digits (machine language) or two or three alphabetic letters (sometimes numbers) arranged as a mnemonic (assembly language) used to specify what the instruction is intended to accomplish; the op code is usually stated as a verb. Example assembly language mnemonics are:

   SUB—subtract

   MPY—multiply

   STR—store (into memory)
3. *Operand(s).* A number(s) or an address(es) expressed in binary (machine language) or hex (assembly language) used to further describe what the instruction will accomplish; the operand can usually be interpreted as the object upon which the verb in the op code is to take action. In assembly language, the operand(s) can be an alphanumeric symbol instead of the actual address or numerical constant.

Convention dictates that the arrangement of these three components of an assembly level computer instruction is in the order shown, using one line per instruction. In the space to the right of the instruction, the programmer may add an optional comment for reference in explaining the instruction. At this point, some example complete instructions will be interpreted for illustration:

CHK1	LDA	ROBOT	The operation mnemonic LDA in this instruction means "load into the A-register." The memory location designated by the symbolic address "ROBOT" contains the data to be loaded into the

		A-register. The symbolic address of the instruction is CHK1.
SLA	7	The contents of the A-register are shifted left seven bit positions. The highest order seven digits are lost. Binary zeroes are added on the right to fill the seven digit positions gained on the right.
JNZ	ROUT1	Jump to ROUT1 for the next sequential instruction if the value of the A-register is not zero.
EI		Enable interrupt; this permits the CPU to be interrupted to perform a special operation whenever the occasion arises as signaled by an interrupt condition.
DI		Disable interrupt; the CPU will ignore interrupts henceforth (until an EI command is encountered) while it performs some high-priority operations.
JMP	IDLE	Jump unconditionally to the memory location symbolically named "IDLE" for next instruction.

The above examples explain only a small subset of the complete assembly language instruction set for a typical microprocessor. However, the reader should be able to visualize the form of assembly language programming from these examples. An example will now use these example assembly language instructions to illustrate an application of microprocessors to robots and manufacturing automation.

Example 15.2
Assembly Language Emergency Routines

In the previous example four logic status variables from an industrial robot work station were of interest in a microprocessor-based automation system. Suppose a logic 1 in these four robot status variables represented the following, respectively:

R_1—Someone or something has intruded into the robot work envelope during operation.

R_2—The robot hydraulic reservoir has exceeded the high temperature limit.

R_3—The discharge chute for finished parts is full and is backing up to the robot release station.

R_4—The in-feed hopper (vibratory bowl) for parts to be presented to the robot for processing is empty.

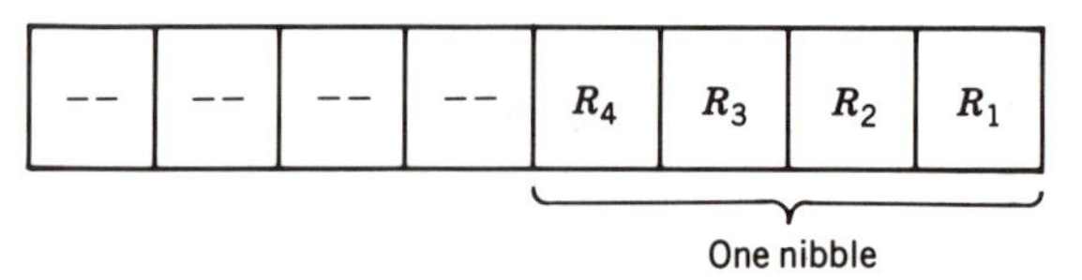

FIGURE 15.8 Robot status variables contained in the low-order bits of memory location ROBOT.

The four bits of data from the robot station represent one nibble (one-half byte) of data. Suppose these data are stored in a microprocessor RAM location symbolically called "ROBOT." Further assume that the microprocessor has a word-length of one byte and that the nibble of interest is in the lower order half of the byte as shown in Figure 15.8. Write an assembly language routine to check for each of the robot conditions R_1 through R_4 and jump to emergency routines ROUT1 through ROUT4, respectively, to service the conditions as required.

Solution

Let us name this assembly language routine "TEST" so that every time it is desired to examine the input conditions R_1 through R_4 for alarm conditions, the program can be directed to memory location TEST for execution. We will assume that each alarm routine ROUT1 through ROUT4 has an appropriate exit either back to test other alarm conditions or to await future interrupts as desired. The following assembly language program with comments supplied will execute alarm routines ROUT1 through ROUT4 as required:

TEST	DI		Disable interrupt. We want no CPU interruptions while we service alarm conditions R_1 through R_4.
CHK1	LDA	ROBOT	Load the contents of RAM address "ROBOT" into the A-register.
	SLA	7	Get rid of high-order bits.
	JNZ	ROUT1	If the A-register is not 0 ($R_1 \neq 0$), jump to alarm routine ROUT1.
CHK2	LDA	ROBOT	Load ROBOT into A-register again.
	SRA	1	Get rid of R_1.
	SLA	7	Get rid of high-order bits to select R_2 this time.
	JNZ	ROUT2	Jump to ROUT2 if $R_2 \neq 0$.
CHK3	LDA	ROBOT	Load ROBOT back into the A-register again.
	SRA	2 }	Select R_3 this time.
	SLA	7 }	
	JNZ	ROUT3	Jump to ROUT3 if $R_3 \neq 0$.
CHK4	LDA	ROBOT	Load ROBOT into the A-register again.
	SRA	3 }	Select R_4.
	SLA	7 }	
	JNZ	ROUT4	Jump to ROUT4 if $R_4 \neq 0$.
	EI		Enable interrupt.
IDLE	LDA	ROBOT }	This is really a do-nothing routine simply to give the microprocessor something to do while it waits around for an interrupt.
	JMP	IDLE }	

As can be seen from Example 15.2, assembly language programming is a detailed and tedious task when compared with BASIC or C. However, the manipulation of logic bits of information is inherently a detailed operation because logic variables are binary variables. Although it is convenient to use compiler languages for data-processing applications for computers, any wide departure from binary can create as many problems as it solves when dealing with *logic variables* as in applications of industrial robots and manufacturing automation. In addition, assembly language can sometimes provide some computational economies because of its close relation to machine language.

The microprocessor assembly language statements shown in this chapter are exemplary of all microprocessor assembly languages. Once the automation engineer has programmed one microprocessor, he or she can easily program others by referring to the instruction set for the particular microprocessor being programmed for the small differences that do exist between assembly languages. It is not the purpose of this text to delve into all the detailed differences of microprocessor assembly languages, or even to teach a single microprocessor language because that would tend to make the reader machine-specific. Rather, this chapter's purpose is to expose the automation engineer to the extraordinary potential of microprocessors and to dispel some of the mystery surrounding their construction, their programming, and their use. With an understanding of these basic microprocessor principles, let us now examine the ways that industry is actually using microprocessors in robots and manufacturing automation.

MICROPROCESSOR APPLICATIONS

The application of microprocessors in automation is in three general categories: original equipment manufacturers (OEM), microcomputers, and special-purpose controls. Each of these categories will now be explained.

Original Equipment Manufacturers (OEM)

It sounds contradictory to refer to an "original equipment manufacturer" as an "application" of microprocessors. But this is a term widely used in the field of industrial automation, and the robotics and automation engineer needs to understand its place in the automation movement. The low cost and high capabilities of the microprocessor make it a very attractive piece of hardware to build right into a new equipment design. In this category of microprocessor application, the equipment designer anticipates all modes in which the automated system will operate and all environments to which the system must react. The programming is virtually always in assembly language, and the predominant memory medium is nonvolatile, usually ROM or PROM. The investment in detailed programming is easily justified by the generally large numbers of identical microprocessor applications envisioned for a single model of automation equipment.

Figure 15.9 displays a piece of automation equipment, designed and built by Artran, Inc., and used in the poultry processing industry. As originally designed, the equipment was controlled by a commercial-model programmable logic controller, as studied in Chapter 13. The PLC was housed in the control box in the foreground in Figure 15.9. After several of the units had been sold and the utility and reliability of the product had been proven, ARTRAN engineers redesigned the equipment to control it with a microprocessor instead of a PLC, at a substantial savings in manufacturing cost. This can be seen as a viable design strategy—in the prototype design, use a commercial programmable logic controller with its simpler, ladder diagram approach to setting up the logic control of all of the variables associated with the application. Then, after the machine is proven and perfected, switch to a microprocessor controller to save in unit manufacturing costs.

It is easy to find automation examples of OEM applications of microprocessors among commercially available products. Microwave ovens, automobiles, watches, stereo systems, and children's toys have all been given the mystique of computer wizardry by equipping them with these low-cost microprocessor chips. The industrial automation equipment manufacturer has had a bonanza in the promotion of equipment outfitted with

FIGURE 15.9 ARTRAN Leg Transfer Machine was originally designed with a PLC controller and later was enhanced by using a microprocessor controller. (Photo courtesy, ARTRAN, Inc.)

microprocessor chips. Although the chip itself is inexpensive, it has added so much value to large, expensive industrial equipment that manufacturers have been able to achieve higher profit margins by incorporating the microprocessor into the original equipment design. Examples of industrial equipment that have had their capabilities greatly increased by the incorporation of microprocessors into their design are industrial timers, machine tools with sophisticated safety interlock systems, specialized test equipment, and laboratory apparatus. Small firms with spectacular success stories have been spawned by automation engineers who understand both their application industries and how to program and use microprocessors in creative ways to solve their industry's problems.

Often the experienced automation engineer who really understands what industry needs is too removed from the new technology to capitalize upon its potential. On the other hand, the bright young computer experts often do not understand the real needs of industry and spend their time constructing microprocessor games in order to exercise their programming prowess. When these two talent areas come together, however, either as a single person or a partnership, the potential can be very fruitful.

The greatest drawback to the OEM use of microprocessors is that the control and programming strategy frequently can be ill-chosen. It is very difficult for the automation equipment designer to anticipate all the ways the equipment will be used and the backgrounds and characteristics of its human users. Perhaps in no other time in history has the field of human factors or ergonomics in design been so important. The more sophisticated that control systems have become, the more they have befuddled the user, when exactly the opposite should be true. Equipment designers have too often designed equipment that seems easy for them to understand because they have spent months studying the system and its characteristics, but the equipment is still bewildering to the user. The power exists in modern-day, low-cost microprocessors to program into the system ample diagnostics to make the system forgiving and user-friendly.

Microcomputer Applications

Sharply contrasted with the OEM category of applications is the use of microprocessors as the heart of a computer and data-processing system. As a microcomputer, the microprocessor usually needs large amounts of RAM and is supported by convenient programming software, usually on floppy disk. Microcomputers are usually programmed by the user in BASIC or one of the higher level computer languages, not assembly language. ROM is usually used to house operating systems, interpreters, or compilers, not automation application programs.

When a microprocessor is used as a microcomputer system, it is best to think of it as a computer, not an integrated circuit chip. The programs it will execute will be many and varied. It may be wired on-line to an industrial process or perform process optimization calculations off-line. The applications of microprocessors as microcomputers closely parallels the automation applications of computers in general, studied in Chapter 14.

Special-Purpose Controls

The third category of automation applications of microprocessors falls somewhere between the extremes of the other two. This third category holds the greatest potential for an automation engineer working within a plant to apply automation and process cost reduction ideas as they arise. Programming is usually in assembly language but may be in BASIC or a process control language such as FORTH. The program is usually stored in an EPROM or EEPROM, recognizing that the program is of sufficient utility to dedicate a microprocessor to its execution, but the opportunity remains open for future adaptation or complete reconfiguration. The system may not be equipped with I/O peripherals, such as keyboards, CRT screen displays, or printers, as are seen with microcomputer systems.

Sooner or later in the development of sophistication on the part of the robotics and manufacturing automation engineer, he or she becomes dissatisfied with the purchase of preconfigured automation equipment that is either unmatched to the application the automation engineer has in mind or is overdesigned for the application, making necessary the purchase of unwanted hardware. Also the sophisticated automation engineer understands the profit margin being realized by the OEM user of microprocessors when the microprocessor could alternatively be applied on a custom basis in the development of special-purpose controls in-house. This knowledge is characteristic of the seasoned automation engineer who understands the movement and knows how to apply robots and automation at their most efficient level. It is appropriate therefore for this chapter to be near the end of this book.

SUMMARY

The microprocessor is the fundamental element in the construction of manufacturing automation equipment, although it may not be so recognized because of its obscure size and invisible function. But an engineer's knowledge of robots and manufacturing automation is not complete until he or she understands the construction, programming, and practical application of these tiny integrated circuit chips. The developments in capability and reduction in cost of these tiny IC chips has surpassed any technological breakthrough of our age. It is no wonder that the subject of robots and manufacturing automation, a primary beneficiary of this technological development, is in a tremendous growth phase.

All computer devices need a central processing unit (CPU), some memory media, and a means for data transfer. Memory can have volatile or nonvolatile characteristics and can be capable of read/write (RAM), read-only (ROM), or various shades between the two: PROM, EPROM, and EEPROM. Each has its place in robots and manufacturing automation applications.

Data transfer is by data buses and utilizes AND logic, the principles of which were studied in earlier chapters in this text. Data transfer to and from the world external to the microprocessor is by serial or parallel input/output.

Although automation engineers need to understand microprocessor programming, this does not mean that they must be experts in assembly language codes. There is a need for assembly language coding in automation applications, and the automation engineer should understand that need, but the actual detailed coding is machine-specific and should be done with the help of a programming codebook for the particular microprocessor being programmed. Some example code was presented in this chapter along with an explanation of the shorthand for assembly language coding—hexadecimal.

Applications for microprocessors vary from use in original equipment (OEM) at one extreme, to use in microcomputer systems at the other extreme. In the middle is a third category—special purpose controls—which has the most potential for productivity improvement and automation within a manufacturing plant. The automation engineer's education is not complete until he or she can custom-develop special purpose robots or controls to solve problems that have been identified within their own manufacturing plants by using microprocessors.

EXERCISES AND STUDY QUESTIONS

15.1. Explain any difference in meaning between the terms *microcomputer* and *microprocessor*.

15.2. A certain improvement in the production process for a wafer containing 200 IC chips reduces costs of wafer production 30 percent and increases the wafer yield from 5 percent to 80 percent. How will this affect the cost of the IC chips?

15.3. Suppose the IC chips in Exercise 15.2 were designed for use in an industrial robot controller and had represented 60 percent of the $300 manufacturing cost of the controller prior to the process improvement. After the process improvement, what percent of the total cost of the robot controller would the IC chip represent if the other components of the controller cost remain unchanged? Suppose 30 percent of the $300 controller cost represented profit. By how much could the robot controller price be reduced after the IC chip process improvement, provided that the profit percentage be held constant at 30 percent.

15.4. A certain microprocessor has ROM addresses 0000 to 9000 and RAM addresses 9001 to FFFF. What is the memory capacity of this microprocessor, and how much memory is allocated to ROM and to RAM, respectively?

15.5. A potential industrial robot application has been examined, and two alternatives are to be compared. The first alternative employs a commercially available robot that costs about $50,000 and comes equipped with a general-purpose microcomputer for which a high-level robotics language, such as VAL, is included as standard equipment. The second alternative represents a hardware expenditure of $30,000 including a microprocessor controller for which an EPROM will have to be programmed in assembly language. The assembly language program development is expected to cost $20,000, which is five times the cost of program development of the commercially available robot. If performance and other characteristics are equal, which alternative would you favor if the application can be expected to remain fixed indefinitely? Which alternative would you favor if the application is expected to need reprogramming approximately once per year?

15.6. The largest constant that a certain microprocessor can hold in its memory is decimal 255. What is the number expressed in hexadecimal? What is the word length of this microprocessor? The largest memory address for this microprocessor is FFFF. The memory capacity of this microprocessor is how many bits, how many bytes, and how many nibbles?

15.7. A microprocessor is equipped with a parallel I/O interface device. How many pins are required for the transmission of data points of value up to hex FF? If the device had been a serial I/O arrangement, how many pins would have been required?

15.8. A microprocessor-based automation system continuously monitors eight logical inputs from an industrial robot. The eight inputs are received in parallel, one nibble at a time, via a 4-bit register that is commanded by the microprocessor to alternate the register every half-second between inputs 1 through 4 and inputs 5 through 8. The microprocessor does this by outputting a logic variable (P), which takes on the value logic 1 when inputs 1–4 are to be received and logic 0 when inputs 5–8 are to be received. Designate the 4-bit input register as logic variables A_1 through A_4 (low order to high order) and the input variables as I_1 through I_8 (low order to high order).

(a) Write the Boolean logic expressions of variables A_1 through A_4 as functions of inputs I_1 through I_8 and P.

(b) Draw a logic diagram that represents the parallel input system of this microprocessor.

(c) When $P = 1$, the input register reads hex C; when $P = 0$, the input register reads hex 9. What inputs from the industrial robot are logic 1?

15.9. A 1-byte parallel I/O register reads hex FB. How many of the eight I/O bits read logic 1?

15.10. Example 15.2 assumed a priority of alarm conditions, R_1 to R_4 in that order. Suppose it is determined that the conditions "discharge chute full" and "in-feed hopper empty" are both more urgent than the condition "hydraulic temperature high." Revise the assembly language program of Example 15.2 to represent the revised priorities.

15.11. The highest memory address in a certain microprocessor is FFFF. What address comes just before 5000? (Hint: The answer is *not* 4999.)

15.12. A certain microprocessor stores program instructions in 8-bit words for the following subroutines:

	Memory Address	
Symbolic name	START	END
TEST	9F99	A007
ROBOT	A022	A100
CHK	0050	0A00
TOOL	BE12	BF00

Which subroutine is longest? How long is it (in bytes)? Which subroutine is shortest? How long is it (in bytes)? List the subroutines in the sequence in which they appear in the microprocessor's memory.

CHAPTER 16

COMPUTER INTEGRATED MANUFACTURING

At this point let us review what we have studied about robots and automation. We began by examining products and processes and by qualifying the designs of both for automation. We have become familiar with the devices and components used to build automated processes, including the devices that handle, orient, and feed piece-parts to machines. We have studied the role of NC machines and their outgrowth—industrial robots. We have seen how important the computer, programmable logic controller, and microprocessor are in the control of application hardware to accomplish specific goals in the enhancement of efficiency and effectiveness of a single machine or work station. We have considered the linking together of automated processing stations in both synchronous and asynchronous production flow lines. In isolated situations we have considered the possibility of informational links, such as the linking of computerized product design data with the process, that is, CAD/CAM. In all of this we have maintained a manufacturing perspective, that is, we have attempted to solve manufacturing problems and, where necessary, we have employed computers to facilitate these solutions.

The approach we have taken in studying this subject parallels industry's perspective: finding problems or processes that can be improved by using automation and then applying available hardware and software to these problems. There is another approach, however, and that approach focuses upon the computer as the center of control of the entire factory. This approach does not stop at computerization of the fabrication and assembly processes, but also encompasses information flow for production control, quality, maintenance, material handling, and inventory control in a totally integrated system. This is computer integrated manufacturing, or CIM.

Let us first acknowledge that various meanings are associated with the term CIM, or CIMS, for computer integrated manufacturing systems. For some people CIM includes off-line procedures for using the computer to assist people to better understand and control a factory. In this broad sense almost any practical use of the computer to analyze data in a factory could be considered to be a part of CIM. But as we move toward the twenty-first century, CIM is taking on a more narrow meaning: the on-line computer control and linking together of all functions in a manufacturing plant. This type of computer integrated manufacturing is an ambitious undertaking and demands new priorities and a new perspective of manufacturing. For instance, in the new perspective, the flow of information is considered at least as important as the flow of material. The processing of information by the computer is considered as important as the processing of material by machines.

A conventional manufacturing plant is a widely distributed layout of materials and processes under the control of a whole group of people. It is not uncommon for material and work-in-process to become lost as it is transferred from department to department and from person to person. In such a situation it may become as important for a centralized point of information to know the status of the material and processes in the plant as it is for the material to actually be processed.

It may be difficult to accept the priority of information in a manufacturing plant, and indeed it is possible to overdo the emphasis upon computers and information processing. But perhaps the following scenario which can be related to our personal lives will bring the problem into focus. Consider the tools and materials available to you in your own garage at home. Can you really find what you need at the time you need it to do a job? Have you ever found it easier to go to the store and buy new material or even tools to accomplish a job, because it has not been worthwhile to spend the time trying to find the materials or tools that you know you already have? Carrying the analogy a step further, you may begin a task only to remember that you forgot to pick up a replacement part when you were at the store, replace a broken blade, sharpen a tool, or buy gasoline for the lawn mower. The missing ingredient in your garage at home is information—the knowledge of where everything is, when it is needed, the state of repair of your tools, and whether you need any additional materials or tools to do a job.

Multiply the problems of your own garage at home a few thousand times and you have the complexity of the conventional manufacturing plant. Many of the employees spend most of their time communicating and interacting with other employees in an attempt to find out the status of machines, processes, materials, and orders that must be filled. According to the computer-integrated-manufacturing perspective, much of this human activity is wasted. What is needed is an information processing system that ties all of these needs together in an on-line, integrated system—CIM.

The emphasis in CIM is upon the communication linkage made by the computer between the various software subsystems that drive the manufacturing operations and support functions. This is no small problem because the computerization of manufacturing functions has as many origins as the functions themselves. Thus, the design engineering function has its CAD software, the NC machining function has its CAM software, the production and inventory control function has its MRP (material requirements planning) software, and the process planning function has its CAPP (computer-assisted process planning) software.

It was logical that CAD and CAM were the first functions to recognize a need to be linked for the purpose of sharing data. Both require meticulous coding to fully describe piece-parts. The marriage of CAD and CAM is resulting in significant savings in the encoding of detailed part characteristics to provide a common data base for the products manufactured by the facility.

Successes with CAD/CAM have generated interest in bringing the process planning, production control, material handling, shop floor control, and quality control functions within the integrated system. To accomplish this task we must face the problem of interfacing a conglomeration of systems and equipment into one organized CIM.

INTERFACING

The biggest barrier to the implementation of computer integrated manufacturing is the collection of barriers that exist between the machines and systems themselves—the problem of interfacing. It would have been nice if all of the automation developments by vendors and users had followed an orderly standard so that machines would be compatible and would link together in a way that all understood, but the piecemeal development of automation breakthroughs has not supported such an orderly development. Indeed, the computer technology to integrate the entire factory system had not been discovered in the early days of numerical control, CAD/CAM, robots, and automation. Still, all of these separate machines and developments must be interconnected in a CIM system, and this is the subject of interfacing.

In Chapter 15 we learned how information is passed between microprocessors and automation equipment by using serial or parallel I/O devices. There is now a standard for serial communications, RS-232-C, established by the Electronic Industries Association (EIA). RS-232-C is the most popular serial interface convention in existence, and designers and suppliers of computerized equipment, or equipment that could conceivably be connected to a computer, are typically building RS-232-C interfaces into their products. RS-232-C is widely used to connect computer peripherals, but in a factory environment where equipment may be widely distributed, it may be appropriate to use a cousin to the RS-232-C—the RS-422.

For computer-to-computer interfaces or for systems requiring massive amounts of rapid data transfer, serial interface is inadequate. Remembering the comparison between serial and parallel I/O, the principal advantage of parallel interface is speed. The standard for parallel interface is IEEE 488,* which has a maximum data transfer rate of 1 megabit/sec. The IEEE 488 is more than an interconnector; it is a bus. This is to say that through control lines it can manage the transfer of information between several devices, up to 15 to be specific. There are a total of 24 pins in the IEEE 488 interface, 8 for data, 8 for ground, and 8 for control.

Despite the fact that the IEEE 488 interface can be used to network up to 15 devices, both the RS-232-C and the IEEE 488 are basically interconnector devices from one device to another. If there are just two devices, these interconnectors are ideal because each device needs only one interconnector plug. But as you add devices to the system, if all the devices need to communicate with each other, the number of interconnectors needed increases rapidly. For 3 devices, 6 interconnectors are needed; for 4 devices, 12 are needed. In general, the number of interconnectors needed can be calculated from the following formula:

$$\text{number of interconnectors} = C_2^n = \frac{n!}{2(n-2)!}$$

where n = the number of devices to be connected

Also the devices must be in close physical proximity to each other. It is straightforward to see that ordinary point-to-point interconnection is not the answer to the problem of computer integrated manufacturing. Thus there has arisen the need for networks, the subject of the section that follows.

Local Area Networks

Local area networks (LANs) are the primary means of interconnecting devices in an automated system to achieve computer integrated manufacturing. Not all networks are "local," and the technology for networking factory information systems had its origins

*Commonly known as the general-purpose interface bus (GPIB).

in the telephone industry. But the physical distribution of a manufacturing facility, or a university or office, is appropriate for local area network technology.

Star Networks

The simplest layout of a network, from a conceptual standpoint is a star arrangement, as indicated in Figure 16.1. This arrangement is analogous to the hierarchical control arrangements for CAD/CAM discussed in Chapter 5. In network terminology, the central control is called a "server" because its network role is to make the interconnection between the various robots, NC machines, and automation devices that need to communicate. A principal disadvantage of the star arrangement is the level of dependence upon the server to handle all communications. Despite its conceptual simplicity, the star arrangement is not a popular type of LAN for computer-integrated manufacturing.

Ethernet

The most popular type of LAN, as of the early 1990s, is Ethernet. Ethernet uses a simple coaxial cable as a bus. This may seem difficult to comprehend, considering the concepts of data buses as we studied them in Chapter 15. How can control of the messages be managed when the only interconnector is a simple two-wire cable? The answer is through software. Each device on the network has a "transceiver" responsible for encoding each message the device is sending and for decoding each message it receives. Each batch of information, called a "packet," contains all of the source and destination information to enable it to be properly received by only the device for which it is intended.

Routing is one problem to be handled by the software; timing is another. Remember that the bus is merely a coaxial cable. Suppose two devices want to put a message on

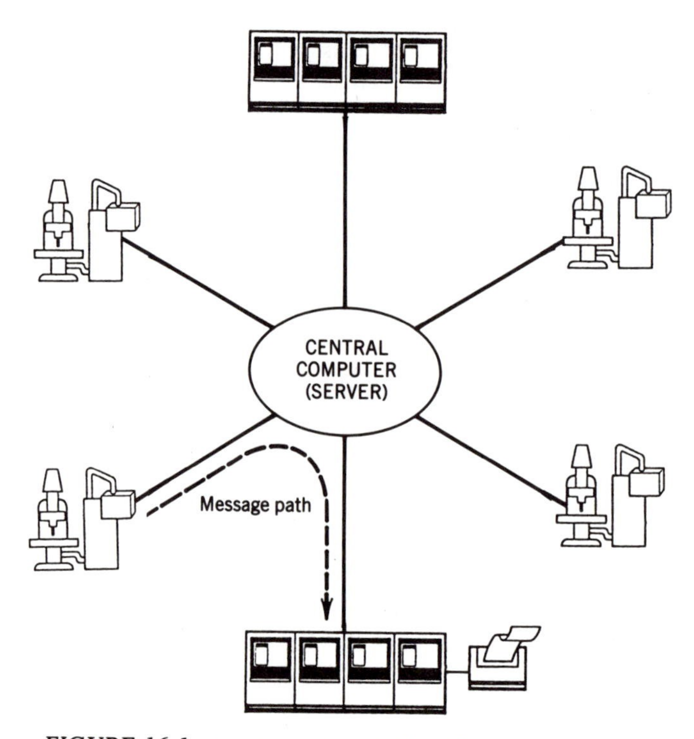

FIGURE 16.1 Star layout for a LAN. All messages must be routed through a central server.

the Ethernet at the same time. The answer is that the messages actually collide on the Ethernet! A system or protocol called "carrier sense multiple access with collision detection" (CSMA/CD) handles this problem. When messages collide, both of the sending devices are notified by the CSMA/CD, which jams the entire network so that all devices will stop transmitting. After waiting an appropriate, but random, amount of time, device transceivers can again attempt to transmit messages.

Ethernet is an ideal choice for office computer networks. Each computer can simply be tapped into the coaxial cable via its transceiver. The message collisions are handled without any perceivable interruption of the users and a typical data transmission speed for Ethernet, as of the early 1990s, was 10 Mbps. There are restrictions, such as the distance between individual taps on the bus. If it is desired to connect two networks, a "repeater" can be used that makes the two buses become a single bus. The physical arrangement of an Ethernet is shown in Figure 16.2. The figure includes a repeater for interconnecting two buses to form one Ethernet.

Token Rings

Another strategy for constructing a LAN avoids the problem of message collisions by providing a message carrier that travels around to all devices connected to the network. This message carrier is called a "token." The token is merely a bit pattern that is packaged with the data to permit the package to be recognized by each device on the network. Each device has a repeater that receives the token, determines whether the data is intended to be received by its device, and then retransmits the token to continue to the next device. The reason this type of LAN is called a "ring" is that the token must continually operate

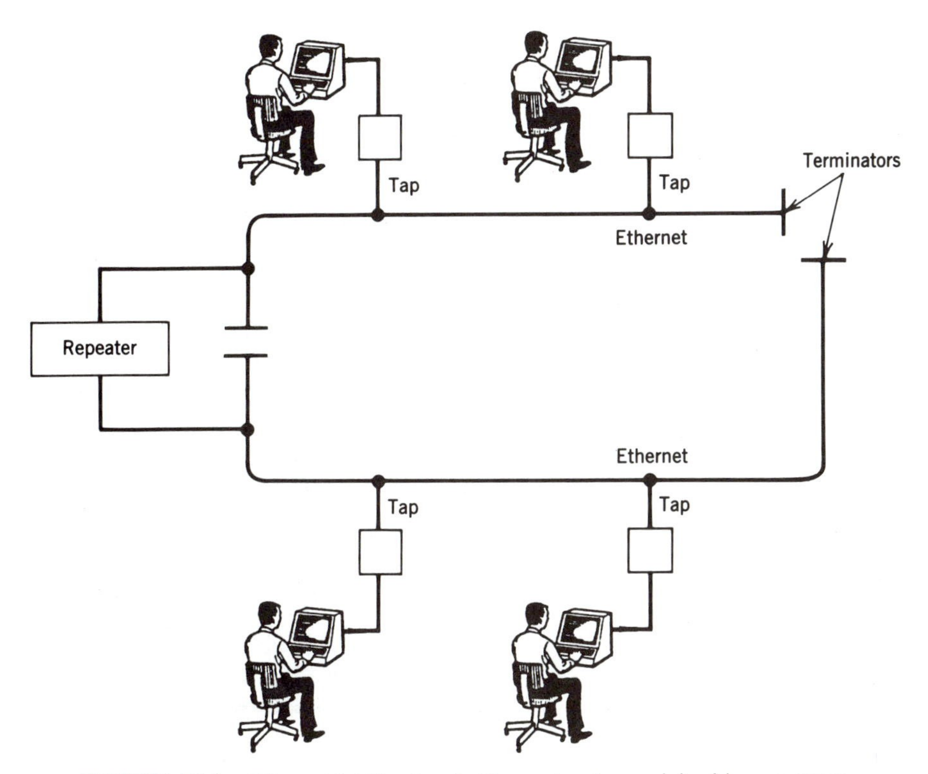

FIGURE 16.2 Ethernet LAN network. Two networks are joined by a repeater.

in a loop, as shown in Figure 16.3. The token travels around to every device in the loop during every cycle, regardless of which or how many devices are intended to receive a given message. Indeed, even the device that sends the message receives that message back again to verify that the transmission proceeded satisfactorily. Only when the token has made the complete loop back to the sender is it ready to carry another message, from either the same device or from another one it encounters in its route around the loop. Once a device encumbers the token with a message, no other devices can send a message until the token has made the complete loop.

One disadvantage of the token ring is that each device must have a repeater. Also, the strategy is slower than with Ethernet, because messages can be sent only by the device that has the token, and only if the token is free at the time. The speed of token rings has been increasing in recent designs, even to the point of surpassing the speed of Ethernet. This development has enhanced the capability of token ring networks to handle the data flow requirements of factory automation.

Token Buses

As the name implies, token buses are a combination of the concepts of token rings and the Ethernet bus. In the token bus strategy, there are two routing modes for the token,

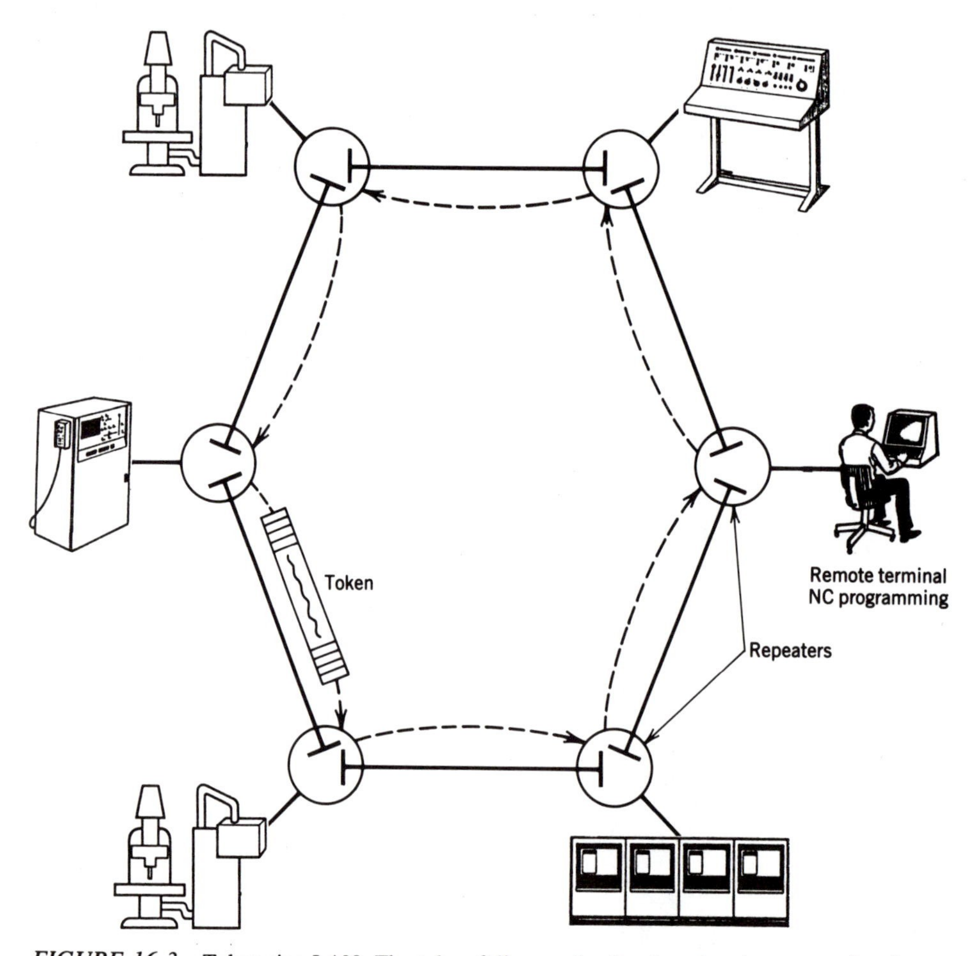

FIGURE 16.3 Token ring LAN. The token follows a fixed path and makes a complete loop.

depending upon whether the token is carrying a message or is simply empty and looking for a device that wants to send a message. If the token is carrying a message, that message has an address; hence, the message is received directly by the device on the bus, without waiting for it to be relayed through the system by the successive operations of the repeaters of each of the devices, as is done in the token ring arrangement. However, if the token is not carrying a message, it must make the rounds through the entire network, visiting each device in case it has a message to send. Since the connections to the bus are not in a physical ring, some method of routing must be used to assure that each device gets a chance at sending a message. The method is to use a chain-linking method that is identical to the chaining of data in a file in a computerized data storage and retrieval system. Thus, every device connected to the bus knows to whom to send the empty token next, and every device receives the empty token from one and only one device. The scheme is illustrated in Figure 16.4.

Packets

In our discussion of the four principal arrangements for LANs we have used the word "packet" to represent packages of data to be put on the network for delivery to one or more devices. In a method analogous to time-sharing, large blocks of data are divided into manageable chunks before being committed to the network. Large blocks of data require more time for transmission and increase the probability of collisions on a bus.

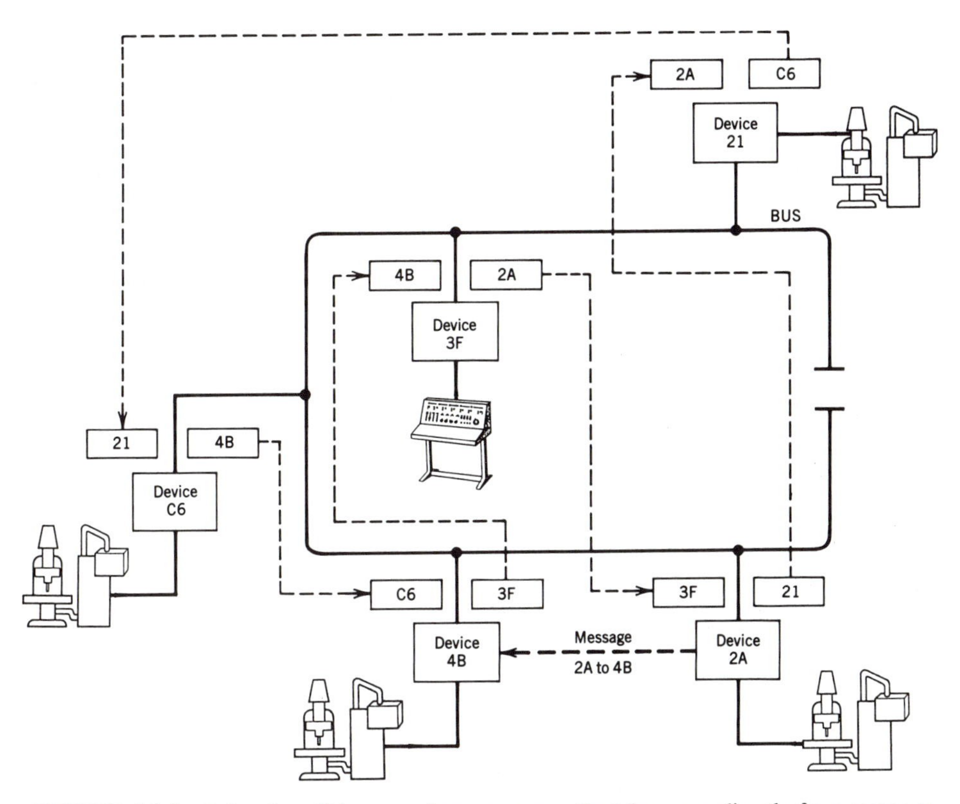

FIGURE 16.4 Token bus. When carrying a message, the token goes directly from source to destination. When the token is empty, it makes the complete loop, following the address chain while it looks for new messages to send.

Also there is always the possibility that a random source of noise will garble a transmission and cause an error. Only one error in a message is enough to be picked up by fault detection software, and the entire block of data must be retransmitted. At the other end of the spectrum, if messages are sent only in tiny blocks, the overhead of attaching the necessary routing header, start and end blocks, and other network information, if any, unnecessarily complicates and overloads the network. Add to this the problem of relinking all of the small blocks together, a problem that must be solved by the receiving device. The compromise is to parcel the data into standard sizes, and this is the explanation of the word "packets," a term that is often used to describe data blocks being handled by LANs.

Gateways

Suppose an organization already has an Ethernet for the computers in its offices and wants to use a different "topology," such as a token ring or a token bus, for the shop floor information system. Is it possible to link the two systems together in a plantwide computer integrated manufacturing system? Possible, yes, but it is expensive. What is required is a dedicated computer that is connected to both systems and is programmed to translate messages from one system to the other. Such computers are called "gateways," and thus we have network interfacing as well as device interfacing.

Standards

If readers are confused by the jargon of tokens, packets, rings, and gateways, they can perhaps take comfort in the realization that they are dealing with a very confusing conglomeration of communication schemes. Not since the erection of the Tower of Babel has there been such a cacophony of unintelligible attempts at communication. There is a natural motivation to want to standardize, but standardization is a difficult goal in this environment.

ISO/OSI

A recognized standard is the familiar abbreviation ISO/OSI, which represents International Standards Organization Open System Interconnect model. The term "model" is a careful choice of words, because the ISO/OSI is more of a model than a standard. ISO/OSI merely organizes the problem into successive layers to be addressed by the interface designer, but does not nail down the standard to be followed in any of the levels. The lowest level in the ISO/OSI model is "physical," which deals with the electrical and mechanical connections, such as RS-232-C and IEEE 488. The second level is "data link," which is involved with software for routing and error detection, among other functions. The point is that the ISO/OSI "standard" merely categorizes the selections to be made. It does not specify what these selections must be. It can be seen that the ISO/OSI standard is really a model for the way in which automated system designers can describe their interface designs, but there is still a great deal of latitude on the part of vendors to chose the particular interface design they think is best.

MAP

Moving a little closer to an actual standard is Manufacturing Automation Protocol (MAP), originated by General Motors. The term "protocol" means a generally accepted set of rules to follow, and thus MAP is still not a hard standard in the conventional sense of the term "standard." MAP adopts the ISO/OSI for describing the system in identified layers, but MAP is more specific than ISO/OSI. For instance, at the physical layer described earlier, MAP has adopted the token bus arrangement. MAP is not fixed; it continues to evolve, and true standardization in the field of interfacing for computer integrated manufacturing will likely remain a goal into the twenty-first century.

Our purpose in this exploration of the details of the interfacing for computer integrated manufacturing systems has been to gain some understanding of the challenges this ambitious goal has before it. The responsible automation engineer, and industrial manager as well, should not be naive about the problems to be faced when contemplating a thrust for computer integrated manufacturing. We now explore another concept that is closely related to CIM: flexible manufacturing systems.

FLEXIBLE MANUFACTURING SYSTEMS (FMS)

> In today's Baskin-Robbins society, everything comes in at least 31 flavors [ref. 74].

These are the words of John Naisbitt in his best-selling book *Megatrends*. Naisbitt continues:

> There are 752 different models of cars and trucks sold in the United States—and that's not counting the choice of colors they come in. If you want a subcompact, you can choose from 126 different types. In Manhattan, there is a store called Just Bulbs, which stocks 2,500 types of light bulbs—and nothing else. Its most exotic bulb comes from Finland and emits light that resembles sunshine. Today, there are more than 200 brands (styles, as the industry calls them) of cigarettes on the U.S. market.

What Naisbitt has observed is the tremendous diversity of options available to the consumer of today, something all of us know about, but its appearance has been gradual, and most of us have not stopped to think about it. What Naisbitt has not stated is *why* these options have appeared. In large part, the flexible production made possible by the computer and microprocessor has given birth to the consumer options we are now seeing. And since the options are there, we are growing increasingly accustomed to exercising them, which in turn stimulates the proliferation of more options on the production side.

The hard automation of the 1940s and 1950s was the ideal production style for that age because almost everyone wanted the same things: a new Ford or Chevrolet, standard light bulbs, and an RCA radio or TV. Such an environment generated stable demand for incredible volumes of identical product—volumes that justified the expensive and long-term commitment to hard automation. But in today's multioption market there is much greater need for flexible production that can be used on a variety of products, models, and styles. Planners for new facilities now want their machines, robots, and material handling subsystems to be integrated into a *flexible manufacturing system* (FMS). The key word is "system." FMS takes advantage of the flexibility of robots, NC machine tools, industrial logic controllers, and microprocessors in creating an overall flexible system.

Before going any further it is imperative to reemphasize priorities by stating that an overall flexible manufacturing system is unachievable without the flexible machines, robots, and automation subsystems described in earlier chapters of this book. And prerequisite to the robots and manufacturing automation subsystems are the objectives of tolerance control, part design for orientation and handling, station reliability and maintainability, and mechanization of the work station. So, before launching into the deep waters of flexible manufacturing systems, the engineer must be sure that the subject plant's processes, robots, material handling, and manufacturing automation subsystems are truly capable of handling this approach.

However, while developing the manufacturing automation capability of a plant, it is worthwhile to keep in mind the vision of a completely automated, flexible, total manufacturing system. Such vision will motivate the planner to build or specify flexibility into the individual manufacturing work stations whenever economically feasible. It is the same vision that causes the engineer to select a general-purpose robot for a single application instead of selecting a special-purpose parts manipulator, if the economics can reasonably permit this choice. It is exactly the approach that was used by the Singer Company in the robot implementation discussed earlier, in Chapter 9. A rigid parts handler was a feasible alternative and was almost equivalent in cost. But the general-purpose robot was selected instead so that the firm could aim for a totally flexible manufacturing system in the future.

At the current stage of development, flexible manufacturing systems are still more concept than reality. But there are a few examples, especially in the computer and electronics products fields, of working flexible manufacturing systems. FMS is also making headway in the machined-metal products industry, in which small quantities of each model are manufactured. An example is large pipeline valves and fittings. Industries of this type are equipped with substantial numbers of NC machines and are amenable to organization into a flexible manufacturing system. Some representative examples of flexible manufacturing systems are useful as models that all automation engineers can study in preparation for implementing the concept in their own plants when the appropriate time comes.

At the General Motors Hydramatic-Warren Plant, a flexible manufacturing system has been installed in which conveyors serve 22 robots in four interconnected machining lines [ref. 37]. The product—bulky, 30-pound aluminum transmission cases—is typical for an application of FMS. Figure 16.5 diagrams the layout of the FMS, which covers an area of 220,000 square feet. The FMS is credited with eliminating 91 percent of the arduous manual labor formerly required in the machining area. The robots selected for this FMS were the pick-and-place variety, and due to their low cost were expected to pay for themselves in one year.

The General Electric Company risked $16 million on the FMS concept to rebuild a locomotive plant in Erie, Pennsylvania, to regain a competitive edge [ref. 39]. The system (see Figure 16.6) was designed to handle motor frames—bulky, heavy metal castings (150 to 2500 pounds) with lengthy, close-tolerance machining sequences. Note the similarity between this application and the General Motors FMS application.

The General Electric FMS application was built around a shuttle car operating along a 212-foot-long track. The shuttle car system is directed by programmable logic controllers under the general supervision of a host computer. The shuttle car system moves pallets each of which is identified by a binary number. Photoelectric sensors read each pallet's identification number to verify its position at each spur along the track.

The General Electric FMS system is reported to have cut the time required to machine a motor frame by 87.5 percent with an overall worker productivity increase of 240 percent. At the same time, capacity was increased by 38 percent while using 25 percent less floor space.

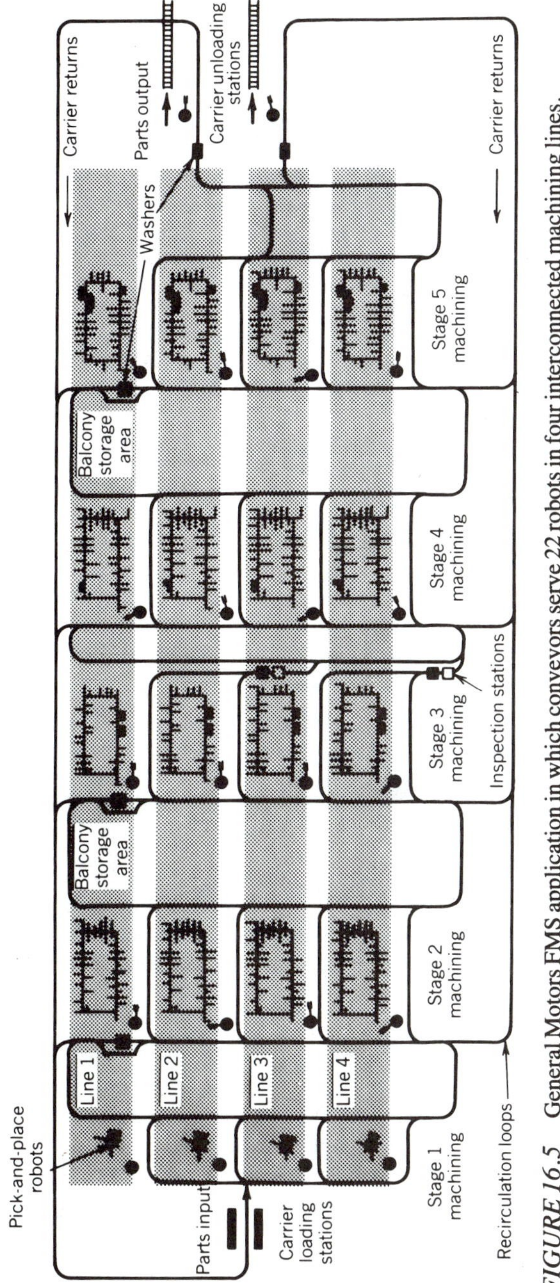

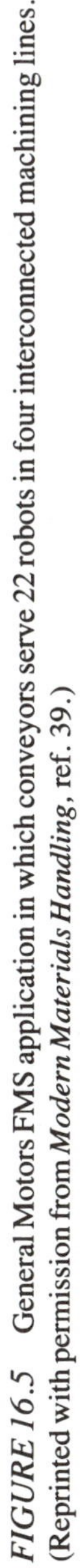
FIGURE 16.5 General Motors FMS application in which conveyors serve 22 robots in four interconnected machining lines. (Reprinted with permission from *Modern Materials Handling*, ref. 39.)

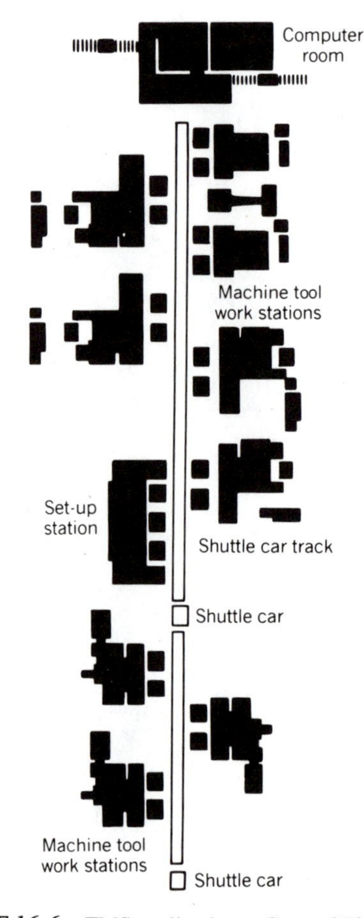

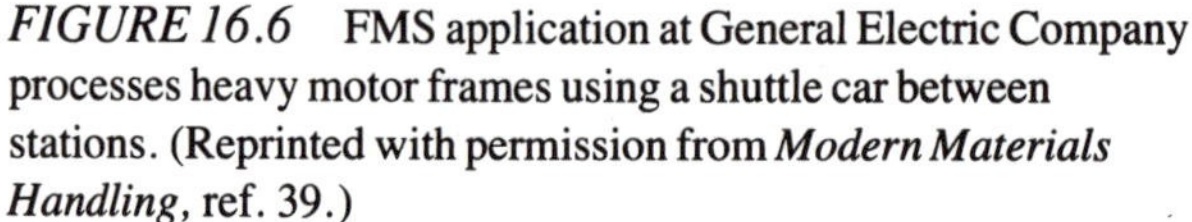

FIGURE 16.6 FMS application at General Electric Company processes heavy motor frames using a shuttle car between stations. (Reprinted with permission from *Modern Materials Handling*, ref. 39.)

Random Production

A curious twist brought on by the flexible manufacturing approach is that the problem has become part of the solution. Remember that the reason for designing a flexible manufacturing system was to permit a wide variety of short-run products and models to be processed by the same manufacturing system. Such an objective forces the selection of very flexible material handling equipment to serve the processing machines, which themselves are of varying degrees of flexibility. Once the flexible system has been installed, however, varying or randomizing the product and model mix can actually *enhance* the efficiency of the system! An example will illustrate.

EXAMPLE 16.1
FMS Random Production Efficiency

In a simplified example for illustration, a certain flexible manufacturing system can accept shop orders and manufacture any of three products A, B, and C in any order upon demand. The system

TABLE 16.1 Estimated Production Volumes for Each of Three Products Manufactured by the Flexible Manufacturing System of Example 16.1, and Operation Times (Including Handling) for Each Product at Each Processing Station in the System

	Products		
	A	*B*	*C*
Estimated production volume	100	50	25
Operation times (including handling):			
Milling machine	8 min	30 min	6 min
NC turning center	22 min	3 min	16 min
Inspection station	6 min	5 min	25 min

consists of a milling machine, an NC turning center, and an automatic inspection station. A matrix of operation times is shown in Table 16.1 for each product and each manufacturing station (machine). The material-handling system has zero storage between stations, but it is completely flexible, and handling time can be considered to be a negligible portion of the operation time specified in the matrix.

(a) How long would it take to complete the specified production volume if the production is scheduled in the conventional way (i.e., produce all 100 of product *A*, then all 50 of product *B*, and finally all 25 of product *C*)?

(b) Suppose that the three products were mixed so that production proceeded according to the following sequence: [*A-B-A-C-A-B-A*]-[*A-B-A-C-* · · ·] etc. How much time would be required to produce the total estimated production volume for products *A*, *B*, and *C* using the mixed approach?

Solution

It is logical to assume that all three machines in the system can be operating at the same time. However, a given machine can process only one piece at a time. Assuming these conditions, the total production time using the conventional strategy in part (a) would be tabulated as follows, using the notation A_i to represent the *i*th item of product *A*:

		Time (min)
A_1: milling machine		8
A_1–A_{99}: NC turning center	99 × 22	2178
B_1–B_{50}: milling machine	50 × 30	1500
C_1: milling machine		6
C_1: NC turning center		16
C_1–C_{25}: inspection station	25 × 25	625
	Total	4333 min

Note that the following operations were omitted because they were performed concurrently with other operations and thus did not contribute to the total production time:

A_2–A_{100}: milling machine
A_{100}: NC turning center
A_1–A_{100}: inspection station
B_1–B_{50}: inspection station
B_1–B_{50}: NC turning center
C_2–C_{25}: milling machine
C_2–C_{25}: NC turning center

Question (b) presents a product-mixing strategy that should result in more efficient utilization of the flexible manufacturing system because of the balancing of machine loads. Figure 16.7 uses a time chart to schedule the various machine operations using the product mix *A-B-A-C-A-B-A*. Note that the cycle begins to repeat itself at time 136. Therefore the time required to perform the cycle *A-B-A-C-A-B-A* in continuous production would be 136 minutes. To complete the full lot of required product, 25 full cycles would be required. The final cycle would require 14 additional minutes to complete the final operations for the last item of product $A(A_{100})$, resulting in a total cycle time of 150 minutes. Therefore, the total production time under the mixed product strategy would be

$$24 \times 136 \text{ min} + 150 \text{ min} = 3414 \text{ min}$$

The savings in production time using a mixed-product strategy instead of grouping into product batches would be

$$\begin{aligned} 4333 - 3414 &= 919 \text{ minutes} \\ &= 15^1/_2 \text{ hr} \end{aligned}$$

This represents an improvement in production efficiency of:

$$\frac{919}{4333} \times 100\% = 21\%$$

There is a subtlety about Example 16.1 that should be mentioned before going on. When analyzing manufacturing systems such as this one, it should be remembered that a new unit of product cannot enter a station until the item previously occupying that station has proceeded to its next station. The point is obvious for cases in which the station is still busy processing that previous item, but even if its processing is complete, the previous item may not be able to advance because its next station may be occupied. The reader may recognize this as station blocking, a concept introduced in Chapter 4. Station blocking is a problem to be considered when planning flexible manufacturing systems with zero storage between stations. In Example 16.1 the milling machine station was blocked from time 52 to time 68 and again from time 76 to time 84. Therefore the milling operations for products A_3 and B_2 were delayed, respectively, even though the milling process for the previous time was complete. In Figure 16.7, contrast the idle periods just described with time period 122 to 136 in which the milling machine station is *not* blocked. The milling process for item A_5 could very well have begun at time 122, but it was not shown that way in Figure 16.7 so that the new "cycle start" could be shown. Note that because item B_3 would have been blocked from the milling operation until time 144 anyway, there is no shortening of cycle time by beginning item A_5 before time 136.

In the real world of manufacturing, one would usually not have the liberty of selecting a regular, preset sequence of product such as was specified in part (b) of Example 16.1. But the beauty of such a flexible manufacturing system is that *random* mixing of product through the system is usually better than organization into product lots, such as was specified in part (a) of Example 16.1. So the flexible consumer demand described by Naisbitt in *Megatrends* is creating a production capacity for which flexibility is not only possible, but is even *desirable* from an efficiency standpoint! How different this new industrial philosophy is from the conventional mass production philosophy of only a decade or so ago.

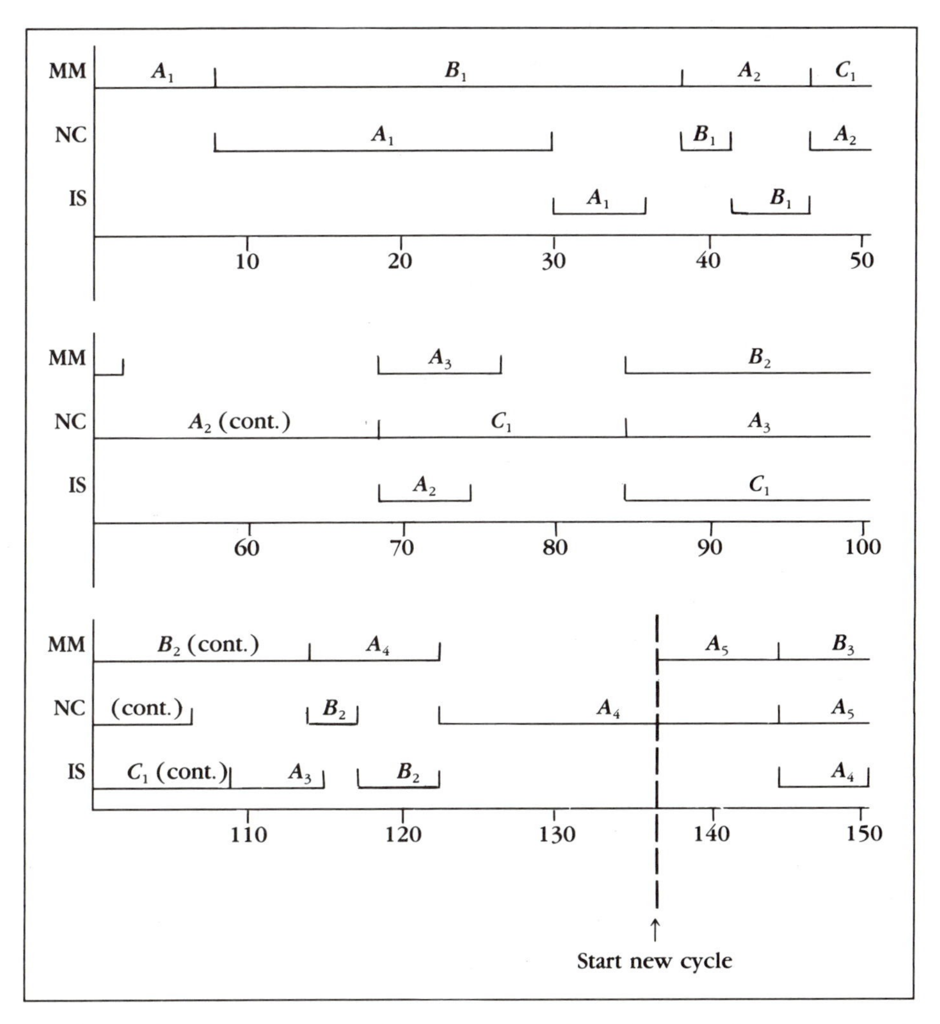

FIGURE 16.7 Scheduling of FMS processes for a product sequence [*A*-*B*-*A*-*C*-*A*-*B*-*A*] [A-B-A-C . . .] etc., as specified in Example 16.1, part (b).

GROUP TECHNOLOGY

To preserve the production efficiencies afforded by producing similar but not identical products on the same machines, a technique known as *group technology* (GT) has emerged. It was practiced in the Soviet Union in the 1950s [ref. 68] and has received recent attention in all parts of the world.

GT Layout

In group technology the emphasis upon part similarities has resulted in the identification of part "families." Machines can be laid out in a fashion that permits production economies for these families. The result can be a significant improvement in efficiency over the traditional process layout in which machine types are grouped together into departments. A natural question to ask is, "What is the difference between group technology and the old idea of *product layout* as an improvement over process layout?" Product layout is geared for mass production of a single product or model, just the type of situation that is disappearing due to today's multioption marketplace. Group technology is indeed closely related to product layout and can be considered an attempt to take advantage of single-product layout economies when in fact the products manufactured are not identical. The result is several product flow lines that may have common elements.

Figure 16.8 illustrates a group technology layout for 13 part families being processed through 24 different operations. Note that the layout is neither product nor process oriented. A product layout would have consisted of 13 independent lines that employed no common machine stations except for the robot welding and assembly operations that merged parts into subassemblies and assemblies. A process layout would have physically grouped machines into departments, such as drill press department or milling machine department. The only operations one might call departments in Figure 16.8 would be foundry, plating, and packaging/shipping.

Note that some operations accommodate several families whereas others are specialized. Also note that only seven part families are shown reaching packaging and shipping, whereas 13 families emerged from the foundry. The reason is that the robot welding and assembly stations merged parts into subassemblies and assemblies. Also note that the number of machine stations is reduced considerably over a corresponding product layout due to the combining of work stations for parts with similar manufacturing characteristics. This can be seen in Table 16.2, which tabulates the number of machine stations (including foundry, plating, and packaging/shipping) had the layout been product oriented. The table counts stations beginning with the production sequence for the topmost part family path in Figure 16.8. The total of 65 is considerably higher than the 24 station operations observed in the group technology layout of Figure 16.8. In a real application of group

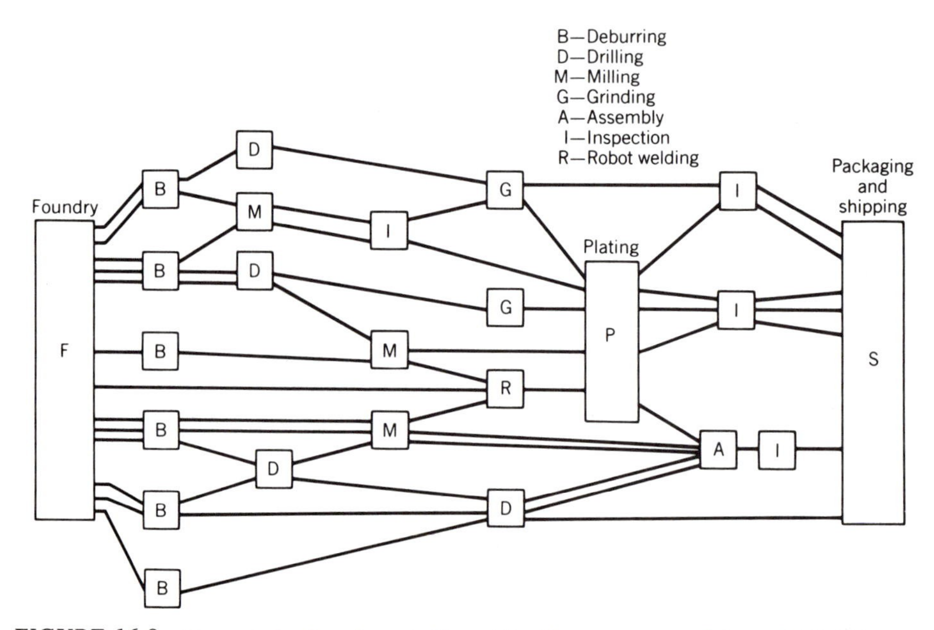

FIGURE 16.8 Group technology layout: 13 part families are processed through 24 different operations.

TABLE 16.2 Product Layout Station Count

Part Family	*Station Sequence*			*Count*
1	F–B–D–G–I–S			6
2	F–B–M–I–G–P–I–S			8
3	F–B–M–I–P–I–S			7
4	F–B–D–G–P–I–S			7
5	F–B–D–M–P–I–S			7
6	F–B–M– }			3
7	F– }	R–P– }		3
8	F–B–M }			3
9	F–B–M– }			3
10	F–B–D–M– }		A–I–S	7
11	F–B–D–D– }			4
12	F–B–D– }			3
13	F–B–D–S– }			4
Total				65

technology, the savings would not likely be so dramatic because foundry, plating, and packaging/shipping operations would likely already be grouped. Still, the group technology approach would save 19 stations because of the grouping of product families through the other machine stations.

Part Identities

Another facet of group technology is the grouping of part identities, especially through part numbers. Attaching logic significance to each digit of the part numbers is of benefit to the design process, especially when computerized. Thus, designers can retrieve part designs from the computer data base by referencing part numbers, even if they are unfamiliar with the exact design and function of the part for which the design is being retrieved. But from the logical part number system designers can be sure of retrieving designs that are at least similar to the design they are seeking. The result is that existing designs may be copied as is or perhaps imitated with only slight modification.

The price paid for using group technology to place interpretive significance upon each digit of the part number is much larger part numbers. Graves [ref. 40] states that some group technology systems for assigning part numbers are known to utilize as many as 36 digits per part number. Such a system almost completely describes a part by means of the part number alone, but the number is so long it is virtually impossible to remember and is useful only as a computer storage and retrieval aid.

The resolution of a part numbering system is a function of the number of digits and the number of possible values for each digit. Thus, a 36-digit part numbering system has the following resolutions:

1. Numeric digits only: 10^{36}
2. Alphanumeric digits: 36^{36}

A part numbering system of 36 digits has tremendous resolving power, enabling a part design to be almost completely described by the part number alone. But a 36-digit part number is cumbersome to use.

There are two systems of digit identification in group technology: hierarchical and chain-type. In hierarchical systems the interpretation of succeeding digits depends upon the selection of higher ordered digits. In the chain-type systems the meaning of each digit position is fixed. For example, consider a classification system for industrial robots. The first digit might represent the axis configuration, the second the power source, the third the capacity, and the fourth and succeeding digits other characteristics of the robot. In a hierarchical system, an articulating hydraulic robot of capacity 25 pounds might be designated:

311XXX . . . X

The third digit is 1 because a 25-pound capacity might represent the low end of the spectrum of capacities for articulating hydraulic robots. In a chain-type numbering system, however, the number might be:

317XXX . . . X

In the chain-type system the interpretation of the third digit is fixed and does not depend upon the selection of the first and second digits. Therefore the "capacity" digit must accommodate any robot configuration and power source from the smallest, table-top electric models to the huge Cartesian hydraulic models. Therefore in the chain-type system a digit 7 is selected to designate a 25-pound capacity robot, a value somewhere in the midrange of robot capacities, all robots considered. From this example it can be seen that the hierarchical system has an advantage in pinpointing the description of the item. But the chain-type system is much easier to interpret because digit decoding remains fixed. Even when a computer is used, the complexity of coding and decoding hierarchical part numbers can be cumbersome.

One way of understanding group technology is to recall the familiar game of Twenty Questions. Most readers will recall that the object of this game is to isolate the identity of an object by asking 20 carefully chosen yes or no questions, given only that the item's composition is prestated to be vegetable, animal, or mineral. This can be recognized to be equivalent to a 21-digit part code in which the first digit is tertiary (vegetable, animal, or mineral) and the other 20 codes are binary (yes or no). The system is hierarchical because the interrogator has the liberty of phrasing each succeeding question based on answers to previous questions. The resolution of the system is:

$$3 \times 2^{20}$$

Such a resolution is minuscule compared to a 30-digit group technology classification; yet anyone who has played Twenty Questions realizes that the resolving power of this simple party game is quite impressive.

GT Economics

Group technology achieves economies in plant layout, data files, and the identification of products, but it also saves hardware costs. One of the most important cost savings from group technology is the reduction in the number of jigs and fixtures necessary for the manufacturing process. Group technology calls attention to the fact that two otherwise unrelated products may fit the same fixture if key geometrical specifications are identical.

Although the basic concepts of group technology have been practiced for decades, there is much still to be learned about its potential benefits. New areas of cost benefits are still being discovered, but the quantification of these benefits remains an objective of group technology research.

SUMMARY

In this chapter, while attempting to maintain our manufacturing perspective, we have explored the possibilities of integrating the entire system of robots and manufacturing automation within a facility—the concept of CIM. As the name implies, the computer is the pivotal element in such a system. The architects of CIM perceive information flow within a plant to be as important as material flow.

The principal barrier to CIM is interfacing. Interfacing on a direct device-to-device level is a reasonably well managed problem that involves hardware and a limited amount of software. But as the number of devices grows, and the needs for each device to communicate with each of the other devices in the system also grows, the problem of communication and message traffic becomes very complex. Ethernet, the most popular of various physical layout schemes and strategies for dealing with the control of message traffic, is an example LAN, the primary method of interfacing in CIM. Standards such as MAP are emerging, and the ISO/OSI is the principal model.

Related to CIM is the topic of FMS. The flexibility in manufacturing that has been spawned by flexible, programmable automation, the principal subject of this book, is creating a demand for more product flexibility which, in turn, is calling for more flexible manufacturing systems. FMS is more an ideal than a reality in most industries, but this chapter has explored some examples in which FMS has been found to be practical.

Group technology is a technique that is not new but is finding new application in the optimization of the production of flexible manufacturing facilities. Group technology takes advantage of product and piece-part similarities to save steps in the design and manufacturing process. An instrument of the group technology method is the part identification number, in which meaning can be attached to each digit to make the classification scheme automatic. A well thought-out part numbering scheme has benefits in preparing the plant for computer integrated manufacturing.

Computer-integrated manufacturing is a far-reaching goal; few plants have reached it, and many will perhaps never reach it. But in the selection of robots, NC machines, industrial logic control systems, and system hardware and software, there is nothing wrong with keeping the general automation of manufacturing facilities on course toward a totally integrated CIM system. With CIM systems, we are getting close to the ultimate in manufacturing automation. A totally integrated CIM system can be called an "automatic factory." As the ultimate in automation and computerization of manufacturing facilities, CIM is a logical end point for our study of robots and manufacturing automation. But before we close this book we have one more topic to consider: the ethics of what we are doing.

EXERCISES AND STUDY QUESTIONS

16.1. What is the maximum data transfer rate of the IEEE 488 bus expressed in bytes? In nibbles?

16.2. Calculate the number of direct connectors required to directly interconnect a group of 20 factory automation machines.

16.3. In a LAN network arranged in a star configuration with a computer server in the center of the star, how many connectors would be required to connect 20 factory automation machines? (Interpret the term "connector" to mean either of the two plugs or other attachment mechanism on each end of a connector cable.)

16.4. Explain why a block of data to be transmitted is not as large as the sum of the data packets into which it is divided before it is placed on a network.

16.5. Explain why an Ethernet is typically faster than a token ring.

16.6. Explain the relationship between MAP and the ISO/OSI standard.

16.7. In a token bus arrangement an industrial robot controller has an arbitrary address C24, a predecessor address 4AC, and successor address 2B3. A second robot controller is purchased and is arbitrarily assigned address F16. How could the addresses be rearranged to accommodate the new robot on the LAN?

16.8. What is an advantage to employing a token on a bus? (*Hint:* The correct answer is *not* to get a fare discount to ride the bus!)

16.9. A group technology parts classification has part numbers consisting of six numerical and four alphabetic digits. What is the resolution of this part numbering system?

16.10. A hierarchical scheme is used in the part numbering system for a six-numeric digit group technology classification system. How many category descriptions must be defined for this system? If the scheme had been fixed instead of hierarchical, how many category descriptions would have been needed?

16.11. What is the resolution of a parts numbering system consisting of eight numeric digits? What would be the resolution of the system if the digits were alphabetic instead of numeric? What would be the resolution if the digits were alphameric?

CHAPTER 17

ETHICS

At some point, all of us should begin to ask some ethical questions:

Is there a limit to the amount of production we should turn over to robots and manufacturing automation?

Will industrial robots put people out of work?

If robots begin to make more robots, can robots eventually take over too much production?

Do companies ever elevate automation to the position of being an end in itself?

Do people sometimes serve machines, instead of vice versa, as intended?

UNEMPLOYMENT

Easily the most significant question concerning the ethics of automation is whether automation causes unemployment. The public image of automation often casts it in the role of the villain underlying unemployment, as in the cartoon in Figure 17.1. But there is definitely a Theory X and a Theory Y on this issue with viewpoints as opposite as Douglas MacGregor's original Theory X and Theory Y as applied to management [ref. 49]. With regard to *automation,* Theory X states that automation *causes* unemployment, and Theory Y states that automation *prevents* unemployment.

Theory X

There is no question that the principal purpose of robots and manufacturing automation is to make processes more efficient. Robots are usually justified on the basis of labor cost savings, and this means that robots are doing jobs that were formerly performed by human workers. Even in new processes for which no human worker has been employed, the use of a robot means that one or more human worker(s) will not be employed in that job. The replacement rate is not one-to-one either. Robots can often perform an operation at speeds equal to or faster than the human operator and can work three shifts per day, seven days per week, which is the equivalent of four human workers. Even considering downtime for maintenance, a robot can often replace more than one worker.

FIGURE 17.1 Public sentiments about automation. (*Chicago Tribune*—New York News Syndicate, Inc.)

Opponents to the Theory X of automation justify robots as job-makers, not job-destroyers, by claiming that workers are needed to build and maintain robots. Although skill levels and job types may change, the supporters of robots rationalize that the number of jobs remains the same or even increases. But this argument seems to contradict the principal objective of robots and automation: to increase efficiency. The Theory X types say: If it were really true that as many or more people are required to build and maintain robots as were formerly required to do the jobs the robots do now, where would be the improvement in system efficiency?

There are, of course, many reasons for employing robots and manufacturing automation other than the drive for higher efficiency. Improving quality, enhancing safety, and removing tedious, fatiguing, or boring tasks are all worthwhile objectives, but are they worth the price of sacrificing workers' livelihoods? The automation engineer must face these arguments when confronted by either their own consciences or the attacks of automation's opponents. To put the problem in perspective, automation engineers should ask themselves: How would I explain to the worker whose job is about to be replaced by a robot that his or her job is "undesirable" and should be eliminated?

In summary, the opponents of automation see robots and automation as job-destroyers, and the arguments that new jobs take the place of the old ones or that undesirable jobs should be eliminated anyway simply does not counter their belief that automation causes unemployment.

Theory Y

But there is another line of reasoning supporting automation as a job-*creator* or at least a job-*saver*. In the face of intense competition worldwide, is it really possible to sustain a job indefinitely when that job can more efficiently be done by a machine? Without automation, whole industries and all jobs in these industries can be lost to competition. Case Study 17.1 will bring the issue into focus.

EXAMPLE 17.1 (CASE STUDY) Automation in the Steel Industry

In 1951 U.S. Steel Corporation employed 301,328 workers. In 1983 the company employed only 98,722 workers, a decline of about 67 percent. In the same period tonnage produced and shipped by U.S. Steel declined from 24.6 million tons to 11.0 million tons [ref. 115]. Ignoring other products made by U.S. Steel, what was the steel production per employee at U.S. Steel in 1951, and what was it in 1983?

Solution

$$1951: \quad \frac{24.6 \text{ million tons}}{301{,}328 \text{ people}} = 81.638 \text{ tons per person}$$

$$1983: \quad \frac{11.0 \text{ million tons}}{98{,}722 \text{ people}} = 111.424 \text{ tons per person}$$

These results are conservative because U.S. Steel was more diversified in 1983 than it was in 1951. Hence, the productivity improvement for the production of steel products at the U.S. Steel Corporation would have been even more dramatic if the total numbers of employees who produce products other than steel had been known and removed from consideration in the calculation. In U.S. Steel's case and in countless others, the savings from productivity improvements in three decades prevented the firms' collapse before foreign competition. The savings from these productivity improvements exceed total profits, and if the savings had not been there, the company would not have survived.

If we look at the larger picture and include a count of steel workers worldwide, we would undoubtedly find that automation is reducing the number of jobs required to sustain a given level of output of steel. But since automation is here, failure to use it by any one steel company is almost surely to result in failure of the company and elimination of all jobs in that company.

Although the steel industry was selected to illustrate the point, it is not difficult to find parallels in other industries. The textile industry is a good example. Without automation, textile industries in countries such as the United States would have little chance of survival against industries employing low-cost labor in foreign countries.

The foregoing argument seems to imply that automation is a necessary evil because others are using it. This suggests negotiated worldwide limitations upon automation just as the world military powers negotiate arms limitations. Such a proposition would hardly be workable, but even if it were, it would not be advisable because the argument for automation as a job-creator goes deeper than merely its utility as a weapon against competition. If productive power is saved by the efficiencies of automation, that power can be applied to other needed products and services or more of the same products and services. For instance, suppose an automation breakthrough were able suddenly to cut the price of automobiles in half; more families would decide that they could afford and could use an additional automobile. Or if they need only one automobile, they would have funds for a bigger house or other products that would increase the number of jobs in these other areas.

Early in the twentieth century, an automobile was as extravagant a luxury as the personal airplane is today. But Henry Ford's assembly lines and Detroit-style automation made automobiles available to the average person and resulted in huge demands for automobiles in the 1920s. Along with the huge volumes came millions of jobs for persons employed by the automobile and related industries. Today automation breakthroughs are increasing volume of sales of other products and at the same time creating more jobs due to increased volumes, just as seen in the automobile industry in the 1920s.

LABOR'S ATTITUDE

Most readers can recall an era when labor and management were diametrically opposed on most issues, and bitter strikes were commonplace. As recently as the 1980s nationwide strikes still occurred in the United States; an example is the well-known air traffic controller strike of 1981. The subject of automation was often an issue between management and labor, and everyone presumed that management wanted more automation and that labor naturally opposed it.

The 1990s has been witness to a blending of the objectives of management and labor. Leadership has come from Japan, where for centuries the culture has recognized the dignity of labor and where management and labor have always seen each other as partners, to be mutually respected. Indeed in Japan the lines that divide labor and management are either blurred or considered not worthy of note. Now, in the United States, companies are attempting to create a corporate culture in which every worker is a part of the management team and has a sense of ownership of the firm. A popular style employee name badge carries the title "employee/owner." This sense of ownership often has substance, as whole firms have been reorganized and actually sold to the employees, especially companies on the verge of bankruptcy. In such companies the old stereotyped roles of management and labor are forever changed. With these changed roles has come significant changes in attitudes toward automation.

In recognition of Theory Y with respect to automation, many workers now see automation as vital to their firm's well-being and, in turn, to their own job security. Criticism of automation by fellow workers has lost some of its former popularity, and peers are seen to be showing a cool reception to jokes casting automation as the villain. The result is a factory team more committed to the common goal of meeting worldwide competition.

In summary neither Theory X nor Theory Y as applied to automation fully explains the problem. It is true that automation eliminates jobs, but it does not follow that it creates unemployment. In fact, logic points to the opposite.

AUTOMATION INTEGRITY

The greatest harm is generated when the feats of automation are exaggerated. Too often automation engineers in their enthusiasm exaggerate the efficiencies to be gained by implementing their pet projects in robots and manufacturing automation. The result is marginal or wasted capital investments in equipment, which in reality is unable to generate enough production cost savings to justify its initial investment cost. The sad part is that a worker's job is often sacrificed for an industrial robot or automated system of such marginal utility. The consequence is lost jobs with no real gain in efficiency and no reduction in total product costs. No one wins in this situation, and enthusiasm for automation can be severely blunted for future projects.

MURPHY'S LAW

"If anything can go wrong, it will." Almost everyone is familiar with Murphy's law, but its origins are obscure. There is ample reason to believe that Murphy was an automation engineer. Automation projects are particularly susceptible to Murphy's law for several reasons. First of all, there is the strong and constant motivation, perhaps *demand,* to find some way to improve productivity and reduce costs. The rewards can be sensational, and the penalty for failure to automate can be critical. Therefore many ideas will be tried, and it is inevitable that some of these ideas will be marginal. Second, the pursuit of automation is almost always a launching into the unknown. When new processes and equipment are tried, there is always a generous measure of unexpected occurrences. Examples of this were seen in Chapter 9 in the discussion of robot implementation.

The possibility for unexpected problems has ethical ramifications. When planning and justifying a new robotics project, allowance should be made for unexpected problems to avoid premature commitment to a project that will not be worthwhile in the long run. Further consideration should be given to unexpected safety hazards that may arise. Safety is always a concern in initial test and check-out. But safety can also be a problem in operational phases for new automated systems because workers may not be aware of failure modes and potential hazards.

LEGAL HAZARDS

Society has not been very forgiving of automation equipment that has gone awry and injured someone. The notion of an industrial robot harming or even killing a worker is repugnant to society, rare as it may be. Fatalities have occurred both in the United States and in Japan. In one case in the United States a jury awarded the family of a victim $10 million to compensate for the fatal injury of a factory worker [ref. 48]. The worker had climbed up on a storage rack served by a mechanical arm. The attorney for the worker's family said:

> The question, I guess, is "Who serves who[m]?" I think we have to be very careful that we don't go backwards to the kinds of notions we had during the industrial revolution that people are expendable.

If Murphy's law is insufficient motivation to exercise care in an automation project, the social ramifications and the threat of a $10 million lawsuit should be sufficient motivation.

SUMMARY

The ethical considerations of the robots and manufacturing movement cannot and should not be avoided. The principal question of ethics is whether automation causes unemployment or not. This issue has a Theory X and a Theory Y. Productivity statistics appear to support the Theory Y viewpoint that automation saves jobs that would be lost to worldwide competition without the productivity improvements automation makes possible.

It is easy to be overly optimistic about the success of a robotics or other automation project and fail to consider Murphy's law, safety, downtime, and other potentially negative consequences of the decision to automate. Many projects are undertaken in such optimism resulting in the burdening of productive facilities with costly but inefficient, dangerous, or unreliable systems. The resultant loss of jobs on the part of workers and failure of the project to improve the production process result in a no-win situation.

The subject of robots and manufacturing automation arouses strong public sentiments. A measure of this emotion can be seen in the enormous legal settlements now on record for fatal accidents involving robots.

The tremendous potential for robots and manufacturing automation presents temptations to ignore ethics in the pursuit of sensational applications. Robots are now in the spotlight, but their long-range acceptance by the public will be largely dependent upon the ethics of the engineers who specify and install them.

EXERCISES AND STUDY QUESTIONS

17.1. What is Theory X and Theory Y for automation? Does automation cause or prevent unemployment? Support your answer.

17.2. Name some ways an automation engineer can take unethical actions. What are the consequences?

17.3. A certain industry produces a single product and while employing 350 employees in 1980 produced 56,000 units. Gross sales for these units produced a revenue of $29 million for the firm that year of which $3.2 million represented net profits. In the period from 1980 to 1985 the firm embarked upon an ambitious plan to install robots and manufacturing automation systems in their production process. In 1985 the firm's annual production had swelled to 68,000 units and the number of employees had been reduced to 280 as a result of automation improvements to the process. Gross sales in 1985 were $46 million with net profits of $4.1 million. Did robots and automation cause unemployment in this firm? Justify your position by explaining any assumptions you believe to be reasonable. Support your position with calculations.

REFERENCES

1. *Accident Facts,* Chicago: National Safety Council, 1981.
2. "Adapting a Robot Hand to Specialized Functions," *NASA Tech Briefs,* Vol. 11, No. 6, June 1987.
3. Advanced Robotics Corporation Form No. 00099-001, 777 Manor Park Drive, Columbus, Ohio 43228.
4. Aidlin, S.S., and Aidlin, S.H. "The Developing Art of Feeding Plastic Bottles," *Automation* (now *Production Engineering*), Vol. 17, No. 1, January 1970.
5. Amstead, B.H., Ostwald, Philip F., and Begeman, Myron L. *Manufacturing Processes,* 7th ed. New York: John Wiley & Sons, 1979.
6. *AML/2™ Language Reference,* International Business Machines Corporation, 1986.
7. Asfahl, C. Ray. "Curriculum Dilemmas for Robotics Educators,"*T-H-E Journal,* Vol. 11, No. 7, April 1984, p. 98.
8. Asfahl, C. Ray. "A Mathematical Model for Robot Machine Loading Analysis," *International Journal of Robotics and Automation,* Vol. 4, No. 2, 1989.
9. Asimov, Isaac. *Foreward to Handbook of Industrial Robotics.* New York: John Wiley & Sons, Inc., 1985.
10. Aumiaux, M. *The Use of Microprocessors.* Translated by Anne Hutt. Chichester: John Wiley & Sons, 1979.

11. Bailey, J. Ronald. "Product Design for Robotic Assembly," *Proceedings,* 13th International Symposium on Industrial Robots and Robots 7, Vol. 1. Dearborn, Mich.: Society of Manufacturing Engineers, 1983.
12. Baker, James A. "Factory Automation: Industry's Survival Kit," *High Technology,* February 1984.
13. Baker, James A. "Robots: A General Electric Perspective." Paper presented to Society of Manufacturing Engineering executives, Chicago, Ill., September 30, 1981.
14. Barrett, Craig R. "Semiconductor Manufacturing—The Past and the Future," IEEE/SEMI International Semiconductor Manufacturing Science Symposium, Piscataway, N.J., 1989.
15. Baumol, William J. "U. S. Industry Lead Gets Bigger," *The Wall Street Journal,* Vol. 86, No. 56, March 21, 1990.
16. Bernardon, Edward. "Robots in the Apparel Industry," in Richard C. Dorf, Editor, *International Encyclopedia of Robotics,* Vols. 1–3. New York: John Wiley & Sons, 1985, p. 30.
17. Bock, Gordon. "Limping Along in Robot Land," *Time,* Vol. 130, No. 2, July 13, 1987, p. 46.
18. Bolles, R., and Cain, R.A. "Recognizing and Locating Partially Visible Objects: The Local-Feature-Focus Method," *International Journal of Robotics Research,* Vol. 1, p. 57–82, 1982.
19. Bolles, R., and Paul, R. "The Use of Sensory Feedback in a Programmable Assembly System," Stanford Artificial Intelligence Laboratory, Memo AIM-220, October 1973.
20. Boothroyd, G., and Dewhurst, P. *Design for Automatic and Manual Assembly.* Amherst, Mass.: Boothroyd and Dewhurst Manufacturing Consultants, 1983.
21. Boothroyd, G., Poli, C., and Murch, L.E. *Automatic Assembly.* New York: Marcel Dekker, 1982.
22. Butow, Steve; Kent, Stan; Major, Janet; and Matthews, Anthony. "The Get Away Special," *Robotics Age,* Vol. 5, No. 5, September–October 1983.
23. Cameron, Arnold C. "Machines for Driving Screws," *Automation* (now *Production Engineering*), Vol. 19, No. 1, January 1972.
24. Cannon, D.L., and Luecke, G. *Understanding Microprocessors.* Dallas: Texas Instruments, 1979.
25. Chang, Tien-Chien, Wysk, Richard A., and Wang, Hsu-Pin. *Computer-Aided Manufacturing.* Englewood Cliffs, N.J.: Prentice-Hall, 1991.
26. Cooper, K.W., Jr. "Your Computer vs. Rubik's Cube," *Microcomputing,* October 1982.
27. Correale, Anthony. "Physical Design of a Customer 16-Bit Microprocessor," *IBM Journal of Research and Development,* Vol. 26, No. 4, July 1982.
28. Dario, P., et al. "Piezoelectric Polymers: New Sensor Materials for Robotic Applications," *Proceedings,* 13th International Symposium on Industrial Robots and Robots 7, Vol. 2. Dearborn, Mich.: Society of Manufacturing Engineers, 1983.
29. DeGarmo, E. Paul; Black, J. Temple; and Kohser, Ronald A. *Materials and Processes in Manufacturing,* 6th Ed. New York: Macmillan, 1984.

30. den Hamer, H.E. *Interordering, a New Method of Component Orientation*. New York: Elsevier Scientific Publishing Company, 1980.
31. Diesenroth, Michael P. "Robot Teaching," in S.Y. Nof, *Handbook of Industrial Robotics*. New York: John Wiley & Sons, 1985, p. 352.
32. Dodd, John. "Robots, the New Steel Collar Workers," *The Personnel Journal,* September 1981.
33. Dorf, Richard C., Editor. *International Encyclopedia of Robotics,* Vols. 1–3. New York: John Wiley & Sons, 1985.
34. Dyer, R.D., et al. "Robot to Solve Rubik's Cube," *Robotics and Industrial Inspection, Proceedings of SPIE,* Vol. 360, August 1982.
35. Emerson, C. Robert. "Automation Needs Prompt Different Approaches For Systems Analysis," *Industrial Engineering,* Vol. 17, No. 6, June 1985, p. 28.
36. Engelberger, Joseph F. *Robotics in Practice: Management and Applications of Industrial Robots*. New York: AMACOM Division of the American Management Associations, 1980.
37. "Flexible Automation Saves Us Time and Effort," *Modern Material Handling,* Vol. 38, No. 14, October 6, 1983, p. 52.
38. Flynn, Raymond W. "7th Annual Programmable Controller Update," *Control Engineering,* Vol. 37, No. 2, February, 1990, p. 65.
39. "GE's New FMS—Regaining a Competitive Edge," *Modern Material Handling,* Vol. 38, No. 12, September 6, 1983, p. 48.
40. Graves, Gerald R. *Some Considerations for Group Technology Manufacturing in Production Planning*. Doctoral dissertation, Oklahoma State University, 1984.
41. Groover, Mikell P. *Automation, Production Systems, and Computer-Aided Manufacturing*. Englewood Cliffs, N.J.: Prentice-Hall, 1980.
42. Groover, Mikell P., et al. *Industrial Robotics Technology, Programming and Operations*. New York: McGraw-Hill, 1986.
43. Gurumoorthy, B. "CAD-model Based Programming of Robots," *Journal of the Institution of Electronics and Telecommunication Engineers,* Vol. 5, No. 4, July–August 1989, pp. 218–221.
44. *Heathkit® User's Manual for the HERO Robot Model ET-18,* Benton Harbor, Michigan: The Heath Company, 1982.
45. House, Karen E. "The '90s and Beyond," *The Wall Street Journal,* Vol. 213, No. 15, January 23, 1989.
46. *Introduction to Microelectronics*. Grand Rapids, Mich.: Industrial Training, 1981.
47. Jablonowski, Joseph. "Aiming for Flexibility in Manufacturing Systems," *American Machinist,* Vol. 124, No. 3, March 1980.
48. "Jury Awards Family $10 Million," *Northwest Arkansas Times,* Fayetteville, Ark. (Associated Press article), August 14, 1983.
49. Kinnucan, Paul. "How Smart Robots Are Becoming Smarter," *High Technology,* Vol. 1, No. 1, September–October 1981.
50. Kwei, Edwin C.S. *Evaluation of Process Improvement Alternatives for a Frozen Okra Process*. Unpublished masters degree research report, University of Arkansas, Fayetteville, 1971.

51. Kye, S.J., and D. Elford, "Animal Positioning, Manipulation and Restraint for a Sheep Shearing Robot," *Proceedings of the First International Conference on Robotics and Intelligent Machines in Agriculture,* American Society of Agricultural Engineers, St. Joseph, Mich., 1984.
52. Lancaster, M. Carol. *Robotics—The New Industrial Revolution.* Unpublished student paper, University of Arkansas, Fayetteville, November 1981.
53. Lane, Jack E., and Mariotti, John H. "Workplace Handling System Design Targets on Productivity as Main Objective," *Industrial Engineering,* Vol. 13, No. 4, April 1981.
54. Langston, Marcus, *Electro-Optic Force/Pressure Sensor and Transducer.* University of Arkansas Center for Technology Transfer Technical Bulletin 90-1, January 1990.
55. Lapedes, Daniel N., Editor. *Dictionary of Scientific and Technical Terms,* 2nd ed. New York: McGraw-Hill, 1978.
56. Low, Paul R. Keynote Address at International Industrial Engineering Conference, San Francisco, Calif., May 1990.
57. MacGregor, Douglas. *The Human Side of Enterprise.* New York: McGraw-Hill, 1960.
58. McManus, G.J. "Labor Eases Automation Stance," *Iron Age,* March 1966.
59. McWhorter, Gene. *Understanding Digital Electronics.* Dallas: Texas Instruments Learning Center, 1978.
60. *Machine Guarding—Assessment of Need* (NIOSH 75-173), U.S. Department of Health and Human Services (formerly HEW), June 1975.
61. *Making Our Robots and Your Process Work as One.* Columbus, Ohio: Advanced Robotics Corporation, 1983.
62. Marbach, William D.; Rogers, Michael; and Conant, Jennet. "Working at the Wafer's Edge," *Newsweek,* August 22, 1983.
63. Mehrotra, Rajiv, and Grisky, William I. "Shape Matching Utilizing Indexed Hypotheses Generation and Testing," *IEEE Transactions on Robotics and Automation,* Vol. 5, No. 1, February 1989, p. 70.
64. Miller, Irwin, and Freund, John E. *Probability and Statistics for Engineers,* 3rd ed. Englewood Cliffs, N.J.: Prentice-Hall, 1985.
65. Miller, Richard K. *Robots in Industry: Applications for Metal Fabrication.* Madison, Ga.: SEAI Institute, 1982.
66. *Minimover-5 User Reference and Applications Manual.* Mountain View, Calif.: Microbot, 1980.
67. Mitrofanov, S.P. *Scientific Principles of Group Technology* (English translation from Russian). National Lending Library for Science and Technology, 1966. (Original text published in 1959.)
68. *Model 510 Programmable Controller User's Manual.* Johnson City, Tenn.: Texas Instruments, February 1981.
69. Moder, Joseph J., Phillips, Cecil R., and Davis, Edward W. *Project Management with CPM and PERT,* 3rd ed. New York: Van Nostrand Reinhold, 1984.
70. Molander, Thony. "Routing and Drilling with an Industrial Robot," *Proceedings,* 13th International Symposium on Industrial Robots and Robots 7, Vol. 1. Dearborn, Mich.: Society of Manufacturing Engineers, 1983.

71. *MotionMate Robot System Installation and Operation Manual*. Akron, Ohio: Schrader-Bellows Division, Scovill, December 1981.
72. Mowforth, Peter, "The First International Robot Olympics," *AI Expert,* Vol. 6, No. 2, February, 1991, p. 56.
73. Mutter, Randolph F. "Effective Interfacing Through End Effectors," *Proceedings,* 13th International Symposium on Industrial Robots and Robots 7, Vol. 1. Dearborn, Mich.: Society of Manufacturing Engineers, 1983.
74. Naisbitt, John. *Megatrends*. New York: Warner Books, 1982.
75. Nof, Shimon Y., Editor. *Handbook of Industrial Robotics*. New York: John Wiley & Sons, 1985.
76. Nof, Shimon Y., "Robot Ergonomics: Optimizing Robot Work," in S.Y. Nof, *Handbook of Industrial Robotics,* Chap. 30. New York: John Wiley & Sons, 1985, pp. 549–604.
77. Nof, S.Y., Knight, J.L., and Salvendy, G., "Effective Utilization of Industrial Robots—A Job Skill Analysis Approach," *AIIE Transactions,* Vol. 12, No. 3, September 1980, pp. 216–225.
78. Nof, Shimon Y., and Lechtman, Hannan. "Robot Time and Motion System Provides Means of Evaluating Alternate Robot Work Methods," *Industrial Engineering,* Vol. 14, No. 4, April 1982, p. 38.
79. Oppenheim, Irving, and Motazed, Behnam. "Robotics in Construction," *CIT: The Magazine of the Carnegie Institute of Technology,* Vol. 10, No. 1, Fall/Winter 1990.
80. Ottinger, Lester V. "Robotics for the I.E.: The Automated Factory," *Industrial Engineering,* Vol. 14, No. 9, September 1982.
81. Prenting, Theodore. "Why Automatic Assembly Needs IE's." *Automation* (now *Production Engineering*), Vol. 15, No. 12, December 1970.
82. Pressman, Roger S., and Williams, John E. *Numerical Control and Computer-Aided Manufacturing*. New York: John Wiley & Sons, 1977.
83. Quinlan, Joseph C. "Welding with Robots—Is It for You?" *Tooling and Production,* March 1983.
84. Rakowski, Leo R. "Robots: An Aye for an Eye," *Machine and Tool Blue Book,* Vol. 85, No. 4, April 1990, pp. 30–34.
85. Redford, A.H., K.G. Swift, and R. Howie, "Product Design for Automatic Assembly," *Proceedings,* 2nd International Conference on Assembly Automation, Brighton, U.K. May 18–21, 1981, pp. 129–142.
86. Remick, Carl. "Robots: New Faces on the Production Line." *Management Review,* American Management Association, May 1979.
87. *Results with Robots,* product literature, Prab Robots, 1982.
88. Rodgers, Robert C. "What's News in PCs?" *Production Engineering,* Vol. 28, No. 11, November 1981.
89. Rogers, P.F., and Boothroyd, G. "Designing Slot Orienting Devices for Vibratory Bowl Feeders," *Automation* (now *Production Engineering*), Vol. 19, No. 1, January 1972.
90. Rosato, Pat John. "Robotic Implementation—Do It Right," *Proceedings,* 13th International Symposium on Industrial Robots and Robots 7. Dearborn, Mich.: Society of Manufacturing Engineers, April 1983.

91. Rosenfeld, Azriel, "History of Vision Systems," in Richard C. Dorf, Editor, *International Encyclopedia of Robotics,* Vols. 1–3, p. 1877. New York: John Wiley & Sons, 1985.
92. "Rubik's Robot," *Science 82,* Vol. 3, No. 6, July–August 1982.
93. Russell, Marvin, Jr. "Odex I: The First Functionoid," *Robotics Age,* Vol. 5, No. 5, September–October 1983.
94. Russell, R.A. "Closing the Sensor-Robot Control Loop," *Robotics Age,* Vol. 6, No. 4, April 1984.
95. Sadowski, Randall P. "History of Computer Use in Manufacturing Shows Major Need Now Is for Integration," *Industrial Engineering,* Vol. 16, No. 3, March 1984, p. 34.
96. Sanz, Jorge L.C., et al. "Industrial Applications of Vision Systems," in Richard C. Dorf, Editor, *International Encyclopedia of Robotics,* Vols. 1–3, p. 1885. New York: John Wiley & Sons, 1985.
97. Schmitt, Neil M., and Farwell, Robert F. *Understanding Electronic Control of Automation Systems.* Dallas: Texas Instruments Learning Center, 1983.
98. *SMART™ I Programmer's Manual.* Pittsburgh: Aerotech, February 1982.
99. Smith, Randall C., and Nitzan, David. "A Modular Programmable Assembly Station," *Proceedings,* 13th International Symposium on Industrial Robots and Robots 7. Dearborn, Mich.: Society of Manufacturing Engineers, 1983.
100. Standard EIA RS-267-B, Electronic Industries Association. Standard available in its entirety from Electronic Industries Association, 2001 I Street NW, Washington, D.C. 20006, (202) 457-4900.
101. Stauffer, Robert N. "An Exercise in Robot Press Loading." *Manufacturing Engineering,* Vol. 86, No. 2, February 1981.
102. Staugaard, Andrew C., Jr. *Robotics and AI: An Introduction to Applied Machine Intelligence,* Englewood Cliffs, N.J.: Prentice-Hall, 1987.
103. Stoffel, Jane M., "TI Gains Momentum in the Low-End PLC Market," *Control Engineering,* Vol. 36, No. 7, June 1989.
104. Susnjara, Ken. *A Manager's Guide to Industrial Robots.* Englewood Cliffs, N.J.: Prentice-Hall, 1982.
105. Tanner, William R. "Industrial Robots Today," *Machine and Tool Bluebook,* March 1980.
106. Tanner, William R., Editor. *Industrial Robots,* Vol. 1 and 2. Dearborn, Mich.: Society of Manufacturing Engineers, 1979.
107. Thornton, G.S. "Integrated Knowledge-Based Design and Assembly Robot Programming in the 'Design To Produce' Project," *Proceedings,* 10th International Conference on Assembly Automation, Kanazawa, Japan, October 23–25, 1989, pp. 157–171.
108. TISOFT Programming Package (Tutorial Diskette). Texas Instruments, Johnson City, Tenn.
109. Trevelyan, James P. "Sensing and Control for Sheep-shearing Robots," *IEEE Transactions on Robotics and Automation,* Vol. 5, No. 6, December, 1989, pp. 716–727.
110. Ullrich, Robert A. *The Robotics Primer.* Englewood Cliffs, N.J.: Prentice-Hall, 1983.

111. *Unimate Industrial Robot System Planbook.* Product literature. Unimation-Westinghouse (no date).

112. *Unimate Puma™ Mark II Robot-500 Series, Vol. II—User's Guide to VAL™.* Danbury, Conn.: Unimation-Westinghouse, 1983.

113. U.S. Congress, Office of Technology Assessment. *Making Things Better: Competing in Manufacturing,* OTA-ITE-443. Washington, DC: U.S. Government Printing Office, February 1990.

114. U.S. Department of Commerce, Bureau of Economic Analysis. *Business Conditions Digest.* Washington, DC: U.S. Government Printing Office, September, 1989.

115. U.S. Steel Corporation. *Annual Report,* 1983.

116. Vanderbrug, Gordon; Wilt, Donald; and Davis, Jim. "Robotic Assembly of Keycaps to Keyboard Arrays," *Proceedings,* 13th International Symposium on Industrial Robots and Robots 7. Dearborn, Mich.: Society of Manufacturing Engineers, 1983.

117. "Vertical Fin Skin Robot," *Production Technology Bulletin 85-09.* Fort Worth, Texas: General Dynamics Fort Worth Division, 1985.

118. Waldman, Harry. "The Programmable Controller: Continuing to Grow in Use," *Manufacturing Engineering,* Vol. 91, No. 2, August 1983.

119. Warnecke, H.J., et al. "Application of Industrial Robots for Assembly Operations in the Automotive Industry," *Proceedings,* 13th International Symposium on Industrial Robots and Robots 7. Dearborn, Mich.: Society of Manufacturing Engineers, 1983.

120. *Western Electric Engineer,* Vol. 27, No. 2, 1983.

121. Witriol, Norman M., and Cowling, David H. "Illuminescent Industrial Robotic Vision Systems." Conference on Lasers and Electro-optics, IEEE, New York, 1989.

122. Zuech, Nello, and Dunseth, Jim. "Vision Systems, Theory," in Richard C. Dorf, Editor, *International Encyclopedia of Robotics.* New York: John Wiley & Sons, 1985, p. 1926.

APPENDIX A

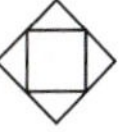

TABLE OF POPULAR, COMMERCIALLY-AVAILABLE PROGRAMMABLE LOGIC CONTROLLERS

Manufacturer	*Model*	*Capacity* *I/O*	*Capacity* *Pgm. Mem.*	*Scan Rate*	*Relay Ladder Logic?*	*High-Level Languages?*	*Analog Capability?*
Allen-Bradley							
	SLC 500	72	1K	6 ms	Yes	No	Yes
	PLC-5/25	1920	21K	2 ms	Yes	Yes	Yes
	PLC-3	8192	2M	2.5 ms	Yes	Yes	Yes
Eagle Signal							
	Eptak 100	16	250 steps	25 ms	Yes	No	No
	Eptak 220 Series 2	896	40K	0.75 ms	Yes	Yes	Yes
	Eptak 7000	2048	48K	0.75 ms	Yes	Yes	Yes
Eaton Corp. Cutler-Hammer							
	CRA 28	28	1K	7 ms	Yes	No	No
	CPU 20	224	4K	1.5 ms	Yes	Yes	Yes
	CPU 50	512	8K	.9 ms	Yes	Yes	Yes
GE Fanuc							
	Series ONE JR.	96	700	40 ms	Yes	No	No
	Series FIVE	2048	16K	1 ms	Yes	Yes	Yes
	Series SIX PLUS	8000	64K	0.8 ms	Yes	Yes	Yes
Honeywell							
	620-06	192	18K	2 ms	Yes	Yes	No
	620-10	1024	4K	10 ms	Yes	Yes	No
	620-35	4096	32K	2.5 ms	Yes	Yes	Yes
Mitsubishi							
	F1-12M	32	1K	12 ms	Yes	No	Yes
	F2-60M	120	2K	7 ms	Yes	No	Yes
	A3MCPU	2048	60K	2.25/ 0.2 ms	Yes	Yes	Yes
Modicon							
	Micro 984	120	6K	5 ms	Yes	No	Yes
	984-685	3968	16K	3 ms	Yes	No	Yes
	984B	18532	64K	0.75 ms	Yes	No	Yes

Manufacturer	Model	Capacity I/O	Capacity Pgm. Mem.	Scan Rate	Relay Ladder Logic?	High-Level Languages?	Analog Capability?
Omron							
	C28K	84	1194	10ms	Yes	No	Yes
	C120	256	2.6K	15 ms	Yes	No	No
	C2000H	2048	38K	0.4 ms	Yes	No	Yes
Siemens							
	S5-100U	128	2K	70/ 1.6 ms	Yes	Yes	Yes
	S5-115U	2048	48K	30/3 ms	Yes	Yes	Yes
	S5-155U	8192	896K	1.4 ms	Yes	Yes	Yes
Square D							
	SY/MAX 50	256	4K	8 ms	Yes	No	Yes
	SY/MAX 400	4000	16K	0.7 ms	Yes	No	Yes
	SY/MAX 700	14K	64K	1.3 ms	Yes	No	Yes
Texas Instruments							
	TI 510	40	256	16.7 ms	Yes	No	No
	TI 315	74	700	20 ms	Yes	Yes	Yes
	TI 435	640		0.49 ms	Yes	Yes	Yes
	565	8192	384K	2.2 ms	Yes	Yes	Yes
Westinghouse							
	PC-500	128	1K	70 ms	Yes	No	Yes
	PC-503	256	10K	2 ms	Yes	Yes	Yes
	HPPC-1700	8192	224K	1 ms	Yes	No	Yes

CASE STUDY 13.4 PLC SEQUENCE CONTROL OF PICK-AND-PLACE ROBOT

Outputs	
20	Rotate base c.c.w.
21	Rotate base c.w.
22	Raise arm
23	Lower arm
24	Extend arm
25	Retract arm
26	Grasp (close gripper)
27	Release (open gripper)

Series of Actions	*Cumulative Time (sec)*	*Time Taken (sec)*	*Order of Outputs*
Extend arm	0–2	2	24
Lower arm	2–3	1	23
Grasp	3–3.5	0.5	26
Lift arm	3.5–4.5	1	22
Retract arm	4.5–6.5	2	25
Rotate c.c.w.	6.5–10	3.5	20
Extend arm	10–12	2	24
Lower arm	12–13	1	23
Release	13–13.5	0.5	27
Lift arm	13.5–14.5	1	22
Retract arm	14.5–16.5	2	25
Rotate c.w.	16.5–20	3.5	21

2 Counts/sec

Use .5 sec intervals ∴ $\dfrac{20 \text{ sec}}{.5 \text{ sec/interval}} = 40$ intervals

Time (sec)	*Count Number*	*Action*	*Time (sec)*	*Count Number*	*Action*
0	0	SET 24		21	
	1		11	22	
1	2			23	
	3		12	24	SET 23 RST 24
2	4	SET 23 RST 24		25	
	5		13	26	OUT 27 RST 23
3	6	OUT 26 RST 23		27	SET 22
	7	SET 22	14	28	
4	8			29	SET 25 RST 22
	9	SET 25 RST 22	15	30	
5	10			31	
	11		16	32	
6	12			33	SET 21 RST 25
	13	SET 20 RST 25	17	34	
7	14			35	
	15		18	36	
8	16			37	
	17		19	38	
9	18			39	
	19		20	40	RST 21
10	20	SET 24 RST 20			

PLC Code

STR	SHF	0	ENT				
AND	NOT	TMR	SHF	6	0	0	ENT
TMR	SHF	6	0	0	ENT		
SHF		5	ENT				
STR	TMR	SHF	6	0	0	ENT	
STR	CNT	SHF	6	0	1	ENT	
CNT	SHF	6	0	1	ENT		
SHF	4	0	ENT				
STR	SHF	6	0	1	ENT		
SHF	0	ENT					
SET	SHF	2	4	ENT			
STR	SHF	6	0	1	ENT		
SHF	4	ENT					
SET	SHF	2	3	ENT			
RST	SHF	2	4	ENT			
STR	SHF	6	0	1	ENT		
SHF	6	ENT					
OUT	SHF	2	6	ENT			
RST	SHF	2	3	ENT			
STR	SHF	6	0	1	ENT		
SHF	7	ENT					

PLC Code (continued)

```
SET   SHF   2     2     ENT
STR   SHF   6     0     1     ENT
SHF   9     ENT
SET   SHF   2     5     ENT
RST   SHF   2     2     ENT
STR   SHF   6     0     1     ENT
SHF   1     3     ENT
SET   SHF   2     0     ENT
RST   SHF   2     5     ENT
STR   SHF   6     0     1     ENT
SHF   2     0     ENT
SET   SHF   2     4     ENT
RST   SHF   2     0     ENT
STR   SHF   6     0     1     ENT
SHF   2     4     ENT
SET   SHF   2     3     ENT
RST   SHF   2     4     ENT
STR   SHF   6     0     1     ENT
SHF   2     6     ENT
OUT   SHF   2     7     ENT
RST   SHF   2     3     ENT
STR   SHF   6     0     1     ENT
SHF   2     7     ENT
SET   SHF   2     2     ENT
STR   SHF   6     0     1     ENT
SHF   2     9     ENT
SET   SHF   2     5     ENT
RST   SHF   2     2     ENT
STR   SHF   6     0     1     ENT
SHF   3     3     ENT
SET   SHF   2     1     ENT
RST   SHF   2     5     ENT
STR   SHF   6     0     1     ENT
SHF   4     0     ENT
RST   SHF   2     1     ENT
      END OF CODE
```

PHONEME CHART

Phoneme Code	*Phoneme Symbol*	*Duration (ms)*	*Example Word*	*Phoneme Code*	*Phoneme Symbol*	*Duration (ms)*	*Example Word*
00	EH3	59	jacket	20	A	185	day
01	EH2	71	enlist	21	AY	65	day
02	EH1	121	heavy	22	Y1	80	yard
03	PA0	47	no sound	23	UH3	47	mission
04	DT	47	butter	24	AH	250	mop
05	A2	71	made	25	P	103	past
06	A1	103	made	26	O	185	cold
07	ZH	90	azure	27	I	185	pin
08	AH2	71	honest	28	U	185	move
09	I3	55	inhibit	29	Y	103	any
0A	I2	80	inhibit	2A	T	71	tap
0B	I1	121	inhibit	2B	R	90	red
0C	M	103	mat	2C	E	185	meet
0D	N	80	sun	2D	W	80	win
0E	B	71	bag	2E	AE	185	dad
0F	V	71	van	2F	AE1	103	after
10	CH*	71	chip	30	AW2	90	salty
11	SH	121	shop	31	UH2	71	about
12	Z	71	zoo	32	UH1	103	uncle
13	AW1	146	lawful	33	UH	185	cup
14	NG	121	thing	34	O2	80	for
15	AH1	146	father	35	O1	121	aboard
16	OO1	103	looking	36	IU	59	you
17	OO	185	book	37	U1	90	you
18	L	103	land	38	THV	80	the
19	K	80	trick	39	TH	71	thin
1A	J*	47	judge	3A	ER	146	bird
1B	H	71	hello	3B	EH	185	get

Phoneme Code	*Phoneme Symbol*	*Duration (ms)*	*Example Word*	*Phoneme Code*	*Phoneme Symbol*	*Duration (ms)*	*Example Word*
1C	G	71	get	3C	E1	121	be
1D	F	103	fast	3D	AW	250	call
1E	D	55	paid	3E	PA1	185	no sound
1F	S	90	pass	3F	STOP	47	no sound

*T must precede CH to produce CH sound.

D must precede J to produce J sound.

EXAMPLE

Suppose it is desired to program an industrial robot to speak the following message in the event a sensor detects that someone has intruded upon its work envelope:

"Danger! Stand back!"

The message can be synthesized using the following sequence of the standard phonemes taken from the phoneme chart:

PHONEMES:	1E	06	21	0D	1E	1A	02	3A	3E	3E	3E
PHONICS:	D	A1	AY	N	D	J	EH1	ER	. . . pause . . .		
PHONEMES:	1F	2A	2F	00	0D	1E	0E	2F	00	19	
PHONICS:	S	T	AE1	EH3	N	D	B	AE1	EH3	K	

INDEX